Multifunctional Agriculture

A Transition Theory Perspective

Multifunctional Agriculture

A Transition Theory Perspective

Geoff A. Wilson

www.cabi.org

CABI is a trading name of CAB International

CABI Head Office
Nosworthy Way
Wallingford
Oxfordshire OX10 8DE
UK

CABI North American Office
875 Massachusetts Avenue
7th Floor
Cambridge, MA 02139
USA

Tel: +44 (0)1491 832111
Fax: +44 (0)1491 833508
E-mail: cabi@cabi.org
Website: www.cabi.org

Tel: +1 617 395 4056
Fax: +1 617 354 6875
E-mail: cabi-nao@cabi.org

A catalogue record for this book is available from the British Library, London, UK.

A catalogue record for this book is available from the Library of Congress, Washington, DC.

ISBN: 978 1 84593 256 5

Printed and bound in the UK by Cromwell Press, Trowbridge, from copy supplied by the author.

Contents

Part 2 From productivist to post-productivist agriculture … and back again?

Part 3 Conceptualising multifunctional agricultural transitions

List of tables and figures

Tables

Figures

Acknowledgements

Ideas contained in this book can be traced back many years and are, therefore, bound to leave a long trail of intellectual and other debts. In particular, a book that focuses on thoughts, ideas and theories can not be successfully written without the help of many individuals who have critically commented on various aspects of the book.

My biggest debt of gratitude is to my wife Olivia Wilson who supported this book project throughout, had to endure my long 'absences' while writing, who was asked regularly to comment on lines of thought developed in the book, and who gave crucial editorial advice on the final draft. Special thanks are also due to Clive Potter who provided invaluable constructive criticism on conceptual issues developed in Part 3, to Karlheinz Knickel who critically commented on approaches used in the book, and to Jonathan Rigg for comments on issues related to agricultural change in developing countries. One of my biggest debts of gratitude is to Jacqui Dibden who not only unearthed important materials on multifunctionality, but also commented critically and enthusiastically on ideas developed here. I also wish to thank Peter Shanahan for crucial help with bibliographical searches. During stages of writing, this book also benefited from intellectual support and critical comments from various academics and friends including, in particular, Raymond Bryant, Rob Burton, Richard Perkins, Chris Cocklin, Michael Redclift, James Sidaway, Peter Wilson and Jochen Kantelhardt. Thanks also to my PhD student Rob Hopkins whose work on energy descent pathways has informed some of the conceptual ideas developed in Part 3.

Ideas for this book were also tested at various conferences, seminars and workshops, and I am grateful to all those who commented critically and asked probing questions. This includes the many people who attended my inaugural lecture in 2004 (University of Plymouth, UK) in which the idea of a multifunctionality framework was first exposed to critical scrutiny, but also the 2004 Environmental Economic Geography workshop at the University of Köln (Germany), the 2004 conference of the International Geographical Union (Glasgow, UK), the 2004 Anglo-German Rural Geographers meeting at the University of Exeter (UK), the rural research workshop at the Berlin-Brandenburgische Akademie (Berlin, Germany) in 2005, and the 2006 Anglo-French Rural Geographers meeting in Vichy (France) where ideas from Chapters 9 and 10 were exposed to critical scrutiny for the first time.

A book like this can not be written without the supportive environment of research-led institutions that also reward 'blue sky' research not linked to the generation of overhead income. I, therefore, both wish to thank the Department of Geography at King's College London (where the first part of this book was written) and the School of Geography at the University of Plymouth for providing academically stimulating work environments and, in particular, for enabling me to find the time beyond the duties of teaching and administration to successfully complete this book

I am also indebted to staff at the cartographic unit of the School of Geography (University of Plymouth), in particular Brian Rodgers who produced the 26 figures and who patiently addressed regular suggestions for changes to these figures as the conceptual framework gradually unfolded. I would also like to thank my editor Nigel Farrar for help with editorial aspects, but also for rigidly enforcing a maximum length for this book which has led to a much sharper and more succinct storyline. I am also indebted to Tracy Ehrlich for help with preparing the camera-ready copy.

Finally, I also wish to acknowledge the beautiful landscape of south Devon where I and my family live, which has not only invited me to take relaxing breaks along the coastal footpaths while writing, but which has also greatly inspired me to think about the challenges of agricultural and rural transitions in a strongly multifunctional rural setting.

List of abbreviations

AEP	Agri-environmental policy
BML	Bundesministerium für Ernährung Landwirtschaft und Forsten (Germany)
BSE	Bovine spongiform encephalopathy
BVEL	Bundesministerium für Verbraucherschutz, Ernährung und Landwirtschaft (Germany)
CAP	Common Agricultural Policy (EU)
DEFRA	Department for the Environment, Food and Rural Affairs (UK)
EEC	European Economic Community
ESA scheme	Environmentally Sensitive Area scheme (EU)
EU	European Union
FAO	Food and Agriculture Organisation
GAEC	Good agricultural and environmental condition
GATT	Global Agreements on Tariffs and Trade
GDR	German Democratic Republic
GM crops	Genetically modified crops
IMF	International Monetary Fund
IUCN	International Union for the Conservation of Nature
MAFF	Ministry for Agriculture, Fisheries and Food (UK)
NCOs	Non-commodity outputs
NGOs	Non-governmental organisations
OECD	Organisation for Economic Cooperation and Development
p/np	productivist/non-productivist
p/pp	productivist/post-productivist
RDR	Rural Development Regulation (EU)
RSPB	Royal Society for the Protection of Birds (UK)
SAPs	Structural Adjustment Policies
SFPs	Single Farm Payments (EU)
UK	United Kingdom
URAA	Uruguay Round Agreement on Agricultural Trade
USA	United States of America
WTO	World Trade Organisation

Chapter 1

Introduction

This book has emerged out of two key arenas of concern that have influenced recent research and thinking on conceptualisations of agricultural and rural change. The first is a growing dissatisfaction with the uncritical and weakly theorised use of the notion of 'multifunctionality' in contemporary debates on agricultural change. The last 20 years or so have seen the use of this term in a wide variety of contexts, spanning a wide spectrum of proponents from policy-makers to rural stakeholder groups and from politicians to non-governmental organisations (NGOs). Yet, none of these debates has shed sufficient light on *what* the notion of multifunctionality implies, *who* the beneficiaries should be and *how* it ought to be put into practice – in other words, the notion of 'multifunctionality' has remained undertheorised and poorly linked to wider debates in the social sciences. The second issue revolves around a growing dissatisfaction with the proposed transition from 'productivist' to 'post-productivist' agriculture. Although there is growing criticism of this postulated transition, it is assumed by many that agricultural systems *have* moved to a post-productivist era, and that we are now firmly embedded in this new 'post-ism' that has left the legacy of environmental degradation and agro-business orientation of the previous productivist model behind. Yet, many studies have highlighted that there is little empirical evidence to support the notion of a transition towards post-productivism.

The aim of this book is to bring these two seemingly separate concerns together by (re)conceptualising agricultural change from a transition theory perspective. First, I will analyse to what extent the reductionist notion of a proposed transition towards post-productivist agriculture holds up to scientific scrutiny and propose a *modified productivist/non-productivist model* that better encapsulates the complexity of contemporary agricultural and rural change. Second, I will *theoretically anchor* the notion of 'multifunctionality' within these debates and will, for the first time, conceptualise multifunctionality as a transitional process of agricultural/rural change embedded in a spectrum bounded by *productivist* and *non-productivist* actor spaces. I will highlight that only by linking these different arenas of investigation can the notion of multifunctionality begin to make sense and, eventually, become a robust and tangible *normative* concept to be used by decision-makers at various spatial scales to protect, shape and change contemporary agricultural and rural spaces. Third, I also wish to use the new framework as the basis for a *plaidoyer* against economistic interpretation of multifunctionality simply as an 'externality' issue, against the predominant productivist agri-business model that continues to dominate agriculture in advanced economies and that increasingly influences agricultural practices in the developing world, and against the globalisation of agro-commodity chains that, in my view, weakens global agricultural multifunctionality pathways.

1.1 A changing agriculture

Ever since humans embarked on the domestication of plants and animals some 10,000 years ago, agriculture has gone through fundamental changes. Such changes have been associated with changing technologies, changing attitudes, and with changing food markets and consumer demand. The invention of the plough about 3000 years ago, for example, allowed the planting of previously unusable areas, resulting in the rapid spread of agriculture into large and often densely forested regions. Mechanisation of agriculture from the early 19th century, together with the increasing globalised nature of agro-commodity chains, meanwhile, enabled much more efficient and labour-saving forms of agricultural production, resulting in 'industrial' agriculture often characterised by mass production of uniform and standardised food products (Mazoyer and Roudart, 2006). Such changes, and many others that agriculture has experienced over thousands of years, have been fundamental and arguably even 'revolutionary'. However, some researchers are arguing that these processes pale into insignificance when compared with changes that have occurred since the Second World War, and that developments over the past 50 years have been more dramatic and far-reaching than anything that affected agricultural production in previous millennia (Mannion, 1995). Some commentators even argue that we are facing the end of conventional 'agriculture' that had, as its sole purpose, the production of food and fibre, and that a new agricultural regime may be emerging that has much wider purposes, including the 'production' of nature and new spaces for leisure (Braun and Castree, 1998), and that sees the modern farmer as a person who is as much an environmental manager as a producer of food and fibre (Marsden, 1999a).

The purpose of this book is to analyse this recent *transition* from agriculture as a producer of food and fibre to that of agriculture as a producer of what has been termed *multifunctional* products and spaces – a transition that goes well beyond traditional understandings of what 'agriculture' is about. I will argue throughout this book, that we may be at a crucial threshold in human interaction with the countryside and agricultural production, as, for the first time, a new *consciously orchestrated* multifunctional agriculture may begin to take shape in both the developed and developing world. As Marsden (2003: ix) emphasised, "to say that the nature of agriculture, and its role in rural development is at something of a crossroads is both to understate and to reaffirm many of the debates that have been articulated in both academic and policy-making circles for more than a decade". Before analysing the possible emergence of *multifunctional agriculture*, we will need to investigate the nature and pace of this possible transition. The underlying theme of this book will, therefore, be focusing on understanding 'transition', both from a theoretical and practical vantage point. I will place particular emphasis on recent debates surrounding the notion of a possible transition from a *productivist* to a *post-productivist* era over the past 50 years – debates that have particularly dominated theorisations of agricultural and rural issues in the English-speaking world since the late 1980s. I will argue throughout that the conceptualisation of multifunctional agriculture is only possible when considered against the background of debates on the transition to post-productivism.

Although 'multifunctionality' in agriculture has been discussed for decades as an emerging issue, these discussions have focused largely on policy change, macro-economic considerations and the economics of farming, and have remained relatively uncritical and, in my view, atheoretical. As a result, 'multifunctionality' currently means different things to different people (similar to the notion of sustainability) and has, so far at least, only had limited policy relevance as it lacks a thoroughly grounded widely acceptable conceptualisation. I will, therefore, argue that only by contextualising multifunctionality in the context of debates on the transition from productivism to post-productivism will it be possible, first, to understand what multifunctional agriculture is about, and, second, to anchor the notion of multifunctionality theoretically in the context of agricultural change. If this is done successfully, then I hope that this book will also contribute towards making the notion of 'multifunctional agriculture' a more conceptually grounded – and therefore more policy-relevant – term than it is at present.

1.2 Agricultural systems in transition: from productivism to post-productivism to multifunctionality?

Over the last 50 years or so, global agriculture has witnessed profound changes in food and fibre production, actor spaces, policy frameworks, food regimes, ideologies, and impacts on the environment. Food production has increased dramatically so that today more than six billion people can, in theory at least, be adequately fed. Many agricultural systems have also seen dramatic structural upheavals, including the rapid decline in agricultural workforce in most advanced economies, with the concurrent destabilisation and restructuring of many rural communities. This has been accompanied by the substitution of subsistence farming patterns with cash crop production for export in many regions of the developing world, and the concurrent dismantling of often sustainable traditional farming systems. This has also led to an overall increase in environmental degradation linked to over-intensive agriculture. As Pretty (2002: xi) reminded us, "something is wrong with our agricultural and food systems", best exemplified by nearly 2 billion hectares of land worldwide now classified as heavily degraded (Wilson and Juntti, 2005). In recent years, much theoretical and conceptual work has attempted to understand these changes, both in the developed and developing world. Indeed, in the Anglo-American context, the last 20 years have seen the emergence of some of the most interesting and challenging theoretical debates about the nature, changes and future trajectories of modern agricultural and rural systems from a variety of economic, social, political and environmental stances.

The most powerful theoretical concept to emerge has been the notion that modern agriculture has moved from a *productivist* to a *post-productivist* era (Cloke and Goodwin, 1992; Marsden *et al.*, 1993). Holmes (2006: 143) argued that the concept of the post-productivist transition is "currently the only overarching conceptualization of the rural transition". From an advanced economies perspective, the productivist era has been described as lasting from the end of the Second World War to about 1985. It has been broadly characterised as a period when the main preoccupation of agriculture was maximum food production to ensure national or regional self-sufficiency, as a time when agriculture held a central 'hegemonic' position in society, and as an era characterised by a small but powerful and tight-knit agricultural policy community. In addition, productivism has seen a 'strong' state with predominantly top-down policy-making structures, and with farming techniques that have often relied on the application of high external inputs and the use of heavy machinery that have caused severe environmental degradation in intensively farmed regions. In developing countries, the 'productivist' era is often allied to a concern with achieving 'food security'. In the 1960s and 1970s in particular, and reflecting the general Malthusian pessimism of the time, notions of food security in developing countries were couched in highly productivist terms (Wilson and Rigg, 2003).

For some advanced economies, it has been argued that the transition to a *post-productivist era* began in the 1980s and has lasted to the present day (Mather *et al.*, 2006). As Chapter 6 will highlight, post-productivism has generally been seen as the 'mirror-image' of productivism. Thus, for advanced economies, post-productivist agriculture has been characterised by: a reduction in the intensity of farming through extensification, diversification and dispersion of agricultural production; an associated move away from agricultural production towards 'consumption' of the countryside; the loss of the central position of agriculture in society characterised by 'contested' countrysides; a widening of the agricultural community to include formerly marginal actors at the core of the policy-making process; and a weakening of the state role in policy-making powers with a more inclusive model of governance that also includes grassroots actors. Simultaneously, farming techniques in the post-productivist era are seen to be more in tune with environmental protection through reduced application (or total abandonment) of external inputs (e.g. organic farming). During the post-productivist era the main threats to the countryside are generally perceived to be agriculture itself rather than urban or industrial development.

Issues surrounding the transition from productivism to post-productivism have also recently been highlighted by work that has attempted to assess how applicable this transition is outside the UK. Seminal works include Holmes' (2002, 2006) and Argent's (2002) analyses of the applicability of the concept of post-productivism in the Australian context, Jay's (2004) work in New Zealand, and recent work on the 'transferability' of the concept to the European situation (Wilson, 2002). Reference to 'productivism and 'post-productivism' is now burgeoning in the context of agricultural and rural research in advanced economies to such an extent that for some it is becoming the accepted orthodoxy for conceptualising recent agricultural change (e.g. Ilbery and Bowler, 1998). Post-productivist theorisations of agricultural change have risen to such prominence that they have prompted Roche (2003, 2005), in recent reviews of 'rural geography' in *Progress in Human Geography*, to situate research on 'rethinking post-productivist rural spaces' as one of the key themes preoccupying rural scholars at the beginning of the 21st century. As this book will amply demonstrate, there is an inherent danger in uncritically accepting such seemingly 'straightforward' and 'linear' concepts of societal change (see Chapter 7 in particular), echoing Evans *et al.*'s (2002: 325) critique that "given the discussion in rural geography on the use of dualistic notions of Fordism and post-Fordism, it is rather surprising that a similar debate and critique has not been forthcoming in relation to the notion of a shift from 'productivism' to 'post-productivism'. The commonality of the term's usage is matched conversely by minimal theoretical contouring". For developing countries, meanwhile, Sen's (1981) work echoed debates that emerged in advanced economies surrounding the notion of a transition to post-productivism. Sen's work on famine, for example, led to increased nuancing and locally grounded definitions of 'food security', emphasising household food security and food systems. Arguably, by the end of the 1990s a 'post-productivist' view may have emerged for some developing countries that emphasised the multiple and shifting ways with which individuals and households achieve food security. Thus, food security in the South has moved from the global and the national scale to the household and the individual, from a food to a livelihoods perspective, from objective indicators to subjective perception, and from production to 'consumption' of the countryside (Wilson and Rigg, 2003).

In this book, I will refer to this postulated transition as the **productivist/post-productivist transition model** (hereafter the 'p/pp transition' or 'p/pp model'). Debates on the possible applicability of this model have undoubtedly added an interesting new conceptual dimension to agricultural/rural research – a field that was, for a long time, seen as having relatively static theorisations of agricultural and rural change (Cloke, 1989; Buttel *et al.*, 1990). Marsden (2003: 100) argued that these debates have shown "the utility of theory for guiding research, policy and more effective interpretation of the contemporary rural economy". In particular, debates on the p/pp transition have brought together a wide array of researchers from different disciplinary backgrounds including rural studies, geography, history, sociology, economics or social psychology, to name but a few (e.g. Buttel *et al.*, 1990; Van der Ploeg, 2003). Yet, several years after the first widely publicised discussions of a shift towards post-productivism, and at a time when further profound changes in agricultural/rural arenas at local, national and global scales are taking place, it is time to re-evaluate existing conceptualisations (Goodman, 2004). Such a re-evaluation is particularly important in light of the almost ubiquitous (and relatively uncritical) use of the term **'multifunctional agriculture'** in both academic and policy-related literature. As Durand and Van Huylenbroek (2003: 16) suggested, "multifunctionality has been introduced in recent years at different occasions as a leading principle and new paradigm for the future development of agriculture and rural areas". Similarly, Marsden (2003: 88) argued that "the farm as a multi-functional business is emerging in many rural regions". Yet, as Chapter 8 will highlight, the notion of multifunctionality has only relatively recently (early 1990s) been used by researchers as a key concept for understanding the complexity of agricultural change. Current understandings of multifunctionality are *reductionist* and based on relatively narrow *economic* and *policy-based* approaches predicated on structuralist interpretations of agricultural and rural change. McCarthy (2005) rightly asked whether the notion of multifunctional rural geographies is 'radical' or merely 'reactionary'.

I will argue that the narrow structuralist interpretations of multifunctionality have suffered from *discursive insularity* that has confused rather than clarified what multifunctionality could be about. 'Multifunctional agriculture' is still understood mainly as a policy-led *process describing* current agricultural trends, rather than as a *concept explaining* agricultural change, and is still largely embedded in *structuralist* theory rather than informed by *normative* concepts (Andersen *et al.*, 2004). Clark (2003: 225) suggested that "no formal meaning of multifunctional agriculture exists, which clearly poses difficulties for researchers analysing this new ... paradigm". Delgado *et al.* (2003: 28) also emphasised that "the concept of multifunctionality is still being formed ... Even the different countries supporting it do not interpret it the same way". Similarly, Di Iacovo (2003: 122) argued that the "multifunctionality of agriculture is not yet totally explored", while I recently suggested that the "concept of 'multifunctionality' currently remains poorly conceptualised, and more work is needed in the future to further sharpen this possible alternative/extension to the productivist/post-productivist model of agricultural change" (Wilson, 2005: 117). Losch (2004: 338) argued that the notion of multifunctionality "is never tackled head-on", and that it is "becoming simultaneously a subject of negotiation, a foil or a reference point for the formulation of alternatives". As a result, Knickel and Renting (2000: 512) argued that researchers "must improve their understanding of multifunctionality", while Buller (2005: ii) suggested that "what is missing is a more holistic evaluative framework for assessing the broader multifunctional contribution of agriculture". Such concerns have not only come from English-speaking commentators. Lardon *et al.* (2004: 6; my translation) argued from a French perspective that "multifunctionality is a concept that is not easy to grasp", while Rapey *et al.* (2004b: 49; my translation) emphasised that the characteristics of multifunctionality have often been defined in very general and variable terms, depending on disciplinary biases and research objectives, highlighting that "we need to simplify the meaning, characterisation and representation of multifunctional agriculture".

While there is an *implicit* assumption that 'multifunctionality' is a relatively clear concept that both describes and explains processes of agricultural change, there are few *explicit* explanations and definitions of multifunctionality. Garzon (2005: 1) emphasised that the "multifunctionality of agriculture has been a question debated at international level for more than two decades and yet is not a generally accepted notion". McCarthy (2005: 778) also emphasised "that land use is necessarily multifunctional is hardly a novel idea; the challenge is rather to theorise the emergence and significance of contemporary articulations of 'multifunctionality'". Clark (2003: 247) similarly emphasised that, currently, "there is no definable blueprint of the characteristics of multifunctional [farm] businesses", while Peterson *et al.* (2002: 423) referred to the "continuum of views on multifunctionality, generating an impasse that will be difficult to resolve". Further, Vatn (2002: 312) argued that "the concept of multifunctionality seems to have somewhat different meanings in the literature", while Holmes (2006: 145; emphasis added) suggested that "only over the last two or three decades have Western, market-oriented modes of rural occupance revealed a marked trend towards *overt recognition* of multifunctionality". The result has been a fragmented notion of multifunctionality largely predicated on its relevance for *policy-based* decision-making, characterised by short-termism, weak theorisation and pragmatism. There continues to be a pronounced lack of *critical theoretical debate* about 'multifunctionality' in social science research, and we are currently left with an understanding of the term that has not yet gelled into a *coherent workable framework* for fully *understanding* contemporary multifunctional agricultural and rural change.

This book echoes both Cloke and Goodwin's (1992) earlier suggestion that there is a need to theorise the complexity of agricultural/rural change in a more satisfactory manner and Delgado *et al.*'s (2003: 28) call for giving multifunctional agriculture "a content that is politically possible, socially suitable, and economically efficient". It further addresses Marsden's (2003: 142) recent concern that "while ... there is a recognition ... about the integrative and holistic nature of the new processes of rural change, the approaches thus far have yet to theoretically develop an approach which begins to guide a clearer understanding of the processes which are making things different in the ... countryside". Similarly, it takes into account Buttel's (2001) call for more theoretical innovation in agricultural/rural research, away from the flurry of rich empirical work that, although important, has tended to create a relatively random set of micro-empirical studies, thereby possibly losing sight of the 'big picture'. Finally, it also addresses Knickel and Renting's (2000: 526) call that "the increasing recognition of the complexity of rural development processes should be accompanied by the adoption of more multidisciplinary, holistic approaches". Current shortfalls in our understanding of multifunctional agriculture emphasise that we urgently need a *new* concept of multifunctionality that is conceptually and theoretically better anchored in current debates on agricultural change than has hitherto been the case. In particular, we need a model of multifunctionality that is *holistic* and goes beyond mere economic and policy-based understandings, and that is applicable not only in a European context but also *globally*. Academic and scientific debate needs to *re-appropriate* the notion of 'multifunctionality' from policy-makers who have used the term in rather cavalier fashion, expose it to more thorough theoretical analysis, and reconceptualise it into a notion that can be used to *explain* what is happening in the countryside.

1.3 Aims of the book: conceptualising multifunctional agriculture

The aims of this book are to, first, shed a critical light on the p/pp transition model, and to analyse whether 'post-productivism' is the best theoretical framework to conceptualise contemporary agricultural change. Second, this will be used as a theoretical and conceptual springboard for analysing the concept of a transition towards multifunctional agriculture that may provide a more robust 'alternative' concept describing contemporary agricultural change. As this book will highlight, several issues have emerged that threaten the robustness of the p/pp transition. It has been argued that the bipolar assumption of the transition does not fully encapsulate the diversity, non-linearity and spatial heterogeneity that can be observed in modern agricultural systems (Wilson, 2001, 2002). Instead, I will suggest that the notion of *multifunctional agriculture*, rooted in a revised *productivist/non-productivist* model, may be a more appropriate concept to describe and comprehend contemporary agricultural/rural trajectories. I will argue for a significant reframing of conceptualisations of agricultural change and for the dispelling of the myth that modern agriculture is now well and truly post-productivist. In particular – inspired by visionary approaches to multifunctionality by authors such as John Holmes (Australia), Karlheinz Knickel (Germany), British researchers Terry Marsden, Jules Pretty, Julian Clark and Clive Potter, US researchers James McCarthy and Gail Hollander, and French commentators Hélène Rapey, Bruno Losch, Sylvie Lardon and Patrice Cayre – I will challenge existing understandings of multifunctional agriculture that see it merely as an economic or policy-based *process* and suggest, instead, that multifunctionality should be understood as an overarching normative *concept* that both describes *and* explains contemporary agricultural change.

The analysis will be based on what has been loosely termed *transition theory* (Pickles and Smith, 1998; Rotmans *et al.*, 2001). Through the lens of this theory we will be able both to identify the inherent weaknesses of the p/pp transition model and anchor the notion of multifunctionality theoretically as a crucial part of, and possible alternative concept to, this postulated transition. As Holmes (2006: 159) argued, "the concept of a multifunctional rural transition invites positioning within current theory on the role of place and space in contemporary society". Similarly, Goodman (2004: 10) reminded us that many current academic debates "overlook the complexities of transition, its uneven spatial and temporal intensity, and the possibility that processes of change may not engender convergence". Chapters 2-4 will highlight that transition theory provides a particularly useful framework to assess whether modern agricultural systems have moved towards post-productivism and/or multifunctionality, as it places the nature, pace and processes surrounding 'transition' at the heart of the investigation. Transition theory will help us understand how agricultural systems have changed over the last 50 years, why these changes have occurred, and the form that this transition has taken. It will also help us understand whether a *full* transition to post-productivism has occurred and whether this transition has been linear and homogenous. I, therefore, wish to take theorisations of agricultural change beyond the confines of mono-disciplinary approaches, and the lens of transition theory provides an ideal *multi-disciplinary* platform from which to assess, re-evaluate and reconceptualise current thinking on agricultural change.

One of the aims of the book will be to go beyond UK/Euro-centric conceptualisations of multifunctionality and to investigate whether and how theorisations such as post-productivism and multifunctionality may also find applicability in the developing world. I wish to explore how easily the concept of multifunctionality can be 'exported' to the situation in the South, partly as a response to the scarcity of previous efforts to link theoretical themes of agricultural change between the advanced and the developing world (Wilson and Rigg, 2003), and partly inspired by Pretty (1995, 2002) whose seminal work on regenerating global agriculture is one of the few examples that elegantly moves between both the advanced and developing world without losing sight of the dramatic changes facing agriculture, farmers and rural society in any location on Earth. Although transition theory has often been combined with *scenario building* based on notions of path dependency and transitional trajectories, it is not the aim of this book to engage into a large-scale discussion of future scenarios, but Chapter 10 will include a discussion of the challenges faced by different decision-makers about the best management strategies for future agricultural multifunctional transitions. I also hope to contribute to contemporary thinking about societal transitions as a whole. In particular, I wish to address the question how current debates on the transition to post-productivism in agriculture and rural society compare with other 'parallel' (and, at times, not so parallel) debates on transitions from 'isms' to 'post-isms'. How can we conceptualise transition as both a process of social and environmental transformation? Here, my background as a *human geographer* is vital, as the discipline of geography – possibly more than any other social science discipline – enables insight into a variety of sociological, economic, political, cultural, environmental and spatial dimensions that form key building blocks of conceptualisations of the p/pp transition and multifunctional agriculture.

As the title of this book implies, the focus will be largely on *agricultural* rather than *rural* multifunctionality. This is deliberate, although I will also analyse repercussions for rural areas, in particular when discussing conceptualisations of multifunctionality in Chapter 9. I acknowledge that it is increasingly difficult to conceptually separate these two terms, especially if we wish to extend discussions beyond the UK where understandings of the 'rural' can differ considerably from UK-centric interpretations. Throughout, I will consider the difficulty in applying notions linked to the p/pp transition model and multifunctionality to *both* the 'agricultural' and 'rural' realms. The rural focus will be particularly evident with regard to some of the key transitions that underpin notions of post-productivism and multifunctionality, such as changing ideologies towards agriculture, changing spaces of agricultural actors (who synonymously are also rural actors), agricultural policies (that cannot be easily separated from rural policies), and food regimes and agro-commodity chains (that can rarely be understood in their entirety by focusing on agriculture alone). Multifunctionality will not only be conceptualised in light of the p/pp transition model for agricultural change, but also with regard to multifunctionality issues beyond the farm gate.

I acknowledge that the terminology 'multifunctional agriculture' is far from ideal, particularly as its emphasis on 'agriculture' may indicate (to English-speaking readers) an over-emphasis on agricultural production and, therefore, on traditional 'productivist' notions. Nonetheless, if the notion of multifunctionality is to be broadened beyond advanced economies, the more neutral terminology of multifunctional *agriculture* (as opposed to rural) may be more appropriate. The starting point of any critique of the p/pp transition model and for any conceptualisation of multifunctionality has to be *agriculture*, as it is within this framework that these terms were initially conceptualised (Cloke and Goodwin, 1992; Ward, 1993). Agricultural activity is still one of the key vectors for ensuring a close and lasting connection between society and the land(scape) and continues to be an important type of land use in most parts of the world (e.g. over 50% of the total territory of the European Union is classified as 'agricultural land'). When referring to 'agriculture', I will use the traditional definition of agriculture as the economic sector in charge of *food and fibre production*[1] with its associated actors, institutions and politics (Le Heron, 1993). Yet, it is also this seemingly 'precise' definition of 'agriculture' that lies at the heart of the investigation in this book. Chapters 7-10 will show that the seemingly 'clear' boundaries of what constitutes 'agriculture' – and, concurrently, the distinction between 'agricultural' and 'rural' – are increasingly blurred (Woods, 2005). In Chapters 9 and 10, for example, I will argue that a conceptual non-productivist territory that lies 'beyond agriculture' may exist within a broader spectrum of multifunctional agricultural pathways, and that the notion of 'agriculture' itself can never be completely divorced from notions of *productivism*.

This book should, therefore, be of interest to a wide audience including both agricultural *and* rural researchers, but particularly to those who are interested in issues of agricultural change, agricultural/rural theory, and, at a broader scale, in theorisations of societal change and transition. The novel theoretical framework adopted here is designed to provoke further debate and comparative enquiry integral to theory building from any disciplinary vantage point. It is also hoped that this book will provide an effective intellectual groundwork, following on from seminal publications on the p/pp transition and multifunctionality highlighted above, that will stimulate further theoretical and empirical explorations of agricultural transitions. It will already be obvious to the reader that I want to write about ideas, debates and conceptual approaches, rather than stating 'facts'. Many subject areas will, therefore, only briefly be mentioned, left unfinished or may not be as fully explained as could be. Although I will assume some background knowledge on agricultural change, food networks, agricultural and rural policy, agriculture and conservation issues, and on the role of specific actors and stakeholders in various agricultural and policy-making processes, I hope, nonetheless, that there will be sufficient discussion of pertinent debates to implant further ideas and to stimulate thought, discussion and further reading. As a human geographer, I also particularly wish to emphasise the importance of geography and spatial patterns in the conceptualisation of multifunctional agriculture, and throughout this book the *geography* of productivism, post-productivism and multifunctionality will, therefore, be an important underlying theme.

---

[1] Fibre production refers to agricultural fibre plants (e.g. cotton, flax or linseed), not to the production of fibre from forestry which is usually treated as a separate economic sector (see Robinson, 2004).

1.4 Structure of the book

The structure of the book mirrors the theoretical orientation based on transition theory. It provides a step-by-step approach in three distinctive parts that aims to provide the theoretical background for conceptualising transition and introducing the reader to 'transition theory' (Part 1), discusses debates on the transition from productivist to post-productivist agriculture and deconstructs unilinear assumptions about the nature of this transition (Part 2), and introduces 'multifunctional' agriculture as a concept theoretically embedded in productivist and non-productivist action and thought (Part 3).

Part 1 will begin with an overview in Chapter 2 of what 'transition theory' means and will outline key models of transition. Chapters 3 and 4 will then analyse key debates and theories surrounding the notion of transition. Specific emphasis will be placed on analysing different models of transition, and on highlighting key debates surrounding the possible transition from 'isms' to 'post-isms' in the social sciences (e.g. Fordism to post-Fordism or socialism to post-socialism). In particular, I will show in Chapter 4 that current debates on the transition to post-productivist agriculture share many similarities with debates about 'other' transitions. I will highlight that, parallel to many other debates on transitions, conceptualisations of post-productivism need to take into account issues of temporal non-linearity, spatial heterogeneity, global complexity and structure-agency inconsistency currently lacking in debates of agricultural change.

This will form the basis for understanding why we may need to question the seemingly 'simple' notion underlying the p/pp transition model (Part 2). In order to set the wider context for the debates, Chapters 5 and 6 will analyse issues and debates surrounding conceptualisations of 'productivist' and 'post-productivist' agriculture. These chapters will pay particular attention to several inter-linked dimensions that have formed the basis of conceptualisations of the post-productivist transition. Chapter 7 will then broaden the discussion by linking the debate on the p/pp transition to parallel debates on 'other' transitions discussed in Part 1. The chapter will outline the strengths and weaknesses of the p/pp transition model, and specific emphasis will be placed on analysing the assumptions of linearity, homogeneity, universality and causality. Acknowledging that the p/pp transition model has made a vital contribution to current debates on agricultural transition, I will suggest that post-productivism should not be seen as the 'end-point' of agricultural change, and that, instead, the notion of a productivist/*non*-productivist spectrum of decision-making forms a better non-linear conceptual model for understanding agricultural change.

Part 3 will build on this discussion by arguing that, on the basis of the critique of linear and directional assumptions underlying the post-productivist transition model, the notion of a multifunctional decision-making spectrum bounded by productivist and non-productivist action and thought better encapsulates the non-linearity and heterogeneity of agricultural systems. I will begin Part 3 by critically examining contemporary conceptualisations of multifunctionality (Chapter 8). Chapter 9 will then *theoretically anchor* the notion of multifunctionality in the context of the productivist/non-productivist decision-making spectrum discussed in Chapter 7. This will enable us to investigate in more detail the conceptual boundaries of multifunctional agricultural and rural systems. I will suggest that discussions about multifunctional agriculture have to imply *normative* and *subjective* value judgements and, based on the conceptualisation of weak, moderate and strong multifunctionality, I will argue that the strong multifunctionality model is qualitatively and morally the 'best' model to follow and implement. Chapter 9 will also investigate the geography of multifunctionality with specific emphasis on the multi-layered nature of multifunctional processes from the local to the global. Chapter 10 will analyse issues linked to multifunctional agricultural transitions over time. As the focus will be on transitional processes, there will be strong interlinkages between Chapter 10 and Part 1. Chapter 10 will highlight that it is possible, based on the new multifunctionality framework developed in Chapter 9, to identify the transitional potential of individual farms and individual nation states. The chapter will also investigate the key issues of path dependency, transitional corridors, system memory and transitional ruptures for the understanding of multifunctional transitional processes in both the developed and developing world. The chapter will conclude by discussing how the strong multifunctionality model could be implemented, by whom it should be orchestrated, and will discuss various challenges facing future decision-makers in the quest towards strongly multifunctional transitional pathways. Chapter 11 will conclude the book by highlighting the key arguments of the reconceptualisation of multifunctionality, and by arguing that more empirical work will be needed in future to further substantiate theoretical and conceptual issues of multifunctional transitions.

Inevitably, the complexity of the argument presented in this book will make for lengthy reading, especially for those interested in agricultural/rural issues or multifunctional agriculture and less familiar with transition theory and debates on societal change. Readers who may find themselves getting bogged down in Part 1 (transition theory), or in debates on the transition to post-productivism (Part 2), may wish to skip ahead to Chapter 8 to read immediately about multifunctional agriculture itself. However, the argument in this book is based on a logical 'storyline' that is intended to narrow down the argument from the broad (transition theory) to the intermediate (deconstructing post-productivism) to the specific (conceptualising multifunctional agriculture), in which each component builds on the others and may, therefore, make less sense on their own. As a result, I hope that this book will not only form an important reference point for readers interested in agricultural and rural issues revolving around multifunctionality, post-productivism and the nature and pace of global agricultural change, but that it will also act as a trigger for future theoretically informed empirical work on conceptualising societal transitions in general.

Part 1

Conceptualising transition

The first part of this book comprises three chapters which set the scene for conceptualising the notion of 'transition'. This will form the basis for understanding conceptual and theoretical weaknesses of traditional models of agricultural change linked to 'productivism' and 'post-productivism'. **Chapter 2** provides insights into 'transition theory' – the conceptual framework used in this book – and discusses different models of transition. Eight 'classic' examples of transition from both the social and natural sciences are discussed in **Chapter 3** as a basis for understanding key assumptions underlying transitional processes, and to highlight the pitfalls of dualistic and linear assumptions underlying many debates on transition. A critical review of the eight examples of transitions is then used in **Chapter 4** to highlight the diversity, non-directionality and heterogeneity inherent in transitory systems. This will provide the framework for applying transition theory to the evolution of agricultural systems. In particular, it will form the basis for the deconstruction of the unilinear and binary assumptions underlying the p/pp model in Part 2 of this book, and for theoretically anchoring the notion of agricultural 'multifunctionality' in Part 3. For those interested in *transitory processes* in general – especially as a basis for understanding problems associated with unilinear assumptions underlying the p/pp transition model – Chapters 2-4 will provide a crucial framework within which to contextualise discussions in the remainder of this book. I hope, therefore, that in the following the strikingly parallel debates, critiques and reconceptualisations across various disciplines and fields of investigation analysing transition will become apparent.

Chapter 2

Theorising transition

2.1 Introduction

The aim of this book is to critically evaluate the recently postulated transition towards post-productivism, and to use this analysis as a theoretical and conceptual springboard for conceptualising multifunctional agriculture. Critiques of the p/pp model of agricultural change gain particular prominence if we view debates on the p/pp transition as part of wider theoretical debates about the nature, trajectories and philosophical foundations of other societal transitions. This is particularly true for debates on the possible transition from other 'isms' to 'post-isms' discussed in Chapters 3 and 4. *Transition theory* forms a useful theoretical framework within which to contextualise the approach in this book. This will help us understand the nature, cause and outcomes of the postulated post-productivist transition and to cast a critical eye on the assumptions underlying this transition. Section 2.2 will analyse what 'transition theory' means and how transition has been conceptualised from different disciplinary vantage points. Section 2.3 then discusses six basic models of transition that emerge from critical analysis of existing 'transition studies'. This will provide the conceptual framework for both the deconstruction of the postulated unilinear transition to post-productivism (Part 2) and for theoretically anchoring the notion of 'multifunctionality' (Part 3).

2.2 Transition theory – theorising transition

Transition theory is a general theory at the heart of which lie general principles, patterns and processes that are applicable across different, and apparently unrelated, fields (Pickles and Smith, 1998; Martens and Rotmans, 2002). The increasing importance of transition theory in the social sciences is linked to recent radical changes in conceptualisations of societal change linked to the 'cultural turn'. As Pickles and Unwin (2004: 14) argued, the cultural turn meant that "scholars were now more interested in the contingencies of transitions and the ways in which local people and ways of life shaped the specific outcomes of transition in particular places". Chakrabarti and Cullenberg (2003: 5) argued that 'transition theory' now forms one of the key analytical components in the social sciences, as concerns over development trajectories "have multiplied in recent years given the amount of literature on transition and development as well as the media coverage on the transition process". Theorising transition has also received added impetus through emergent literature on *complexity theory* in which the complex nature of transitional processes (e.g. path dependency; see below) is becoming increasingly evident (O'Sullivan, 2004).

   However, transition theory has not yet gelled into a coherent theoretical framework that spans both the social and natural sciences (Burawoy, 1985; Pickles and Smith, 1998), although some overarching concepts and ideas for theorising transition have been applied across disciplinary and philosophical divides (e.g. Frigg, 2003, on self-organised criticality).

In this sense, transition theory shares common ground with other newly emerging 'theories', such as complexity theory, that are seen by some as 'quasi-theories' rather than fully-fledged theories on their own (Thrift, 1999). As a result, Pickles and Unwin (2004: 11) argued that "for many, 'transition' is better understood as 'transformation', a process of reworking in which the legacies of the past are the resources for the struggle over the construction of whatever is new". Rotmans *et al.* (2002: 3) argued that "there is still not enough scientific proof to legitimise the use of the concept of transitions in a broad social sense. Nevertheless, there are sufficient indications that the concept of transitions is an attractive, useful and helpful aid for figuring out social complexity and coherence". Similarly, Reason and Goodwin (1999) suggested that quasi-theories such as transition theory are still useful as they offer principles and metaphors that may inspire more thorough future theorisation and richer understanding of transitional processes. In the following, I will, therefore, draw on a wide variety of literature on transition from both social and natural science perspectives, and identify specific patterns and parallels surrounding different notions and types of transition. In addition to this book's main contribution towards understanding the transition towards multifunctional agriculture, I also aim to contribute towards refining and further developing the *theory of transition* itself.

Transition surrounds us all, and terms such as 'beginning/end' or 'start/finish' highlight the importance of transition in everyday life. Since the Big Bang (apparently about 15 billion years ago) everything in our universe has been in transition, with time itself as the ultimate and endless transition (Davies, 1995). Human life is also synonymous with transition with its stages of birth, youth, adulthood and death – a seemingly inevitable linearity used in many scientific works attempting to unravel and understand 'macro-concepts' such as the rise and fall of human societies (Spengler, 1992, is probably the most prominent author to apply linear notions in his classic text *The decline of the West*; see also Toynbee, 1962; Diamond, 2006). Time and human life particularly highlight two of the main theoretical underpinnings of transition theory as some transitions have no specific start and end point (e.g. the passing of time itself), while other transitions may have a clear start and end (e.g. human lifespan). Agricultural transition has had a clearly defined starting point (i.e. when the first animals and plants were domesticated some 10,000 years ago), but continues to be in the process of transition with no clearly visible 'end point' – although, and as will be argued in Chapter 6, some see current 'post-productivist' changes in modern agriculture as the possible 'end of agriculture' (Marsden, 1999b). However, the focus of this book will not be to understand the transition in agriculture since its inception 10,000 years ago (see Mannion, 1995, and Mazoyer and Roudart, 2006, for good discussions), but to focus on a small, but arguably crucial, temporal slice of this transition for the time period since about 1950. The most relevant debates on 'other' transitions that will help us understand this recent agricultural transition will, therefore, also be mostly related to theorisations of *recent* societal transitions that have occurred over the past 100 years or so.

Theorisations of such transitions have been led in particular by economic theorists (e.g. Lipietz, 1992), sociologists (e.g. Giddens, 1979) and political scientists (e.g. Pickles and Smith, 1998). *Transition theory* has been strongly embedded within regulation theory (e.g. Dunford, 1998; Smith and Swain, 1998), as both transition and regulation theory share the conception that societal transition phases, i.e. the structuring and restructuring of everyday life, occur "within complex articulations of local, regional, national and globalizing contexts. Histories, political economies, discursive formations, and institutional assemblages and practices each comprise complex articulations of *universalizing* and *particularizing* processes" (Pavlinek and Pickles, 2000: 22; emphases added). As economic theorist Lipietz (1987) argued, these articulations can be seen as structuring moments which *normalise* and *regulate* social, economic and political life. Lipietz (1992) further referred to 'development models' that characterise particular economic and political organisational forms adopted (or imposed by external forces) during different stages of societal transition, including specific 'regimes of accumulation', 'modes of regulation' and 'hegemonic blocks'. These involve power relations in politics, ideology, culture and behaviour that contribute to securing the continuation and stability of a particular 'ism'/'post-ism' – an assertion also relevant for conceptualisations of the p/pp transition.

Rotmans *et al.* (2002: 3) argued that transition can be defined as "a gradual, continuous process of societal change where the structural character of society (or a complex sub-system of society) transforms". Pickles and Unwin (2004: 10) further suggested that "in common usage 'transition' generally implies a change from one known state to another, with a clean break between them". I broadly concur with both definitions, but, as the remainder of this book will demonstrate, the notion of transition as 'gradual' and 'continuous' is problematic, as is the suggestion that there necessarily should be a 'clean break' from one transitional stage to another. I suggest that, from a social science perspective, transition theory should be seen as *a theoretical framework that attempts to understand and unravel socio-economic, political, cultural and environmental complexities of societal transitions (or sub-systems of society such as agriculture) from one state of organisation to another* (or, indeed, multiple 'other' states of organisation; see below). Transition theory suggests that, at times, coherent phases of societal organisation can be identified (e.g. agricultural productivism), while at other times complex and even chaotic transitional characteristics may dominate, leading eventually to a new set of 'structured coherences' (e.g. agricultural post-productivism) (Cloke and Goodwin, 1992). Latour (1993: 10) described transition as a clear "break in the regular passage of time", and argued that it often designates a "combat in which there are victors and vanquished" – in other words, where a change of the 'old guard' with its 'old ideas' is occurring. Transition theory, therefore, often assumes that there are certain key stages or periods in societal transition (but never 'end points'; cf. Davies, 1995), and that any of these stages may, in turn, become the starting point for the next transition – an assumption that will be critically examined throughout this book with regard to the p/pp transition model.

There are differences between the application of transition theory in the social and natural sciences that need to be emphasised here – especially as the discussion below will also briefly touch upon natural science concepts of transition. In its most fundamental form, the notion of transition was arguably first theorised by *natural scientists* (especially physicists) in an attempt to understand transitions in natural systems (Rotmans *et al.*, 2002). While in the natural sciences transition processes are usually *non-anticipatory*, where a system under investigation (e.g. natural landscape change) cannot forecast and adjust for a change in output (Thornes and Brunsden, 1977), human systems are usually *anticipatory*. In the latter, *system memory* is crucial and may lead to a *learning* and *adjustment* phase based on past experience that streamlines transition pathways that may not exist in natural systems. Transitions in social systems are, therefore, *non-deterministic*.

Transition theorists with roots in evolutionary political economy point to the importance of *path dependency* that shapes the nature and pace of societal transitions (Stark, 1992; Grabher and Stark, 1997). Here, the emphasis is largely on institutionalised forms of learning and institutional thickness (Amin and Thrift, 1993) and on the fact that personal choices can be self-reinforcing and, therefore, self-fulfilling. In other words, means may become ends and alternative pathways may not even be considered. Such 'lock-in effects' may be more typical for human systems, and imply that once a transitional pathway has been chosen it may be very difficult to leave this pathway due to various cultural, socio-economic, political and institutional factors (O'Sullivan, 2004). Thus, Rotmans *et al.* (2002: 3) argued that "a transition process is not set in advance, because during a process of change, humans are able to adapt to, learn from and anticipate new situations". Hudson (2000: 301; emphasis added) rightly cautioned for social transitions that "it remains an open question as to whether a revolutionary shift from one path to another can be achieved through *incremental change* and evolutionary reformist modifications to the existing developmental trajectory or whether it requires a *rapid quantum leap* from one trajectory to a qualitatively different one". For human systems, this means that forecasting the effect of transition may be even more complex than it is for natural systems, as the direction of change is influenced by both the passing of time and, often unpredictable, human adjustment strategies (Diamond, 1998, 2006). In addition, and as Chapters 3-6 will show, the direction and pace of transitional pathways can be influenced by policy and other institutional interventions. This suggests that there is no 'theory-neutral' way of both observing and understanding transition. Darwin's work, in particular, has emphasised the 'theory-ladenness' of scientific research, while social scientists have usefully pointed out that the notion of transition is, and can only be, socially constructed.

2.3 Models of transition

In this section, I wish to discuss different models of transition. The purpose is to highlight that transitional models are useful to show that there are many different ways of conceptualising 'transition', i.e. that there is not just one pathway to understand how a transition can occur. Six basic models of transition can be conceptualised on the basis of various studies investigating transition. These models can be represented as two-axis representations that show time on the x-axis and the trajectory of a specific societal or natural change on the y-axis (Fig. 2.1).

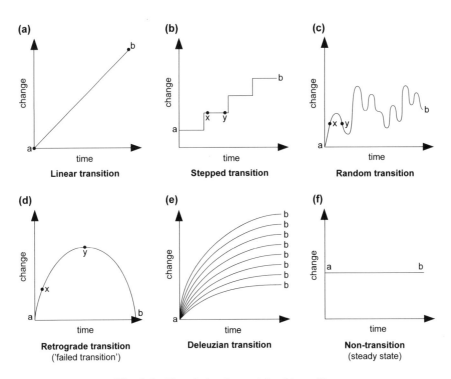

Fig. 2.1. The six basic models of transition

A good starting point for our discussion is the **linear transition model** (Fig. 2.1.a). Here, transition from a given starting point 'a' in time is seen to evolve along a linear trajectory that develops at a steady and regular pace along a seemingly predictable straight line towards 'b'. Because of this predictability, this type of model is *deterministic* (i.e. we can determine the direction of future change) in which the importance of *predictability* is key. This model is closely related to the concept of 'transformationism' or Lyell's notion of 'uniformitarianism' based on the gradual change of an object or its essence (Mayr, 2002). It is also interesting to note that while pre-Christian concepts of time and transition may often be dominated by the non-linear metaphor of the cycle (i.e. ever-returning seasons), modern societies based on Christian beliefs tend to portray the image of time in linear form as a pathway from sin on Earth, through redemption, to eternal salvation in heaven (Eliade, 1959). Notions of linearity are, therefore, closely associated with Judeo-Christian conceptualisations of change (see below), and "the distinguishing feature of ultimate progression has led the way to a new linear concept of time and, with it, a sense of firm beginning" (Hassard, 2001: 133).

Linear transition models are more common in the *natural sciences*, and "linear equations are arguably the most widely used mathematical structure in science" (Frigg, 2003: 622). Common examples of this model relate to biological or physical laws of transition over time, such as the change of a star from one type (white, yellow, red, blue) to another, the gradual rise of a mountain range through tectonics and its subsequent erosion, gradually increasing levels of oxygen in Earth's atmosphere over billions of years, or the gradual cooling of the Earth's surface in the first few hundred million years of its existence. However, even with these seemingly linear transitions, over time the linear pathway can be questioned (e.g. the impacts of large asteroids with Earth which would have substantially disturbed linear processes). It is even more questionable whether linear transitions can *ever* be observed in any domain of the *social sciences* to explain societal change, due to the often unpredictable and even random nature of social development (e.g. Polanyi, 1944).[2] However, some social scientists suggest that societal change can, at times, occur along relatively linear pathways (see Wallerstein, 1979, 1991, or Frank and Gills, 1993, on 'world-systems theory'). As we will see in Chapters 5 and 6, there are some proponents of the p/pp model who have suggested a relatively linear development trajectory from productivist to post-productivist agriculture that has, seemingly, followed a clearly definable temporal trajectory.

The second type of transition is the **stepped transition model** (Fig. 2.1.b). In contrast to the linear model, transition is seen here to involve relatively 'stable' phases (Rotmans *et al.*, 2002, term this the 'predevelopment' phase) that are then suddenly 'overthrown' ('take-off' and 'acceleration' phases) and replaced by a new 'stable' phase ('stabilisation' phase) that, in turn, will eventually be overthrown through revolutionary change. Similar to the linear model, it is also a *deterministic* model in which the 'direction' of change is predictable (i.e. the direction of change is usually into one direction, i.e. 'upwards'). In the natural sciences, this line of reasoning has been particularly popular among 'catastrophe theorists'. Kay and Schneider (1995: 52) argued that "catastrophe theory describes the change in systems over time. It predicts that systems will undergo dramatic, sudden changes in a discontinuous way ... The general insight from catastrophe theory is that the world does not always change in a continuous and deterministic way". Catastrophe theorists suggest that many of Earth's landforms have not been formed by gradual processes, but by sudden catastrophes or 'discrete events' (e.g. meteorite impacts or sudden natural dam bursts) followed by phases of little and gradual change (Thornes and Brunsden, 1977). There are parallels here with the notion of *self-organised criticality* that originated in physics to explain organisational patterns in natural systems based on the notion of a self-organised critical state (or threshold) beyond which a 'new' discrete level or pathway of development is reached (Frigg, 2003). The stepped transition model has been particularly popularised by Kuhn (1970), who used this concept to describe the structure of scientific revolutions. Kuhn argued that scientific paradigms evolve uninterrupted for a certain period of time until they are challenged to such an extent by new ways of thinking so that they are replaced by a new (and arguably more 'elevated') paradigm that would, in turn, dominate thinking for a certain time. Again, we will see in our discussion in Chapters 3 and 4 that some debates in the social sciences (notably

---

[2] Key examples of unpredictable and non-linear transitions in societal development include most 'revolutions', such as the French Revolution in 1789, the Paris Commune of 1871, the Russian Bolshevik Revolution in 1917 and China's Cultural Revolution of the 1960s and 1970s (Chase-Dunn and Hall, 1997).

those on recent economic transitions) suggest that social change may often occur along such 'stepped' development trajectories (see also notion of 'bifurcation' developed by Wilson, 1981, that discusses the shift from a linear relationship to a periodic or even chaotic one as applied to urban systems; or Bracken and Wainwright's, 2006, notion of 'equilibrium' in geomorphological transition).

The **random transition model** (Fig. 2.1.c), meanwhile, is closely associated with the notion of 'chaos'. Contrary to the linear and stepped models, transition is seen here as unpredictable, random and non-directional (Gleick, 1988). It is both a *probabilistic* and *non-deterministic* model that suggests that the future can *not* be predicted. Chaos theory argues that "change in any dynamic system is ultimately not predictable, because individually small interactions between components accumulate" (Kay and Schneider, 1995: 52). This model can be seen as a critique of the unidimensional and directional linear and stepped transition models, although the random transitional model is still based on a somewhat linear trajectory from point 'a' to point 'b'. Examples from both the natural and social sciences relate to development paths that are highly unpredictable (e.g. some theories of the development of the universe or the currently wide-ranging debates about global warming may be cases in point), and, arguably, many societal developments will fall under the umbrella of such random, unpredictable and non-directional transition trajectories (Polanyi, 1944; but see also Wallerstein, 1979, 1991). Discussions about 'Kondratief long-wave cycles' may have repercussions for the conceptualisation of random transitional processes, as increasing evidence suggests that what may initially appear as a random transition may show previously 'invisible' long-term patterns (Bracken and Wainwright, 2006).

The random transition model shows parallels with our fourth model, the **retrograde transition model** (or 'failed transition') (Fig. 2.1.d). In this model, although a transition has taken place over time along a specific development trajectory, points 'a' and 'b' may be situated at the same development 'level' after a certain time has elapsed (i.e. no real change has occurred).[3] We will see that the notion of 'failed transition' is becoming increasingly prominent in social science debates on transition, especially concerning the perceived failure of economic transition from socialist to market-led economies in countries of the former Soviet Union (see Chapter 4).

The fifth type of transition is less well established in the literature and less clear-cut than above examples. I term this the **Deleuzian transition model**, after Deleuze and Guattari's (1987) notion of a 'thousand plateaus' of societal development (see also Frank and Gills, 1993) which shows a complex conceptualisation of transition (Fig. 2.1.e). As a *non-deterministic* model (but also not quite probabilistic), this model interlinks with social science debates on 'co-evolution' that argue for increasingly complex interlinkages and pathways of space, place and human territoriality (Norgaard, 1994; Graham, 1998), with Harvey's (1996: 260-261) notion of 'cogredience' as "the way in which multiple processes flow together to construct a single consistent, coherent, though multi-faceted time-space system", and with Amin's (1990) notion of multiple development paths in an increasingly 'polycentric' world. This model argues that it is impossible to identify one transitional pathway. Instead, and starting from temporal point 'a', we may have a thousand (or more) possible transitional pathways to arrive at point 'b'.[4] Some authors, therefore, argue that transition should not only be conceptualised as a directional movement from one point to another, but that a transition itself can act as a trigger for specific action, akin to a development with a life of its own (Deleuze and Guattari, 1987). This theorisation suggests that, once embarked on a pathway, the outcome of that transition is indefinable and that any of the multiple pathways may ultimately lead to the 'final outcome'. Yet, individual

---

[3] Although Figure 2.1.d depicts the retrograde transition as a smooth linear development, there is no reason to argue why this type of transition may not also be 'stepped'.

[4] Again, there is no specific reason here why the parallel transitional trajectories in the Deleuzian transition model have to be linear. As with the retrograde transition model, transitional pathways in the Deleuzian model may be stepped or, indeed, even retrograde.

pathways necessarily should not be seen as closed systems, as *complementarities* may exist between different (and closely related) trajectories. With the notion of Deleuzian transitions we arguably leave the realm of the imaginable and enter the notion of 'parallel universes' where parallel transitions may occur simultaneously, each with its own distinct and universe-specific trajectory. However, what this model usefully highlights is that understanding transition may be based on different and multiple vantage points, world views and interpretations – i.e. that one and the same natural or social science development issue may be interpreted differently depending on who does the interpretation. We will have to bear Deleuze and Guattari's cautionary viewpoint in mind when discussing *specific* productivist, post-productivist and multifunctional agricultural transitional pathways with seemingly *specific* outcomes in subsequent chapters of this book. In particular, Chapters 9 and 10 will present various conceptual models of multifunctional agricultural pathways that will highlight that the transition towards multifunctional agriculture shows many similarities with the Deleuzian transition model and that, indeed, this model may be the closest in depicting the 'reality' of agricultural transition overall.

The final model is the **non-transition model** (or steady state model) (Fig. 2.1.f). This model highlights that in any discussion of 'transition' as a concept, there needs to be room for the notion of 'non-transition', i.e. where over the timespan between points 'a' and 'b' nothing changes. Although there is no such thing as a time-independent process, there can be processes whose effects are negligible over very long time periods (Davies, 1995). Examples of this will be impossible to find in the social sciences, as human society can not stand still and 'something' changes all the time (although such change may not conform to linear or stepped transition models as outlined above). It may be easier to think of relevant examples of non-transition from the natural sciences (where time periods analysed are often much longer when compared with human systems). For example, dying stars or dying planets may stay 'the same' for billions of years (i.e. the core has cooled; there are no surface processes). However, even in these time scales of seemingly unchanging eons, the odd catastrophic meteorite impact may cause (at times substantial) ripples in our seemingly 'flat' line of non-transition, thereby creating a transitional pathway more akin to the stepped transition model (Fig. 2.1b), albeit with a non-transitional 'plateau' that may last billions of years.

The model of non-transition should, therefore, be seen as a *hypothetical model* that may never be applicable in the 'real' world. Indeed, the *transitory* and *directional* passing of time itself can be seen as the basis of the most elementary form of 'transition' (Davies, 1995), and as Thornes and Brunsden (1977: 2) succinctly argued, "time is distinguishable [from other processes] by possessing the property of intrinsic direction and in the macroscopic sense being irreversible". Time goes forward and what has passed is 'lost' in the sense that any system *can not* alter past events in its feedback or system memory controls.[5] This is the reason why the discussion of transition models above has not considered *cyclical transition models*, as societal transitions can never revert back to a former stage of societal development but always imply a forward or directional development over time (Nietzsche, 1887/1967). It is here that the notion of *system memory* is important – a notion that is particularly relevant in human systems where 'memory' (i.e. knowledge, experience, accumulated wisdom) can be passed on from generation to generation or from actor to actor. Any human system will be at its specific starting point in a transition precisely because of the history of decision-making trajectories *preceding* that starting point. In other words, a system carries with it the memory – or, in a more negative sense, the 'baggage' – of previous decision-making trajectories. In human systems in particular, *history matters* and transitional

---

[5] Einstein's theory of relativity has, of course, challenged the notion of linear time, and has highlighted that all transitions are *relative* events depending on the viewpoint of the observer. In addition, human *perception* of time is often not linear. Anthropological studies have shown that the notion of *linear time* is more closely associated with industrial societies, the use of clocks, and Cartesian modes of rational thinking, while many tribal societies live in *cyclical times* that are closely associated with the the cyclicity of natural processes such as the onset of the rainy period or the passing of the seasons (Possemeyer and Killmeyer, 2005).

path dependency means that system trajectory is a function of past states (O'Sullivan, 2004; see also Chapter 10).

In addition to the six transition models discussed here, several model variations are also possible. A stepped transition may take the shape of an s-curve, rather than depicting sudden and abrupt changes, and may, therefore 'lie between' the linear and stepped transition models (Fig. 2.2). *Combinations* of different models should also be considered, for example a transition may start as a linear transition but may develop Deleuzian transitional pathways in later stages (Fig. 2.3). Some argue that pathways of time can be best described in this manner, with 'past time' depicting a linear transition that has already occurred and that is irrevocable, but where 'future time' may take any possible Deleuzian pathway depending on specific actions occurring in 'present time' (Grolle, 2004). The seminal work by Butler (1980) on transitional stages of tourist resort developments – starting with relatively linear transitional phases of exploration, involvement, development and consolidation and then branching out into possible parallel pathways ranging from 'decline' to 'rejuvenation' – is a case in point.

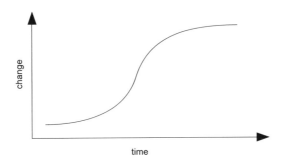

Fig. 2.2. S-shaped stepped transition model

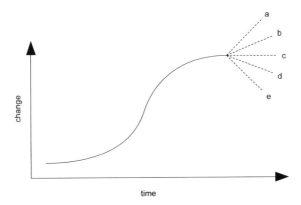

Fig. 2.3. Linear-Deleuzian transition (Source: author; after Butler, 1980)

As the word 'model' implies, the six types of transition models outlined in Figure 2.1 are *idealised representations of transition* that can only ever be approximations of 'reality'. As Morgan and Morrison (1999: 28) argued, "we do not assess each model based on its ability to accurately mirror the system, rather the legitimacy of each different representation is a function of the model's performance in specific contexts". A critical examination of assumptions underlying these models reveals two key points. First, although Figure 2.1 shows the development trajectories along a 'positive' y-axis (i.e. > 0), the notion of transition should be *value-free* and could, therefore, also be 'negative'. All models shown in Figure 2.1 will, thus, also have equally valid *negative mirror images* where the direction of the transition from temporal points 'a' to 'b' may point *downwards* (not shown in the figure for ease of representation). It is only when a specific transition is *interpreted* by humans that it becomes value-laden and directional (e.g. biological evolution theory; see Chapter 4), which results in the usual portrayal of transition as a 'positive' development trajectory (i.e. with a line going 'upwards'). This means that *transition* can not always be separated from the notion of *regression*, and that any discussion of transition should also take into account the possibility of a *downward tendency* and concurrent reduction in the x-axis 'value' between points 'a' and 'b' over time. It is, therefore, important to bear in mind that post-productivism (if it exists) may not necessarily be 'positive' compared to productivism (but see Chapter 9).

Second, the time frame of analysis (i.e. the temporal distance between points 'a' and 'b' in Fig. 2.1) may influence the interpretation of the nature, pace and direction of a transition. In most models time is *quantized* into imaginary (shorter) sections for ease of analysis, or because data or information may only be available for certain time periods. For example, the retrograde transition model (Fig. 2.1.d) may only be a *subset* of the random transition model (Fig. 2.1.c) if certain time points 'x' and 'y' are only selected for analysis. In a similar vein, if too short a time period is selected for analysis between time points 'x' and 'y' in the stepped transition model (Fig. 2.1.b), a long period of relative stability could be (wrongly) interpreted as a linear transition or non-transition – in other words that the short micro-transition time frame selected for research may have neglected to take into account the 'revolutionary' macro-transitional changes that may lead to a stepped transition over time. Linear transition trends may, therefore, at times also appear as a *subset* of a stepped transition. Finally, a subset of the retrograde transition (Fig. 2.1.d) between points 'x' and 'y' could be interpreted (again wrongly) as a linear transition. This is also relevant for our discussion of the p/pp transition and multifunctionality in Parts 2 and 3, as the chosen time frame for analysis (i.e. the past 50 years) may mean that we will only investigate a small 'blip' in what is a much longer transitional pathway that may well have a different long-term shape and direction (see in particular Chapter 10). Choosing an appropriate time frame for analysis of a specific transition is, therefore, crucial in order to identify large-scale transition patterns (macro-transitions).

2.4 Conclusions

This chapter has highlighted that *transition theory* forms a useful theoretical framework within which to contextualise our analysis of the p/pp transition and multifunctionality. The analysis has helped highlight underlying difficulties of conceptualising the meaning of transition and has shown that any interpretation of transition is inevitably a value-laden *social construction* of 'reality'. The six basic models of transition that emerge from critical analysis of existing 'transition studies' have highlighted the complexity of understanding transitional processes, and will form a useful basis for contextualising our discussion of the postulated transition from productivism to post-productivism and multifunctionality in agricultural systems. Chapter 3 will now apply transition theory to specific debates in both the social and natural sciences and will highlight some of the parallel (and not so parallel) debates surrounding the postulated shifts from one state of organisation to another.

Chapter 3

Transitions: social and natural science debates

3.1 Introduction

In the following two chapters, I wish to draw on the wealth of existing literature on 'transitions' to discuss specific debates that have been developed to understand shifts from one organisational state to another. The focus of analysis will be largely within the social sciences, but I will also draw on debates in the natural sciences (biological evolution theory), as patterns of transition transcend the boundaries between disciplines and, indeed, between the social and natural sciences. Chapters 3 and 4 set the wider context for debates on a possible shift from agricultural productivism to post-productivism by discussing parallel (and not so parallel) debates regarding *theories of transition*.

With reference to transitions in agriculture, "more work is needed on linking debates on productivism/post-productivism in agriculture/rural society with those debates in other 'post-...isms' that, at first glance, show some striking parallels in their questioning of directional and unambiguous shifts from one organisational state to another" (Wilson, 2001: 97). Similarly, Marsden (1999a: 504) suggested that "understanding the balance of economic, social and environmental processes which shape the contemporary countryside ... requires far more than the rigidly sectorialized forms of knowledge which have characterised rural research in the post-war period. A synergy between previously discrete knowledge bases is now needed". Such a synergy will help us, first, to understand that debates on the theoretical elements of p/pp transition model are paralleled by equally contested assumptions about 'other' societal transitions. In Chapters 3 and 4 I will place particular emphasis on debates surrounding transitions from 'isms' to 'post-isms', as the possible shift from productiv**ism** to **post**-productiv**ism** can, and should not, be understood in isolation. This has also been recognised in other debates on transition. Casterline (2001: 17), for example, argued for the demographic transition (discussed below), that this transition "has not occurred in isolation from other societal changes. Indeed, in their determination to describe and explain the profound demographic changes of the past two centuries, demographers can be faulted for failing to appreciate the scale of the other significant societal transformations that have occurred during this same time period". Other transitional debates, therefore, will help us provide a framework for deconstructing the implied linearity of the p/pp model in Part 2.

Analysing conceptualisations of other transitions will highlight that terms such as 'productivism' have long been used to conceptualise wider economic issues ranging from Fordism and 'productivist' ideologies underlying Marxist conceptions (Derrida, 1994), to 'liberal productivism' advocated by neo-liberal 'left-wing' EU governments (Giddens, 1998), to concepts of 'industrial productivism' criticised from a 'post-industrial' or 'post-materialist' vantage point (Peck and Tickell, 1994). Thus, any discussion of the p/pp transition would remain simplistic and devoid of wider context if it was not embedded in wider theorisations of other 'isms' and 'post-isms'. Chapters 3 and 4 will not give a complete analysis of existing debates on all types of transitions discussed in the literature. Instead, I will pick selective examples that are particularly useful in highlighting possible issues and debates surrounding unilinear transition models.

Section 3.2 will give insights into contemporary debates surrounding four transitions from 'isms' to 'post-isms' that are of particular relevance for our investigation in Part 2. Section 3.3 will expand this discussion by investigating four other 'classic' transitions (ranging from demographic to evolutionary transition). The aim will be to highlight that many parallel themes permeate debates on transitions – both in the social and natural sciences – and that these debates can be used to inform contemporary discussions on the transition towards post-productivism and multifunctionality in agriculture discussed in Parts 2 and 3. Section 3.4 will discuss why many conceptualisations of transitions have been based on relatively simplistic unilinear, binary and homogenous assumptions about the rate and pace of change. This will form the basis for understanding four key 'fallacies' underlying many transitional models discussed in Chapter 4.

3.2 Transitions from 'isms' to 'post-isms': insights into the debates

In this section I wish to look at the postulated transition from Fordism to post-Fordism, as well as the transitions to post-modernism, post-colonialism and post-socialism. The examples here are used to illustrate patterns and processes identified in discussions of 'transition' that may serve as useful starting points for our analysis of the p/pp transition. Although the focus will be on transitions that have *explicit* implications for debates on the post-productivist transition in agriculture (in particular Fordism to post-Fordism or modernism to post-modernism), I also wish to investigate two other debates (socialism to post-socialism; colonialism to post-colonialism) that, at first glance, have no explicit link with agricultural productivism/post-productivism, but that may *implicitly* yield valuable insights that may be useful for subsequent discussions in this book.

3.2.1 Fordism to post-Fordism

The postulated transition from Fordism to post-Fordism is an obvious starting point for two reasons. First, many authors have highlighted the striking parallels that exist between debates on post-Fordism and discussions of the post-productivist transition in agricultural systems (Cloke and Goodwin, 1992; Goodwin and Painter, 1996). Many have, for example, suggested that post-productivist agricultural systems are a direct expression of post-Fordist modes of accumulation (Philo, 1993; Halfacree, 1997b). Yet, Evans *et al.* (2002) argued that given the manifold discussions on the use of dualistic notions of Fordism and post-Fordism, it is surprising that a similar critique has not been applied to the p/pp transition. Second, much has been written on the post-Fordist transition by academic and populist writers, from a variety of disciplinary backgrounds and philosophical vantage points. Debates on post-Fordism highlight that there is some consensus that a transition to post-Fordist modes of accumulation *has* taken place in most advanced economies. However, arguments in the literature suggest that recent economic transition has *not* moved in a linear fashion but that it has often followed the *stepped transition model* (see Fig. 2.1.b above). As Chapter 4 will discuss, there are serious doubts whether the post-Fordist transition can be conceptualised in simplistic dualistic terms, and whether a less directional process of transition may be closer to economic 'reality' – cautionary notes crucial in our critique of the p/pp transition.

One of the main theorists of recent economic transitions is French economist Lipietz. In his 1992 book *Towards a new economic order*, Lipietz viewed economic development as moving from a relatively stable period of 'grand compromise' on the basis of an agreed 'development model' and shared world views (e.g. Fordism) to 'major crises' which induce transitions to another societal state. Lipietz (1992: x-xi) argued that during the era of the 'grand compromise' "there are arguments about sectional interests and about improvements to the [existing] model … but the model itself, the overall design, is not challenged: we accept the promises of the model as a practical ideal, as the aim of our social activity, whether economic or political. In these periods, utopias offering change exist, but they are far removed from the everyday … practice of those who advocate them". Key to maintenance of the 'grand compromise' is a robust mode of regulation, which involves all mechanisms that control conflictual action and thought of individuals towards the collective principle of Fordist accumulation. Fordism, in this view, is usually conceptualised as a labour process model that began at the turn of the century, but that gathered particular speed in advanced economies after 1945. This model has mechanisation within firms at its core, and involves a process of rationalisation of production based on hierarchical separation of a thinking elite, organisers of production (e.g. engineers) and operatives carrying out production (manual workforce). Characteristic of the Fordist era are mass production with increases in productivity, a proportionate share of 'value added' (increases in real wages), and a relative stability in the profitability of firms together with full employment – often associated with the 'productivist model' (note the similarity with 'productivism' in agriculture), the 'Golden Age' and with what may be loosely termed the 'American way of life'. Fordist production is also associated with changes in employment relations through 'temporal rigidification', as it is now time rather than skill or effort that becomes of paramount concern (Hassard, 2001). These Fordist characteristics are also linked to social legislation covering minimum wage levels and annual pay rises, a welfare state system which ensures that the whole population remains consumers even when they are out of work, and availability of financial credit issued by private banks and no longer a function of a nation's available gold reserves.

The Fordist paradigm, therefore, consists of conceptions of technical progress, social progress (largely through greater purchasing power), and progress of the nation state (Gramsci, 1971). The 'grand compromise' during the Fordist era comprises an agreement between those holding the means and the capital for Fordist production and workers' unions who are granted a share of productivity gains accruing from rationalisation. The key to success of this regime is, therefore, an arguably unique and unprecedented social compromise between owners and workers. Fordism has also been closely associated with state control, in particular an active role in 'controlling' the economy and stimulating growth, the growth of the USA as the world's dominant economic power, and acceptance of the US dollar as the international credit currency. In addition, a supranational institutional framework needs to be in place (developed after the Second World War) that ensures global economic stability by regulating international trade, financial markets and currency regulation (Marglin and Schor, 1991). Institutional control is important as it forms an overarching framework that regulates ideologies, policies, production techniques and actor networks (Jenson, 1989). These institutionalised controls can be top-down (state-determined laws, policies, budgets), at grassroots level (e.g. collective agreements), or 'semi-public' (agreements with both top-down and grassroots elements of control), and occur within a paradigm that has a 'common core' which accepts variants (within certain limits). Shared world views are critical to the conceptualisation of stable development periods, as they shape agreement on a certain way of life in society (defining terms such as 'moral', 'normal', 'desirable' or even 'terrorist'), and they constitute what Lipietz (1992) termed the 'societal paradigm' which duplicates ideologically and behaviourally the development model.

The *transition period*, or period of 'major crisis', meanwhile, is characterised by the challenging of the aims, rules and promises of the grand compromise, as it is now seen unworkable economically and rejected politically and socially (Hudson, 2000). The Fordist paradigm was hit by crisis from many sides at once during the 1960s: lower profitability of the Fordist production model; internationalisation of markets and production compromising nation state control and national modes of regulation, as well as questioning the concept of nation states as 'isolated planets'; worker dissatisfaction (alienation from work); calls for more individual autonomy; and growing reservations about top-down administrative solidarity (Gertler, 1988). During the 1960s productivity gains began to fall in many developed countries, and while increases in real wages continued to grow, the profitability of firms declined (Webber and Rigby, 1996). When price rises outstripped wage increases and purchasing power, and with increasing replacement of labour by fixed capital, demand fell and economic activity slowed, exacerbated in the early 1970s by the first global oil crisis. At the same time, multinational corporations spread their operations over entire continents and moved into developing countries (Leonard, 1988), with management of growth becoming less amenable to government control. This was accompanied by post-Fordist tendencies of agglomeration that facilitated interaction between increasingly vertically disintegrated industrial functions. While the Keynesian 'safety net' of the Fordist era prevented a complete meltdown of advanced economies during the early 1970s, it increased state indebtedness leading to higher taxation which, in turn, led to a further drop in profitability of investment. "In the end, the very legitimacy of the welfare state and welfare benefits was called into question, and with it, the whole Fordist compromise" (Lipietz, 1992: 16).

The Fordist regime is usually seen to have lasted from about 1945 to the mid 1970s. Lipietz (1992, xi) termed these periods the 'crossroads of history', "when initiatives for change override the dead weight of routine; they are open periods, where the outcome remains uncertain". These transition periods can be crises of hegemony, where elites, and the social groups that sustain them, can no longer offer a world view acceptable to the whole of society. A new model inevitably comes up against resistance, Lipietz argued, not only from the 'old guard' but also from others offering alternative trajectories of development. Often, proponents of the new model – or new paradigm – are branded 'alternative' or 'radical', until the new societal model (e.g. a 'post-ism') is accepted as the new practical ideal and the ultimate goal of social activity. The transition towards a new model is characterised by the gradual acceptance of a particular model (a new 'grand compromise'), which then leads to stabilisation of the new model over a (stepped) transitional period at the end of which the rationale of the new model is no longer challenged (see also Kuhn, 1970). In this sense, 'terrorists' become 'politicians', and 'radicals' become 'mainstream' – i.e. our specific use of discourses lends 'credibility' to different transitional stages.

Many European thinkers (e.g. Harvey, 1989; Amin, 1994) have argued that the crisis in the Fordist paradigm was largely initiated by the first great anti-Fordist mass movement in 1968 (in France alone nine million people went on strike). Such 'revolt' was spurred by better education and greater self-awareness among workers and a desire for greater work 'satisfaction' and 'dignity' among those who produced the goods of the Fordist 'Golden Age'. Increasingly, the environmental side-effects of Fordist modes of accumulation were also called into question. Transitional pressures were exacerbated by the rise of the 'liberal-productivist' model of the Thatcher-Reagan era in the late 1970s and early 1980s, which influenced advisory and regulatory bodies such as the Organisation for Economic Cooperation and Development (OECD), the International Monetary Fund (IMF) and the World Bank (Hudson, 2000). This transitionary model emphasised the productivist techno-economic imperative, fragmentation of social existence (world markets as the operating environment weakening the welfare state), 'civil society' taking over responsibilities previously part of the welfare state, and increasing social polarisation. The most important problem was the lack of solutions to the crisis of the Fordist labour-process model (Amin, 1994). In the post-Fordist era, casualisation, in particular, has further removed workers from their jobs, with concurrent reduced worker interest in the fight for productivity and quality – often referred to as the 'neo-Taylorist' model of subcontracting (Gramsci, 1971). Neo-

Taylorism tries to eliminate the residual and hidden involvement of unskilled workers and low-level employees. This is exacerbated by macroeconomic crises of overproduction and periodic slumps ('boom and bust' cycles), leading to high government spending with low taxation, periodically resulting in stock market crashes based on loss of investor confidence in what are increasingly seen as financially unsustainable markets.

While dimensions of Fordism are relatively easily identified, definitions of *post-Fordism* are less clear – a pattern we will encounter throughout our analysis of transitions (including the p/pp transition). Recent researchers have argued that the term 'after-Fordism' may be more appropriate than 'post-Fordism' (Peck and Tickell, 1994). Characteristics of post-Fordism include the dilution of the welfare state to boost profits and, therefore, investment (Gertler, 1988), and a free market approach, often termed liberal *productivism* (Lipietz, 1992), characterised by reduced labour costs, increase in casual work, subcontracting, and the relocation of production into the developing world. Liberal-productivism differs from the Fordist paradigm in that it is no longer an 'organicist design', although it is just as hierarchical in excluding the vast majority of people from the operation of 'free' enterprise. Others place particular emphasis on post-Fordist modes of accumulation as vertically disintegrated production systems (e.g. Murray, 1992). Lipietz (1992) and Amin (1994) highlighted the weaknesses of the Fordist development model as the 'crisis of work', the misguided 'consumption model', the crisis of the welfare state, and the environmental crisis stemming from relatively uncontrolled 'productivist' industrial output and pollution. Post-Fordism is also characterised by the contradiction between the increasingly internationalised nature of production and modes of regulation that remain largely based on nation state frameworks (Harvey, 1989). Post-Fordism is also closely associated with new forms of solidarity within society (at sub-national level at least), based on rejection of the monetarist approach in favour of subsidies for self-organised activities with an agreed utility and, arguably, a move towards forms of grassroots democracy and governance that are more 'organic' and less delegative than those of Fordism.

In the post-Fordist era, firms opt for less mass-production techniques which mobilise skills of line workers, and encourages dialogue between machine design, maintenance and production – going hand in hand with improved relations and communication between 'embedded' firms in the same production process (Grabher and Stark, 1997). The new 'grand compromise' of post-Fordism, therefore, involves employees feeling that their long-term interests are tied to the firm and that they will get something in return for their allegiance. In contrast to the Fordist 'wage society', a rebargaining of the wage relation takes place where employees are all individual entrepreneurs, and where all staff are involved in the management of flows within the firm, and even between firms. This often goes hand in hand with partnership and strategic alliances between firms (economies of scale) to pool research and coordinate specialisms (Grabher and Stark, 1998). Simultaneously, a new post-Fordist 'contract' emerges between cleaner modes of industrial production and environmental sustainability (Welford, 1996). Indeed, many argue that post-Fordism is a reaction to Fordist modes of accumulation because "the natural tendency of firms is to deplete their resources or overwhelm them with waste" (Lipietz, 1992: 52). Post-Fordist environmentally friendlier production is often forced upon firms by pressure from social movements (since the late 1960s) that manage to get environmental laws and regulations introduced – but thereby increasing costs of production and further aggravating the Fordist 'supply-side-crisis' discussed above. At the extreme end of the post-Fordist spectrum lie, arguably, 'post-industrial' processes in which Fordist modes of *production* have completely given way to post-Fordist modes of *consumption* of industrial landscapes. This is evident in the Ruhr area in Germany, where rapid deindustrialisation of the coal mining and steel industry has led to the conservation of some of the last remnants of Fordist production systems, best expressed through the preservation of the 'Gasometer' (giant gas container) near the city of Essen as a viewing tower and exhibition space for tourists and visitors.

Post-Fordist transitional models suggest a degree of directionality, but Chapter 4 will highlight that there are many criticisms of the implied linearity and dualistic notions in the stepped economic development model underlying conceptualisations of the transition to post-Fordism – criticisms that we will need to bear in mind when discussing the postulated agricultural p/pp and multifunctional transitions.

3.2.2 Socialism to post-socialism

Similar theoretical assumptions about the nature of societal and economic transition have been adopted for the transition from socialism to post-socialism in countries which emerged from the political and economic disintegration of the Soviet Union in 1991 (Russia, Ukraine, Estonia, Georgia, etc.). Theorisations of the transition to post-socialism are highly relevant for our discussion in this book. First, debates have highlighted many of the fallacies inherent in 'traditional' transition models – issues we will discuss in detail in Chapter 4. Second, it is in research associated with the transition to post-socialism that *transition theory* has been conceptualised and used most extensively in the social sciences to describe recent processes of societal change. In particular, the groundbreaking work by Pickles and Smith (1998) has greatly influenced current debates on transition theory, and has also acted as an important trigger for the selection of the conceptual framework adopted in this book (see also Pavlinek and Pickles, 2000). Transition theory has been appropriated to such an extent by those analysing the post-socialist transition that Smith and Pickles (1998: 2) referred to the theorising of this transition as "mainstream transition theory", while Swain (2006) argued that a 'transition industry' has emerged that involves a wide array of stakeholders whose purpose has been to realise economic transition in Central and Eastern Europe. As with post-Fordism, *regulation theory* has been widely used to understand the nature and pace of the transition to post-socialism. Smith and Pickles (1998: 13) suggested that regulation theory conceptualisations of how "political economic forms are stabilised or enter into crisis through the coupling of forms of accumulation and mechanisms for their regulation is a valid lens through which one can begin to understand transition".

The transition from socialism to post-socialism is seen as one of the classic *recent* (i.e. last few decades) societal transitions. It, therefore, shares a similar time frame with the p/pp transition model discussed in Part 2. The *era of socialism* is seen to typically involve adoption (forced or voluntary) of state socialist politics and modes of production, with a mode of social regulation characterised by a one-party system and top-down economic planning (Pavlinek and Pickles, 2000), or in the words of Przeworski (1995) a system of 'modernisation without internationalisation'. The most prominent example was the former Soviet Union from 1917 to 1991, but socialism was also imposed onto, or voluntarily adopted by, many other countries that had close ties with the Soviet regime (e.g. Cuba, North Korea, North Vietnam, Albania, etc.). The transition to post-socialism occurred through several interlinked processes and socio-economic upheavals (Bradshaw and Stenning, 2001). First, the dismantling of the socialist ideology was made possible by *decolonisation* of former Soviet satellite states in 1991 (e.g. Georgia, Kazakhstan, Uzbekistan). Second, the gradual process of *democratisation*, characterised by a transition from totalitarian to post-totalitarian rule, was crucial for providing a socio-political basis for the transition. Third, *economic liberalisation* ('from Marx to the market') entailed moving from state ownership of the means of production in centrally planned economies to price liberalisation, privatisation of state-owned enterprises, end of state monopoly over trade, and new property rights for citizens. Fourth, *globalisation* enabled (in theory at least) the inclusion of newly emerging post-socialist states into the global world economy with modernisation via internationalisation (Przeworski, 1995). This led (again in theory) to other globalising forces through inclusion of former socialist countries into the global capitalist system such as free trade, a widening of national and international investment, and, arguably, the emergence of more liberal policies in newly formed democratic societies (Stiglitz, 2002). As Van Hoven *et al.* (2004: 1) argued, "transition was a powerful tool, created as an idea largely by neo-liberal economists in North America and Europe. It was eagerly embraced by politicians in Central and Eastern Europe, and was seen by many people living there … as a remedy for their problems". Some argue, therefore, that the result of the transition to post-socialism has been a fundamental reorganisation of material life within former socialist countries, a transformation of geopolitical relations on a global scale, and a major ideological/discursive shift in policy implementation (Smith and Pickles, 1998). However, Chapter 4 will highlight that there are many debates arguing that a transition to post-socialism has only occurred *in theory* but not *in practice*.

Conceptualisations of the transition to post-socialism are interlinked with different transition models discussed in Chapter 2. Different schools of thought view the transition to post-socialism from different vantage points, resulting in highly differing interpretations of the nature, pace and causes of the transition. After the break-up of the Soviet Union, politicians and economists faced hard choices about 'the best way forward', with the key argument revolving around the *pace of transition*. On the one hand, some argued that if privatisation did not occur rapidly (within a few years), thereby creating stakeholders with a vested interest in capitalism, countries that had been under the influence of the Soviet Union would quickly revert back to socialist ideologies. This school of thought advocated the strategy of *shock therapy* (Sachs, 1990, 1992), with a concurrent belief in the *stepped transition model* (Fig. 2.1.b) as the ideal form of transition that would allow socialist structures to be replaced quickly by new (arguably post-socialist) politics and modes of production. This was amplified by the writings of Fukuyama (1992), who argued for the apparent historical 'victory' of capitalist democracy over communism. On the other hand, some argued that if the transition occurred too quickly, the reforms could be a disaster resulting in economic failures compounded by political corruption, thereby opening up the way to a backlash from either the extreme left or right (Stiglitz, 2002). This *gradualist approach* was akin to ideals enshrined in the more gradual *linear transition model* highlighted above (Fig. 2.1.a). Gradualists believed that the transition would be more successful by moving at a slower and more 'linear' pace. The 'shock therapy' approach was strongly advocated by both the US Treasury and the IMF, who provided large sums of money for rapid political and economic restructuring. However, at the beginning of the 21st century it appears that countries that adopted the gradualist approach (e.g. Poland, China) may have been more successful in the transition to post-socialism than those adopting the shock therapy approach (e.g. Russia). Indeed, Russia has seen a large increase in poverty since 1990 (Stiglitz, 2002), although recent high oil and gas prices have led to an improvement in the economic situation.

This debate highlights that the transition to post-socialism is complex and that it can not easily be explained and understood by unidimensional models of transition. Transition theorists with roots in evolutionary political economy have pointed to the importance of 'path dependency' in shaping the nature and pace of post-socialist transitions (Grabher and Stark, 1997). They have argued that the 'memory' inherent in many social systems to a certain extent streamlines transition pathways (Smith and Pickles, 1998). Grabher and Stark (1998), for example, suggested that transitions arise out of the particular trajectories or paths taken by specific countries or regions – in other words, that transitional systems have an in-built memory, or legacy, that may define the transitional trajectory. Similarly, Smith and Swain (1998) argued that an understanding of the diversity of local responses to transition is partly dependent on the way in which previous legacies of socio-cultural, political and economic relations are 'reworked' within the possible framework of these relations. Some elements of the post-socialist transition may, therefore, follow the *Deleuzian transition model* (see Chapter 2), based on the existence of 'a thousand plateaus' of possible parallel transitional pathways, depending on individual path dependency and memory at a given transitional starting point in time. The complexity of debates on the transition to post-socialism, and that many believe that *no* full transition to post-socialism has yet occurred, also highlights the importance of the *retrograde transition model* (or 'failed transition' model) (Fig. 2.1.d). Manser (1993), in particular, applied the concept of 'failed transition' to the post-socialist transition in Central and Eastern Europe, arguing that none of the transitional pathways used by former socialist countries has led to a 'new' and 'better' final outcome (see also Dunford, 1998, for the Russian case). Post-socialism has been characterised by economic collapse, an onslaught on labour, and social and political disorientation that has made the contemporary situation worse in economic terms than during the end of the communist regime (e.g. the economic output of Moldova is 30% lower than a decade ago; Ukraine's GDP is only one third of what it was 10 years ago; Stiglitz, 2002).

Many debates on the transition to post-socialism suggest a certain degree of directionality – similar to the transition to post-Fordism outlined above – but Chapter 4 will analyse in detail the criticisms of implied temporal linearity, spatial homogeneity, global universality and structural causality in conceptualisations of the transition to post-socialism.

3.2.3 Modernism to post-modernism

The postulated transition from *modernism to post-modernism* is another arena of investigation that has direct relevance for our discussion, as it has been argued that agricultural post-productivism may be an expression of a post-modern society (Philo, 1993; Halfacree, 1997b). Further, some argue that linear and unidimensional conceptualisations of societal change may be a particular expression of *modernist* grand narrative and trajectory linked to 'directional' industrial progress (Macnaughten and Urry, 1998; Hassard, 2001). Yet, a note of caution is necessary when considering the transition to post-modernism. As Chapter 4 will discuss, an argument could be made that this transition may only be a *hypothetical assertion* expressed solely in philosophy and social science theorisation[6] (Lyotard, 1984; Latour, 1993), leading Bertens (1995: 10) to suggest that post-modernism is an "exasperating term" surrounded by "massive but also exhilarating confusion". There is a considerable body of work that discusses the 'visible' and more tangible expression of the postulated post-modernist transition 'on the ground' – especially through arts and architecture (e.g. Berman, 1982), but, since the 1980s, also describing a specific approach to philosophy and a banner under which some seek to attack the *ideological* underpinnings of modern society – i.e. post-modernism as 'method' through deconstruction (Dear, 1986). As Latour (1993: 10) suggested, "the adjective 'modern' designates a new regime, an acceleration, a rupture, a revolution in time" – key words that embody the notion of transition. Debates on the transition to post-modernism can, therefore, inform our discussion of the transition to post-productivist agriculture as well as shed further light on possible transitional pathways based on our six key models of transition discussed in Chapter 2.

Modernism is embedded in the concept of 'modernity', which has several meanings, including the process of rationalisation (where science replaces magic), functionalism, disembedding mechanisms, a belief in progress and order, individualism, liberalism and consumerism. It is also often seen to involve processes of industrialisation, scientification, commodification, technologification and globalisation (Bradbury, 1976). The term

---

[6] A similar point has been made about conceptualisations of the transition from *structuralism to post-structuralism* (see Culler, 1983, and Sarup, 1993, for detail).

modernism is used for the ideology behind these processes[7] and is often associated with *cultural changes* accompanying processes of modernisation (Berman, 1982, 1992), although modernism has also been defined as a social condition or a stage in history (Sarup, 1993). Giddens (1997) argued that modernism is a Western project that may have started as far back as the end of the 15th century, although the era of modernism has been largely associated with late 19th and 20th century developments in Western European architecture, literature and the arts (especially between 1890 and 1930), challenging both the realist and romantic period of the 19th century and responding to a series of major crises in capitalist modernity (Nederveen Pieterse, 2004). Modernist artists sought to escape from the notion that art may be a direct reflection of the world, opening art (in particular painting) to the possibilities of abstraction, ambiguity and uncertainty (Lunn, 1985). It is, thus, with regard to modern art and literature that conceptualisations of 'modernism' are most strongly associated, especially regarding the 'new' (early 20th century) art forms of surrealism and cubism that overturned hitherto existing conventions. Similarly, modernist architecture has become a well-used categorisation for a purist architectural style (mainly between 1890 and 1930) focused on function and order and using 'modern' materials such as concrete (Gregory, 1994). Modernism has also been linked to changing ideologies and ideas in the scientific and academic world, especially through the 'new' formation of the social sciences that brought with it new perceptions of society.

As with other 'post-isms', *post-modernism*[8] is conceptualised as a critique of modernism and the modern era (Lyotard, 1984), but there is little agreement about the key components of this 'post-ism' (Jencks, 1987). Bertens (1995: 5), for example, argued that post-modernism is "either a radicalization of the self-reflexive moment in modernism, a turning away from narrative and representation, or an explicit return to narrative and representation ... and sometimes it is both". Similarly, Simon (1998) argued that geographers, sociologists, planners and others struggle to integrate the various meanings of post-modernism, although post-modernism can be seen as a 'distinct' period succeeding the modern era, identified by different forms of economic, social and political organisation and behaviour. Others argue that, in contrast to the modern era with its arguably privileged position for artists and observers, the post-modern vantage point emphasises pluralism, and an openness "to a range of voices and perspectives in social enquiry, artistic experimentation, and political empowerment" (Johnston *et al.*, 2000: 620). Hassard (2001), in particular, suggested that post-modernism is predicated on the view that symbolic boundaries between art/high culture/the academy and everyday life and popular culture are dissolving. This suggests that post-modernism brings with it more open and fluid social identities compared with the traditionally fixed structures of the modernist era. Modernism is also seen by many to be associated with space-time compression on the basis of reconceptualisation of spatial and societal factors on the back of economic recessions, radical transformations of Western cities, rapid industrialisation, changing gendered spaces, problems associated with fading colonial empires, and the rise of the computer and internet (Harvey, 1989). Der Derian (1988: 189) suggested that post-modernism "defies the grand theories or definitive structures which impose rationalist identities or binary oppositions" – thereby emphasising the post-modern critique of unilinear and dualistic conceptions of societal change explored in Chapter 4. The notion of post-modernism is, therefore, closely associated with a key phrase in Marx and Engels' (1848/1972: 338) famous *Manifest* that argued that revolution and disturbance characterise the 'bourgeois epoch' from all earlier times, and where "all fixed, fast-frozen relationships, with their train of venerable ideas and opinions, are swept away, all new-

---

[7] As opposed to 'modernity' which is usually associated with a condition or manifestation of modernism (Simon, 1998).

[8] 'Post-modernism' is not the same as 'post-modernity'. The latter is usually seen as the historic period associated with post-modern economic and socio-cultural processes and is, therefore, a subset of post-modernism (Lyon, 1994).

formed ones become obsolete before they can ossify. All that is solid melts into air, all that is holy is profaned…". Indeed, the phrase 'all that is solid melts into air' is seen to represent post-modernism *per se*, as a reflexive modern frame of mind undermines modernity's own credibility and legitimacy. Post-modernism critically reviews the modernist project and questions modern narratives, leading Harvey (1996: 315) to argue that modernism and post-modernism are "dialectically organised oppositions within the long history of modernity", while Giddens (1991) characterised modernism as the separation of time and space (social relations are extended to include global systems), disembedding mechanisms, and institutional reflexivity.

Again, it is with the arts and architecture that the most tangible assessments of post-modernism can be found, partly because arts and architecture share *visible* expressions of change, in contrast to more 'abstract' post-modernisms related to academic enquiry or reconceptualisations of spatial and societal factors. There is some agreement, for example, that post-modern architecture often reflects regional traditions and generally 'fits in' with its surroundings, is more heavily decorated than 'modern' buildings, depicts building facades that are usually broken into diverse forms and surfaces, and uses an architecture that reflects the needs of users (Ellin, 1996). Simon (1998) argued that post-modernist architecture is characterised by time-space disconnection and decontextualisation of specific purpose-built sites and locations, best expressed through theme parks (e.g. Disneyland), pleasure domes and other buildings and sites for international package tourists. Johnston *et al.* (2000: 621) further argued that "in these features there would be a conscious rejection of the abstraction, universalism, and historical erasure brought to architecture and design by the modern movement". Similarly, some agreement exists about 'post-modern' changes in writing that aim at re-positioning the 'traditional' role of authors in terms of race, culture, class or gender, but that, at the same time, are seen to reduce the possibility to provide an adequate understanding of places, people and 'the other' (Olsson, 1991).

We see in these debates some elements of our transitional models mentioned above. That some suggest that modernism can be neatly delineated temporally (i.e. 1890-1930) suggests that the transition to post-modernism may be akin to the *stepped transition model* (Fig. 2.1.b). Post-modernism can be interpreted as providing a sudden new set of circumstances that may have rapidly replaced modernist structures and patterns after 1930 (Bradbury, 1976). However, a similar case can easily be made for those advocating that the transition to post-modernism has been much 'smoother' and *linear* (Fig. 2.1.a), with a more seamless transition towards post-modern characteristics, patterns and processes (Giddens, 1991). Similarly, and as Chapter 4 will discuss, many arguments can be found to underscore the assertion that no full transition towards post-modernism has occurred yet (i.e. *retrograde or 'failed' transition* as per Fig. 2.1.d) (e.g. Latour, 1993).

3.2.4 Colonialism to post-colonialism

As with the transition to post-socialism, debates on the transition from colonialism to post-colonialism are embedded in political, economic and historical discourses. This transition has been viewed from many different vantage points, with considerable differences between analysts from developed countries (vantage point of colonial powers) and those from developing countries (i.e. those who 'suffered' from colonialism) (Bauer, 1976; Wolf, 1982). There are two specific reasons why debates on this transition are highly relevant for our discussion. First, the transition to post-colonialism is another crucial and long-standing discussion within the wider framework of transitions from 'isms' to post-isms'. Debates started as early as the first moves of former colonies towards independence, while academic debates on post-colonialism gathered particular pace in the 1970s and 1980s (Abdel-Fadil, 1989; Childs and Williams, 1997). Second, notions of '*neo*-colonialism' (see below) have added an interesting conceptual and theoretical critique of the implied linearity of the colonial/post-colonial model, and will be relevant for our discussion of the fallacies inherent in transition models discussed in Chapter 4. Some theoretical links also exist between debates on colonialism and modernism (see above), as, within the modernist paradigm, colonialism was seen as the 'modern' expression of the state in its quest for control over non-European societies and cultures (Fieldhouse, 1981).

There is considerable debate surrounding the beginning of the era of *colonialism*, as the notion of colonial powers can be traced back to the Greek colonial system and beyond. Writers from the North often argue that the era of colonialism began in the 15th century, with European expansion first to Africa, then Asia and the New World. This phase of European colonialism was led by Spain and Portugal (and later by other European powers such as France and the UK). The 'first phase' of colonialism (about 1470-1800) was mainly concerned with short-term wealth accumulation in form of territories, gold and silver, slaves and natural resource exploitation, while the 'second phase' (about 1800-1914) entailed a more permanent settlement with the establishment of political and economic (and, at times, cultural) roots in colonial territories embedded in an expansionary world capitalism (Wolf, 1982). This second phase was based on almost complete European hegemony over world trade, finance and shipping, with one-quarter of the globe's land surface distributed (and re-distributed) among only a handful of colonial powers in just a few decades (Rodney, 1972). The viewpoint 'from the North' usually sees the period of colonialism as a *benign* force of economic modernisation (a transition in itself from pre-modern to modern; see above), and as a form of territorial expansion where certain forms of rule were established and maintained by sovereign powers over a 'subordinate' culture often perceived as 'alien' or 'exotic' (Bauer, 1976). Often, colonialism went hand-in-hand with colonisation, i.e. the physical settlement of people from the sovereign power. Colonialism has been strongly associated with political and legal domination over an alien society, coupled with economic, political and military dependence and racial and cultural inequality – the latter often artificially exacerbated by ruling elites linked to the colonial power. Colonialism was also associated with exploitation of natural resources and unsustainable environmental management practices (Wilson and Bryant, 1997). The view from the South and that of Marxist critics of colonialism (e.g. Rodney, 1972), meanwhile, sees colonialism as en era of *violent conquest and plunder*, as the subordination and domination of the poor through the establishment of labour regimes instituted to promote cheap commodity production by ruling elites from the North, and as an era of wholesale destruction and dependency (Chomsky, 2000). Here the focus is more on issues of imperial hegemony, and less on the myth of improved local participation and empowerment.

The notion of *post-colonialism* describes the era that follows 'after colonialism' between the 18th and 20th centuries, when former colonial territories gained independence (e.g. late 18th century for the USA; early 19th century for the Spanish colonies in South America; 20th century for African colonies). Yet, some argue that it took until World War II for 'real' colonial independence (e.g. India in 1949), emphasising substantial temporal shifts over at least two centuries in the transition to post-colonialism. Particularly after 1945, many anti-colonial movements were burgeoning in the remaining colonial territories, coupled with an outdated imperial system that proved increasingly uneconomic and ungovernable. The transition to post-colonialism, therefore, could be relatively quick, but at times politically complex and only implemented after organised revolts (e.g. Kenya), wars (e.g. Mozambique, Yemen), or constant lobbying for independence (e.g. India). Many see this transition, therefore, as a typical example of a *stepped transition*, where one form of colonial organisation was superseded quickly by a new form of post-colonial organisation. However, and as will be analysed in Chapter 4, there have been 'many colonialisms' (including new ones; cf. Chomsky, 2000), and it is difficult to pinpoint a specific time period associated with post-colonialism (Childs and Williams, 1997).

Since the 1980s, post-colonialism has also emerged as a theoretical approach that theorises *underdevelopment* in the context of the 'North-South debate' in light of the colonial experience of many 'Third World' territories. This approach attempts to re-inject the relevance of 'culture' and 'cultural factors' in a post-structuralist attempt to reinterpret the impacts of colonialism (Mohan and Stokke, 2004). This theoretical enquiry has relied heavily on the analysis of various *discourses* of colonialism and the assessment of how power works and is expressed through language. It has led, in particular, to the acknowledgement that many territories formerly conceptualised as 'post-colonial' show, in fact, many facets of 'neo-colonialist' discourses – structures that suggest that the former colonial politics, institutions, actors and networks have only rarely (if ever) been superseded by truly post-colonial structures (see Chapter 4). This line of thought, therefore, rejects the notion of a stepped transition towards a post-colonial era, but argues instead that the *retrograde, or 'failed', transition model* better describes the current plight of many former colonial territories.

3.3 Demographic, technological, environmentalist and evolutionary transitions

When discussing transition at a broader level, we need to go beyond 'isms' and 'post-isms'. I, therefore, also wish to investigate debates on 'other' transitions where the emphasis has focused on the *transition itself*. This will provide a wider framework for contextualising the post-productivist transition in Part 2. Inevitably, examples chosen will be selective, and to incorporate debates on all possible 'transitions' would go beyond the scope of this book. My essential concern is the relationship between debates on the transition to post-productivist and multifunctional agriculture and some strands of recent social theory. A major criterion for selection of transitions discussed here is, therefore, their heuristic potential for explaining/understanding processes that can inform discussions in Parts 2 and 3. The objective is to show the effectiveness of these debates in highlighting similar problems, processes and issues that will also be encountered in conceptualisations of the p/pp transition and multifunctional agriculture, and to shed light on underlying assumptions and weaknesses of theories of transition from the social and natural sciences.

The four 'classic' transitions of *demographic transition*, the *transition towards environmentalism*, *technological transition* and *evolutionary transition* will be used here to shed further light on common transitional patterns from different disciplinary vantage points, and to illustrate that most (if not all) of these postulated transitions find resonance in the six transition models analysed in Chapter 2. These debates will form an important building block for our discussion in Chapter 4 of what I term the *four fallacies* inherent in many transition models. I have selected these four transitions because of their relevance for recent transitional trends spanning the last 40 years or so (e.g. transition to environmentalism), and their portrayal of what can be termed a 'classic' and often cited transition (e.g. demographic transition model; technological transition). I have also included a discussion of evolutionary transition to highlight that transition theory permeates the social *and* the natural sciences (indeed an argument could be made that transition theory was originally developed in the natural sciences; see Chapter 2), especially because evolutionary theories have also been applied to analyse transitions in human systems (e.g. cultural evolution; the evolutionary theory of meaning; social Darwinism; socio-biology; etc.), and because the theory of biological evolution has seen some of the most challenging recent debates on the general notion of 'transition'.

3.3.1 The demographic transition model

The demographic transition model is one of the 'classic' and best established transitional models in the social sciences (Caldwell, 1976). As Hirschman (2001: 116) argued, "relative to most of what passes for theory in the social sciences, [demographic] transition theory is a remarkable achievement". Debates on this transitional model have shared many similarities with those surrounding theorisations of transitions from 'isms' to 'post-isms'. Some even argue that the notion of 'transition' in the social sciences was to a large degree based on the demographic transition model (Rotmans *et al.*, 2002). Debates on demographic transition are, therefore, the most long-standing in transitional research (e.g. Thompson, 1929; Davis, 1945) and, as a result, warrant closer scrutiny.

The demographic transition model is a generalised model that describes how birth rates and death rates of human populations change over time (Kirk, 1996). As it argues that there are four *distinctive stages*, the demographic transition model is a typical example of a *stepped transition model* (see Fig. 2.1.b). The first stage is termed the 'high stationary stage' with high birth and death rates. Total population levels during this stage are usually low, and population growth is generally defined by high mortality rates due to diseases, famines or wars (Chesnais, 1992). Birth rates tend to be relatively stable, while population growth/decline tends to be defined by a fluctuating death rate (e.g. episodic crop failures or natural disasters). The second transitional change, the 'early expanding stage', sees growth in population due to a rapidly falling death rate and a still relatively high birth rate. Key drivers for this stage are better sanitation and medical care and improved nutrition. The third stage is termed the 'late expanding stage' in which the death rate stabilises at a relatively low level, and where birth rates start to decline rapidly. There is still substantial debate about the underlying reasons for the fall in birth rates at this stage (Szreter, 1993), although there is general consensus that the rise of urbanisation, with associated improved levels of education and changing attitudes towards children (e.g. children no longer necessary to support older population), are key drivers for a reduction in birth rates. The final stage, the 'low stationary stage', is the mirror image of stage 1 with high overall absolute population numbers but low death and birth rates and low population increase. The demographic transition model is particularly useful in highlighting *geographical* and *temporal* diversity during the transition – key issues that will also be of particular relevance when discussing the p/pp and multifunctional transitions in subsequent chapters.

Agriculture-based societies are more likely to be situated in stages 1 and 2, while industrial societies usually fall into stages 3 or 4 (Boserup, 1993). The literature increasingly highlights, however, that world fertility is declining steadily (Bulatao and Casterline, 2001). World population growth rates are down from a peak of over 2% in the 1960s to only 1.2% in the early 21st century, with projections for 2050 as low as 0.3% annual growth rate (United Nations, 2003). This led Cleland (2001) to suggest that the world is experiencing the 'globalisation of fertility decline', while Chesnais (2001) argued that we now witness the post-transitional demographic stage as a new regime of permanent disequilibrium. Although global population is still likely to increase by 2050 due to the 'population momentum' that built up in the last decades of the 20th century, the drastic falls in fertility have led many commentators to conceptualise a new Stage 5 in the transition model, characterised by population decrease where death rates *exceed* birth rates (Day, 1992). Although in some regions population decrease is counter-balanced by net immigration gains, leading to a temporary stabilisation or even increase of populations (e.g. some countries of the EU; USA), this development may suggest that world population is in a 'post-transitional' stage (Bulatao and Casterline, 2001). As Chapter 4 will highlight, these recent developments increasingly challenge aspects of the 'traditional' demographic transition model, and the notion of 'post-transition' interestingly suggests that there may be a conceptual territory 'beyond' transition.

3.3.2 Technological transition

Debates on technological transition form one of the key pillars of transition theory discussions. Technological transition is a particularly pertinent issue for this book, as both 'post-productivism' and 'multifunctionality' can be interpreted as *innovations* (both in practical and ideological terms) that may follow similar transitional patterns to those described for technological transitions. In Chapters 5-7 we will, therefore, pay particular attention to technological transition to contextualise the p/pp transition. Technological transition surrounds us all and can be witnessed continuously, with the best known recent examples ranging from rapid changes in communication technologies, computers, music, to energy production (Grübler, 1998). Parayil (1999: 1) argued that "if there is one phenomenon that distinguishes the 20th century from all the previous ones in human history, it has to be the spectacular changes that have been taking place in the technological knowledge base of modern and modernizing societies". Technological transition theories, therefore, form a key component of any debate attempting to contextualise theories of transition.

The first arena of research enquiry on technological transition is about *temporal patterns of change*. Research on technological transition was particularly sparked by Kuhn's (1970) notion of paradigm shifts, with earlier studies often treating technological change as an expression of scientific change (Parayil, 1999). Research on technological transition was also influenced by Darwinian evolutionary theory, with some treating technological change as similar to biological evolution, resulting in technological transition perceived as a cumulative and gradual process (DeGregori, 1985). Darwin's evolutionary transitional model (see below) gave analysts a new historiographic idea to reformulate evolutionary change into a potentially powerful model to explain technological transition (Smith, 1993). This linear development sequence of technological change is usually seen as starting with an invention (often linked to functional failure or limitations of the old technology), and then moving through distinctive 'phases' including technology transfer and development, the growth of the new technology (and solving of associated problems), the acquisition of momentum during which the system gains 'direction' and 'velocity' (see also discussion of diffusion below), and finally the 'mature' stage during which a technological transition may begin to disintegrate only to be replaced by a new invention (Hughes, 1976).

In his seminal work on 'mechanical inventions', Usher (1954) emphasised the continuity of technological process, claiming that technological change is a cumulative process in which smaller changes and inventive activities accumulate over a period of time (the 'cumulative synthesis model'). This model suggests that great technological developments are the cumulative synthesis of a large number of minor achievements, where technological transition is the end result of several minor inventions and other procedural technological activities cumulating over a period of time (Parayil, 1999). For Usher, technological transition was a cumulative process of inventions (stepped transitions) similar to evolutionary transition, characterised by bit-by-bit modifications of designs that may remain unchanged over long time periods. 'Cumulative' in this context means that technological change is irreversible, and that ideas, practices, theories and laws from the past inform the development of newer technologies (system memory). This body of work, therefore, suggests that technological transition is fundamentally influenced and moulded by its antecedents – an idea similar to the notion of path dependency highlighted in Chapter 2. In a similar vein, White (1967) identified key cumulative (and 'stepped') technological innovations in human society as the increased use of iron, the replacement of oxen by horses, the harnessing of natural power, and the modernisation of the plough as major causes for rapid technological transition. Echoing Kuhn, Constant (1980) used the example of changes in aircraft propulsion (e.g. from propeller to jet-propelled aircraft) to suggest a stepped cumulative technological transition, marked by specific 'technological revolutions' that occur because of functional failures in previous designs. By discussing various examples of technological change, ranging from the cotton gin, steam power, the electric motor, to barbed wire, Basalla (1988) took these ideas further by arguing that technological change can only be a continuous and cumulative process characterised by *evolutionary* change.

Emphasis is, therefore, often placed on the fact that technological transitions perform a similar role to both evolutionary transition and scientific paradigms, taking into account both continuities (linear pathways) and discontinuities (stepped transition). In this literature, technological transition is generally defined by ever changing production possibilities. Within a specific technological paradigm (e.g. in our current oil-based global economy), the efforts and the technological imagination of those driving technological 'progress' are focused in very precise directions while being 'blind' to other technical possibilities (e.g. the hydrogen-fuelled car). The classical theory argues that technological change is driven either by 'demand-pull' or 'technology-push' factors or by both occurring simultaneously, while the purpose of technological evolution is seen to be improvement of human living conditions – hence the attractiveness of linear technological transition models that aim towards an 'improved' end-point of development. However, both Grübler (1998) and Parayil (1999) provided an important note of caution by arguing that most models and theories of technological transition are based on *successful* inventions and innovations and, may, therefore, neglect crucial assessment of equally important *failed* innovations (akin to our notion of 'failed transition' in Fig. 2.1.d). Further, and as will be discussed in detail in Chapter 4, notions of relatively linear technological transition pathways are increasingly challenged. Although economic forces are often seen as a key driver for technological change, such one-directional explanations have been questioned for their oversimplified description of often complex drivers of technological change (Dosi, 1982). The prime trigger for technological change is often seen as a specific need by society for a new technology (e.g. transport), triggered by increased affluence and/or education, with scientists or inventors 'responding' to that need by producing the required technological change. But Grübler (1997) argued that a technological innovation may be so radical a departure from existing solutions that it may create its own market niche – in other words, innovation may spawn further innovation (and demand) on a new technological transition pathway.

In a seminal study on theorising technological transition, Dosi (1982) suggested various explanations for understanding technological trajectories. Key to Dosi's argument is that technological transition should not just be seen as an economic 'demand-pull' process, but that this transition is characterised by complex interactions of knowledge transfer (both practical and theoretical knowledge), know-how (including cumulative system memory), methods, procedures, experience of successes and failures, and physical availability of devices and equipment. New technologies are selected through complex interactions between economic factors (e.g. profit, new markets) and institutional forces (e.g. interests of existing firms, influence of government institutions). Most importantly, empirical evidence shows that unsolved technological difficulties do not automatically imply a change in technological pathway – in other words, *system memory* plays an important role in, at least partly, defining some technological trajectories (Ausnubel and Langford, 1997). Dosi's (1982) approach, therefore, highlighted the role of *continuity* and *discontinuity* in technological transition, whereby 'incremental' versus 'radical' innovations can be reinterpreted as part of 'normal' technological transition pathways that often resemble a stepped transition. System memory explains why technological transition pathways are often hard to change, but also allows for the existence of multiple parallel (Deleuzian) pathways – or, in Sahal's (1985) words, *innovation avenues* – whose ultimate 'success' is difficult to predict. Nonetheless, evidence suggests that synchronisation can exist between different technological pathways, emphasising possible interdependencies and cross-fertilisation between different trajectories (Grübler, 1997, 1998). The evolutionary character of technological change is, therefore, largely hinged on the assumption that 'selection' of a particular technology is based on a multitude of possible (Deleuzian) directions of development according to the momentum generated by the technological trajectory on its specific development pathway.

In this context, debates on the notion of 'technological leapfrogging' are of particular interest. As Perkins (2003: 177; emphasis added) suggested regarding diffusion of industrial technology in the South, "developing countries need *not* pass through the dirty stages of industrial growth that marred the past of today's developed countries. Instead, they may be able to bypass these by *leapfrogging* straight to modern, clean technologies [e.g. information and communication technologies or biotechnology]". Thus, certain stages in seemingly linear transition processes may be 'left out' (Felipe, 2000). This suggests that transitions may be characterised by *hiatuses* and *gaps* – hiatuses that may be particularly apparent in 'deep' leapfrogging scenarios (e.g. in Asian 'tiger' economies) that call for the development of more radical technologies than 'shallow' leapfrogging pathways (e.g. India). As an interesting variant to the stepped transition model (leapfrogging as a 'large' step bypassing intermediary steps), we will revisit the notion of leapfrogging in Chapter 7 when discussing the possibility for agricultural systems to move straight from 'pre-productivism' to post-productivism without 'having to pass' through the transitional productivist phase.

The second area of investigation on technological transition concerns *spatial diffusion of technological innovation* (Rogers, 1995). These debates on the geography of technological change are particularly interesting for understanding the spread of innovations. Authors often refer to 'phases' of innovation diffusion not dissimilar to arguments we will encounter in Chapters 9 and 10 about diffusion of different multifunctionality pathways in agriculture. Cumulative diffusion is often depicted through a semi-linear model (similar to the s-curve model in Fig. 2.2.) showing a progression from 'early adoption', to 'take-off', to full adoption (Rogers, 1995). The first phase is a period of invention of new technology – usually in advanced economies, although patterns are beginning to shift (e.g. China, India) – whereby firms are exploring consumer reaction to technological inventions. If a new technology is successfully adopted by a large number of consumers, the second phase sees the spatial 'spread' of this innovation into new markets (Parayil, 1999). Grübler (1997) showed how the growth of different types of infrastructures in the USA (canals, railways, roads) all followed the s-curve trajectory (but spaced apart by about 50 years). The reason for the 's-shape' (as opposed to a linear pathway, for example) is linked to both the role of information diffusion and the reduction of uncertainty in individual adoption processes (learning and acceptance curve).

Early studies often argued for linear patterns of diffusion from one source (usually located in advanced economies) to the rest of the world. However, recent research has shown that technological diffusion is much more complex and uncertain (Rogers, 1995). Some authors suggest that for some technologies (e.g. cyber-technology) the shift may ultimately be felt by *every* citizen of the globe based on the notion of 'global village' (see Virilio, 1993, for an extreme technological utopianist view of this 'global homogeneity' hypothesis). Grübler (1998) further highlighted that in today's world diffusion of innovation does not necessarily entail a spread from advanced economies to the developing world (e.g. the motor car), but may also entail reverse patterns in which new technologies are 'exported' from developing to developed countries (e.g. new computer technologies from India). He also emphasised that spatial diffusion of technological innovation can take very different shapes and forms, depending on the nature and timing of the innovation, existing demand (which can vary substantially between societies) and other socio-economic factors. Lack of knowledge often acts as a barrier to successful uptake of technology-heavy innovations

(Parayil, 1999). Thus, spatial technological diffusion patterns are becoming more complex in a globalising world, with innovation flows often reverted or rapidly changed – processes that will also be key in explaining the transition towards multifunctional agriculture at the global level (see Chapter 10).

3.3.3 The transition towards environmentalism

I have chosen to discuss the possible transition towards 'environmentalism' as part of our general discussion of transitions for two key reasons. First, discussions about this transition attempt to highlight societal changes that have taken place over the past 50 years or so – a similar time frame that I have selected for analysis of the postulated transition to post-productivism and multifunctionality. Second, the transition towards environmentalism emphasises changes in attitudes and perceptions of individuals and stakeholder groups, highlighting that transition does not necessarily imply tangible and 'measurable' parameters but that it also occurs 'in the mind', often associated with intangible shifts in opinion (see also discussion of post-modernism above). This will be particularly relevant for our discussion in Chapters 9 and 10 on changing ideologies, attitudes and identities of different actors in the transition towards multifunctional agriculture. In addition, debates on the transition towards environmentalism have been criticised both for being heavily biased towards industrial societies and for assuming that generalisations can be made about the applicability of transitory tendencies to a large proportion of individuals in society – misconceptions we will also identify for conceptualisations of agricultural post-productivism in Chapter 7.

The transition towards environmentalism refers to changes in action and thought towards the environment. 'Environmentalism', thus, highlights a concern that the environment should be protected from unsustainable human activities and where unspoilt nature is seen to hold important intrinsic values (Carson, 1962; Milton, 1996). Such changes in world views are part of a much broader history of cultural and economic development and changing human-environment interactions. For example, the spread of the Judeo-Christian religion is often associated with a view that humans are separate from nature and that nature exists solely for human use (Glacken, 1967; Wilson and Bryant, 1997), while Passmore (1980) suggested that this ethos of domination over nature even pre-dates Christianity and is rooted in ancient Greek and Roman civilisation. The history of intensifying human (ab)use of the environment is largely associated with the triumph of the *technocentric* world view and the spread of global capitalism. Disillusionment with adverse environmental effects may have promoted a transition towards more *ecocentric* world views (Inglehart, 1977; O'Riordan, 1995). It has been argued that this transition began with the emergence of the environmental movement in the 1960s and 1970s strongly linked to the anti-Vietnam and the 'hippie' movement in the USA and to growing resentment among the younger generation of advanced economies in the 1970s against global capitalism and its environmentally unsustainable development ethos (Schumacher, 1973). This was exacerbated by neo-Malthusian 'doom-and-gloom' scenarios of the 1960s and 1970s that suggested that runaway resource exploitation was resulting in anthropogenic environmental destruction that threatened to breach 'absolute limits' of sustainability (Meadows *et al.*, 1972). This grew into a mainstream and increasingly politicised movement in the 1970s and 1980s, culminating in the first 'green' political parties in countries such as New Zealand (from 1972) and Germany (from 1983) and, subsequently, in most industrial nations (Dobson, 1995; McCormick, 1995).

Yet despite the seemingly linear evolution towards 'greener' thinking, the transition towards environmentalism is complex and covers a whole spectrum of different views ranging from 'deep green ecocentrics' (e.g. Devall and Sessions, 1985), arguing for a cessation of environmentally destructive human activities, to technocentrics or cornucopians who question whether an environmental crisis even exists (Lomborg, 2005; Simon and Kahn, 2005). As Chapter 4 will highlight, we, therefore, only witness partial elements of a *linear transition* from environmentalism in the 1960s to the present day. There are increasing questions about the implied linearity of the transition towards environmentalism, the spatial coherence of the movement (both within and outside advanced economies), and how far the movement has permeated all segments of a multi-layered society.

3.3.4 Biological evolution theory

With this final example of a 'classic' transition, I wish to leave the social sciences arena and delve into what I believe are some of the most interesting, and intriguing, debates regarding concepts of 'transition' linked to *Darwinian evolutionary transition* theory. Indeed, any discussion of 'transition' would be incomplete without some consideration of this body of work – a theory that, arguably, has changed the way humans think about themselves and the world more profoundly than any other debate on transition. Biological evolution theory addresses one of the most important transitions of all, that of evolution of life on Earth from primitive to highly complex organisms, and as Mayr (2002: vii) argued, "evolution is the most profound and powerful idea to have been conceived in the last two centuries". I will investigate basic assumptions of biological evolutionary theory, in particular the relatively unilinear 'classical' Darwinian evolutionary model and recent challenges to that model. I wish to demonstrate that intellectual debates on the nature of transitions are not only restricted to the social sciences, but that many theories of transition in the natural sciences also show striking parallel critiques and re-evaluations that, ultimately, can be used as a basis to re-think and reconceptualise unilinear transition models of agricultural change. Indeed, some have argued that evolutionary transition shows striking parallels with technological transition (e.g. Dennett, 1995), emphasising that transitional patterns cross the perceived divide between natural and social sciences.

Biological evolution theory is a particularly powerful transition theory as it allows comparisons to be made with virtually all 'transition models' discussed in Chapter 2 and is, therefore, vital in informing our subsequent discussion of the p/pp transition. While Mayr (2002: 234) argued that "evolution means directional change", Dennett (1995: 21) suggested that "in a single stroke, the idea of evolution … unifies the realm of life, meaning, and purpose with the realm of space and time, cause and effect, mechanism and physical law". We see elements of linear transition in many debates of evolution (e.g. Lamarck, 1809; Darwin, 1859), stepped transition (Bak and Sneppen, 1993; Mayr, 2002), Deleuzian transition (Darwin, 1859; Eldredge and Gould, 1972), random or chaotic transition (Kauffman, 1993), as well as retrograde transition (Dennett, 1995; Mayr, 2002). However, evolutionary theory crosses the divide between disciplines, and since the second half of the 19th century has also been applied to evolutions in linguistics, philosophy, sociology, economics (e.g. evolutionary political economy) and other branches of thought (e.g. evolutionary network theory).

Evolutionary theory before Darwin was based on the concept of 'essentialism', i.e. that the world was perfectly designed by God (directional teleological thinking or finalism). Although essentialism implied that each class of animals had existed in its current form since creation by God, some weakened versions of essentialism between the 17th and early 19th century allowed for gradual linear change (transformation). Typical theories of unilinear evolution included, for example, Bonnet's 'great chain' of life (Bonnet first used the term 'evolution' in 1745), Lamarck's (1809) linear evolutionary pathway based on the belief in the inheritance of acquired characteristics from 'primitive' organisms to humans that postulated a steady evolutionary rise to perfection and, in particular, Linnaeus' (1751) taxonomy based on the assertion that 'nature does not make leaps'. Even today, those adhering to the notion of 'creationism' (e.g. some traditionalist and conservative factions of US society) continue to believe in a purposeful God-driven evolutionary pathway with humans as the 'logical' end-point of creation. In many ways, linear thinking inherent in these theories can be seen as an expression of Cartesian dualism that puts humans 'above' animals and that, as a result, could only conceptualise humans as the end-product of a directional chain of events linked to Judeo-Christian notions of human evolution (see Section 3.4). Thus, pre-Darwinian evolutionary models postulated essentially straight phyletic lineages (Mayr, 2002).

Darwin's (1859) original theory of evolution was a radical break from essentialism. Darwin recognised that evolution is *continually taking place*, and most scholars today acknowledge that an evolutionary transition *has* taken place (see Ridley, 1996, or Futuyama, 1998, for good philosophical discussions on this issue). Darwin's theory was based on the assumption of a slow and gradual transition in species development (gradualism) akin to the linear transition model shown in Fig. 2.1.a. Darwin's theory suggested the descent of all organisms from common ancestors (branching evolution) and the gradualness of evolution. Darwin's great contribution lies in his proposition of branching evolution based on the 'tree of life' that allows for both the temporal positioning of successful species on the same 'level', as well as allowing for whole branches of the evolutionary tree to die out and disappear without successors (Dennett, 1995). In his thinking, Darwin was strongly influenced by the concept of uniformitarianism that emphasised gradual changes in Earth's geology over time. Biological change is seen in Darwin's model as driven by selection for small genetic mutations that operate more or less uniformly at all times and in all places based on both the theory of natural selection (as opposed to Lamarck's essentialism) and the theory of common descent as typical theories of causation. Thus, "a population or species changes through the continuous production of new genetic variation and through the elimination of most members of each generation, because they are less successful either in the process of nonrandom elimination of individuals or in the process of sexual selection" (Mayr, 2002: 83). In this view, individuals with the highest probability of surviving and reproducing successfully are the ones best adapted, owing to their possession of a particular combination of attributes, leading to a continuing and (more or less) linear change in the genetic composition of every species population – a process usually referred to as 'survival of the fittest'. The resulting process *necessarily* leads in the direction of individuals in future generations who tend to be better equipped to deal with problems of resource limitation faced by individuals of the parent generation. Nevertheless, Darwin also acknowledged that a truly linear evolutionary transition was probably not possible (he lacked knowledge of genetic inheritance) and, depending on the time scale of analysis, also acknowledged that evolution could progress by 'very small steps'.

Fossil evidence with its substantial gaps in phyletic series, however, could never support gradualism and has rather suggested a stepped transition model with, in geological times, 'sudden' appearances and disappearances of species (see Chapter 4). The fossil record is, therefore, characterised by discontinuities (fossils are usually not connected with their ancestors by a series of intermediates) that have made it difficult to find empirical support for Darwin's notion of gradual and linear evolution – a problem exacerbated by the fact that Darwin's notion of speciation (the development of distinct species) can only be *retrospectively* identified (i.e. we can not 'observe' the development of new species) and that there is no way to mark the 'birth' of a species. This led Eldredge and Gould (1972) to question the classical Darwinian transition model, with Gould (1992a) particularly challenging the effectiveness of Darwinian theory to explain diversity. Eldredge and Gould's new concept was based on the notion of *punctuated equilibrium* as a typical stepped transition model. This suggests that saltational evolutionary tendencies (i.e. 'sudden jumps' or 'fits and starts') dominate over gradualist evolutionary pathways, based on the assumption that catastrophic events over the Earth's history led to massive extinctions enabling new animal and plant groups to develop. This theory suggests longer phases of species stability (stasis) followed by rapid change (punctuations) triggered by catastrophic events such as large meteorite impacts resulting in rapid climate change and massive species extinction (Bak *et al.*, 1994). During such events, highly specialised species (e.g. dinosaurs) are more vulnerable than more 'flexible' species (e.g. small mammals). The most dramatic of these catastrophes probably occurred around 250 million years ago through the impact of a large asteroid, a change in climate, and/or changes in chemical composition of the atmosphere, with estimates that over 70% of all species became extinct (a similar event occurred 65 million years ago leading to the extermination of dinosaurs).

Eldredge and Gould's (1972) model on the re-establishment of new species after such catastrophes is based on the principle of 'survival of the first' (i.e. species able to occupy free niches will survive and adapt) based on the notion of 'saltation' (sudden jump) characteristic for the stepped transition model. For example, the 'Cambrian explosion' saw the simultaneous emergence of many new species followed by long periods of stagnation. Their stepped model suggests that if a new species is successful and becomes effectively adapted to a new niche, it may subsequently remain unchanged for millions of years. Gould (1982) suggested that the constraints of inherited form and development pathways channel any selective change down permitted paths, where the channel itself represents the primary determinant of evolutionary transition (path dependency). Gould argued, therefore, that the constraints themselves become the most interesting aspect of evolution, and as soon as the appropriate level of adaptedness had been acquired by a species the rate was reduced drastically. This stepped transition model, based on long periods of 'stasis', therefore, has partly supplanted Darwin's old notion of 'survival of the fittest' (Dennett, 1995).

A further conceptualisation of evolutionary transition that builds on Gould's model has been provided by the 'Bak-Sneppen model' that uses the concept of *self-organised criticality* as a basis for understanding remaining inconsistencies in evolutionary transition (Bak and Sneppen, 1993). This model also challenges the Darwinian notion that evolution is a linear process, and draws on recent research that highlights that large extinction events are just part of a more general pattern of regular extinction events that can be expressed as a power law (Frigg, 2003). In the model, a specific 'fitness' is attributed to every species, and selection is taken to act at the species level through mathematical modelling of species interaction. This led Bak and Sneppen to suggest that evolution may be a process of 'self-organised criticality', where extinction can be seen as an avalanche where the extinction of a given species causes the extinction (or birth) of other species. In other words, the species with the lowest fitness goes extinct and is replaced by a new species with a random fitness (also affecting other species). The model generates a relatively clear stepped transition, as there are periods where almost nothing happens, while suddenly a large number of extinctions occur once the system has reached 'criticality' (Bak and Sneppen, 1993). The model particularly demonstrates that there is not necessarily a contradiction between Darwin's evolutionary model and Gould and Eldredge's punctuated equilibrium model of evolution as a stop-start transition (Bak, 1997). Thus, although the Bak-Sneppen model may not tell us much about the underlying processes of evolution, it shows that Darwin's and Gould/Eldredge's seemingly contradictory transition models can be compatible.

This discussion of evolutionary transition highlights the complex nature of debates on transitional processes. Chapter 4 will shed further light on how assertions of seemingly unilinear transitional pathways have been increasingly criticised by scholars from different disciplinary vantage points. The chapter will highlight how this existing body of work on 'transition theory' is beginning to gel into a common critical framework based on four key fallacies inherent in many debates on transitional patterns and processes – four fallacies that will be crucial in guiding our critique of the p/pp transition model of agricultural change.

3.4 Transition theory and Cartesian dualistic thinking

Chapter 2 has highlighted that transition theory can be seen as a theoretical framework that helps unravel socio-economic, cultural and environmental complexities of societal (and other) transitions from one state of organisation to another. Yet, as Sections 3.2 and 3.3 have highlighted, underlying the analysis of different transitions are often relatively simplistic assumptions about the *linearity* and *directionality* of these transitions. Before we can critically examine the fallacies of transition models in Chapter 4, we need to consider underlying philosophical and cultural parameters that underlie the often implied linearity and directionality of transition models. In this section I will briefly speculate why, based on *dualistic thinking*, the human mindset often seeks reductionist, unilinear and simplistic explanations to understand 'transition'. I will outline the importance of *dualistic thinking* embedded in Western thought as a basis for understanding problems inherent in many transition models – problems that also dominate conceptualisations of the p/pp transition model.

Many researchers and philosophers have suggested that world views and philosophies in Western society are largely driven by *dualistic thinking* (Sayer, 1991; Gerber, 1997). Dualism implies that social processes and transitions are seen from a specific, arguably narrow and reductionist, viewpoint, in particular through the lens of various binaries defined by opposites, such as 'good-evil', 'mind-body', 'developed-undeveloped', or 'matter-anti-matter' (Bookchin, 1992; Morgan and Morrison, 1999). Derrida (1994) criticised Western philosophy for its 'logocentrism', arguing that Western society's belief structures are based on a series of unequally valued binaries. In particular, Western intellectual history has a tradition of *Cartesian dualism*, expressed as an opposition between 'culture-nature', 'mind-matter' and 'subject-object' (Culler, 1983). Many commentators relate this dualism to Christian dualistic values of 'heaven-hell' and 'God-Devil'. White's (1967) seminal work emphasised that Christianity *legitimised* the dualism between humans and nature, resulting in the belief that humans operate *outside* of nature and that there is, therefore, no contradiction in exploiting nature for the purported betterment of human society in the name of God. Yet, non-Christian societies have also often adopted dualistic philosophies and world views.[9] Dualism is, therefore, by no means restricted to how the Western world attempts to understand societal change but also has to be seen as a 'natural' part of being human. Fundamentally, humans are shaped by the natural dualism of 'birth-death' that shapes the conceptualisation of life as a dualistic and temporally linear spectrum.

Such dualistic framing will inevitably have important implications for conceptualisations of transitions, and they may explain our 'obsession' with seeking relatively simple models (see Fig. 2.1.) to explain complex transitions. In the context of agricultural transitions, Pretty (2002: 151) argued that "a persistent problem is that the dualistic modes of thought go very deep. We have learned them well, and find it difficult to shake them off". Smith and Pickles (1998: 5), therefore, suggested that "there is a need … for an alternative set of conceptual frameworks on transition". The quest for an understanding of transition can be seen as an expression of dualistic imagination seeking to understand the above-mentioned 'grand compromise' necessary to move to the next 'step' of economic or social development (Kuhn, 1970; Lipietz, 1992). A basic understanding of dualism, and its inherent assumptions about the nature, pace and trajectory of transitions, is, therefore, crucial if we are to understand conceptualisations of societal transition in general, and ideas and concepts on the possible transition to post-productivist and multifunctional agriculture in particular.

---

[9] Probably the best-known non-Western example is Chinese Taoism with its conception of 'yin-yang', based on the assumption that all things and social expressions have an 'opposite' and form an unpenetrable unity that also finds expression in human health or illness (Luczak, 2001).

As we will see in Chapter 4, debates surrounding possible transitions have, therefore, brought unease about the implied 'simplicity' inherent in traditional transition models. Many theoreticians – from the rural research arena to other social (or natural science) disciplines – have begun to question a variety of assumptions underlying transition models. Deleuze and Guattari (1987), in particular, criticised both Western dualistic imaginations based on simplistic two-point unilinear transitions leading from 'one point' to 'another' and the use of 'single signifiers' in descriptions of Western economies, for what they thought are often 'a thousand plateaus' of multiples and hybrids transforming themselves among independent and interlinked multiple transitional pathways – an argument used for conceptualisation of the *Deleuzian transition model* in Fig. 2.1. Bookchin's (1992: 24; emphasis added) work has also acted as a powerful critique of reductionist thinking and antagonistic dualism as he argued (for biological debates) that "in nature, balance and harmony are achieved by ever-changing differentiation, by ever-expanding *diversity*. Ecological stability, in effect, is a function not of simplicity and homogeneity but of *complexity* and *variety*". Poster (1989: 9) went further by arguing that "linearity and causality are the spatial and temporal orderings of a now-bypassed modern era", suggesting that theorisations of linear transition models, such as shifts from 'isms' to 'post-isms', may belong to an 'old-fashioned' era embedded in modernist thought.

Debates on dualism are, thus, important as they help us contextualise and understand important philosophical and structural barriers for the broadening and critical analysis of transitional concepts in society. In this book, I do not claim to be able to 'jump beyond' the boundaries of Cartesian dualistic thinking – in particular as I am a product myself, like most people from developed countries, of the Cartesian life world and philosophy that have influenced scientific investigation for centuries. However, I will propose a new framework of multifunctionality based on a *spectrum* of decision-making that acknowledges the inherent pitfalls of dualistic thinking – a critique that will permeate the analysis of transition in the remainder of this book.

3.5 Conclusions

Chapter 3 has drawn on the wealth of existing literature on 'transitions' to analyse specific debates about shifts from one organisational state to another. By analysing postulated shifts from 'isms' to 'post-isms' and other transitions, this chapter has explained basic assumptions underlying such transitions as well as highlighting the applicability of specific models of transition discussed in Chapter 2. This will help us better understand underlying principles surrounding the possible shift to post-productivism discussed in Part 2, and will enable the reader to recognise the close interlinkages and parallel arguments that exist in investigations of transitional processes. The chapter also highlighted why many conceptualisations of transitions have been based on relatively simplistic unilinear, binary and homogenous transitional assumptions based on Cartesian dualistic thinking that underpins most transitional debates. Critically engaging with these debates in Chapter 4 will help us provide a framework for deconstructing the implied linearity of the p/pp model and for understanding four key 'fallacies' underlying many transitional models. This, in turn, will form the basis for introducing a new notion of 'multifunctionality' theoretically anchored in a *revised concept* of productivism and post-productivism in Part 3.

Chapter 4

Reconceptualising transition: the complexity of transitory systems

4.1 Introduction

Chapter 3 has highlighted the different and multi-faceted debates surrounding various transitional models in both the social and natural sciences. In this chapter, I wish to investigate in more detail problems associated with relatively simplistic assumptions inherent in many transition models. The examples of transition in Chapter 3 have already indicated that notions of *linearity*, *directionality* and *inevitability* underlie many transitional debates. I will highlight that all eight transitions discussed in Chapter 3 have been criticised for their often simplistic and unilinear assumptions, and that many examples can be found to support the notion that supposedly 'simple' and 'linear' transitions have instead been characterised by complexity and unpredictability. As the discussion of dualistic thinking in Chapter 3 highlighted, cynics may argue that the entire notion of a 'post-ism' is already based on a fallacy of directionality, as the transition to a 'post-ism' is often portrayed as *inevitable* and imbued with a certain (often 'positive') direction. This may be because 'post-isms' tend to be conceptualised in the 'negative' as what *they are not* rather than what *they may be*. The prefix 'post' may merely denote something which comes *after* another thing, and does not necessarily mean its *opposite* in a temporal and spatial sense. As novelist Margaret Atwood (1988: 86) succinctly argued, "post this, post that ... everything is post these days, as if we're all just a footnote to something earlier that was real enough to have a name of its own".

I will show that transitions are more complex than simplistic transitional models would let us believe. Transitions, while not necessarily 'chaotic' in Gleick's (1988) terminology, are usually characterised by non-linearity, heterogeneity, complexity and inconsistencies. I will argue that these simplistic assumptions form four key 'fallacies' underlying transitional debates: first, issues related to transition as a *linear* temporal process from one stage of development to another (the *fallacy of temporal linearity* discussed in Section 4.2); second, that we can generalise transitional patterns and processes *spatially* (the *fallacy of spatial homogeneity*; Section 4.3); third, that transitional processes apply equally to *advanced economies and developing countries* (the *fallacy of global universality*; Section 4.4); and, fourth, that *all actors and stakeholder groups* are on *the same transitional pathways* (the *fallacy of structural causality*; Section 4.5). These categorisations echo Rotmans *et al.*'s (2002) suggestion that transition can be conceptualised through both time (temporal directionality) and scale or space (spatial homogeneity/global universality). I argue that additional conceptualisation of the fallacy of structural causality adds a vital ingredient to debates on transition and reflects recent post-structuralist (another post-ism!) changes in research focus on issues of governance, actor and stakeholder roles and networks, and state/non-state actor interactions (cf. Goodwin and Painter, 1996; Jessop, 1998).

This chapter will suggest that, instead, the notions of **temporal non-linearity** (i.e. that 'post-isms' do not necessarily follow sequentially from, and even replace, 'isms'), **spatial**

heterogeneity (i.e. that not all geographical localities are affected in similar ways by transitions), **global complexity** (i.e. that transitional trends identified in developed countries may not apply to developing countries) and **structure-agency inconsistency** (i.e. that 'structure' and 'agency' are not equally affected by transitions) better encapsulate the complex processes that can be observed in societal transitions. This discussion will provide the crucial argumentative framework for our analysis of agricultural transition, in particular as understanding the 'fallacies' in conceptualisations of 'other' transitions will help us recognise the strengths and weaknesses of both the p/pp transition model and the transition towards 'multifunctionality' discussed in Parts 2 and 3. It can be hypothesised that the four fallacies inherent in 'other' transitions may also be apparent in debates on agricultural transitions.

4.2 Temporal linearity or non-linearity?

The notion of *temporal non-linearity* is a direct critique of those advocating that transitions follow temporally linear patterns. It suggests that transitions are *not* necessarily linear, directional and inevitable. Temporal non-linearity implies that processes of transition usually do not follow an easily definable linear transitional pathway over time, and that, instead, transition, if it occurs at all, is characterised by a complex inter-weaving of 'development stages' that may, at times, also mean a move 'backward'. Rotmans *et al.* (2002: 9) argued that "transitions are characterised by alternative pathways [i.e. Deleuzian in our terminology] of slow and fast dynamics, yielding a strong non-linear behaviour", and that "a transition is the result of developments in different domains [and] can be described as a set of connected changes, which reinforce each other but take place in several different areas ... A transition can be seen as a spiral that reinforces itself; there is multiple causality and co-evolution caused by independent developments" (Rotmans *et al.*, 2002: 3). Although it is acknowledged that some form of transition may be taking place in a given system, the specific form that this transition takes is often temporally irregular, unpredictable and, most crucially, may be characterised by individual components of the transition moving along the transitional pathway at *different speeds*. Hallinan (1997), therefore, argued that social change is often 'highly non-linear' and 'discontinuous'. Rotmans *et al.* (2002) further argued that while economic transitions, for example, may take place in the short term (e.g. through election of a new government), demographic change requires at least two generations (see below), while full environmental recovery from degradation may take centuries – referred to as a 'hybrid mixture of fast and slow dynamics' where different transitional time scales overlap and influence each other. A 'successful' transition, therefore, only occurs once the *slowest* components of the system have completed the transition – a process that, in some instances, may never be completed (see examples below). On the one hand, the notion of temporal non-linearity is a critique of the often simplistic assumptions underlying the linear and stepped transition models shown in Fig. 2.1. On the other hand, it supports the more 'fuzzy' temporal notions inherent in the random transition model (i.e. no obvious temporal linearity apparent; see Fig. 2.1.c), the retrograde transition model (i.e. a transition may 'fail'; Fig. 2.1.d) and the Deleuzian transition (Fig. 2.1.e) where 'a thousand plateaus' of possible transitional pathways suggest that different components of a transition may occur in different temporal horizons. In the following, I will look at various examples from the critical literature on transition, with specific emphasis on the eight examples of transition discussed in Chapter 3.

A good starting point is the discussion of temporal non-linearity in the transition towards *post-modernism*, as it is in these debates that the implied temporal linearity of the concept of 'post-isms' is arguably most problematic – partly because, as we saw in Chapter 3, post-modernism is so vaguely conceptualised and defined. Many have argued *against* such a linear and directional shift, in particular Latour (1993) who suggested that 'we have never been modern', while Johnston *et al.* (2000: 620; emphasis added) argued that "the break between modern and postmodern genres [is not] all that clear" and that "a fixation with these and other *lines of continuity* … have led some theorists to deny any significant break between modernism and postmodernism". Similarly, Dear (1986) argued that the notion of post-modernism with its "erratic and chameleon-like form renders it susceptible to quick dismissal", while Newman (1985: 1) argued that the very term of post-modernism "signifies a simultaneous continuity and renunciation, a generation strong enough to dissolve the old order, but too weak to marshal the centrifugal forces it has released". In particular, there is considerable debate about the *temporal onset* of post-modernism, best expressed by Van de Kaa (2001: 294-295) who argued that "some see it [post-modernism] having emerged following the Enlightenment of the seventeenth and eighteenth centuries, others place it around 1875. Others speak about 'the last few decades', the 1980s and 1990s even". Berman (1982), thus, emphasised that it is difficult to define when the modern, let alone the post-modern, period began. The most trenchant critique has, interestingly, come from Umberto Eco (1984: 65-66), famous author of the *Name of the rose*, who, in a postscript to his book, suggested that "unfortunately, 'postmodern' is a term *bon à tout faire*. I have the impression that it is applied today to anything the user of the term happens to like". Eco concluded that post-modernism should not be regarded as a chronologically defined term but as a heuristic 'category', not dissimilar to Jencks' (1984: 5) suggestion that "defining our world today as Post-Modern [sic] is rather like defining women as 'non-men'. It doesn't tell us very much, either flattering or predictive. All it says is what we have left – the Modern world, which is paradoxically doomed, like an obsolete futurist, to extinction".

Nederveen Pieterse (2004: 62), therefore, argued that the recent 're-thematisation' of modernism suggests the "continuing interest in modernisation thinking" – a type of thinking that has *not* been fully replaced by post-modern action and thought. As a result, he suggested the idea of "multiple paths of modernisation" (Nederveen Pieterse, 2004: 63) akin to Deleuzian transitional development pathways. This was further echoed by Gaonkar (2001) and Eisenstadt (2002) who suggested that different societies create their own modernities and post-modernities, all operating in a non-linear fashion and at a different pace. This has been particularly articulated by Latin American writers who argued for the coexistence and interspersion of pre-modernism, modernism and post-modernism (e.g. Hösle, 1992; Vargas, 1992). Berman (1982: 17), thus, argued that the experiment of modernism has "shattered into a multitude of fragments", and that the idea of modernity "conceived in numerous fragmentary ways, loses much of its vividness, resonance and depth … As a result of all this, we find ourselves today in the midst of a modern age that has lost touch with the roots of its own modernity". Berman mentioned the example of 'collage' as an artistic technique that is meant to be a central indicator for post-modernism, but reminded us that this art form was already used frequently during the 'modern' era. Similarly, many commentators have argued that the so-called 'post-modern' repositioning of authors – especially concerning race, gender and class – has occurred many times in the history of writing, and that, again, a direct temporal transition from one writing style to another can not be supported empirically (Barnes, 1996). If we are unsure about the nature and meaning of 'modernism', it may be futile to attempt to conceptualise a temporally linear transition towards 'post-modernism' to emerge from the ashes of modernism – a concern echoed by Harvey (1989) who, like Latour (1993), rejected the notion of a 'post-modern culture' as an 'art of surfaces' that simply masks underlying socio-political and cultural changes.

Many authors would, therefore, argue that there is little evidence of a linear transition, as modern and post-modern action and thought can be found *simultaneously* (Lunn, 1985; Gregory, 1994). The transition to post-modernism may only be a *hypothetical assertion* expressed in philosophy and social science theorisation (Lyotard, 1984; Harvey, 1989). Thus, authors such as Taylor (1996) argued that advanced economies are still firmly embedded in the 'late phase' of the modern world – in other words, that humanity has not even embarked upon the transitional trajectory towards post-modernism. This was echoed by Taylor (1996: 210) who suggested that what is commonly described as post-modernism "does not define a new society as its 'post' appellation implies; rather it is merely a 'phase' in the development of high culture in the modern world-system". King (1995) went one step further in challenging the implied linearity of the post-modern transition by arguing that 'relabelling' is required: Northern industrial capitalism of the 19th and early 20th centuries (hitherto the era of modernism) should be renamed 'pre-modern', while the 'modern' should become what has Euro-centrically been called the 'post-modern'. In other words, there should be a "respatialization of the modern" as a way to overcome the problem "that what some have labelled 'postmodern' culture pre-dated what they have labelled 'modern' culture" (King, 1995: 120-121). In a similar vein, Leontidou (1993) argued for Southern Europe that post-modernity has been characterised by a transition from pre-industrial to post-industrial *without* widespread industrialisation having been experienced. As Simon (1998: 235) argued, "this challenges the conventional linear-stage approach to development as well as views of the postmodern as being strictly epochal". We will see striking parallels between this argument and that of conceptualising post-productivist agriculture in Part 2.

Simplistic assumptions underlying the fallacy of temporal linearity in the transition from *socialism to post-socialism* have also been criticised. While it is often assumed that *all* countries that were part of the former Soviet hegemonic block (including Russia) have gone through post-socialist transitional pathways at the same pace (Hall, 1996), several years after the beginning of this postulated transition there is substantial debate about the linearity of this transition. Pavlinek and Pickles (2000: 29), for example, argued that "detailed case studies identifying different national and regional pathways from state socialism ... challenge notions of smooth and linear transitions". It is argued by many that the transition of post-Soviet countries towards capitalism or democracy should not be seen as a predetermined trajectory – probably best expressed in the differing proposals between 'gradualists' and 'shock therapists' for the temporal sequence of transition (see Chapter 3). Kalb *et al.* (1999: 11), therefore, referred to this transition as "a more diffuse and pluralistic notion of post-communist transformation". Those advocating temporal non-linearity in the post-socialist transition point towards the emergence of policy reform long *before* the end of the Soviet Union. As Pavlinek and Pickles (2000) showed for environmental regulation, struggles over environmental issues had been occurring in socialist countries throughout the 1980s based on a rapidly deteriorating environment. Critics also point towards continuing existence of parallel modes of accumulation between socialist and capitalist modes in former countries of the Soviet Union. State socialism did not operate in a vacuum, as labour process models and regimes of accumulation mirrored those in developed capitalist countries, especially as state socialism could justify itself only through economic success compared with the capitalist economies (Pavlinek and Pickles, 2000). The result was similar 'productivist' rationality after 1945 in both socialist and capitalist societies and the emergence of *parallel pathways* of Fordist accumulation models.

Socialism, therefore, never completely overrode capitalist modes of accumulation (Altvater, 1993, 1998). Johnston *et al.* (2000: 624), therefore, suggested that the "transition to either or both of capitalism and democracy should not be considered as predetermined or automatic ... We need to recognise that [post-socialist countries] are not necessarily all travelling in the same path: they are in effect assuming multiple trajectories". Indeed, in some post-socialist countries (e.g. Poland) different transitional pathways were used *simultaneously* to smooth the transition. Shock therapy was used for bringing down hyperinflation, while a gradualist approach was adopted for privatisation of state-owned enterprises (Vickers and Yarrow, 1991). Similarly, evidence from China suggests that while gradualism has been used for the transition from state to privatised enterprises, this approach was less successful in changing the authoritarianism of the ruling Communist Party (Stiglitz, 2002). In other words, different components of the transition towards a free market and democratisation in China have taken different temporal trajectories (i.e. the economic transition occurred much earlier than the political one), highlighting that different social arenas often have different temporal transitory dimensions. There is, therefore, no single economic path to reconstruction in post-socialist countries, and there are multiple alternatives between state socialism and liberal free market capitalism, resulting in a diversity of temporally non-linear path-dependent trajectories towards post-socialism (Stark, 1992; Grabher and Stark, 1998). Smith and Pickles (1998: 2), therefore, argued that "transition is not a one-way process of change from one hegemonic system to another. Rather, transition constitutes a complex reworking of old social relations".

Two possible alternative transitional pathways are, therefore, suggested by those attempting to theorise the transition to post-socialism. First, the notion of a 'Third Way' represents a reworked modernisation theory which recasts the transition so that it leaves room for an alternative, not necessarily temporally and spatially unilinear path of development (Smith and Pickles, 1998) – not dissimilar to Sayer's (1995) 'radical political economy' of post-communism. This crucially echoes debates on other shifts from 'isms' to 'post-isms', in particular the notion of 'neo-colonialism' as a non-linear outcome of the classical colonialism/post-colonialism transition model (see below) and the notion of 'territorialisation' and 'multifunctionality' for the discussion of non-linear shifts in the transition of agricultural systems discussed in the remainder of this book. Second, proponents of temporal non-linearity in the post-socialist transition have been instrumental in advocating the *retrograde transition model* (or 'failed' transition; see Fig. 2.1.d) as an alternative concept (e.g. Dunford, 1998). This critique argues that no 'measurable' transition has taken place (i.e. points 'a' and 'b' in Fig. 2.1.d remain at the same level of 'development'), thereby challenging the notion of a linear, directional and inevitable shift from one transitional stage (socialism) to another (post-socialism). Stiglitz (2002: 162), therefore, argued for Russia that "instead of a smoothly working market economy, the … transition led to a disorderly Wild West", and that the imposition of an 'alien' framework of rules and regulations has made a meaningful transition towards post-socialism impossible. This has meant that the transition has been marred by the highly problematic implementation of a set of unsuitable and incompatible policies involving economic liberalisation, marketisation and democratisation. Pavlinek and Pickles (2000: 294), even posited that it is, in theory, possible to think of state socialism as "some kind of minor interruption in a 'normal' process of capitalist development or as a mere aberrant form of capitalist exploitation" – a suggestion we need to keep in mind when discussing the possibility of agricultural productivism as an aberration in an essentially non-productivist era (see Chapter 7).

Debates on the transition from **colonialism to post-colonialism** have also highlighted problems of temporal linearity. While *all* former colonies are now often described as being in the post-colonial era (Childs and Williams, 1997), some have questioned whether a post-colonial era has at all superseded a colonial period (Abdel-Fadil, 1989; Hall, 1996). In particular, proponents of *dependency theory* suggest that, under current economic and political global regimes, 'equal' economic development of developing countries (including former colonies) is difficult, if not impossible (Yearley, 1988). Here, it is argued that economic and political structures of former colonies are shaped by the often involuntary relationship these countries still have with their erstwhile colonisers, and that this *artificial dependence* creates underdevelopment in many so-called 'post-colonial' nations. Thus, decolonisation has "not resulted in meaningful economic or political independence"

(Johnston *et al.*, 2000: 95) – in other words, colonial and post-colonial elements in former colonial territories continue to exist side-by-side. Many studies on the transition to post-colonialism have found that 'post-colonial' development strategies are ultimately designed to make 'them' more 'like us', with development experiences in the North still commonly regarded as the 'ideal' model. This is particularly evident through the continued dependence of many former colonies on primary exports, the many dependent and corrupt political elites linked to former colonial powers, and the perpetuation of colonial discourses among many new leaders of former colonial territories. Simon (1998), therefore, emphasised how pre-colonial, colonial and post-colonial traditions, institutions and practices still lie side-by-side.

This has led critics to argue that work on the transition to post-colonialism has failed to take into account decentred economics, indigenisation, local stakeholder participation and burgeoning democratic processes as the 'true' indicators of a new post-colonial world (Sidaway, 2003). Johnston *et al.* (2000: 613) suggested that "an understanding of colonialism and its successor projects has to grasp the complicated and fractured histories through which colonialism passes from the past into the present". Hall (1996), therefore, argued that the notion of post-colonial transition (both in time and space) should be seen as an uneven, serialised and *temporally non-linear* process of decolonisation, a process that clearly questions the binary concept of the traditional colonialism/post-colonialism dualism. Commentators such as McClintock (1992) go even further by arguing that the entire notion of a shift to 'post-colonialism' is flawed, as it has continued to place Europe as the core subject of 'colonial history' in a Euro- and Anglo-American-centric and, arguably, colonialist conceptualisation of the world. Indeed, the very notion of post-colonialism conjures undertones of *unidimensional colonial thinking*.

Many commentators have, therefore, argued that 'post-colonialism' is not the best term to describe contemporary developments in the developing world, with the notion of 'neo-colonialism' emerging as a powerful critique of temporal linearity (Abdel-Fadil, 1989; Said, 1993). Neo-colonialism is seen as a means of political and economic control exerted in particular by powerful Northern states (especially the USA, EU and Japan), also termed the continued 'development of underdevelopment' (Mohan and Stokke, 2004). Although post-colonial conceptualisations assume that dominated states have become seemingly 'independent' over time, the argument of the neo-colonial school of thought is that political and economic systems of most (if not all) developing countries continue to be controlled from 'outside'. This embeddedness of most developing countries in *neo-colonial global capitalist structures* is particularly evident through international trade conventions (e.g. the Lomé Convention between the EU and over 60 African, Caribbean and Pacific states), neo-colonialist aid packages aimed (to some extent at least) at increasing the dependency of receiver states (Peet, 1991; Dicken, 1998), and financial dependency created through extensive loans by the IMF or the World Bank (Stiglitz, 2002). The notion of neo-colonialism implies, therefore, that there has *not* been a temporally linear development from 'colonialism' to 'post-colonialism', but that, instead, the economic, political and socio-cultural trajectories of developing countries have been much more complex than a simple temporally linear model can ever describe. 'Hybridity', therefore, is a more accurate descriptor for the era 'after colonialism' where, according to Bhabba (1990), hybrid trajectories act as intercultural brokers in the interstices between nation and empire, producing non-colonial or anti-colonial counter-narratives from the nation's margins. At times, therefore, less developed (but decolonialised) countries show elements of 'neo-colonial' development – akin to the model of a retrograde (or 'failed') transition.

There are similar concerns about issues of temporal linearity in debates on the transition from ***Fordism to post-Fordism***, as there is often an assumption that Fordism has *inevitably* given way to post-Fordism (Amin, 1994; Dicken, 1998). However, evidence from many advanced economies has increasingly shown that Fordist and post-Fordist modes of accumulation began to *overlap* during the 1970s and 1980s, and that the temporal sequence of a smooth shift from Fordism to post-Fordism – towards a new 'grand compromise' in Lipietz's (1992) words – was never fully achieved. This prompted Peck and Tickell (1994) to reject the notion of 'post-Fordism' and to suggest the alternative concept of 'after-Fordism' that argues that a new regime of accumulation, embedded in the notion of post-Fordism, has not yet emerged. Similarly, Cloke and Goodwin (1992) and Goodwin and Painter (1996) argued that Fordist and post-Fordist modes of regulation occur *simultaneously* in spatial, temporal and conceptual terms. Hoggart and Paniagua (2001: 42), therefore, suggested that there is "regular questioning of the prevalence of so-called post-Fordist production practices". In addition, many have argued that liberal-productivism (characteristic of post-Fordism) has remained as hierarchical as Fordism itself (Amin, 1994; Goodwin and Painter, 1996), with Lipietz (1992: 25; emphasis added) acknowledging that "the monetarist and neo-liberal principles of the early 1980s have lost their prestige, *but they are by no means politically discredited*". Conclusive evidence for a temporally linear shift to post-Fordism is lacking and it is, indeed, difficult to point towards a contemporary region or society that could be described as 'post-Fordist' in the true sense of the term.

Writings on **technological transition** also provide plentiful ammunition for challenging the fallacy of temporal linearity in transitional processes. Although there has been a considerable body of work that has suggested relatively unilinear directional transitional pathways for technological evolution (e.g. Usher, 1954; see Chapter 3), authors such as Rogers (1995) have suggested that *temporally non-linear* technological innovation adoption tends to be the *norm* in most societies. This differential pattern is usually referred to using a five-point innovation adoption spectrum that shows substantial differences in the pace of adoption, ranging from 'innovators' (highest innovation adoption scores and earliest adopters), 'early adopters', 'early majority', 'late majority' to 'laggards' (lowest adoption scores and latest adoption). This suggests that different segments of society adopt innovations at different speeds (if at all), depending on socio-economic characteristics, personality-related variables, homophily (similarities between interconnected individuals), communication behaviour, diffusion networks, critical mass, or the perceived or real need for innovation. Grübler (1997: 27), therefore, argued that the process of technological transition "is not gradual and linear but is instead characterised by long swings and discontinuities", and that technological transitions "remain discontinuous in time and heterogeneous in space" (Grübler, 1997: 29). He used the example of the temporal global spread of the car, arguing that initially it diffused quickly in the early 20th century (i.e. in a few decades) because it was replacing the horse, but that further diffusion was much slower as it depended on infrastructure developments, changing settlement patterns, restructured service activities and income growth. In this context, Parayil (1999: 8) suggested that "technological change [is] an open-ended process without a clear trajectory". Similarly, Graham (1998: 180) argued that "the very notion of technological 'impact' … is problematic, because of its attendant implications of simple, linear, technological cause and societal effect". Smith (1993), in particular, argued that linear technological transition models are simplistic and ahistorical and that they only capture *successful* innovations – highlighting that the 'silent', 'unseen' and 'unsuccessful' technological innovations (that make up the vast bulk of innovations) may well follow non-linear transitory pathways. These debates are crucial for our analysis of the p/pp transition, as they highlight that, viewed from a presentist perspective, transitions often appear as a selective and linear process, while in reality non-linearity, randomness and even 'chaotic' patterns may be the norm.

Debates are, however, complex. Although most would now criticise the implied linearity in many technological transition models, many also argue that technological change can be a *cumulative* process that relies on *system memory* and *path dependency* in the achievement of technological innovations (Basalla, 1988; Parayil, 1999). Most technological transitions show that a new development by no means displaces 'old' forms of technology – i.e. once a technological pathway has been developed, it is difficult to leave this pathway due to market forces, consumer lethargy and social and psychological factors (e.g. one may get 'used to' a certain type of technology irrespective of arguably 'better' alternatives). Constant's (1980) analysis of changes in aircraft propulsion is a good example: although jet-propelled planes have now largely replaced propeller-driven aircraft, they have not completely replaced propeller technology – suggesting the existence of parallel technological pathways that are not mutually exclusive. Similarly, the invention of the car did not displace all other modes of transportation (walking, cycling, trains, etc.), nor did the invention of the CD player completely displace cassette or record players, while the invention of the computer and word processing did not displace other forms of writing and printing (Grübler, 1998). Thrift (1996a), therefore, argued that new technologies often act as a *supplement* rather than a *replacement*, with innovations like the telephone, fax and computer used to *extend* the range of human communication and not to *substitute* them.

Strassmann (1959) argued that old technologies often co-exist with new ones, and that *total* technological change (i.e. complete replacement of one technology by another) only takes place when the rate of obsolescence occurs with unforeseen rapidity. Economic, political, institutional, scientific and social factors combine to operate as selective devices of technological transition operating within a large set of often Deleuzian pathway possibilities. As a result, Dosi (1982: 160) argued that technological transition should be conceptualised as an 'outward' movement (i.e. non-linear) that includes a cluster of pathways in multidimensional space "along something like a cone, rather than the movement along ... a smooth curve". However, critics also acknowledge that inventions are often linked to complex feedback mechanisms in which the market may play an increasingly important role once the invention is marketed – i.e. the awakening of new consumer needs linked to technological change such as the perceived need for the latest computer or mobile phone technology. In this sense, technological transition can be seen as an autonomous (or quasi-autonomous) process in which it is impossible to explain why, at a given point in time, an invention takes place. Parayil (1999), therefore, argued that technological transition may best be described as an *ortholinear* or *directional variation*, characterised by a selective retention process adapted to a sequential process of variation and selection, while Dosi (1982: 161) concluded that "technological paradigms and trajectories are in some respects metaphors of the interplay between continuity and ruptures" in transitional processes.

Similar criticisms apply to the seemingly temporally linear transition towards *environmentalism*. The starting point for debates rests on the assumption that 'environmentalism' as a nature-society philosophy is much older than the wave of environmental concern over the last 40 years or so (Glacken, 1967). Indeed, writings by early American 'environmentalists' such as Thoreau (1859) or Marsh (1864/1965) showed that environmental thinking was present in the New World at a time of most severe environmental degradation (Wilson and Bryant, 1997). Indeed, Marsh (1864/1965: 280) already argued in the mid-19th century for North America that "we have now felled forest enough everywhere ... Let us restore this one element of material life to its normal proportions". Parallel to this, some of the oldest environmental NGOs were created in the mid- to late 19th century to lobby for the protection of endangered species, with the Society for the Protection of Birds (established in the UK in 1881) or the Natal Game Preservation Society (established 1883 in South Africa) being good examples. Yet, although the revulsion of the urban middle classes in 19th century Europe and North America against the social and ecological effects of industrial development largely spawned what we today call the 'conservation movement', a large proportion of 19th century society was still preoccupied with survival and maximum exploitation of the environment based on utilitarian principles (Pepper, 1984; Brimblecombe, 1987).

Since the 19th century, utilitarian and conservation-oriented attitudes have existed side-by-side, suggesting multiple (and often parallel) pathways of temporal evolution of environmentalism (Wilson and Bryant, 1997). While segments of society – in advanced economies and also increasingly in developing countries – show growing disenchantment with the environmental degradation that is often a practical outcome of the adoption of utilitarian attitudes, a large proportion of contemporary society still continues to be staunchly utilitarian (Dobson, 1995; McCormick, 1995). Although many societies have seen the increasing political importance of green parties and environmental NGOs, they have also seen the growing importance of often highly utilitarian stakeholders such as transnational corporations or international financial institutions (Wilson and Bryant, 1997; Forsyth, 2003). The temporal non-linearity of the transition towards environmentalism has been further challenged during the 1990s and 2000s, as it is increasingly argued that, although environmental and green thinking has now become firmly embedded in mainstream politics, the environmental movement has lost momentum in the new consumer-oriented generations of the post-1990 era (Barr, 2003, 2006). There is, therefore, increasing evidence that the 'green movement' has lost ground, and that a *reversal of attitudes* may be taking place with today's younger generations being less environmentally aware or interested than their parents. Some argue, therefore, that the transition to environmentalism may only have been a brief ideological 'fashion' and that recent trends show signs of a *retrograde transition* (see Fig. 2.1.d) (McCormick, 1995; Lomborg, 2005).

There have also been many critical voices about the seemingly temporally linear directionality of the ***demographic transition model***. It can be argued that the notion of temporal directionality is particularly embedded in this model, as it is a typical example of a 'directional model' that allows only one possible linear transitional pathway (Szreter, 1993; Mason, 2001). As Casterline (2001: 18) argued, "the pace of transition is a relatively unexplored topic, despite its short- and mid-term demographic consequences and despite the deep impression that the apparent rapidity of transition has left on many scholars". Cleland (2001) further argued that the 'classic' demographic transition model rigidly assumed that the transition would *always* be characterised by falling death rates, an ensuing period of rapid natural increase, a lagged decline of birth rates and an eventual return to population equilibrium (see Chapter 3). This dilemma was echoed by Hirschman (2001: 117) who argued that "much of the controversy … arises from assumptions that all fertility transitions follow a similar path". Yet, there is increasing evidence that not all societies neatly follow the temporal sequence suggested in the Euro- and Anglo-American-centric demographic transition model. Starting with early critiques (e.g. Davis, 1945), researchers began to argue that individual stages of the model may be by-passed completely, or, more realistically, that they take completely different forms in different contexts (Kuijsten, 1996). Casterline (2001: 26), therefore, suggested that we should acknowledge the existence of *multiple demographic transitions* and that there also is "substantial regional variation in the timing of the onset of transition".

Ireland is often cited as a key example of a country that has not complied with the linearity of the demographic transition model (Casterline, 2001). Ireland's population went very quickly from stage 1 in the model (high birth and death rates) to stage 4 (low birth and death rates) or even 5 (low death rates and even lower birth rates) in only a few decades, thereby 'compressing' or even 'by-passing' stages in the model. An explanation may be that Ireland never witnessed an 'industrial era' which, in many other developed countries, was associated with the gradual shift from stages 1 to 3 (Mason, 2001). At the other extreme are countries that have not yet 'left' stage 1. Although such countries are increasingly in a minority in an age of rapidly declining global fertility rates, examples include many countries in sub-Saharan Africa where the average total fertility rate at the beginning of the 21st century still stands at 5.2, with Uganda (7.1) and Somalia (7.25) having the world's highest fertility rates (United Nations, 2003). This led Cleland (2001) to argue that these countries may still be in a 'pre-transitional' stage. Further, some world regions that had been firmly embedded in stage 4 of the demographic transition model, currently show tendencies of *increasing* death rates, with Eastern Europe as a particular case in point. Other countries such as Sweden, meanwhile, are witnessing *rising fertility levels* (from 1.6 to 2.0 births/woman) thereby questioning whether they should still be classified as within stage 4 of the model (Bongaarts, 2001). Indeed, "a substantial fraction of the population of the globe resides in countries where the fertility transition is far from complete" (Casterline, 2001: 19). By the early 1990s, 'only' 72 countries (out of 200) were judged to have begun the fertility transition towards zero growth, and only the UK has most closely conformed to the 'classic' demographic transition model (Kates, 1997). The latter is not surprising considering that the model was largely developed on the basis of the UK experience (an important point we will also discuss in Chapter 7 for the implied Anglo-centrism in the p/pp transition model). While the transition from high birth to low death rates took about 150 years in the UK and was closely linked to stages of industrialisation, birth rates declined in France much earlier (i.e. in pre-industrial times) while many other European countries (e.g. Eastern Europe) saw a compression of stages 3 and 4 of the transition in only a few decades (Kates, 1997; Mason, 2001).

As Chapter 3 highlighted, recent world population trends with rapidly declining fertility in most areas (e.g. Spain with a fertility rate of only 1.15; Germany 1.3; Japan 1.45) do not seem to easily 'fit' the relatively simplistic assumptions of the demographic transition model. Casterline (2001: 32), therefore, argued that "the alleged rapidity of fertility declines is a lynchpin of the case made by many scholars against conventional demographic transition theory", while Bongaarts (2001: 260) suggested that "conventional demographic theories have little to say about the level at which fertility will stabilize at the end of the transition". Further, the temporal sequencing of the model has been criticised for failing to take into account the increasingly important effects of *migration* on population patterns and birth and death rates (Ogden and Hall, 2004). Indeed, immigration is an issue that is playing an increasingly important part in defining national population growth (e.g. the USA currently has 1.1 million net migration gain/year; United Nations, 2003), and highlights that the basic components of the 'traditional' demographic transition model (birth and death rates) no longer suffice as explanations for population trends (Greenhalgh, 1988). Thus, Hirschman (2001: 118) argued that "the portrayal of demographic transition theory as a universal, unilinear, ahistorical model of modernization and fertility decline is too simplistic".

The issue of temporal non-linearity is arguably most complex regarding *evolutionary transition*. That this body of knowledge has been predicated largely on deductive theorisation has meant that temporal non-linearity has been considered in great detail. Contrary to most societal transitions, biological evolution theory, therefore, can be used to highlight issues regarding *both* temporal linearity and non-linearity. While both Lamarck (1809) and Darwin (1859) argued for temporal linearity and directionality in evolutionary transition, later refinements of Darwin's theory have argued for temporal non-linearity. Some models (e.g. Gould's notion of punctuated equilibrium) directly challenge the notion of temporal directionality in Lamarck's and Darwin's evolutionary models (Dennett, 1995; Mayr, 2002) – also sparked by increasing criticisms of the uncritical application of Darwinian linear transitional models to social Darwinism and the notion of 'survival of the fittest' within human societies (e.g. Spencer, 1870; Hoy, 1986). Thus, according to Gould (1982), evolution is both non-gradual and, most often, at a 'dead stop'. The starting point for Gould's critique is that Darwin's model can not explain catastrophic changes in Earth's biota, such as extinction of the dinosaurs 65 million years ago. Those who continue to adhere to Darwinian notions of evolutionary transition invoke external causes, such as asteroid impacts, climate change or volcanic eruptions, to account for 'irregularities' in what could be seen as a smooth and linear process. Irrespective of the causes of mass extinctions, during certain periods in Earth's history up to 90% of all organisms were exterminated, thereby severely disrupting linear processes of evolution (Mayr, 2002). Gould (1992b: 21) emphasised "the unpredictability of the nature of future stability, and the power of contemporary events … to shape and direct the actual path taken among myriad possibilities", thereby highlighting the similarities between evolutionary transition and Deleuzian transitional pathways. Gould (1989) also argued that evolution often takes the form of a *random* non-linear transitional pathway. Indeed, if we were to wind the tape of life back and play it again and again, the likelihood would be infinitesimal of humans being again the product of evolutionary change – in other words, for whatever reasons evolution embarked on one specific and unrepeatable evolutionary pathway out of millions of alternative Deleuzian possibilities.

However, recent critics of Gould highlight that his 'new theory of evolution' may just be a question of scale. Indeed, what Gould termed stepped transitional processes based on 'punctuated equilibrium' may well continue to fit into a relatively linear Darwinian tree of life, or, in Dawkins' (1986: 242) words, "the origin of a new kind of animal in 100,000 years or less is regarded by palaeontologists as 'sudden' or 'instantaneous'" – an important point that highlights that understanding the *scale* of investigation of any transition is vital in order not to mistake small transitional steps as representative for a more complex larger whole (see also Chapter 2). Dennett (1995: 284-285), therefore, argued that "orthodox Darwinism was already a theory of punctuated equilibrium. Even the most extreme gradualist can allow that evolution could take a breather for a while, letting the vertical lines extend indefinitely through time until new selection pressure somehow arose". Dennett (1995) made another important observation that also highlights the interconnectedness between theories of transition across disciplinary divides, by suggesting that Gould's ideas of punctuated equilibrium may have been strongly influenced by Marxist interpretations of the world around him. Following a visit to Russia, Gould particularly emphasised the difference between gradualness of reform and the suddenness of revolution, arguing that Marx had been right about the validity of the larger model of *punctuational change*. This has opened the door for the interesting juxtaposition of notions of *evolutionary change* (i.e. akin to Darwin) punctuated by *revolutionary transitions* (i.e. Gould's model).

Bak and Sneppen's (1993) evolutionary model based on self-organised criticality of evolutionary systems is another critique of implied temporal linearity in Darwinian thinking. Recent critics (e.g. Dawkins, 1995) have argued that evolution may not even be a directional process (regressive evolution). Therefore, evolution does not necessarily mean 'progress' (i.e. akin to the notion of 'retrograde' or 'random' transition highlighted above) and may be subject to what Dawkins terms 'random genetic drift', or to what Kauffman (1993: 26) described as "contingent historical accidents". The most cited example of non-linear evolutionary processes is that of 'convergent evolution', where similar evolutionary processes (e.g. the development of a structure like the eye or the wing) originated numerous times and independently in different kinds of organisms (Maynard Smith, 1986). Similarly, 'primitive' organisms that evolved billions of years ago continue to thrive in some ecological niches and have not been displaced by new species, while bacteria – the earliest organisms on Earth – are probably the most successful of all organisms with a total biomass that may well exceed that of all other organisms combined. As Mayr (2002: x) argued, there have been "constant disagreements about the causes of evolutionary change … and about whether evolution [is] a gradual or discontinuous process". Similarly, Maynard Smith and Szathmary (1995) suggested that the evolution of life should be seen as a story of 'numerous transitions' and that 'saltation' (i.e. a sudden change or 'jump' in evolutionary processes) may be a better description of evolution than a linear transitional pathway. Dawkins (1986: 73) went a step further by arguing that "the actual animals that have ever lived on Earth are a tiny subset of the theoretical animals that could exist. These real animals are the products of a very small number of evolutionary trajectories through genetic space". Thus, debates continue as to whether evolution can be seen as a teleological (goal-directed) and deterministic process, or whether it is non-teleological and non-deterministic (and, therefore, non-linear) based on the fact that 99.99% of all evolutionary lines have become extinct (Mayr, 2002).

4.3 Spatial homogeneity or heterogeneity?

The notion of spatial homogeneity *implicitly*, rather than *explicitly*, underpins many transitional models, largely because the immediate spatial context of a postulated transition is only rarely specified and discussed (i.e. the geography of transition has often remained unexplored). The notion of *spatial heterogeneity*, meanwhile, challenges the assumption that transitions are spatially homogenous – a notion largely underpinned by *geographical research* into the unevenness of spatial diffusion processes (e.g. Thrift, 1996a; Massey, 2001). Spatial heterogeneity implies that not all geographical spaces from local to global follow the same pathways of transition. Some localities may not embark on a transitional pathway at all, and it is easy to over-emphasise the *mobility* of people in simple all-encompassing assumptions about place transcendence (Thrift, 1996b). Although most spaces will undergo some form of transition, transitional patterns will be *geographically highly complex*. The notion of spatial heterogeneity is, therefore, a critique of simplistic assumptions about geographically homogenous transitional patterns, and emphasises recent trends in disciplines such as human geography that place increasing emphasis on the individual (as opposed to the state), the region (as opposed to the country) and cultural diversity as part of the 'cultural turn' (as opposed to cultural homogeneity). The notion of spatial heterogeneity, therefore, echoes the Deleuzian transition model, emphasising that different territories can embark on very different transitional pathways.

Most of the criticism of spatially homogenous transitory processes has occurred within debates on the ***post-socialist transition***. It is now commonly accepted that post-socialism did not completely supersede, nor eradicate, socialist modes of accumulation, cultures of production and mentalities in countries influenced by the Soviet Union. Smith and Pickles (1998: 10), for example, cautioned in their book *Theorising Transition* that "many of the contributors to [this book] are also concerned with the important argument that treating post-communist Eastern Europe as a whole fails to recognise the ever-present diversity of some 27 states and 270 million people". Similarly, Pavlinek and Pickles (2000: 30) argued that "without a clear understanding of uneven development, geographical variability and geographic scale, analyses of post-communist transitions fail to address the social and environmental complexities of state socialist and transitional models of development". While urban areas have tended to more readily adopt post-socialist patterns of consumption (e.g. rapid adoption of consumerism or capitalist share dealing), many rural areas in Eastern Europe remain still firmly embedded in socialist action and thought (e.g. expectations about the role of the state in helping rural communities; resistance to adopt the Western model of 'family farms' in East Germany to replace large production cooperatives modelled on the socialist agrarian system; cf. Wilson and Wilson, 2001; see also Meurs and Begg, 1998, for Bulgaria). Similarly, marked cultural differences between Russia and other newly independent states of the former Soviet Union continue to exist. Johnston *et al.* (2000: 624) argued that "for Russians the choice of identity is often seen as between focusing either on democratic state-building or on empire rebuilding, the latter including the re-colonisation of the post-Soviet borderlands". While some Eastern European states have built 'proto-democracies' modelled on Western democratic states (e.g. Baltic states, Czech Republic,

Ukraine), others have adopted more authoritarian forms of governance (e.g. southern post-Soviet states).

Post-socialism can, therefore, not be equated with a united spatial identity, and a 'territorialisation' of the post-socialist model is emerging with large regional differences between post-socialist countries. Stark (1992: 18; original emphasis), therefore, argued that "East Central Europe must be regarded as undergoing a *plurality of transitions*. Across the region, we are seeing a multiplicity of distinctive strategies; within any given country, we find not one transition but many occurring in different domains – political, economic, and social – and the temporality of these processes is often asynchronous and their articulation seldom harmonious". Pavlinek and Pickles (2000: 21) concurred by arguing that "there is already clear evidence ... of the regionally differentiated nature of national responses to transition. Certainly, few would now contest that the experience of transition has been quite different for Central European countries on the one hand and those of south-eastern Europe on the other. But even this geographical distinction eludes the intensity of competition over policy, and the geographically specific nature of these struggles at all scales." Stiglitz (2002) mentioned China as one of the clearest examples of a 'geographically polarised transition', where pronounced spatial differences in transitional pathways between rapidly westernising and developing urban areas (e.g. Pudong) and relatively under-developed remote rural areas are particularly pronounced.

Post-socialist transitions operate at different, often spatially heterogeneous, scales and paces, and each region has its own history and geography with often differing political and historical contexts (Pavlinek and Pickles, 2000). As Hörschelmann (2002: 52) argued, "post-socialist transformation is ... a complex process that fits uneasily into pre-given categories and disrupts an ordering logic that divides between a Western postmodern 'us' and 'the rest' of the world". Similarly, Smith (1995, 1996) argued that there is ample empirical evidence to suggest that, in the context of Eastern Europe, different countries have followed different transitional pathways from state socialism, and in some cases this transition has led to regional fragmentation. Thus, Pickles and Unwin (2004: 20) argued that post-socialist transformation processes "have been shaped by the complex geographies of regional ecology, economic, and socio-political variation across the region. Place and region have shaped the transformation" – in other words, 'geography matters' in conjunctures of local and wider forces which can not be logically derived from universal theoretical models such as 'post-socialism'. The path-dependency model developed by Stark (1992) for the post-socialist transition has been particularly useful for understanding complex transition issues, precisely because of its sensitivity towards multiple outcomes of transitions both in *different social realms* and at *different geographical scales* (Grabher and Stark, 1997; Altvater, 1998). Although post-socialist transitions in individual countries share some general features (e.g. collapse of one-party system and central planning; processes of democratisation; marketisation and relative economic decline [transitory recession]), this does not hide the large geographical and 'mental' differences that accompany the transition. Pavlinek and Pickles (2000: 29), therefore, argued that the post-socialist transition is "an experience of spatial and temporal unevenness ... [and] we need to appreciate fully the geographical and temporal variability of the transition".

The spatial homogeneity hypothesis has also been challenged in recent debates on **technological transition**, in particular in work analysing differential diffusion patterns of innovations. Hägerstrand (1952, 1967) was among the first to analyse how spatial distance affects the spread of innovation. Using diffusion simulation techniques, he highlighted that the 'neighbourhood effect' was crucial for the pace and success of innovation diffusion and that it explained often highly uneven patterns of diffusion. Using historical and contemporary examples, Grübler (1997) similarly highlighted that diffusion of technological innovations can take very different forms, ranging from the example of the 'Captain Swing' movement in

1830s England that successfully resisted (at least for a few years) the introduction of mechanical threshing (with workers smashing the new machines), to the spatially differential adoption of new technological devices in 20th and 21st century Europe (the current spread of broadband technology is a case in point) with some regions resisting innovations much longer than others. He also mentioned the example of the spatial diffusion of railway networks around the globe as a particular challenge to the spatial homogeneity hypothesis, as areas with late development of railway infrastructure (e.g. the west coast of North America) never 'caught up' with early centres of railway development (e.g. east coast of North America), as other technologies (namely cars) had taken over mass public transport by the time these remoter areas were opened up by railway lines. There were also substantial differences in the spatial diffusion of railway technology in Europe, where the German rail network, developed several decades later than in the UK, has remained one-third less dense than the British one. The spread of the car across the globe is another key example. Countries that adopted widespread use of the car early (e.g. the USA) continue to have the highest densities (USA = 600 cars/1000 people), while 'latecomers' such as Japan (widespread car use only since the 1950s) 'only' have about 300 cars/1000 people.

Recent studies also emphasise the importance of *cultural resistance* to technological innovation, ranging from outright rejection of 'Western' technology among indigenous tribes wishing to preserve their traditional lifestyles, to the religious sect of the Amish in the USA who have successfully resisted adoption of any kind of 'modern' technology such as televisions and cars (Rogers, 1995). Rogers also mentioned examples of the resistance of Peruvian households towards public health-related innovations due to specific cultural beliefs, or Korean women's adoption and non-adoption of contraceptive methods due to specific 'village norms'. In a different realm, there is increasing evidence that new 'cyber-technologies' (largely linked to virtual spaces such as the internet and computers) also have a pronounced spatial dimension suggesting substantial spatial heterogeneity (Warren, 2006), with Graham (1998: 173) arguing that "cyberspace is, in fact, a predominantly metropolitan phenomenon" still largely bypassing 'rural' areas around the globe (see also Castells, 1996). Currently, France may be the most extreme example of cyber-technological spatial heterogeneity, with Paris accounting for 80% of all investments in telecommunications infrastructure in France. These studies highlight the importance of *compatibility* of technological innovations with existing values, past experiences and needs, and resulting *divergence* in innovation transitions. A spatially homogenous diffusion of technological innovation is, therefore, the exception rather than the norm. Indeed, adoption of new ideas in a given region may be influenced by current market needs, cost, fashion, and familiarity, amongst many other factors. Graham (1998: 181; emphasis added), therefore, concluded that "new technologies are inevitably enrolled into complex social power struggles, within which both new technological systems and *new material geographical landscapes* are produced".

Parayil (1999) suggested that some technological innovations are not meant to be 'equally' available around the globe. Using examples ranging from the first widespread use of the cannon (which gave the Portuguese total command of ocean routes in the 15th century) to current technology linked to nuclear missile guidance systems and weapons of mass destruction, he argued that it can be in the interest of nation states *not* to divulge technological secrets (e.g. Cold War era; era of global terrorism). Similarly, patent rights held by the country where an invention originates often act as a barrier for 'equal' spatial diffusion of a new technology. Diffusion of technology is, therefore, essentially linked to diffusion of knowledge systems. For this reason, technological diffusion can not only be about superficial diffusion of technological artefacts, but it also needs to consider whether *value and knowledge systems* of affected societies have adapted to the new technology – a key reason for slow and spatially heterogeneous technological diffusion in many parts of the world. Linked to this is that some technologies (e.g. high-speed rail, plane travel, high-speed computer linkages) may lead to 'tunnel effects' which bring certain spaces and places closer together, while pushing adjacent areas further away (Graham, 1998). Unequal access to technologies can, therefore, mean that the centres of two geographically distant cities (e.g. 'global' cities like London and New York) may be closer to each other than to their own geographically close peripheries.

The notion of transition to *post-modernism* has been equally criticised concerning spatial patterns and processes (Bradbury, 1976; Berman, 1992). In particular, commentators increasingly advocate the existence of *multiple* modernisms (Nederveen Pieterse, 2004), similar to Amin's (1990) notion of multiple modernist and post-modernist pathways in an increasingly 'polycentric' world. Authors such as Singh (1989) have highlighted the different pathways that modernism has taken in the South and East Asian context, while Tominaga (1990) showed evidence that Japanese modernism and post-modernism have followed a different path from that of the West. Li (1989) and Sonoda (1990) made similar cases for modernist pathways in both Taiwan and China that differ markedly from any experienced in the European context (see also discussion of global complexity below). Van de Kaa (2001) provided a particularly interesting analysis in which different countries are 'positioned' with respect to their 'levels of post-modernity' based on a set of variables used in the *World Values Survey* ranging from attitudinal (e.g. importance of religion), economic (e.g. indicators of post-materialism) to demographic parameters (e.g. low fertility transition score for 'post-modernists'). Although extreme caution has to be taken about interpretation of such survey results, 'clusters' of more post-modern countries can be identified that include, for example, Finland and Sweden at the extreme end of the 'post-modern' spectrum, and countries such as the Netherlands, Denmark, Austria, Germany, Norway, Switzerland, Belgium and France positioned at the post-modern end. Less post-modern (or more modern?) countries were identified as Canada, Spain, the UK, the Czech Republic, Lithuania and Slovenia, while Bulgaria, Belarus, Portugal, the USA, Ireland, Russia and Hungary scored relatively low on the 'post-modern scale'. As with debates on the diffusion of technological innovation, these debates interlink with arguments of monocentrism versus polycentrism, suggesting a multiplicity of different spatially heterogeneous pathways and transitions towards both modernism and post-modernism. The importance of culture has been particularly emphasised as a key driver of differentiated spatial post-modern pathways. Van de Kaa (2001: 297) suggested that if 'post-modernists' from different cultural backgrounds have to demonstrate a clear awareness and conscious acceptance of some of the most significant theoretical and philosophical principles of post-modernism, then "there would in any country only be a small minority of them". Indeed, "the transition reaches countries with different cultural heritages at different times and is, consequently, likely to differ in speed and impact" (Van de Kaa, 2001: 324). As societies become more modern, the emphasis people place on higher-order needs increases differentially depending on cultural norms. As Gellner (1992) argued, different cultures will seek different post-modern self-expressions and will focus differentially on their own well-being and on actions they perceive as giving meaning to their lives – meanings that differ markedly between societies and cultural and ethnic groups (also *within* nation states). As a result, spatial heterogeneity in adoption of post-modern action and thought becomes one of the *defining characteristics* of the post-modern project.

The *post-Fordist* literature also provides ample material for a critique of the spatial homogeneity thesis. Debates on post-Fordism are often based on the general applicability of the model to *any* region or area in the developed world (e.g. Lipietz, 1992), rather than discussing the 'geography of post-Fordism' that may highlight large, and relatively unexplored, spatial differentiation in the adoption of post-Fordist modes of accumulation both *within* and *beyond* advanced economies (e.g. O'Connor, 1989; Peck and Tickell, 1994). Yet, many commentators have criticised the underlying assumption of spatial homogeneity in post-Fordist theory, and have highlighted that there have been large national and regional variations. For Europe, Hudson (2000: 299), for example, argued that "Fordism diffused unevenly over space and through time". Thus, many industrial regions in developed countries are still dominated by Fordist modes of production characterised by 'the worship of machines and hierarchy', and by a relative contempt for line workers' skills and union demands. According to Lipietz (1992), Fordist patterns and processes also continue to be evident for vast numbers of women and for workers in smaller subcontracting firms. Fordist and post-Fordist modes of production, therefore, often exist side-by-side. Any visit to recently deindustrialised regions in Europe (for example the Ruhr area in Germany), highlights that so-called 'post-industrial' landscapes still contain Fordist elements of production (e.g. continuing coal mining and steel production in the Ruhr albeit at a limited scale) that lie immediately adjacent to 'post-industrial' landscapes in which most evidence of the former industrial past has been eradicated.

In a similar vein, most of the literature on the transition from *colonialism to post-colonialism* highlights that the colonial experience involved *multiple* forms of resistance and adaptation to colonial rule, and that each colony had its own specific trajectory of colonial, post-colonial and neo-colonial developments based on economic, political and socio-cultural factors unique to that territory (Said, 1993; Sidaway, 2003). The transition to post-colonialism is, therefore, characterised by a pronounced spatial heterogeneity with different expressions of anti-colonial movements and decolonisation processes (Bauer, 1976). For example, in the process of post-colonialism former colonies have responded in very different and highly complex ways to pressure exerted by the IMF and the World Bank, and "not all states have responded in the same way, nor have they had the same capacities to respond to international pressures" (Smith and Pickles, 1998: 11). Colonial spaces can, therefore, be seen as 'heterogeneous spaces' where parallel stages of colonialism, post-colonialism and neo-colonialism may co-exist (Childs and Williams, 1997). Indeed, post-colonial spaces are "complex, fractured and foliated, and can rarely be reduced to any simple binary geometry" (Johnston *et al.*, 2000: 614).

The fallacy of spatial homogeneity is also evident in debates on *environmentalism*. Most arguments revolve around broad-based spatial generalisations about the adoption of new environmental thinking, especially within advanced economies (Devall and Sessions, 1985; Lovelock, 1995). Yet again, debates suggest a much more complex picture. Wilson and Bryant (1997), for example, suggested that environmental attitudes vary from region to region, with specific groups around the world depicting *specific* attitudes towards the environment. They illustrated that many hunter-gatherer societies,[10] for example, tend to show greater 'respect' for the environment than many other stakeholder groups. Hunter-gathering societies, and to some extent also small-scale subsistence farmers or nomadic pastoralists, often attach more spiritual significance to their environments than other groups in society (Denslow and Padoch, 1988; Durning, 1993). These stakeholders articulate their often complex relationships with nature through rituals, oral histories or cultural artefacts (e.g. masks, totem poles) and often attach – consciously or sub-consciously – specific 'taboos' to the over-use of specific environmental resources, thereby protecting the environment (in particular game animals or valuable plants) in sustainable ways for future generations. Spatially uneven tensions have been best expressed in the clash between utilitarian-minded pioneer farmers and indigenous populations around the globe, where the environmentally destructive attitudes of the former often led to the destruction of the environment the latter depended on for survival (Williams, 1989; Wilson, 1992).

Debates about *recent* trends towards environmentalism emphasise an equally spatially varied picture. Many studies have shown that within advanced economies, the adoption of

---

[10] Hunter-gathering is by no means only a fringe activity. In countries such as India, for example, it is estimated that tens of millions of people still live what could be described as 'hunter-gathering lifestyles'.

environmentalist or green thinking has been spatially uneven (Dobson, 1995), with some countries or regions adopting elements of green thought more readily than others. Current discussions about the highly differential rate of recycling in the EU are a case in point (Barr, 2003, 2006), as are uneven participation of green political parties in national politics within Europe, North America and the Antipodes (Johnston, 1996). Many studies suggest a correlation between socio-economic status and the extent of 'greenness' (Lowe and Goyder, 1983; Wilson, 1992, 1996), although many exceptions can be found as adoption of conservationist action and thought is influenced by a complex array of factors involving educational levels, income or urban/rural residency in a given territory (McAllister, 1994). Further, continuing 'pioneering' processes in many parts of the remaining tropical rainforest areas (especially in Amazonia and in Indonesia) highlight that extreme utilitarian and short-term environmental attitudes continue to persist at the periphery of the 'post-modern' world. These attitudes may be in stark contrast with conservation-oriented views held by environmental NGOs attempting to prevent further forest destruction, and with environmentally sustainable lifestyles of indigenous hunter-gathering and shifting cultivation societies severely affected by unsustainable forest clearing (Bryant and Bailey, 1997). These tensions, therefore, result in a multifaceted array of different stakeholder perceptions towards the environment located in close geographical proximity. As long as such pronounced regional variations exist in the adoption of green thinking, *spatial heterogeneity* in the transition towards environmentalism may be a more accurate depiction of reality.

Spatial homogeneity also underlies basic arguments in the ***demographic transition*** model (e.g. Chesnais, 1992; Leete, 1996). Originally, it was anticipated that all world regions would 'follow the lead' of European countries where the model was developed, and that all societies would 'move through' the four key stages of the transition model at broadly the same pace. Hirschman (2001: 117) suggested that "much of the controversy over the appropriate theoretical framework for fertility change arises from assumptions ... that the same causal variables are present everywhere". As with other transitional models based on Euro-centric assumptions, there is increasing evidence that this is not the case (Rosero-Bixby and Casterline, 1993). Indeed, the original model has been criticised for failing to take into account the complex interaction of many factors (multi-causality) in influencing demographic trends, including, in particular, different cultural, political, religious and socio-economic factors (Hirschman, 1994; Mason, 1997). The case of Ireland has already been highlighted for issues of temporal non-linearity in demographic transitions, and also emphasises highly variable population trajectories within the EU. Eastern and Central Europe, characterised by extremely low fertility rates, also stand out regarding different demographic pathways and relative positions in the demographic transition model. Politics are partly responsible for these low fertility rates (especially linked to the post-socialist transition; see discussions elsewhere in this Chapter). The classic example of politically-guided demographic trends comes from China, where the 'one-child-family-policy' from 1979 led to rapid *falls* in China's fertility rates (from above 3 to only 1.8 today) – possibly faster than in any other country (Feeney and Wang, 1993). An opposite trend could be observed in Iran after the Islamic revolution of 1979 where fertility *increased* based on the weaker socio-political role of women in a fundamentalist state (Casterline, 2001). Religious factors regarding demographic trends continuously come to the surface, possibly best highlighted by the difference between Protestant and Catholic families in Northern Ireland, with the latter having much higher fertility rates. It is generally accepted (although this is also currently changing rapidly) that Catholic countries tend to have higher birth rates than Protestant countries due to anti-abortion and anti-contraceptive policies adopted by the Catholic Church (Potter, 1999). Casterline (2001), therefore, argued that considerable inter-country diversity in the pace of fertility decline has characterised recent spatially-specific fertility trends, and that spatial variations in the pace of this decline are largely explained by changes in social, cultural and economic aspirations and expectations (e.g. achievable economic aspirations are increasingly perceived as undermined by continued childbearing). As a result, Mason (2001: 165) suggested that "differences in family structure and gender stratification between Europe, Asia, and sub-Saharan Africa may help to explain the [highly differentiated] historical sequence of fertility transitions across these regions".

The issue of ageing populations (i.e. where the elderly above 65 make up an increasingly large proportion of the total population) possibly best highlights spatial heterogeneity in the demographic transition. Although there is now a global trend of ageing, the emergence of ageing populations is still spatially uneven. These populations are a particular feature of developed countries (especially Europe, Japan, Australia), but many developing countries also increasingly show signs of rapidly ageing populations. It is anticipated that the number of elderly in the developing world will rise by over 1.2 billion by 2050, with countries like Mexico, Brazil or Thailand soon expected to have an 'ageing crisis', while commentators are beginning to refer to the 'greying of Latin America'. Of particular concern for our discussion is that the traditional demographic transition model does not sufficiently take into account factors such as ageing, in particular as the 'ageing issue' suggests a much more complex situation for stage 4 than simplistic assumptions based on birth and death rates alone suggest. Thus, Casterline (2001: 17; original emphasis) argued that demographers are increasingly liable to "criticism of being overly focused on the *number* of children as the first and foremost among many fertility parameters". This is particularly important as the 'replacement migration' issue is gaining increasing prominence, arguing that if Europe, for example, intended to maintain current 'support ratios' (proportion of workers/dependants) it would need to allow 700 million new immigrants and their descendants into Europe by 2050. Immigration, therefore, increasingly blurs 'traditional' demographic pathways and may, by 2050, be the overriding force defining spatial demographic patterns (Greenhalgh, 1988).

Evolutionary transition, meanwhile, is different from *social science* based transition models concerning the fallacy of spatial homogeneity. Indeed, the *non-existence of spatial homogeneity* underlies many arguments in biological evolution theory, as a key starting point for the theory was precisely to explain the geographical *variations* in species evolution (Darwin, 1859; Mayr, 2002). Contrary to social science theorisations of transition which usually seek to find *inductive* generalisable patterns and processes first, and then attempt to differentiate these patterns spatially later, biological evolution theory (and arguably many other debates on transitions in the natural sciences; cf. Thornes and Brunsden, 1977; Davies, 1995) *deductively* begins by asking the question why certain inconsistencies (be they spatial or temporal) exist and then attempts to generalise from the particular to the general. Biological evolution theory, therefore, has helped explain reasons for spatially heterogeneous distributions of plants and animals based on the history of species *dispersal* from original points of origin. This has helped explain why floras and faunas of Africa and South America are different from each other and why there are usually no mammals on oceanic islands (Darwin, 1859). Since Darwin, the notion of 'discontinuous distribution' in evolution theory, therefore, challenges notions of spatial homogeneity in evolutionary transition. According to the notion of 'geographical speciation' a new species may evolve when a population acquires isolating mechanisms while being isolated from its original parent population (Mayr, 2002). In other words, different evolutionary skills tend to have different 'payoffs' for different subgroups of a population, resulting in these subpopulations to diverge, each one pursuing its favoured type of excellence, eventually leading to the development of a new species (Dennett, 1995). Similarly, the notion of 'adaptive radiation' assumes that phyletic lineages can establish themselves in numerous different niches and adaptive zones and that these 'radiate out' from a common pool of ancestors. In many ways, parallels can be drawn here to the notion of 'innovation diffusion' in social science theorisations (see above) where innovative ideas or behaviours (e.g. related to agricultural post-productivism) 'radiate out' from an original source, thereby emphasising spatially heterogeneous territories into which such adaptive radiation is possible.

4.4 Global universality or complexity?

Theories about social transitions are often strongly rooted in the experiences of advanced economies, and it is (often) simplistically assumed that these experiences can be easily transposed to the situation in developing countries (Young, 1990). However, there is increasing evidence to suggest that transitional tendencies evident in the North can not easily be transposed to the situation in the South. This *fallacy of global universality* is a particular by-product of traditional European- and Anglo-American-centric 'hegemonic' conceptualisations of the world (Shohat and Stam, 1994; Gregory, 1998), and is related to the issue of 'exporting' Euro-centric 'high theory' developed in the context of advanced economies to the experience of developing countries (Simon, 1998). Two important caveats need to be mentioned here: first, the fallacy of global universality exclusively applies to *social science theorisations of transitions*; second, we also need to acknowledge the inherent problems in the binary and dualistic division of the world into two distinctively separate entities (i.e. developed/developing world) – binaries that are increasingly more difficult to defend in a rapidly globalising world where emergent economies such as China are already among the economically most powerful countries on the globe. Debates, therefore, intertwine closely with notions of *particularism* versus *globalism*.

There is much debate about the 'exportability' of theories developed in the context of advanced economies to the experience of developing countries – an issue particularly emphasised in the context of the problems of 'travelling theory' by cultural critic Edward Said (1983). Said made particular reference to what he termed the existence of 'interruptions' and 'irregularities' that often characterise differential North-South differences, which make it difficult to paint the situation in developing countries with the same theoretical brush as that applied to advanced economies. As Said emphasised, a key problem is the Northern-biased view of developing countries as 'the other' through the long-distance lens implicit in terms such as 'the savage', 'the native' and 'the oriental'. These are important issues for the exportability of the notion of post-productivist and multifunctional agriculture to the developing world (see Chapters 7 and 10 in particular). Questions need to be asked whether we should automatically assume that developing countries should follow the same transitional pathways as their advanced economy counterparts. Further, should the assumed transition necessarily follow from an 'ism' to a 'post-ism', or should we leave room for conceptualising 'pre-isms' (i.e. in our case a 'pre-productivist' agriculture), as well as allowing for the possibility that societies in the developing world may move directly from a 'pre-ism' to a 'post-ism' thereby bypassing the seemingly crucial transitory 'middle' stage?

The fallacy of global universality is possibly most apparent in debates on the transition to *post-modernism*. Most discussions focus on post-modernism in the context of advanced economies and have often neglected to discuss whether these notions can be applied to the developing world (Jencks, 1987; Ellin, 1996). This has led Simon (1998) to ask whether the notion of post-modernism is so 'esoteric' and 'self-indulgent' that it may have no relevance to theoretical and practical concerns about development in the South, largely because of Euro-centric perceptions combined with a lack of personal engagement, confidence and curiosity about the applicability of post-modernism beyond the developed world. Simon (1998: 239), therefore, argued that "authors of much of the Northern literature on postmodernity … either explicitly or implicitly assume the global purchase of the constructs". There have been few attempts to apply notions of modernism and post-modernism beyond Western Europe (Kaplan, 1996), although there are some notable exceptions (e.g. Crush, 1995; Marchand and Parpart, 1995). Nederveen Pieterse (2004: 61), thus, suggested that modernism "is a theory of westernization by another name, which replicates all the problems associated with Eurocentrism". Indeed, as both Featherstone (1995) and King (1995) emphasised, theories of modernism strongly reflect the particular experience of successful Western societies. Simon (1998: 220), therefore, posited that "few post-modernists are concerned explicitly with the South, but those who are pay inadequate attention to the ongoing transformation of the social, political, and economic relations within which development takes place, giving rise to contradictions, disjunctures, fragmentations, and rapid changes frequently referred to in a Northern context as characterising postmodernity". He further suggested that the developing world has been inappropriately excluded from postmodern engagements, especially regarding intellectual practice, and that the key explanation of this exclusion may lie in the fact that the fuzzy 'indicators' of post-modernism are "largely irrelevant where widespread modernisation is still an unrealised aspiration and poverty remains widespread" (Simon, 1998: 224) – in other words, where societies may still be largely embedded in a 'pre-modern' era. Such concerns are echoed by critics *from* the South who have advocated anti- or post-developmentalist discourses from an eco-feminist perspective that challenge the notion that post-modern thought, with all its Western values and problems, is effectively a 'desirable' thing for developing countries (Shiva and Bedi, 2002). As Chakrabarti and Cullenberg (2003: 5) argued, these backlashes against the global universality paradigm of post-modernism "question the path of transition and the logic of development underlying the transition process", and that "postmodernism, after all, is a vast milieu constitutive of multiple imaginations in terms of historical phase, existential state, or condition, style, and critique".

Yet, most Northern writers on post-modernism implicitly – and at times explicitly – "assume the global salience and applicability of their constructs" (Simon, 1998: 225). An oft cited example is Habermas' (1987) attempt to develop a unifying theory on the project of modernism that also applies to developing areas. Some studies (e.g. Lunn, 1985; Kaplan, 1996) have discussed post-modernism in the context of developing areas, and there is a growing volume of post-modern writing emanating from the developing world, most notably South America (e.g. Colas, 1994). Yet, most debates have occurred within a European context, leading Bauman (1992) and King (1995) to emphasise the 'northern boundaries' of post-modernism, while Marchand and Parpart (1995) highlighted the *globally complex* and highly varied contributions of various strands of feminism in undermining Northern-biased post-modern conceptualisations. Further, Hassard (2001) highlighted that, despite notions of the global village and global space-time compression, there are still substantial global differences in the accessibility to 'post-modern' artefacts and structures.

Simon (1998) further nuanced the debate by arguing that while the notion of post-modernism as 'epoch' and as 'expression' is difficult to apply to the South, post-modernity as a 'problematic' – i.e. as a different way of doing, seeing and envisioning – *can* be applied to the developing world. He referred particularly to adoption of greater post-modern contextuality in research and writing, and through explorations of the interplay between global, national and local that find similar expressions both in the North and South. However, particularly in view of the highly contentious transition to post-modernism in (some) advanced economies (see above), there are strong indications that there is no global universality for the applicability of the post-modern transition, and no study has yet convincingly shown that notions of this transition can be successfully applied to the context of the developing world. This resistance by many scholars should, in itself, be seen as a possible expression of global complexity of the transition, and, as King (1995: 111) argued, "these shortcomings are the direct outcome of defining the modern and modernism by reference to a particular point in time, but without reference to space".

Similar discourses can be identified for the transition to ***post-Fordism***. As Thrift (1995) argued, the rapidly expanding literatures on globalisation and flexible post-Fordist production in the global economy have highlighted divergent capital formations, development policies, and economic fortunes of different countries and regions *across* the globe. Yet, Simon (1998: 222) suggested that "although the necessity of disaggregating simplistic constructions of globalisation is increasingly being recognised, such perspectives are still dominated unequivocally by Northerncentric worldviews". Both Peck and Tickell (1994) and Webber and Rigby (1996) argued that Fordism has been largely a feature of OECD countries, and that many developing countries were excluded, or excluded themselves, from the Fordist project due to historic, economic, social and cultural reasons. Indeed, rapid growth in newly industrialising countries is still primarily addressed in terms of Western deindustrialisation and international capital mobility (e.g. China) rather than as endogenous transitions of their own (Simon, 1998).

These debates suggest that post-Fordist notions will only rarely apply to the developing world context, and that little consideration is given to possible alternative perspectives in which the focus could be on local world views and indigenous industries that may differ substantially from Northern-centric models (Adedeji, 1993). In particular, notions of linearity inherent in the Fordist model (see Chapter 3) have been criticised for their applicability to the developing areas context, as many developing countries may be characterised by more time-flexible 'task-oriented experiences' and rarely fit 'classical' notions of labour separation. According to Hassard (2001), different cultures are characterised by a melange of conflicting Fordist and post-Fordist time horizons, making it impossible to argue for a specific post-Fordist pattern of work in the developing world. It is here that debates overlap with the implied *temporal linearity* of Fordist and post-Fordist theorisations (see above), as room also has to be made for the possible existence of pre-Fordist modes of production that may continue to dominate production in many developing countries. Can pre-Fordist regions or societies, therefore, be in transition to post-Fordism, or do they 'have to' progress through the Fordist stage before being 'eligible' for the post-Fordist label? Critics, therefore, argue that the resulting possible pattern of simultaneous pre-Fordism, Fordism and post-Fordism across the globe supports the notion of global complexity, and discredits notions of a globally universal transition towards post-Fordism.

Similar debates on the fallacy of global universality can be found in discussions on the ***demographic transition model***. Although the model was "devised with particular reference to the experience of developed countries" (Johnston *et al.*, 2000: 160), developed countries have often undergone transitionary processes that have differed from those in developing countries. The main problem lies in the fact that the demographic transition model was developed almost entirely on the basis of historical developments in Europe and that it is, therefore, based on Europe-specific historical trajectories of change and development. Although there have been numerous attempts at using the model to forecast population development in the developing world, it has attracted considerable criticism as a globally universal model. Evidence shows that each country or region on Earth has undergone its own specific pace of transition, with industrial countries usually starting stage 2 of the transition in the 19th century (or earlier) (see Chapter 3), and having since long been embedded within stage 4 (usually for several decades or longer) or even showing signs of 'post-transitional' population decline typical for stage 5 (Kuijsten, 1996), the latter also referred to as the 'second' demographic transition (Van de Kaa, 1987; Ogden and Hall, 2004). As a result, we witness as many different demographic trajectories as there are countries, with each country, or even sub-region, showing region-specific population trends based on complex interlinkages of socio-economic, political and cultural factors (Watkins, 1987).

Simplistic assumptions about the applicability of the four stages in developing countries have been particularly criticised (Lesthaeghe and Surkyn, 1988). Kates (1997), for example, highlighted that trajectories of birth and death rates in most developing countries have not conformed to the traditional demographic transition model. Thus, "it took a hundred years for deaths to drop in Europe, whereas the drop took thirty years in the Third World" (Kates, 1997: 45). This more rapid pace is now commonly associated with faster than expected societal changes in developing countries, including changing labour needs (i.e. fewer children needed), greater child survival, improved opportunities for women, and access to birth control. In addition, critics have particularly argued that 'movement' from one phase of the transition to another is only possible once certain 'ingredients' are in place – ingredients often modelled on the demographic situation in developed countries. Rotmans and Martens (2002), for example, argued that the shift from stage 1 (high birth and death rates) to stage 2 (high birth rates, declining death rates) is only possible if the availability of primary supplies of food, energy and water is guaranteed, while factors such as education, income levels and women's participation in the workforce are key ingredients for the transition from stages 2 to 3 (low death rates, declining birth rates). Inevitably, in many parts of the developing world the latter ingredients are not (yet) in place, resulting in a spatially heterogeneous patchwork of global regions in different stages of transition. Casterline (2001: 26), therefore, argued that the proposition that later-starting demographic transitions tend to be more rapid "appears to be correct if based on a comparison between the historical European transitions ... It is an incorrect description, however, of the experience of developing countries".

A few examples will suffice to illustrate these large transitional differences. At the 'extreme end' of the spectrum are countries in sub-Saharan Africa with a persistently high fertility rate in 2002 (United Nations, 2003). This, combined with a relatively high death rate, places these countries still firmly in stage 1 of the model (Casterline, 2001). Many other developing regions witness rapid falls in fertility and have, therefore, moved through stage 2 into stage 3, with some countries even showing 'post-transitional' tendencies. Latin American countries are a particular case in point, where Brazil now has a fertility rate of only 2.2 – just above replacement level – while Mexico currently has a fertility rate of 2.5. Even India, one of the most populous countries and until recently perceived as a potential 'population bomb' (Ehrlich, 1970; Ehrlich and Ehrlich, 1990), has seen dramatic falls in fertility rates to 'only' 3.0 by 2002. Similarly, Bangladesh, which had one of the highest fertility rates in the 1970s at over 7.0, now only has a fertility rate of 3.5 (Cleland, 2001). Post-transitional tendencies may be particularly apparent in Cuba (fertility rate of 1.55 well below replacement level) and China with a fertility rate of 1.8. The latter particularly highlights the success of a combination of the politically enforced 'one-child-family-policy' in the 1970s and 1980s and, more recently, the aim of many new middle-class Chinese households for small families (see also Goodkind, 1995, for Vietnam). The key assumption for long-term forecasts about the *completion* of the demographic transition (if one can 'complete' a transition at all?) is, therefore, still highly questionable (Kates, 1997). Johnston *et al.* (2000: 160) concluded, therefore, that there is "widespread doubt about the model's validity and applicability", and that the first stage of the model, in particular, is oversimplified to take into account social and cultural complexity evident in many developing areas (Watkins, 1987). Above examples suggest a much greater *periodicity* for larger scale fluctuations in population increase or decrease than is highlighted in the seemingly 'stable' stages of the model (Coale and Watkins, 1986; Ogden and Hall, 2004). Most contentious, however, is the question whether developing countries will 'automatically' follow the directional progression from stages 3 to 4, and whether a 'new stage 4' with a much more differentiated pattern of birth and death rates may begin to emerge in some developing countries (Leete, 1996). That some developing countries appear to be unable to 'leave behind' stage 1 leads to severe doubts about the utility of the demographic transition model as a *globally universal* concept (Sen, 1994). It is here that the 'ageographical' context of the traditional demographic transition model finds its most acute expression (Watkins,

1987; Sen, 1994), leading Cleland (2001: 72) to argue that "in terms of rates of growth and long-term potential increases in size, the demographic transition in developing countries stands apart from the earlier European transition" – in other words, *global complexity* is the norm rather than global universality.

One of the most interesting debates on the notion of global universality refers to *technological transition*. The 1950s saw burgeoning interest in the effects of technological change on the newly independent 'post-colonial' developing countries by many scientists from advanced economies, especially how modernisation and economic development could be induced by technological transition (Parayil, 1999). It quickly became evident that most developing countries lacked the technological and scientific capital to mobilise the forces required for economic development from within, and that *transferring technology* from advanced economies under an array of developmental assistance became the norm (Escobar, 1995). Thus, technological transition became the main vehicle for 'developing' countries of the Third World from 'outside'. As we will see in Part 2, these debates are particularly relevant for our analysis of the p/pp transition, especially for the introduction of Green Revolution technology into developing countries (Pretty, 1995; Parayil, 1999). More recently, emphasis has been placed on differential drivers for adoption of new technologies in different regions and countries linked, in particular, to economic, political and social factors (Grübler, 1998). Of crucial influence has been the criticism of traditional 'modernisation theory', with a key argument that patterns and processes of modernisation applicable in advanced economies rarely find parallels in the culturally, socio-economically and historically different countries of the developing world. Yearley (1988) strongly opposed the promotion of the Western 'natural' model of modernisation theories for developing countries as *historically unacceptable* and *untenable*. He argued that advanced economies 'developed' because they had the opportunity to exploit developing countries, and that developing countries are, therefore, not in a favourable situation to repeat the historical experience of their colonisers – particularly regarding development and adoption of technological innovations.

Emphasis is often placed on the differential knowledge trajectories and system memories that may prevent adoption (let alone development) of new technologies by subjugated people – especially in the poorest countries. Technological knowledge production in developing countries is seen as an extension of advanced economies' interests, rather than as an autonomous development in itself, although new technologies have often only been transferred to developing countries because they helped military conquest, pacification or 'normalisation' of colonial territories (Parayil, 1999). Recently, the importance of *indigenous culture* and *past experiences* has, therefore, been emphasised in work criticising simplistic assumptions of the global universality of technological diffusion. Lack of recognition of the importance of indigenous culture is seen as a key explanation why certain 'Western' technologies are simply not suitable for many people in developing countries (Rogers, 1995). It is argued that new technologies can only be adopted by indigenous people if they fit somehow into existing belief systems, world views and attitudes (the compatibility attribute) (Lansing, 1991).

Some commentators are beginning to argue that technological innovation does not necessarily have to diffuse from advanced to developing countries, although most suggest that advanced economies have an important 'head-start' when it comes to new technological developments based on economic, institutional and knowledge-based factors (Grübler, 1998). Dosi (1982: 161) argued that "the process of technical change as such is not likely to yield to convergence between countries starting from different technological levels", emphasising that *system memory* is important in defining technological transition pathways. Similarly, Parayil (1999: 62; emphasis added) argued that this concern has become "particularly salient when analysing the impact of modern technology on the Third World and the dynamics of *technology transfer* from the industrialised nations to less-developed nations. The question as to whether the *tenet of symmetry* can be applied to two historically divergent and socially, culturally, and economically disparate systems … is a serious issue". Parayil particularly highlighted that technological diffusion is largely about *diffusion of knowledge*. Thus, incomplete mechanisms of transfer of technological knowledge systems in many aid or development programmes are one of the key reasons why many 'modernisation' programmes have failed. Perkins (2003) further suggested that established technologies, usually emanating from developed countries, tend to have considerable market advantages, making it difficult for *new* technological options to penetrate global markets (especially if they are incompatible with existing technologies). Rip and Kemp (1998), therefore, argued that such 'lock-in effects' are one of the reasons why hydrocarbon-intensive technologies, for example, continue to dominate in developing countries, despite the existing range of alternative low/zero-carbon substitutes. Indeed, Grübler (1997) argued that the 'decarbonisation' of the energy system in technologically least interconnected developing regions of the world may take several decades longer than in advanced economies. Thus, *global complexity* in technological diffusion – whether enforced by the North or not – is a better descriptor of current patterns of technological transfer to the developing world.

As would be expected, debates are complex for notions of global universality in the transition towards **post-colonialism**, in particular as colonialism itself is an inherently Northern-based notion whose conceptual problems have been transposed into conceptualisations of post-colonialism (Said, 1983). Simon (1998) argued that the diffuse literature on post-colonialism connects remarkably little with conventional developmental agendas, although some work has pointed towards spatially heterogeneous restructuring of inequitable colonial inheritances and the highly differentiated cultural politics of identity of groups subordinated and marginalised by colonial practices, official histories, and Northern-biased feminist and environmentalist discourses. He suggested that modern development is not a uniform or smooth process, and that the post-colonial acceptance of modernisation "need not lead to global homogeneity, especially if undertaken with a degree of politicoeconomic and cultural autonomy" (Simon, 1998: 222). Several studies inform debates on the fallacy of global universality in conceptualisations of the transition to post-colonialism. Thus, selective historical reconstructions of a bygone colonial area have been examined, as have post-colonial architectural styles catering especially to international mass tourism (Potter and Dann, 1996; Worden and Van Heyningen, 1996). Other studies have investigated the post-colonial restructuring of cities or the changing life-worlds of rural-urban migrants in countries such as Papua New Guinea, India and South Africa, as well as forms of post-colonial expression such as new national symbols (flags, place-names) or new capital cities (Connell and Lea, 1995; Masselos, 1995). Most studies show that post-colonial form and function has not necessarily superseded colonial (or indeed neo-colonial) historical reconstructions, architectural styles or socio-economic processes linked to city restructuring or rural-urban migration – in other words, the 'post-colonial' world is still characterised by multiple and simultaneous colonial, neo-colonial and post-colonial territories.

The notion of global universality also underlies arguments surrounding the global shift towards **environmentalism**, where it is often assumed that what has been true for transitional tendencies in advanced economies may also be true for many less developed countries (Milton, 1996). There is, therefore, a key geographical component to debates on environmentalism, as ecocentric ideas are now increasingly spreading into developing countries, as more local and global environmental problems emerge. However, as many critics have argued, environmentalism in the less developed world remains 'patchy', and not all regions or stakeholder groups in society have embraced notions enshrined in ecocentric thinking (Princen and Finger, 1994; Wapner, 1995). Bryant (1996, 2001), in particular, has highlighted problems for South-East Asian countries to adopt models of Western-style environmental NGOs to their specific socio-economic and cultural contexts. There is increasing evidence that many environmental NGOs in developing countries are embarking on their *own transitional pathways* towards a 'greening' of society that may qualitatively differ from those adopted by NGOs in advanced economies. Most debates on the global complexity of the transition towards environmentalism revolve around the different cultural and religious underpinnings of these transitions. While the often implied dichotomy between an 'environmentalist North' and 'non-environmentalist South' is increasingly criticised (Wilson and Bryant, 1997; Bryant, 2005), it has been similarly argued that some cultural-religious contexts generally seen as relatively benign regarding anthropogenic mismanagement of the environment have much more complex human-nature relationships than hitherto assumed. Outside the Western cultural context, it has often been suggested that other cultures have provided more propitious conditions for the development of an ecocentric world view. In particular, the role of Buddhism in Asian societies has been associated with an inclination towards a symbiosis of humans with nature (Callicot and Ames, 1989). However, recent research has shown that Asian cultural contexts have not necessarily resulted in more conservation-oriented management of the environment (Bruun and Kalland, 1994). Further, although many see the spread of global capitalism as a possible great 'leveller' (global universality) for environmental attitudes across the globe (i.e. with a tendency towards technocentric ideals based on profit-maximisation), environmental degradation and technocentrism are not necessarily synonymous with capitalism (Walker, 1989). Indeed, proponents of the technocentric world view argue that it is precisely the accumulation of capital across the globe that enables the reduction of environmental degradation through technological solutions (Barbier, 1987; Lomborg, 2005). As with other transitions, the adoption of environmentalist ideas and notions, therefore, is better conceptualised through *globally complex* processes rather than through 'one size fits all' assumptions about the global transferability of green thinking and behaviour.

4.5 Structural causality or structure-agency inconsistency?

The fallacy of structural causality is probably the least well explored of the four fallacies discussed here. This may be because political economy and other structuralist interpretations of the world – approaches that still tend to dominate social science debates – have not left much room to acknowledge that societies are complex entities comprised of multiple and often overlapping stakeholder groups, some of which are more closely associated with the state and the macro-structural level (i.e. state organisations themselves, parastatal organisations, etc.) while others are more closely related with the 'grassroots' level (e.g. farmers, fishers, local decision-makers, etc.). As with the fallacy of global universality, the fallacy of structural causality only applies to theorisations of transitions in the *social sciences* (i.e. it is a fallacy that only concerns the role of human actors in transitional processes). In what could be termed a post-structuralist political economy approach, this fallacy largely hinges on Giddens' (1979, 1984) theory of structuration that attempts to bridge the schisms of macro and micro, actor and structure, and traditional structuralist interpretative frameworks, and that argues that the social sciences should focus their analysis more on social *practices* rather than social *structure*. Human 'agency' is thereby expressed through social systems, beliefs, attitudes and identities, while 'structure' is based on rules, resources, or other exogenous forces (e.g. the wider political economy) influencing actors' actions and thought. Structure is particularly manifested in social systems in the form of the "reproduced relations between actors or collectivities, organised as regular social practices" (Giddens, 1984: 25). The process of structuration is, therefore, seen to incorporate both *agency* and *structure* (duality of structure) which interact with each other to structure a society or social system. Structure is the medium for actions by human agency, as well as the outcome of the reproduction of social practices/actions of the agents who constitute the social system. The pertinence of Giddens' framework has been emphasised through the upsurge in interest on issues of *governance* which has usefully highlighted the multi-layered character of decision-making in human society (Williamson, 1996; Rhodes, 1997) and the crucial importance of understanding 'agency' and how it influences policy, politics and socio-economic processes (Jordan, 1999; Wilson, 2004). Thus, any conceptualisation of transitional processes has to take into account *both* structure and agency (Jessop, 1998; Imrie and Raco, 1999).

As we saw in Chapter 3, transitional models (here only in the context of the social sciences) often imply that both structure (policies, political economy) and agency (actors' behaviour, identities and thoughts) 'move along' the transitional spectrum at the same linear pace and in the same homogenous manner – i.e. there is an underlying assumption of *structural causality* in transitory processes. In other words, transitional research has often treated actor and stakeholder groups as relatively *homogenous* entities – merely reacting to external change – thereby largely neglecting key questions regarding possible differential transitional pathways among different groups in society. However, as the discussion in this section will highlight, most often we find evidence for a disjuncture (temporal as well as spatial) between adoption of transitory pathways among different segments of society – a process I term *structure-agency inconsistency* that will also be investigated for the postulated p/pp transition in Part 2. This process is marked by pronounced differences in the nature and pace of transitory tendencies among different actor groups in pluralistic multi-layered societies, ranging from 'grassroots' actors such as farmers or fishers, to 'intermediate' actors such as regional policy officials or representatives from medium-sized companies and firms, to state and parastatal actors and organisations that may have power and jurisdiction over the entire nation state territory (Wilson and Bryant, 1997).

Different segments of society may, therefore, move at different speeds through a transition. Inevitably, there are tensions between different stakeholder interests, often leading to different outlooks on, and adoption of, transitory 'opportunities' by different groups in society. Readers familiar with recent literature on *hybridity* will recognise that this discussion is closely related to emerging debates on 'hybrid identities' (especially linked to the notion of globalisation as hybridisation), where it is argued that any society or individual is increasingly comprised of multiple and parallel highly differentiated identities that can never form a uniform whole – or, in the context of our discussion in this chapter, that can never form an 'ism' that transcends all segments of society (Nederveen Pieterse, 2004). This leads to the question whether we should conceptualise *a* specific transition (e.g. from 'ism' to 'post-ism') or whether, instead, *multiple transitional pathways* resembling our Deleuzian transitional model (see Chapter 2), adopted differently by segments of society and stakeholder groups, are more likely. Several commentators are, therefore, rightly beginning to ask whether theorisations of social transition may often only be conceptual constructs of an 'intellectual elite' that may describe patterns at the macro-structural level (i.e. the level most affecting that elite), but neglecting transitional forces and pathways at 'other' societal levels (Pickles and Smith, 1998; Stiglitz, 2002).

Debates on the postulated **transition to environmentalism** may provide the most useful evidence for deconstructing simplistic assumptions about structural causality in transitional processes. Although an underlying argument surrounding this transition implies that most sectors of society are beginning to move towards environmentalism, it is increasingly recognised that environmentalism may (continue to) be a preoccupation of the wealthy middle classes in advanced economies who have the time and finances to be able to afford the 'luxury' of environmental thinking and behaviour (Pepper, 1984; O'Riordan, 1989). Indeed, increasing empirical evidence highlights the differentiated nature of adoption of environmentalism, with different sections of society adopting environmentalism more readily than others, resulting in many different forms of 'environmentalism' (Wilson and Bryant, 1997; Barr, 2003). The most important debates were launched by O'Riordan (1981, 1989), who distinguished between segments of society that have remained 'technocentric' by putting human concerns above those of nature (i.e. 'anthropocentric' faith in the application of science and technology to 'fix' environmental problems), and those who have become 'ecocentric', i.e. who have gone through the transition towards environmentalism by emphasising the intrinsic rights of nature and the environment (Wilson and Bryant, 1997). Ecocentrics typically advocate that humans must live within well defined environmental bounds, while 'deep green' writers such as Devall and Sessions (1985), Naess (1989) and Sessions (1994) emphasise the need to manage the environment so that resources are not depleted and criticise contemporary human-environment interactions based on economic growth and global capitalism. There are diverse perspectives within ecocentrism (Jacob, 1994; Pepper, 1996) with radical ecocentrics – reflected in the thinking of the 'deep ecology' or 'Earth First!' movement – calling for a revolutionary transformation of human-environment interactions (Devall and Sessions, 1985; Foreman and Haywood, 1988). In contrast, less radical ecocentrics assert the importance of local-level action based on conservation but with a view towards reform of existing systems rather than complete revolution (Schumacher, 1973; Dobson, 1995).

Technocentrics, meanwhile, would dismiss such arguments as hopelessly idealistic and naïve. As reflected in the works of Simon (1981), Karshenas (1994) and Lomborg (2005), technocentrics do not dwell on environmental problems but emphasise technological and market-based solutions. In contrast to the ingrained pessimism of ecocentrics, technocentrics are confident that top-down solutions will resolve even the worst environmental problems (Wilson and Bryant, 1997). In the technocentric camp there is again a great diversity of perspectives, with extreme technocentrics (or 'cornucopians') suggesting that there are no environmental crises at all, while moderate technocentrics acknowledge the existence of environmental problems, but are confident that technological solutions can be found. This highlights that the transition towards environmentalism occurs at *different pace* for different segments in society, and that some stakeholder groups in society may not enter the transition at all. Wilson and Bryant (1997), in particular, have suggested that different types of stakeholders can be associated with different types of environmental attitudes and world views, illustrating the multi-layered nature of environmental thinking in society. State actors, transnational corporations and financial institutions may be broadly linked to the technocentric world view, while grassroots actors may be broadly associated with the ecocentric world view, although they also acknowledged that most actors' environmental identities are usually exposed to the combined influence of ecocentric and technocentric views. Further, complex attitudinal differences may also be found within individual stakeholder groups. For farmers, for example, Wilson and Bryant (1997: 74) argued that "although many farmers favour … practices based on conservation-oriented attitudes, the growing dependency of most farmers on the global capitalist economy, and the associated reliance on industrial-type agricultural techniques, suggests that, on balance, a majority of today's farmers have to be utilitarian in outlook" (see also Burton and Wilson, 2006). Even within environmental NGOs different 'shades of green' can be depicted reflecting the depth of their attachment to ecocentric world views (Bryant, 2005). Elements of this debate are, therefore, more akin to our above conceptualisation of a *Deleuzian transition*, in which several different transitional 'plateaus' of environmentalism may co-exist simultaneously in contemporary societies.

Similarly, in conceptualisations of the transition towards **post-modernism**, there is an underlying assumption that *all* segments of society will eventually move towards this 'post-ism', with little critical analysis of highly differentiated structure-agency patterns (Berman, 1992; Ellin, 1996). This is particularly perplexing as one strand of the argument for the transition towards post-modernism rests on the assumption that 'post-modernisation' refers to a process of value change, i.e. the transformation from modern to post-modern value orientations among different segments of society (Bertens, 1995). Many have recently challenged assumptions about uniform acceptance of post-modern action and thought among different stakeholder groups. Thus, Nederveen Pieterse (2004) argued that just as there are multiple trajectories of 'globalisation' – i.e. several globalisations affecting different stakeholder groups in many different ways (or not at all) – there are multiple expressions of post-modernism that influence different segments of society in multi-layered ways. Recently, these differences have been exacerbated by the influx of migrants into previously relatively homogenous 'modern' societies. This is echoed by Chakrabarti and Cullenberg (2003: 6; emphasis added) who argued that "the projected singular postmodernity is in fact a *disaggregated* discursive space constitutive of *multiple* standpoints and political positions". In contemporary post-modern analysis, therefore, notions of 'hybridity', 'global mélange', 'heterogeneity' and 'multiplicity' have become crucial keywords, suggesting multi-faceted and complex adoptions and non-adoptions of post-modernism among different stakeholder groups (Lowe, 1991). Van de Kaa (2001: 294) argued that the advent of post-modern society "has had a generation-specific impact: fragmentation, discontinuity, and incongruity are standard". Post-modern fragmentation in society, therefore, does not only occur horizontally *between* different segments of society, but also vertically across generations *within* individual cultural groups. Indeed, authors refer increasingly to stakeholder-group-distinctive plots of music, clothing, behaviour, advertising, body language or visual communication patterns – post-modern multi-ethnic and multi-centric patterns strongly influenced by mass education and global communication (Canevacci, 1992). Post-modernistic outlooks, therefore, question the fundamentals of the 'meaning-giving system' of modernity. As a result, Friedman (1999: 237) suggested that "the current stage is one in which culture has begun to overflow its boundaries and mingle with other cultures, producing numerous new breeds or hybrids", while Appadurai (1999: 230) argued that the various flows of post-modern culture we see depict clear signs of structure-agency inconsistencies in that they "are not coeval, convergent, isomorphic or spatially consistent. They are in relations of disjuncture ... The paths and vectors taken by these various kinds of things have different speeds, different axes, different points of origin and termination, and different relationships to institutional structures in different regions, nations or societies".

Nederveen Pieterse (2004) argued that although the notion of post-modern global cultural synchronisation has not become irrelevant – i.e. suggesting that there is an element of structural causality in the adoption of post-modern thought that *can* be observed – many have overlooked the countercurrents of post-modernism through the impact that non-Western cultures and 'diasporas' are increasingly making on Western post-modern action and thought. It is argued, therefore, that much work overrates the homogeneity of Western 'post-modern' culture and that multiple expressions of post-modernity can be identified in all societies (Lowe, 1991). Some of these expressions can be described as 'assimilationist hybridity' that leans towards the centre, adopts the predominant post-modern canon, while a 'destabilising hybridity' may blur the canon, reverses currents (even 'back' towards modernism) and subverts the centre. Thus, "we find the traces of asymmetry in culture, place, descent", and post-modern society may rather be "subversive of essentialism and homogeneity, disruptive of static spatial and political categories of centre and periphery, high and low, class and ethnos" (Nederveen Pieterse, 2004: 74). Bell (2004: 23), therefore, argued that "the modernism-postmodernism debate ... is one that many have found tiresome – even some of those who are intrigued by matters of theory – and with some justice, given its polarizing tendencies".

Debates on the differential adoption of ***post-Fordist*** modes of production between different segments of society are equally manifold. Societal impacts have particularly taken place at two levels. First, the transition towards post-Fordist modes of production has *indirectly* affected society as a whole, resulting in cleaner, more efficient and worker-friendlier working environments away from mass production. As Chapter 3 highlighted, this was often accompanied by tendencies of agglomeration that facilitated interaction between increasingly vertically disintegrated industrial functions, as well as a move towards forms of grassroots democracy and governance that are more 'organic' and less delegative than those of Fordism. In particular, in the post-Fordist mode of production firms can opt for less technologically sophisticated modes of production, thereby mobilising the skills of line workers, and encouraging dialogue between machine design, maintenance and production – processes that often go hand in hand with improved relations and communication between 'embedded' firms in the same production process (Grabher and Stark, 1997).

Second, in today's post-industrial service economies of the developed world, only some stakeholder groups continue to be *directly* affected by the transition towards post-Fordism. This is particularly true in the service economy of the UK (where manufacturing now accounts for only about 15% of the entire workforce), but is also evident in all other advanced economies where rapid recent de-industrialisation has led to substantial differentiation between those still embedded in Fordist/post-Fordist modes of production and the large sections of society no longer directly involved in manufacturing. It is regarding stakeholders involved in manufacturing that we see increasing differentiation into Fordist and post-Fordist modes of production. Inevitably, discussions here have centred on Marxist interpretations of the particular configuration of class relationships and power that have inhibited the development of post-Fordist modes of production. A starting point has been debates surrounding the grassroots and entrepreneurial levels, where the rebargaining model, characteristic of post-Fordist wage relations (see Chapter 3), does not necessarily apply to all segments of the labour force. As Lipietz (1992: 40-41) argued, "it is quite possible to have, in the same factory or the same office, neo-Taylorism for the unskilled and semi-skilled, and an 'individually negotiated involvement' for more qualified employees ... There can be guaranteed employment and collective bargaining in major firms, while neo-Taylorism is kept for women, immigrants and handicapped people in subcontracting firms and consumer services". This is particularly evident in the emergence of a two-tier society in countries such as Japan and Germany where upper echelons of the labour force retain Fordist social benefits from the past (e.g. big pension schemes) while the rest are 'condemned' to neo-Taylorist casual employment conditions (Aoki, 1990). This may be evidence of structure-agency inconsistency within firms themselves where managers operate on a different post-Fordist transitional trajectory compared to their workers. This situation is complicated by the fact that in ultra-modern highly productive manufacturing environments (e.g. car production in the UK and Japan) traditional 'workers' have been increasingly replaced by 'robots' that work 24-hour shifts without the need to be represented in unions. Some argue that this creates a rift between post-Fordist modes of production for people and ultra-Fordist modes of production for machines – conceptually an interesting territory that suggests a further blurring of the implied structural causality in the traditional Fordist/post-Fordist model of transition.

Debates on structure-agency inconsistencies in post-Fordism have been particularly prominent in development studies. In many developing countries Fordist modes of production continue – or have indeed been reinforced through rapid relocation of industry from the developed world (Leonard, 1988) – exemplified by the textile industry in China. The latter highlights, in particular, the link between continued Fordism and the cost of employment, as the cheaper the labour costs, the more likely it is that Fordist modes of mass-production predominate. With a focus on India, Chakrabarti and Cullenberg (2003) discussed the post-Fordist transition from a Marxist perspective with a focus on class, class formation, and the effects of class differentiation on the Indian economy, especially for class effects produced by the transition of Indian society from one mode of production to another. They argued that adoption of both Fordist and post-Fordist modes of production has varied greatly among different class-related segments of Indian society, especially as 'class' in Indian society is still characterised by relatively homogenous groups of people (e.g. peasant class; see also Patnail, 1987) formulating and executing specific decisions or non-decisions concerning adoption of Fordist and post-Fordist modes of production based on highly differentiated 'caste consciousness'. The Indian 'peasant class', in particular, has been portrayed as a 'backward' class resisting (either forced or voluntarily) adoption of Fordist modes of production (e.g. adoption of certain mass agricultural production practices), while 'elite' caste groups have been described as having more easily appropriated the means of Fordist mass production (e.g. mass textile production) and initiating, in parts, the Indian transition towards post-Fordism based on patterns and processes of Indian 'domination-subordination' structure-agency relationships defined by access to knowledge and information (Chakrabarti and Cullenberg, 2003). Similar structure-agency inconsistencies in the adoption of both Fordism and post-Fordism have been described in other geographical contexts of the developing world (Peck and Tickell, 1994), in particular linked to the crisis conditions that developed in conjunction with the end of Fordist modes of production through ruptures between owners and workers and the lack of balance between aggregate productive capabilities of nation-state economies (e.g. Frank, 1969, for Latin America; Bhabba, 1990 for Africa) and the aggregated purchasing power of workers acting as consumers (e.g. Stiglitz, 2002, for China and South-East Asia).

In a similar vein, those arguing that former colonies have moved towards *post-colonialism* often assume that this transition applies to *all stakeholder groups* in society, often irrespective of their political power, gender and social status. Yet, as both Chambers and Curti (1996) and Childs and Williams (1997) argued, there is much evidence to suggest that, if post-colonialism truly exists (see above), only some segments of society have adopted post-colonial action and thought. In particular, the presence of foreign finance and industrial capital does not only affect the international economic embeddedness of a developing country, but also influences structure-agency relationships by restructuring class relations through the maintenance (and at times establishment) of a 'comprador bourgeoisie' (Dicken, 1998). This suggests that, in many instances, structural elements in society (often linked to powerful elites pampered by international capital) may be characterised by 'neo-colonialist' tendencies, but with grassroots stakeholders having little control over the often detrimental reversal of post-colonial transitory processes. Some of the most interesting debates about structure-agency inconsistency have come from the critical (and not so critical) *popular literature* on post-colonialism. Analysis of contemporary writing, in particular, increasingly highlights a multi-layered structuration of many 'post-colonial' societies into *both* remnant colonial and overtly post-colonial discourses. Jacobs (1996) argued that many contemporary writings on social, cultural and gendered identities and historical reinterpretations from different former colonies around the world show clear elements of *both* colonialist and post-colonialist interpretations, with advocates of colonial or neo-colonial views still dominating many segments of society (often, however, such authors no longer reside in their countries). On the other hand, the most fervent advocates of post-colonial ideas (such as Edward Said, Homi Bhabha, Trin Minh-ha, or Gayatri Spivak, to name but a few) are still a small, albeit growing, minority (Simon, 1998). These tensions are further highlighted in the notion of 'neo-colonialism' that also has important repercussions for assumed structural causality, as neo-colonialism implies both that large segments of former colonial societies continue to adhere to 'colonial' ideologies, action and thought (often the ruling elites), and that control continues to be exerted 'externally' over many societal processes in developing countries (Stiglitz, 2002).

The same is true for debates on the transition to ***post-socialism***, where it is often assumed that *all* segments of society have adopted new post-socialist modes of thinking and capital accumulation, irrespective of the position and power of these groups within the new political and socio-economic structures emerging after the break-up of the former Soviet Union. From the outset, this assertion was challenged as, based on individual life experiences and histories, every citizen in post-socialist countries has experienced the transition in different ways. This was echoed by Van Hoven (2004b) who, based on an extensive qualitative study of different individual transitions among citizens in former socialist countries in Eastern Europe, argued that "all of the people ... lived through their own transitions with their own struggles, challenges and successes. Some people struggled more than others and are still struggling. Others have carved out niches for themselves that permitted impressive successes". Pickles and Unwin (2004) highlighted for countries like Latvia and Estonia that simple factors such as age play a crucial role in citizen's perceptions of transition. In these countries, older people, in particular, still have strong memories of their first period of independence (1918-1939), which is key for shaping their current 'post-socialist' identities. Younger people, meanwhile, brought up entirely during the socialist era view the changes to post-socialism in a completely different light. Some of the older generation, therefore, may perceive the socialist era as a mere 'blip' in an otherwise non-socialist period.

Pickles and Smith (1998), therefore, suggested that some segments of 'socialist' society never were truly socialist. During the attempt to introduce socialism in 1917 Russia, Bolsheviks tried to *impose* communism on a reluctant population, and although the Bolsheviks argued that the way to build socialism was for an elite cadre to 'lead' the masses into the 'correct' transitory path, this was not necessarily what the masses thought best. Similar patterns have occurred in the 'new' post-Communist revolution in Russia, where a newly emerging elite similarly attempted to force rapid post-socialist change on a reluctant population. Research has also shown that in some post-socialist countries attitudes to ethnicity and culture have become more polarised, especially in the Yugoslavian successor states (Pickles, 2000). In other countries such as Bulgaria, ethnic discrimination against minorities such as gypsies worsened (Pickles and Begg, 2000), resulting in uneven transitional pathways among different segments of society. In addition, evidence suggests that women's roles in post-socialist countries has partly worsened, with many women losing their political positions (Regulska, 1998). Thus, post-socialist states have not created a level playing field and a 'win-win' situation for all their citizens. Instead, we witness social dislocation, empowerment of elites and the establishment of powerful crime syndicates that, concurrently, often lead to the disempowerment of stakeholders at the grassroots level (Burawoy, 1996; Smith and Pickles, 1998). This means that stakeholders 'at the top' have often benefited from new post-socialist structures and associated access to new financial opportunities, while grassroots actors have not, which has led to profound increases in poverty and inequality (Stiglitz, 2002). Grabher and Stark (1997, 1998), therefore, argued that *multiple pathways* to post-socialism exist, and that the post-socialist transition is a complex culturally-driven process that finds multiple expressions. They particularly highlighted that the nature of the post-socialist transition is highly dependent upon how networks of connectivity between social and economic actors are transformed and understood, emphasising the importance of historically and socio-economically rooted *path dependency* for individual stakeholder groups. This latter point will be particularly relevant for the remainder of our discussion in this book on the reconfiguration of post-productivist and multifunctional actor spaces, and the relative 'lethargy' of many agricultural systems often precluding transitional change between different stakeholder groups (see Chapter 10).

Structure-agency inconsistencies in the context of ***technological transition*** have been the focus of long-standing debate in the social sciences and form one of the key debates challenging simplistic assumptions about equal adoption of innovations across different stakeholder groups. Such inconsistencies were, for example, a key component in Marx's (1977) classical critique of 19th century political economy. Marx emphasised how rapidly evolving modern technology generated wealth and prosperity, but that most of these benefits only went to a small minority. Marx particularly outlined that the key issue is about who is in control of technological innovations and argued that, in the capitalist system, owners of the means of production are usually the best placed to control and direct these innovations. Bourdieu's (1998) suggestion that society is fragmented between various 'specialist' groups, and that society is bound by a group logic that is sustained by imitating the technological consumption patterns of the upper or elite classes, also lends credence to Marx's argument that technological transition can *never* empower all segments of society in similar ways (see also Latour, 1996). Grübler (1997: 29, emphasis added), therefore, argued that technological transition is "discontinuous in time and heterogeneous in space among the population of potential adopters and *across different social strata*".

Rogers (1995) highlighted that technological diffusion is a kind of *social change and exchange* and that, therefore, it can rarely be seen as a one-way system of 'simple' diffusion form one agent to another. Grübler (1998) further emphasised that technological change can not transcend whole society, as technology cannot evolve without being embedded in appropriate social and institutional contexts – contexts that are often lacking, or poorly developed, among certain segments in society (e.g. those affected by severe poverty or lack of education). Parayil (1999), therefore, equated technological transition with knowledge change, and argued that in no society do all citizens have equal access to information and knowledge. As a result, Rogers (1995: 252) suggested that "individuals in a social system do not adopt an innovation at the same time". In particular, he highlighted that there is a link between technological adoption and diffusion and individual socio-economic status, especially as status is often highly related to an individual's *degree of change contact*. People of lower socio-economic status have, therefore, fewer contacts with other individuals who may encourage them to adopt a new innovation. This explains why not all segments of society – structure and agency – can adopt new technologies at the same time, and that "the same innovation may be desirable for one adopter in one situation, but undesirable for another potential adopter in a different situation" (Rogers, 1995: 12). To a large extent, this explains the cumulative s-shaped innovation adoption curve (see Fig. 2.2.) as, initially, only a few individuals may adopt an innovation (e.g. innovators such as opinion-leaders, village chiefs or household heads). If successful, this adoption rate increases substantially to include a substantial proportion of society, usually with a minimum of 10-25% participation needed during the 'take-off' phase. However, even with the most successful innovations, complete diffusion is almost never achieved (e.g. in advanced economies some households still do not have a television either due to enforced reasons [usually economic] or due to voluntary resistance [ideological reasons]). Thus, Thrift (1996b, 1474) suggested that technologies are "part of a continuing performative history of 'technological' practices, a complex archive of stances, emotions, tacit and cognitive knowledges, and presentations and re-presentations, which seek out and construct these technologies in certain ways rather than others".

In some cases, knowledge is actively withheld by powerful ruling elites who wish to prevent diffusion of technology to the grassroots level (Losch, 2004). Thus, the social construction inherent in the development of any technological innovation shows how powerful social groups try to influence the course of technological transition to satisfy their short-term business interests at the expense of long-term public interests – interests that possibly could have been much better served by other 'grassroots'-friendly technological innovations that never saw the light of day. In the context of leapfrogging technologies, Perkins (2003), therefore, argued that there is a need for *selective intervention* that targets specific technologically-disadvantaged stakeholder groups in supporting the uptake of leapfrog technologies at the grassroots level, mainly through improved cooperative partnerships between state and non-state actors. This means that the appearance of new artefacts in technological diffusion is only the *physical manifestation* of a much more complex process in which different actor groups, although possibly in possession of these artefacts, lack the knowledge and skills to effectively put them into practice. This is best illustrated through examples of agricultural technology diffusion in developing countries linked to Green Revolution technology, high-yielding seed varieties or chemical fertilisers, that have affected different stakeholder groups in very different ways, largely depending on who has wielded enough power within complex actor-networks to obtain the necessary *knowledge* or technological *capabilities* through adaptive learning (Lall, 1992). Rip and Kemp (1998) further argued that the larger the scale of transition, the longer it takes for all involved actors and institutions to adopt a new technology. However, while regimes at the macro-level may often slow down transitional processes (i.e. political inertia), changes in regimes can produce sudden breakthroughs and stimulate rapid transition, also referred to as the 'snowball effect' (e.g. recent regime change in China and the rapid adaptation of new 'Western' technologies). According to Parayil (1999: 157), "social factors and the interests of the actors determine the innovation process". Indeed, political, economic and social interests of dominant groups in society set the stage for technological change and often determine its trajectory. This emphasises the weakness of many models of technological change that have failed to analyse how the *knowledge content* of technology changes, and that, as a result, have placed insufficient emphasis on actors and stakeholders and their social and institutional contexts and structures.

Finally, structure-agency inconsistencies are also apparent in debates surrounding the **demographic transition** model. In particular, the model has been accused of being too simplistic in its assumptions about demographic trends among different segments of society (Bulatao and Casterline, 2001). Here, a variety of cultural, socio-economic, educational and religious factors have been mentioned that explain why different stakeholder groups and population segments have depicted substantially different demographic trajectories (Cleland, 1994). Socio-economic factors are a good starting point to illustrate differences and to highlight path-dependent fertility trajectories common among different segments of society. While wealthy middle-class groups are usually associated with low fertility rates (usually two children but often also falling into the category of 'double-income-no-kids), 'working class' families are often linked to high fertility rates, partly because in many societies children still form important roles as labourers, family supporters and helpers for the elderly (Bongaarts and Watkins, 1996; Cleland, 2001). Casterline (2001) argued that this results in a highly non-linear dynamic, as an increase in the fixed costs of bearing a certain quantity of children induces a shift from *quantity* to *quality*. Hirschman (2001: 118; emphasis added) further suggested that "once the demand for smaller family size was present in many European societies, the new information and change in values spread first *along paths of cultural and linguistic homogeneity*" – in other words within and between specific segments of society. There are interlinkages here with educational factors, as many studies have shown that better educated segments of society tend to have fewer children (more knowledge about contraception; individual freedom valued more highly) (e.g. Santow and Bracher, 1999, on fertility decline among Southern Europeans living in Australia). Further, just as different religious beliefs may find expression in spatially differentiated demographic trajectories (see above), such beliefs also have 'vertical' repercussions within societies (Cleland, 2001). Here we can return to our example of Catholic 'high fertility' families that, even at the beginning of the 21st century, still stand out with higher fertility levels than other religious groups (Potter, 1999). Based on evidence provided by the current critical literature on demographic transition, it is, therefore, difficult to argue for structural causality regarding transitional pathways in society. Indeed, as Cleland (2001: 79; emphasis added) argued, "one possible reaction to this bewildering diversity of circumstances in which fertility has declined is to abandon *simple monocausal* explanations". Instead, stakeholder-specific trajectories are apparent – whether in developed or developing countries – suggesting pronounced *structure-agency inconsistencies* within the demographic transition model.

4.6 Applying transition theory to the evolution of agricultural systems

The discussion in Chapters 2-4 has highlighted that the issue of 'transition' has been critically analysed from different angles and disciplinary vantage points. In Chapter 2, we saw that six basic models of transition can be identified and that conceptualisations of any transition – be it social or natural transitions – bear resemblance to one of these six basic models (or five if we exclude 'non-transition'). However, Chapter 4 has highlighted that substantial criticism has been voiced against the often *unilinear* and *binary* assumptions inherent in many transition models. Critical analysis shows that transitions rarely follow smooth linear pathways and that they are most often characterised by temporal non-linearity, spatial heterogeneity, global complexity and structure-agency inconsistency. The evidence, therefore, suggests that transitional action and thought occurs *simultaneously* along a *spectrum* of decision-making pathways bounded by extreme pathways (e.g. 'isms' and 'post-isms'), rather than being embedded in seemingly temporally linear pathways as traditional transitional models seem to imply. As a result, I have argued that four 'fallacies' can be identified that underpin virtually all debates on transition: the *fallacies of temporal linearity, spatial homogeneity, global universality* and *structural causality*.

The main purpose of Chapter 4 was to highlight that any discussion of a transition towards post-productivist and/or multifunctional agriculture – the focus of the remainder of this book – needs to be *embedded* in the wider context of these 'parallel' debates on transitions, and that we need to consider the possibility of *simultaneous* decision-making pathways that occur along a *spectrum* of constraints and opportunities. In other words, we can not understand the problems inherent in the assumed directionality of the p/pp transition model without, at the same time, acknowledging similar unease and criticism that has taken place for other conceptualisations of transitional processes. In this sense, Part 1 of this book has addressed Holmes' (2006: 159) recent call that "the concept of a multifunctional rural transition invites positioning within current theory on the role of place and space in contemporary society". As Rotmans *et al.* (2002) argued, although individual transitions may have their own temporal and spatial dynamics, it is important to recognise the *intertwined* nature of the various transition – i.e. no specific transition, be it social or natural, can be understood without at least attempting to grasp 'parallel' developments in 'other' transitions. As Habermas (1994) reminded us, the recognition of the complexity of transitions forces us to rethink our ubiquitous 'isms' (i.e. in our case productivism), as all re-conceptualisations of 'post-isms' discussed in this chapter simultaneously re-shape the past and, potentially, the future.

Chapters 2-4 have also highlighted that debates on various transitions have not occurred in a vacuum, and that several important synergies exist between conceptualisations of social transitions and debates on the p/pp transition. This was particularly obvious in the context of post-Fordism, where the notion of industrial 'productivism' has been widely used over the last decades to describe a specific transitory stage during the Fordist regime of production. Further, we will see in Parts 2 and 3 that there are striking parallels between the critiques of post-productivism in agriculture on the one hand, and critiques of classical regulation theory views of the transition to post-Fordist modes of accumulation on the other (e.g. Cloke and Goodwin, 1992; Goodwin and Painter, 1996). Our meaning of, and discussion about the p/pp transition, therefore, has to be firmly embedded in parallel debates surrounding the lack of clear evidence of a shift from Fordism to post-Fordism (Potter and Burney, 2002). Further, in

Part 3 I will argue that the notion of *multifunctional agriculture* can only be fully appreciated within the framework of a possible parallel 'territorialisation' of Fordist/post-Fordist modes of accumulation. Murdoch and Lowe (2003) similarly highlighted how issues of modernism are key to understanding the politics of rural spatial division, with some arguing that agricultural post-productivism may be an expression of post-modern society (e.g. Philo, 1993; Halfacree, 1997b). Chapter 7 will also show that issues raised regarding the transitions to post-socialism and post-colonialism resonate strongly with criticisms of the p/pp transition. Similarly, our discussions of demographic, technological and environmental transitions have uncovered patterns and processes that also show striking parallels with the non-directional and spatially heterogeneous processes of agricultural transition. Finally, debates surrounding evolutionary transition have shown that any discussion of social transitions (such as agricultural change) have to be embedded in much wider debates on the nature and meaning of the concept of 'transition' as a whole, and that many historical and disciplinary biases and erroneous evolutionary transitional assumptions also find parallels in our discussion of the transition towards multifunctional agriculture.

These parallel debates have important repercussions for possible conceptualisations of a transition from productivist to post-productivist to multifunctional agriculture, and, therefore, lead us to ask several key questions based on the four transitional fallacies discussed in Chapter 4: (i) The **fallacy of temporal linearity:** can we assume, as the p/pp transition model leads us to believe, that post-productivist agricultural systems will necessarily supersede productivist systems – i.e. that the transition to post-productivist agricultural regimes follows a linear temporal pattern? Why should the assumed direction of change – from productivism to post-productivism – not be reversed in some circumstances? (ii) The **fallacy of spatial homogeneity:** is it meaningful to suggest that the p/pp transition is spatially homogenous – i.e. that different agricultural regions are affected in similar ways and are simultaneously taking part in the 'transition'? (iii) The **fallacy of global universality:** can we assume that the p/pp transition model can be universally applied to agricultural systems in developing countries – i.e. that the transition can be observed in rural areas in the developed and developing world at the same time? (iv) The **fallacy of structural causality:** is it possible to assert that the p/pp transition has affected all segments of society in similar ways and that all actors are moving along transitional pathways towards post-productivism at the same pace?

It is evident that we will only be able to understand the notion of 'multifunctionality' discussed in Part 3 if we acknowledge that many social transitions – including the p/pp transition – rarely follow linear transitional pathways, only occasionally conform to notions of stepped transitions, but most frequently appear to follow *Deleuzian transitional pathways*. Indeed, Part 3 of this book will argue that the notion of multifunctional agriculture most closely resonates with the Deleuzian transitional model that acknowledges that multiple parallel productivist and post-productivist (better: non-productivist; see Chapter 7) pathways exist that allow for the simultaneous existence of temporal non-linearity, spatial heterogeneity, global complexity and structure-agency inconsistency in agricultural transitions. This will enable us to conceptualise an actor-oriented and spatially 'territorialised' multifunctional agricultural system that is conceptually located between the extreme ends of a productivist and non-productivist spectrum and that comprises elements of *both* productivist and non-productivist transitional pathways (see Chapter 9).

I acknowledge, however, that it is one thing to observe a set of parallel frameworks operating in social theory, and that it is something different to observe whether or not these parallels are more than just parallels, but reflections of *mutually constituted processes*, as discussed in Chapter 3 in the context of Western thinking embedded in Cartesian dualism. The discussion in Chapters 2-4, therefore, also suggests that parallels in intellectual debates surrounding transitions may simply be a *mirror* of social processes that are interrelated in complex ways (e.g. the parallels between post-Fordism and post-modernism have already been noted). We can expect, therefore, that many debates surrounding 'post-isms' and other transitions discussed here will find resonance in our discussion of productivist and post-productivist agriculture in the following chapters. In other words, the *intellectual situatedness* of the issue of agricultural transitions takes on a different, and arguably more significant, dimension if we recognise that the intellectual parallels are a result of social processes that are interrelated in complex ways and that go well beyond discussions of 'agricultural change' alone.

4.7 Conclusions

Chapter 4 has investigated in detail the problems associated with relatively simplistic assumptions inherent in many transition models. Eight examples of transition have been criticised for their often simplistic and unilinear assumptions, and supposedly 'simple' and 'linear' transitions have instead been characterised by complexity, heterogeneity and unpredictability. I have suggested that these criticisms gel into the four typical transitional fallacies of temporal linearity, spatial heterogeneity, global universality and structural causality, and that, instead, temporal non-linearity, spatial heterogeneity, global complexity and structure-agency inconsistency are better descriptors of complex transitional processes at multiple scales and temporal levels. This analytical framework can now be used as a basis for the critique of both the postulated p/pp transition and the transition towards multifunctionality in Parts 2 and 3 of this book.

This is not to say that conceptualisations of transition are futile exercises. Our analysis in Chapters 2-4 has shown that many postulated transitions show clear signs that some transitory processes *are taking place*, although issues tend to be more complex than simplistic linear and unidimensional transitional dualisms seem to imply. We can, therefore, use the discussion in this chapter as a basis to understand possible problems associated with the postulated transition towards post-productivist agriculture, and a critical assessment of this transition will help us to further fine-tune currently existing conceptualisations of agricultural change. Based on the critical discussion of transition theory in Part 1, we now have sufficient baseline knowledge to critically analyse transitional patterns related to agricultural change and, in particular, to cast a critical eye on the binaries 'productivism' and 'post-productivism'.

Part 2

From productivist to post-productivist agriculture … and back again?

Part 1 has provided insights into debates on theorising transition from different vantage points. This forms the basis for deconstructing the implied linearity, spatial homogeneity and structural causality inherent in the agricultural p/pp transition model. The aim of Chapters 5 to 7 is to analyse what is meant by agricultural productivism (**Chapter 5**) and post-productivism (**Chapter 6**), and to critically examine and deconstruct underlying assumptions of the transition towards post-productivism (**Chapter 7**). I will draw on evidence from different disciplines and geographical contexts to illustrate that the p/pp transition has been based on assumptions that are too simplistic and unilinear, and that the reality of agricultural change is more complex and characterised by non-linearity, heterogeneity and complexity. In other words, based on transition theory, I will show in Chapter 7 that no full transition towards post-productivism has taken place. I will acknowledge throughout that notions of productivism/post-productivism go well beyond the issue of agriculture and cannot be fully understood in agricultural terms alone, and that they need to be understood as part of wider debates on transition. This will help us understand that productivism/post-productivism is, in fact, a *spectrum* of different views rather than two easily definable and 'separate' entities on their own. I will highlight that while the concept of 'productivism' remains useful, the notion of 'post-productivism' should be replaced by its true conceptual opposite: 'non-productivism'. This will enable us to understand a multifunctional agricultural 'territory' bounded by the extreme pathways of productivism and non-productivism as a basis for conceptualising multifunctionality in Part 3, and will allow us, for the first time, to *theoretically anchor* the notion of multifunctionality in the context of a refined productivist/non-productivist transition model.

Chapter 5

Productivist agriculture

5.1 Introduction

The aim of this chapter is to highlight debates surrounding the notion of 'productivist agriculture'. We need to fully understand the concepts behind productivism and post-productivism before being able to theoretically anchor and conceptualise the meaning of 'multifunctional agriculture' in the context of the p/pp debate. As Chapter 1 highlighted, in recent years debates on the notion of a possible shift from productivist to post-productivist agriculture have added an interesting new conceptual dimension to agricultural/rural research. Lowe *et al.* (1993: 221) argued that, for the UK context, productivism can be conceptualised as "a commitment to an intensive, industrially driven and expansionist agriculture with state support based primarily on output and increased productivity. The concern [of productivism] was for 'modernisation' of the 'national farm', as seen through the lens of increased production. By 'productivist regime' we mean the network of institutions oriented to boosting food production from domestic sources which became the paramount aim of rural policy following World War II. These included not only the Ministry of Agriculture and other state agencies but the assemblage of input suppliers, financial institutions, R&D centres, etc., which facilitated the continued expansion of agricultural production". Other conceptualisations have broadly concurred with this view, while also emphasising the often *environmentally destructive nature* of productivist agriculture based on maximisation of food production through the application of ever more intensive farming techniques and biochemical inputs (Ward, 1993; Ilbery and Bowler, 1998).

Productivism is, therefore, conceptualised largely on the basis of an *industrially driven agriculture* akin to Fordist modes of production of high *quantities* of food and strongly *supported by the state* through subsidies and a productivist policy regime.[11] Egoz *et al.* (2001) suggested that productivism should be seen as the practice of using farmland to its full potential, creating a 'mechanistic' landscape appearance that reflects the production process. Burton (2004: 200) further suggested that productivist identities of the 'good farmer' are largely based on skills evident in the production of agricultural commodities that could be "visibly assessed from the road by neighbouring farmers". *Maximising production* and *farm modernisation* are, therefore, key components of productivism, as are productivist *institutional structures* (e.g. agriculture ministries, farmers' unions) that aid and abet production maximisation. It is important to note that most definitions of productivism also include reference to productivism as an *era* firmly embedded within 20th century historical development pathways of agricultural change, especially as a reaction towards *food shortages* after the Second World War.

---

[11] The term 'productivist' was used in some agricultural national contexts before it became incorporated into theoretical discourses of rural change (e.g. 'une agriculture productiviste' in the post-war decades in France; Buller, 1999), but the term was usually used to denote a 'productive' agriculture rather than delineating a specific rural/agricultural era.

A review of the literature suggests that productivism and post-productivism have been conceptualised on the basis of seven inter-related *dimensions*: agricultural policies, ideology, governance, food regimes, agricultural production, farming techniques and environmental impacts (Wilson, 2001; Mather *et al.*, 2006). This highlights that a multitude of different characteristics need to be considered to fully understand the postulated p/pp transition, and that focusing on one dimension alone would only provide partial answers. This chapter will be concerned with analysing these seven dimensions for productivism (Section 5.3). However, we first need to briefly investigate the genesis of the terms 'productivism' and 'post-productivism' and highlight who the main proponents of the debates have been (Section 5.2). Section 5.2 will also investigate underlying approaches inherent in conceptualisations of the p/pp transition model and will highlight how these can be linked to the four fallacies inherent in many transitional models outlined in Part 1.

5.2 Approaches underlying conceptualisations of the productivist/post-productivist transition

5.2.1 The genesis of debates

Agricultural productivism and post-productivism are relatively recent terms, with debates surrounding these concepts emerging in academic circles in the UK in the early 1990s. By that time, notions of 'productivism' and 'post-productivism' began to appear in academic publications, spearheaded by some of the key thinkers on agricultural change at the time including, in particular, Terry Marsden, Philip Lowe, Paul Cloke, Richard Munton, Jonathan Murdoch and Neil Ward. Seminal publications include the first mention of the terms productivism/post-productivism in a conference paper by Munton (1990) and in a paper by Symes (1992), as well as the seminal book *Constructing the countryside* (Marsden *et al.*, 1993), in which the scene for a burgeoning debate on the postulated p/pp transition was set that has continued unabated to the present day (e.g. Wilson, 2001; Evans *et al.*, 2002; Mather *et al.*, 2006). Cloke and Goodwin's (1992) article 'Conceptualising countryside change: from post-Fordism to rural structured coherence' also needs to be mentioned in this context, especially as the authors already emphasised the close associations between conceptualisations of post-productivism in agriculture and wider debates on post-Fordism – highlighting that no analysis of agricultural transitions is complete without consideration of debates, patterns and processes in 'other' societal transitions (see Part 1). These debates were taken further in Lowe *et al.*'s (1993) paper on 'Regulating the new rural spaces: the uneven development of land' in which the authors particularly emphasised the spatial heterogeneity of the p/pp transition. Of particular importance were also Ward's (1993) article on 'The agricultural treadmill and the rural environment in the post-productivist era' that put forward several key hypotheses for conceptualisations of agricultural change from a post-productivist vantage point, and Shucksmith's (1993) paper on 'Farm household behaviour and the transition to post-productivism' that looked specifically at post-productivist farm development pathways from a farm household perspective.[12] No analysis of the p/pp transition would be complete without mentioning Halfacree's (1997b) and Halfacree and Boyle's (1998) work on the importance of migration of the middle classes into the countryside and the resulting changes in perceptions and attitudes about the 'rural' and the 'countryside idyll', with wide-ranging repercussions for how we conceptualise agriculture and rural areas. In Chapter 7 I will come back to many of these early conceptualisations of

---

[12] It is interesting to note that parallel discussions on productivism have also occurred in forest science. Mather (2001), for example, linked theoretical debates on agricultural post-productivism with notions of the post-industrial forest, and argued that many of the characteristics that provide the framework for conceptualisations of agricultural post-productivism also apply to notions of the post-industrial forest and vice versa.

the p/pp transition, with a focus on how subsequent work has increasingly challenged the postulated transition towards post-productivism.

Before analysing the different dimensions of productivism highlighted in these (and other) publications in Section 5.3, four key issues need to be analysed that underlie conceptualisations of both productivism and post-productivism. Understanding these issues will help us better grasp problems associated with the postulated p/pp transition, and will also form the basis for our deconstruction of the p/pp transition model in Chapter 7. These four issues include the temporal dimension of the era of productivism, the UK-centrism underlying the conceptualisation of the p/pp transition, the predominance of political economy and structuralist interpretations, and the 'presentist' interpretation of the productivist era.

5.2.2 The temporal dimension of the 'era' of productivism

The timing of the onset of the 'era' of productivism has been the focus of continuing debates. Most of the literature argues that, in the context of the UK (and to some extent other advanced economies; see below), the era of productivism lasted from about the Second World War to the mid-1980s (Lowe *et al.*, 1993; Mather *et al.*, 2006). Some authors have placed the beginning of the post-productivist transition as early as the 1970s in the context of the oil shocks in 1973 and 1979 which gave urgency to resource conservation (Baldock and Lowe, 1996; Halfacree and Boyle, 1998). Clark *et al.* (1997), meanwhile, suggested that the era of productivism ended with the 1992 CAP reforms and the conclusion of the Uruguay Round of the Global Agreements on Tariffs and Trade (GATT) talks. As Chapter 7 will emphasise, others argue that the productivist era is far from over yet (Morris and Evans, 1999; Wilson, 2001). Key issues here are the suggestion that the productivist era may have 'only' lasted from about 1945 to the mid 1980s, i.e. a relatively short period spanning about four decades. As Chapter 10 will suggest, this may suggest that the productivist era may only be a short 'blip' in a much longer sequence of *non-productivist* agricultural pathways.

If we extend this discussion to include the developing world (see below), then the question of timing of the post-productivist transition becomes even more complex. Several interesting questions emerge here. If we accept notions of 'productivism' and 'post-productivism' (which is problematic as Chapter 7 will highlight), then we may also have to acknowledge the existence of a *pre-productivist* agricultural era, possibly characterised by high environmental sustainability, low intensity and productivity, weak integration into capitalist markets and horizontally integrated rural communities. This would mean that most of the world's farmers (most in the developing world but also some in advanced societies) would fall into this category (Chambers, 1983; Pretty, 1995). Does pre-productivism then automatically imply that *all* societies are 'aiming' towards productivism (i.e. implied directionality)? Can post-productivism, therefore, only occur in rural areas that have 'gone through' the productivist phase, or is it possible to conceive of rural areas that leapfrog directly from pre-productivism to post-productivism – in other words, can no developing country, although their agricultural practices may often be environmentally sustainable, be classified as 'post-productivist'? In particular, that some suggest that the productivist era seemingly came to an *end* in the 1980s and was then superseded by the post-productivist agricultural era suggests an underlying assumption of temporal linearity. We will, therefore, need to return to these temporal issues when investigating debates surrounding the postulated p/pp transition in Chapter 7, with particular emphasis on the *fallacy of temporal linearity* outlined for other transitional debates in Part 1.

5.2.3 The UK-centrism underlying the p/pp transition model

The discussion of the seven dimensions of productivism in Section 5.3 will also highlight the UK-centrism of the p/pp transition concept. This is particularly important concerning the notion of 'exporting theory' highlighted in Part 1 (Said, 1983). The UK-centrism of the p/pp concept will also have important repercussions for our analysis of the fallacies of *spatial homogeneity* and *global universality* of the p/pp transition (see Chapter 7). Indeed, "there is an urgency … to develop a conceptual framework for the understanding of post-productivist agricultural regimes that can also be applied in specific [Southern] contexts, particularly as the need for initiatives and policies that emphasise environmentally-friendly agricultural practices has grown as a result of the Green Revolution" (Wilson, 2001: 92). Conceptualisations of productivism and post-productivism have, therefore, largely been developed by English-speaking academics, and have been heavily biased towards the UK experience and history of agricultural development. Indeed, it was UK researchers who developed and popularised the concept of post-productivism during the early 1990s. Thus, Halfacree and Boyle (1998: 6) conceded that "the concept of the 'post-productivist countryside' has been developed mostly in the context of recent changes which have affected what are generally understood to be rural areas of Britain".

Recent attempts have also been made to analyse whether and to what extent the notion of a shift to post-productivist agriculture can be applied outside the UK. The most important debates in this context have revolved around the question of changes in agricultural policies, in particular recent changes to the EU Common Agricultural Policy (CAP) which, it is argued, shows some elements of 'post-productivism' through the greening of agricultural policy discourses (Lowe *et al.*, 1993; Potter, 1998). It is particularly with regard to the development of CAP policies that researchers have ventured beyond the UK framework in their conceptualisations of post-productivism (e.g. Baldock and Lowe, 1996; Clark *et al.*, 1997). I recently discussed the overall applicability of the concept of post-productivism at the EU level and argued that although the implied linearity of the transition towards post-productivism continues to be questioned, there are some interesting parallel developments in ideologies, actor spaces, agricultural practices, food regimes and attitudes across EU countries that share some elements of the UK-centric post-productivist transition (Wilson, 2002). Indeed, much work has highlighted that many similarities exist between agricultural and rural developments in the UK and other Western European countries (Hoggart *et al.*, 1995; Buller *et al.*, 2000) and that similar processes of rural diversity are occurring in different parts of Europe (Van der Ploeg, 1997; Marsden, 1999a). Similarities are also emerging concerning, for example, common EU member state responses to changes in the organisation of food regimes under the World Trade Organisation (WTO) negotiations (Redclift *et al.*, 1999; Potter and Tilzey, 2005), similarities in responses by member states to CAP policy frameworks (e.g. similar adoption of agri-environmental policy structures; Whitby, 1996; Buller *et al.*, 2000), or growing consumer demand for organic products across the EU (Dabbert *et al.*, 2004).

Theoretical notions of a shift towards post-productivist agriculture are also beginning to be discussed in non-European settings. The most prominent example has been a recent theoretical evaluation by Holmes (2002, 2006) which discussed the applicability of the concept of post-productivism to Australia. Holmes (2002: 381; emphasis added) concluded that although some elements of the different 'dimensions of post-productivism' can be applied to the Australian context, new concepts are needed to understand agriculture-environment tensions in the vast Australian outback – in particular theoretical assumptions based on the shift from "productivist agricultural *occupance* to an emerging era of multifunctional rural *occupance*", rather than a complete shift in Australian agricultural systems (see also Mather *et al.*, 2006). Interesting work on the Australian situation has also been conducted by Argent (2002) who examined whether, and to what extent, the notion of a post-productivist countryside can be applied to the Australian context. Both Holmes and Argent highlighted interesting tensions surrounding the notion of a post-productivist Australian countryside, and suggested that more work is needed to shed light on the applicability of the post-productivist concept beyond Europe (see also Smailes, 2002). Further, recent discussions surrounding the notion of 'deagrarianisation' in the South show striking parallels with debates on the shift towards post-productivism in the North (Bryceson, 1997b; Rigg, 2001). I will argue in Chapter 7 that there are some interesting theoretical and empirical overlaps about concepts, ingredients and indicators surrounding 'post-productivism' and 'deagrarianisation'. Chapter 7 will also make the important point, however, that while there may be common elements to the debates, the underlying propelling forces and implications of the two processes are different. Taken together, these theoretical discussions take the debate into 'new' geographical and agricultural spaces well beyond the UK and EU.

Yet, so far little discussion has taken place on whether a shift towards post-productivist agriculture can be observed in a developing world context (see Wilson and Rigg, 2003, for a notable exception in which we explore the 'exportability' of the p/pp transition concept to the developing world). The reason for this lack of research may be that the theoretical debate on post-productivism (and the supporting empirical data) has been based on a variety of assumptions that may only be found in advanced economies, and that these assumptions can not easily be transferred to the South – in other words, that the theory of post-productivism can not easily be 'exported' into socio-cultural, political and economic contexts of the South. Ten to fifteen years after the first conceptualisations of a shift towards post-productivism – undoubtedly injecting a much needed 'new' theoretical dimension to our understanding of contemporary agricultural change – and on the basis of new studies attempting to test the applicability of the concept in advanced economies beyond the UK and Europe, it is, therefore, an appropriate time to ask whether these assumptions can also be applied outside of the context of advanced economies. Broadening the theoretical debate on post-productivism beyond developed countries is particularly important, because, as Chapter 7 will argue, if the notion of post-productivist agriculture does not find applicability at a global scale (i.e. assumption of global universality), 'post-productivism' may be an insufficiently robust theoretical framework to explain broader patterns of agricultural change in the 21st century in any given locality (Wilson and Rigg, 2003). Indeed, in Part 3 I will argue that multifunctionality may be a much more robust framework within which to situate agrarian change in *both* advanced and developing countries.

There is another possible approach to the application of a post-productivist conceptualisation of agricultural and rural change to the poorer world. Namely, that our ideas, constructed as they are on the experience of a very small slice of the world, need to be adapted and developed to conditions outside the countries of the developed North (Wilson and Rigg, 2003). Indeed, if we consider that the great bulk of the theorising and much of the supporting fieldwork comes from the UK and to a lesser extent from Europe and other advanced economies such as Australia, then we are in danger of extrapolating from a population representing only a few percent of the total, and a significantly smaller proportion of the global farming and rural populations. There is an inherently *evolutionary* and *unilinear* undercurrent to the work on rural change in the richer world and its assumed application to the poorer world. This continues to inform our ideas of agrarian transitions and the direction and forms that they take. As I will show in Chapter 7 through selected examples, the multiple experiences of the countries in the South indicate that there are many paths of rural change, some of which resonate with the work on post-productivism in the North, but many more that do not. In Chapter 7, I will, therefore, also attempt to bridge the gap that continues to exist between often 'parallel' theoretical frameworks of agricultural change between advanced economies and the developing world, by arguing that holistic theoretical concepts need to be developed that cross the divide between North and South (see also Pretty, 1995: 2002). Only once we acknowledge that there is as much (or as little) 'productivist/post-productivist' diversity in advanced economies as in the developing world, can we move towards developing theoretical and conceptual models (e.g. multifunctionality) that will help us to better understand global agricultural change.

5.2.4 Political economy and structuralist conceptualisations

As debates on the p/pp transition emerged largely in the early 1990s in the context of the UK, discussions have been dominated by *political economy* and *structuralist* approaches. To some extent, this was inevitable as political economy theorisations have strongly dominated Anglo-American rural research as the overarching paradigm since the 1980s (Morris and Evans, 1999). Structuralist approaches have been challenged recently by the 'cultural turn' in rural studies (Evans *et al.*, 2002), but researchers using cultural approaches have largely shied away from engaging with debates on post-productivism and multifunctionality. As a consequence, many traditional features of the conceptualisation of the p/pp transition have focused on specific actor groups (e.g. policy-makers) or larger structural entities (e.g. 'the state') to the neglect of individuals and their actions. Thus, "the dominant political economy discourse has ... inevitably led to a heavy emphasis on the importance of the state and policies, a strong focus on the importance of macro-economic factors in actor decision-making ... and a heavy emphasis on food production and global market regimes. ... As a result, the farming community has often been viewed as responding almost entirely to *outside* forces, with little acknowledgement of possible changes from *within*" (Wilson, 2001: 85-86; original emphasis). Although structuralist approaches towards conceptualisations of the p/pp transition have been essential to the argument – indeed, one may argue that only structuralist interpretations could have led to the development of the unilinear assumptions underlying conceptualisations of post-productivism – Chapter 7 will highlight that these approaches may have provided only partial answers, and may be one of the reasons why the p/pp transition model has failed to capture the real underlying forces behind agricultural change in the UK and beyond (Evans *et al.*, 2002; Wilson and Rigg, 2003). If the implied linearity in the p/pp transition model was true, it could be hypothesised that both structure (e.g. agricultural policies, rural political economy) and agency (farmers' identities and 'farming culture') should 'move along' the p/pp spectrum at the same pace and in the same manner.

In Chapter 7 I will argue that the predominance of structuralist interpretations of post-productivism has led to an over-emphasis of research on *structural* factors behind the postulated p/pp transition (in particular the heavy focus on policy change as a key indicator), to the neglect of *agency-related* factors that have only recently begun to attract the attention of rural researchers (in particular, farmer identities and attitudes and investigations of decision-making pathways of rural and agricultural actors) (e.g. Burton and Wilson, 2006). This has led to a focus on structural *exogenous* factors or 'indicators'[13] of agricultural change (in particular policy changes; the political economy framework; farmers' economic adjustment strategies to external forces; etc.), rather than agency-related *endogenous* characteristics that may accompany this change (e.g. attitudes, perceptions, behaviour and identities of specific agricultural and rural actors). I will argue that the political economy/structuralist interpretations of the p/pp transition are closely interrelated with the *fallacy of structural causality* inherent in the p/pp transition model – a fallacy that is also apparent in many other debates on societal transitions (see Chapter 4) – because of the unbalanced focus on structural factors that may have shown more evidence of a transition towards post-productivism than rural/agricultural agency (see Chapter 7).

5.2.5 Presentist interpretations of productivism as a form of modernity

It is also important to emphasise that, similar to debates on transitions from 'isms' to 'post-isms' highlighted in Chapters 3 and 4, the evolution of the term 'post-productivism' only

---

[13] The notion of 'indicators' to measure the extent of environmental and ideological change has been increasingly criticised by academics in recent years (see Lowe *et al.*, 1999, and Wilson and Buller, 2001, for detailed critiques of the indicator concept).

brought into being the term 'productivism'. As with other debates on transition we have, therefore, witnessed the *retrospective definition* of the 'productivist' era from a 'post-productivist' vantage point. As Argent (2002: 107) argued in his critique of post-productivism, "this is one of the dangers of [linear] narrative ... because it offers a compelling account of the real world by presenting a particular sequence of events it naturalises, *post hoc* fashion, synchronic and contingent relations". This is highlighted by the fact that during the productivist 'era' (i.e. until about the mid-1980s according to some authors; see above) few actors (if any) were aware that they were 'productivist' (similar to debates on Fordism or modernism, for example). This means that *during* the productivist era no theoretically grounded reference can be found that acknowledged the existence of 'productivist agriculture' – the key reason why the terms 'productivism' and 'post-productivism' only began to appear in scientific debates in the early 1990s.

In addition to the issues of timing, UK-centrism and the predominance of structuralist conceptualisations, presentism can be seen as an additional weakness of the p/pp transition model. This bias is important to consider, as critics may argue from the outset that 'productivism' has been conceptualised from a presentist vantage point – in other words, from a perspective that benefits from the knowledge that agriculture may have 'moved on' from productivism – thereby influencing an 'objective' assessment of what made up this so-called 'productivist era'. In Chapter 7 I will, therefore, also argue that the presentist bias in conceptualisations of the p/pp transition is one of the reasons why assumptions underlying suggestions that advanced economies are now well and truly in a post-productivist era can not be substantiated.

5.3 The seven dimensions of productivist agriculture

In this section I wish to look in more detail at the individual components of what the literature has characterised as productivism.[14] The aim here is to provide insights into the debates on what the literature suggests are the 'ingredients' of productivism. A review of the literature suggests that productivism (and post-productivism; see Chapter 6) has been conceptualised on the basis of seven inter-related dimensions (Table 5.1.): agricultural policies, ideology, governance of rural spaces, food regimes and agro-commodity chains, agricultural production, farming techniques and environmental impacts (see also Evans *et al.*, 2002, for a slightly modified list of indicators). Mather *et al.* (2006) provided a useful overview of the different approaches used to conceptualise productivism and post-productivism, and they particularly pointed to the problem of finding quantifiable 'indicators' for assessing the shift towards post-productivism. These debates highlight that a multitude of different characteristics need to be considered to fully understand the postulated post-productivist transition (Marsden *et al.*, 1993), and that a single focus on specific dimensions would only provide partial answers. In the following I will discuss each of these dimensions in greater detail.

The predominance of political economy/structuralist interpretations of the p/pp transition has meant that conceptualisations of productivist agriculture have placed a large focus on both the evolution and changing discourses of *policies* – best evidenced through the above-mentioned definition of productivism (see Section 5.1) that focuses to a large extent on policy-related aspects of productivism and on institutions (e.g. Ministry of Agriculture in the UK) in charge of policy implementation. An additional reason for the predominant focus on policy is arguably *methodological*, linked to the relative ease with which data and information can be gathered on policy change (e.g. content and discourse analysis based on

---

[14] In the following I will refer to 'productivism' rather than the 'productivist era' as the latter suggests questionable assumptions about temporal linearity (see also Chapter 7).

Table 5.1. Dimensions of productivism (Source: after Wilson, 2001)

Agricultural policies:
- Strong financial state support (Cloke and Goodwin, 1992; Winter, 1996)
- Conservative faith placed in ability of state to plan and orchestrate agricultural regeneration (Marsden *et al.*, 1993; Mather *et al.*, 2006)
- Encouragement to farmers to expand food production (Whitby and Lowe, 1994)
- Government intervention (Marsden *et al.*, 1993)
- Protectionism (Goodman and Redclift, 1989, 1991)
- Price guarantees/financial security for farmers (Potter, 1998)
- Agriculture largely exempt from planning controls (Marsden *et al.*, 1993)
- Security of property rights/land use rights (Whatmore *et al.*, 1990; Marsden *et al.*, 1993)

Ideology:
- Central hegemonic position of agriculture in society (Cloke and Goodwin, 1992)
- Ideological security (Marsden *et al.*, 1993; Halfacree and Boyle, 1998)
- Agricultural fundamentalism rooted in memories of wartime hardships (Newby, 1985; Mather *et al.*, 2006)
- Agricultural exceptionalism (Newby *et al.*, 1978; Newby, 1985)
- Belief in farmers as best protectors of countryside (Newby, 1985; Harvey, 1997)
- Countryside idyll ethos/rural idyll (Mingay, 1989; Hoggart *et al.*, 1995)
- Main threats to countryside perceived to be urban and industrial development (Marsden *et al.*, 1993; Ward, 1993)
- 'Rural' defined in terms of agriculture (Halfacree and Boyle, 1998)

Governance of rural spaces:
- Agricultural policy community small but powerful, tight-knit and with great internal strength (Cox and Winter, 1987; Winter, 1996)
- 'Corporate' relationship between agriculture ministries and farming lobby (Cox *et al.*, 1988; Winter, 1996)
- Relative marginalisation of conservation lobby at fringes of policy-making core (Cox *et al.*, 1988; Hart and Wilson, 1998)

Food regimes and agro-commodity chains:
- Atlanticist Food Order dominated by USA (Goodman and Redclift, 1991; Le Heron, 1993)
- Fordist regime (Goodman and Redclift, 1991; Ward, 1993)

Agricultural production:
- Industrialisation (agri-businesses) (Marsden *et al.*, 1993; Whatmore, 1995)
- Commercialisation (Ilbery and Bowler, 1998)
- Securing national self-sufficiency for agricultural commodities (Lowe *et al.*, 1993; Ward, 1993)
- Intensification (Marsden *et al.*, 1993; Mather *et al.*, 2006)
- Surplus production (Ilbery and Bowler, 1998)
- Specialisation (Ilbery and Bowler, 1998)
- Concentration (Ilbery and Bowler, 1998)
- Increase in corporate involvement (Lowe *et al.*, 1993; Marsden *et al.*, 1993)
- Farmers caught in agricultural 'treadmill' (Ward, 1993)

Farming techniques:
- Increased mechanisation (Ilbery and Bowler, 1998)
- Decline in labour inputs (Lowe *et al.*, 1993; Whitby and Lowe, 1994)
- Increased use of biochemical inputs (Potter, 1998; Pretty, 1998)

Environmental impacts:
- Increasing incompatibility with environmental conservation (Clark and Lowe, 1992; Potter, 1998)

---

published documentary evidence), in contrast to more complex and intangible methodologies linked to investigations of other dimensions of the p/pp transition such as farmer identities and attitudes. Policy analysis has, therefore, provided a relatively 'easy' inroad for many scholars into the complex debates surrounding the p/pp transition (and indeed 'multifunctionality'; see Chapter 8), possibly to the neglect of other 'less accessible' dimensions.

For conceptualisations of productivism, the literature has focused particularly on the CAP of the European Economic Community (EEC) and national legislation from the early 1960s to the mid-1980s. Indeed, the reason why many commentators have argued that the productivist era ended in about 1985 is largely based on *policy evidence* for substantial changes in policy discourses at the time. Thus, the CAP until the early 1980s focused almost entirely on providing a policy framework for increasing agricultural production and encouraging agricultural intensification, possibly best expressed in the British Agriculture Ministry publication *Farming and the nation* (MAFF, 1979) which is usually interpreted as a staunchly productivist policy document in line with the then predominant CAP discourses (Baldock and Lowe, 1996; Marsden, 1999b). Thus, "productivism, reflected in production subsidies and grants, was the cornerstone of the policy framework" (Evans *et al.*, 2002: 323), and "farming grew to follow a 'productivist' model whereby emphasis was placed on maximising food production through the application of intensive production approaches" (Burton, 2004: 195). The policy period after 1985, meanwhile, is generally seen as one of 'greening agricultural policy discourses' (Harper, 1993; Potter, 1998), where the rhetoric of policy changed from *intensification* to *extensification*, from *concentration* of production to *diversification*, and from *high-quantity* production maximisation to production of fewer *high-quality* commodities (Ilbery and Bowler, 1998; Lang and Heasman, 2004). Productivist policies have, however, not been restricted to the EEC (and later EU), but have also been apparent in North America (Potter, 1998) and other advanced economies such as Australia (Argent, 2002; Holmes, 2002) or New Zealand (Jay, 2004).

In the EEC, productivist policy moves to increase production of agricultural commodities have been closely linked to strong financial state (and EEC/EU) support and intervention through farm *subsidies*, *price guarantees*, and *protectionist* and *interventionist* policies that have kept prices for agricultural products artificially inflated since the 1960s, giving farmers a strong sense of financial security (Fennell, 1987; Ritson and Harvey, 1997). As Shucksmith (1993: 466) emphasised, these policies have aimed to "make two blades of grass grow where one grew before". Protectionist and farmer-friendly subsidy policies paid to West German farmers between 1960 and 1980, for example, highlight the ethos of the CAP 'subsidy culture', farm income maximisation, and exaggerated 'artificial' support for farmers in the EEC (Wilson and Wilson, 2001; see also Goodman and Redclift, 1991). Sheingate (2000) referred to this as the creation of an 'agricultural welfare state'. Further, while productivism has been strongly associated with conservative faith placed in the ability of the state to plan and orchestrate agricultural regeneration (e.g. Agricultural Act 1947 [UK]; Loi d'Orientation Agricole 1960 [France]; Australian agricultural policy after 1952), it has, simultaneously, also been characterised by limited state regulation of environmentally harmful agricultural practices (Cloke and Goodwin, 1992), exemption of agriculture from planning controls (Marsden *et al.*, 1993), and the guaranteed security of property and land use rights (Whatmore, 1986; Whatmore *et al.*, 1990) – not only in Europe but also in the context of other advanced economies. This has given farmers freedom to manage their land as they see fit, resulting in farming techniques associated with increased mechanisation and increased use of biochemical inputs progressively incompatible with sustainable environmental management (Knickel, 1990; Mannion, 1995).

Recent debates on the p/pp transition have also highlighted that 'productivist' policies can be identified in many *developing* world contexts. Historical evidence in many developing countries often points towards early productivist policies aimed at boosting agricultural production (Wilson and Rigg, 2003). Probably the most instructive example comes from

China, where the imperial state embarked upon a focused and well-supported attempt to boost production from AD 1000, and possibly much earlier. Bray (1986) highlighted, for example, that in AD 1012 the Song emperor Zhenzong ordered the collection of 30,000 bushels of quick-ripening Champa seed rice to boost production in the Lower Yangzi region.[15] She argued that "the most striking period of development of Southern Chinese agriculture began in the Song period, when the government initiated a series of development policies so sweeping in scope and in result that they may well be compared with the so-called 'Green Revolution' of Asia today" (Bray, 1984: 597). The Song government was printing agriculture manuals of best practice in the 10th century AD, and at the same time introduced tax and credit policies to encourage production (Bray, 1984). Further, the treatment of seed rice with pesticides (such as arsenic) dates from the early Han dynasty (206 BC-AD 220) and commercial fertilisers (e.g. oil cake) were already in widespread use during the Ming dynasty (1368-1644).[16] Other arguably 'productivist' regimes where central structures of authority co-ordinated and directed agriculture through specific 'policies' include Central and South America (the Maya, Aztec and Inca civilisations), South-East Asia (Java, Burma, Cambodia and Thailand; see Reid, 1988) and the Baliem Valley in New Guinea. While these efforts were directed down several avenues, the control of water for irrigation was a central objective, and such co-ordination of effort to achieve stable and high levels of output can be interpreted as 'productivist' in the drive to ensure a stable and maximum supply of food (Wilson and Rigg, 2003; see also Pretty, 1995, 2002).

The second dimension of productivism concerns **agrarian ideologies**, i.e. societal views about the importance and social position of agriculture. In contrast to discussions on policy as a productivist dimension, debates on ideological factors have been less clear-cut, partly due to the less tangible nature of societal perceptions and attitudes. There is, nonetheless, a reasonably coherent body of work that has emerged that is relatively unanimous about the various ideological components of productivism. As Table 5.1 suggests, productivist ideologies have been characterised by a *central hegemonic position* of agriculture in rural society (Cloke and Goodwin, 1992) and a sense of unchallenged *ideological security* for agricultural actors and institutions (Marsden *et al.*, 1993; Halfacree and Boyle, 1998), highlighting that productivist agriculture and food production is seen as occupying a special place in the 'pantheon of traditional conservative values' (Wormell, 1978). Bishop and Phillips (1993) argued that such *agricultural fundamentalism* has been strongly rooted in memories of wartime hardships, with agriculture seen as having a pre-emptive claim on the use of rural land, aptly referred to by Newby (1985) as 'agricultural exceptionalism' (see also Newby *et al.*, 1978). Productivist ideologies hold a strong belief that farmers are the *best 'protectors'* of the countryside (Newby, 1985; Harvey, 1997) and, coupled with notions of the 'rural idyll' (Mingay, 1989; Hoggart *et al.*, 1995), this has led to a conservative vision respectful of private property and traditional agrarian institutions (Halfacree, 1999). As a result, the main threats to the countryside are perceived to be urban and industrial development – not agriculture itself – and the 'rural' is, therefore, mainly defined in terms of agricultural production (Ward, 1993; Halfacree, 1997a). For most advanced economies, Drummond *et al.* (2000) argued that this productivist ideology has long been upheld and reinforced by an alliance of agricultural economists and biological scientists who continue to legitimate the intensification and industrialisation of agriculture.

It is more difficult to find evidence of productivism based on ideological factors beyond the UK – re-emphasising the above-mentioned UK-centrism of the p/pp transition concept. Yet, parallels can, for example, be found between ideological developments in the UK, Germany and Switzerland (Wilson, 2002). For example, the strong and often futile support in areas of former Western Germany to maintain economically struggling small family farms at

---

[15] Champa is an area of coastal central Vietnam (or Annam). It is thought that these indica rices were known to the Chinese for their special quick-maturing qualities (allowing double cropping) as early as the first century AD.

[16] Although chemical fertilisers were not in use in China until the early 20th century.

all costs since the 1960s can be interpreted as the willingness of a society, that sees the survival of small family farming units as part-and-parcel of the German agrarian idyll, to adhere to strongly productivist notions – notions that were severely challenged in the 1990s after reunification of West Germany with the former German Democratic Republic (GDR) and its large agri-business-oriented farm cooperatives (Wilson and Wilson, 2001). In a non-European context, both Holmes (2002) and Argent (2002) suggested that some UK-based elements of 'productivist ideologies' are also applicable to the Australian situation (e.g. central position of agriculture in society after the Second World War; notions of agricultural fundamentalism, especially in the state of Queensland). Argent (2002: 101), for example, argued that "the Australian version of productivism, as in Britain, saw the farm sector installed as a pillar of national economic and social development". However, in the context of both Australia and New Zealand both Jay (2004) and Wilson and Memon (2005) pointed towards substantially different ideological trajectories for the position of agriculture in society, based on the deregulation of agriculture (i.e. withdrawal of farm subsidies in New Zealand in 1984), which led to a repositioning of agriculture in society in which farmers where 'left to fend for themselves' (Wilson, 1994; Losch, 2004). Similar patterns, albeit without the complete abolition of subsidies, can be observed in North America, where re-evaluations of societal perceptions of continuing productivist agricultural practices can be observed (Potter, 1998; McCarthy, 2005).

As Pretty (1995) highlighted, the situation in the developing world is more complicated as agriculture and society are often not as easily separable entities as in advanced economies. Indeed, apart from rapidly growing urban areas, most regions in the developing world can still be characterised as 'rural/agricultural regions' that show characteristics of pre-productivism and where the p/pp transition model may not be applicable (see Chapter 7). In this context we argued that "it is with regard to post-productivist conceptualisations of less tangible agricultural/rural change indicators – such as changing *ideologies* or *attitudes*, or indicators based on culturally constructed complex *societal interactions* (e.g. counter-urbanisation; commoditisation of the countryside; etc.) – that the most challenging questions for the future will lie, and where, currently, the transposition of notions of post-productivism from advanced economies to the South is doomed to fail" (Wilson and Rigg, 2003: 701; original emphasis).

Little work is currently available on the reconfiguration of actor spaces and **governance** as a possible indicator of the transition towards post-productivist agriculture (Wilson, 2004). This echoes Little's (2001: 97) critique of rural geography in that "just as rural geographers were slow to apply theoretical debates on the state in the examination of rural government in the 1980s, so they have shown similar reluctance to engage with recent theoretical and empirical work on so-called new 'governance'" (see also Goodwin *et al.*, 1995; Goodwin, 1998). Nonetheless, many researchers are beginning to engage with issues surrounding the governance of rural spaces, shedding light on the increasingly complex decision-making structures between individual stakeholders in rural societies in both advanced and developed countries (Curtis and Lockwood, 2000; Wilson, 2004). Although early conceptualisations of changing governance did not use the term itself during the early 1990s, much work has shed light on what we today would refer to as changing 'rural governance structures'. Thus, Gilg (1991) and Clark and Lowe (1992), for example, suggested that the productivist agricultural policy community is small but powerful, tight-knit and with great internal strength. In particular, the 'corporate' relationship between agriculture ministries and the farming lobby has been stressed as largely excluding 'other' actors from key agricultural policy decision-making processes. Tilzey (2000) referred to this as 'political productivism'. As highlighted above, this has led to a strong sense of political and ideological security (Cox and Winter, 1987; Winter, 1996), and has marginalised other stakeholder groups, such as the conservation lobby, towards the fringe of the policy-making core (Cox *et al.*, 1986; Hart and Wilson, 1998).

Productivist governance is, therefore, *exclusive* rather than inclusive, and characterised by a few core stakeholder groups wielding a lot of decision-making power. Marsden (2003: 126) argued that "throughout much of the productivist period the agricultural interests … represented something of a stable policy community". This power, it is argued, has stemmed largely from historical developments (e.g. the emergence of agriculture ministries with specific production maximising policy remits) and that, for a long time, grassroots stakeholder groups have been either poorly developed or themselves exclusive (e.g. many farmers' unions that tend to represent the voices of larger and powerful landholders), or only became more powerful actors during the 1980s and 1990s (in particular environmental NGOs and/or the emergence of pressure groups associated with green political parties) (Dobson, 1995; Winter, 1996). Recent research in Australia also tends to support the notion of relatively exclusive productivist stakeholder networks, especially during the 1970s and 1980s, although, as Chapter 6 will highlight, the situation has changed based on more 'inclusive' programmes such as *Landcare* that aim to bring together both state and non-state actors in the management of degraded agricultural land (Wilson, 2004). This suggests that the notion of *inclusive* governance may be closely associated with 'post-productivism' in increasingly pluralistic and multi-layered societies.

It has been argued that productivist *food regimes and agro-commodity chains* have been largely shaped by the 'Atlanticist Food Order' dominated by the USA (Le Heron, 1993; Goodman and Watts, 1997), characterised by mass consumption of agricultural commodities, the expansion of world food trade in a rapidly growing capitalist market, and the adoption of Fordist regimes of agricultural production (Goodman and Redclift, 1991; Cloke and Goodwin, 1992). This has severe repercussions for agricultural production through the industrialisation of agriculture, resulting in both the commercialisation of agricultural holdings increasingly embedded in the 'treadmill' of production and profit maximisation (Ward, 1993), and the emergence of large agri-businesses often poorly rooted in local rural communities (Whatmore, 1995). As productivism has largely been a response to the paradigm of hunger (especially in the wake of the Second World War; see above), the ultimate goal has been to secure national self-sufficiency for agricultural commodities, leading to environmentally harmful intensification (Potter, 1998), and government encouragement for maximum production often resulting in increased surplus production (Wilson, 2001; Robinson, 2004).

The outcome, it is argued, has been regional specialisation of agricultural production and the concentration of farming through the amalgamation of smaller farm units into more efficient larger holdings with associated declines in labour units (Ilbery and Bowler, 1998). This has severe repercussions for the lengthening of agro-commodity chains, characterised by rapidly increasing 'food miles' (distance that food has to travel to reach consumers), and the dislocation of food from its local and regional provenance with associated loss of consumer knowledge about the provenance of their food (Lang and Heasman, 2004). In the UK (and elsewhere), these processes have been particularly associated with the loss of local farmers' markets and the rise in power of a few large supermarket chains that increasingly influence commodity prices, retailing structures and consumer behaviour and preferences (Friedberg, 2003).

Pretty (2002) and Lang (2004) argued that the so-called 'nutrition transition' also needs to be considered as part of conceptualisations of the p/pp transition (see also Popkin, 1998). This is further supported by Murdoch (2002) who suggested that the 'food quality threshold' is conceptually as important as the p/pp transition itself (see also Marsden, 2003). Commentators argue that productivism is associated with the 'conventional food economy', particularly characterised by the consumption of increasingly unhealthy foods leading to severe problems of obesity and diet-related diseases for about one-tenth of the world population (Pretty, 2002). Alarmingly, the obese are increasingly outnumbering the thin in some parts of the world (e.g. Brazil, Chile, Colombia, Tunisia), while 10-20% of Europeans are now classified as clinically obese (Lang and Heasman, 2004). This disconnection of people from food has become a key characteristic of productivism, associated with both the loss of knowledge about food preparation and the shift in perceptions of previously useful plants (for food or medical purposes) now often characterised as 'weeds' – at least in the context of many advanced economies. This has been further exacerbated by the relative lack of customer interest about food quality and the lack of legally enforced food labelling (Dabbert *et al.*, 2004).

Arguably, the use of food regimes and agro-commodity chains is the dimension conceptually most firmly embedded in UK/advanced economies-centric conceptualisations of productivism. Little work is currently available on this dimension from a developing countries' perspective – possibly highlighting that the transposition of this dimension outside advanced economies continues to be problematic (Wilson and Rigg, 2003). The association of food regimes and agro-commodity chains with productivism, in particular the link with the Atlanticist Food Order (cf. Cafruny, 1989), is also one of the key reasons why the timing of the productivist era has been so narrowly placed by some between about 1945 and 1985 – re-emphasising the problems of the predominance of political economy and structuralist interpretations of the p/pp transition highlighted above.

In addition, productivism has had severe repercussions for ***agricultural production***. As the ultimate goal of productivism has been to secure national self-sufficiency for agricultural commodities (see above) this has led to environmentally harmful intensification (Potter, 1998), industrialisation of agriculture with the strengthening of large and globally networked agri-businesses (Marsden *et al.*, 1993; Whatmore, 1995) and the commercialisation of agricultural production practices (Ilbery and Bowler, 1998). As Clark (2005: 476) emphasised, "the agro-industrial model has … divorced agriculture from its scalar economic contexts, and stripped it of its territorially embedded status". Emphasis has been placed on the quantity of production and less on quality, with the result that during the 1970s and 1980s surplus production became an ever-increasing problem in the EEC (Ritson and Harvey, 1997). Many have, therefore, equated productivism with the era of surplus production of agricultural commodities (Ilbery and Bowler, 1998), while Van der Ploeg and Roep (2003: 39-40) argued that "until the early 1990s, scale-enlargement, intensification, specialization, and within sectors, a strong trend towards industrialization, were the parameters that circumscribed developments in the agricultural sector".

Some commentators have also associated productivism with the *specialisation* and *concentration* of agricultural production, particularly in the context of a transformation of 'mixed farming' units into specialised arable or livestock-based agri-businesses (Ilbery and Bowler, 1998). Marsden *et al.* (1993) and Lowe *et al.* (1993) also suggested that this has gone hand-in hand with a rise in corporate involvement, which has increased opportunities for larger economically buoyant farm holdings, while at the same time reducing opportunities for small and economically struggling farm family-based units. Productivist farmers may have become increasingly caught up in often *externally driven path developments* that, in many cases, have predetermined farm decision-making opportunities where farmers often feel that they are losing control over their business decisions (Whatmore *et al.*, 1990; Whatmore, 1995). Ward (1993) aptly described this situation as the productivist 'treadmill' in which farmers have been increasingly caught in a globalised web of commodity export networks and dependency on large agro-chemical companies which has dictated what farmers can grow and how they should go about protecting their animals and crops from pests and diseases (see also Pretty, 1995; Kloppenberg, 1988). As a result, Burton (2004) emphasised how the view of the 'good farmer' over the past few decades has often been shaped around the quality of crops and livestock produced on a farm judged by the two criteria of 'physical appearance' and 'crop yield/animal weight', further epitomised by farmers' penchant for 'clean' and 'ordered' (i.e. productivist) farm landscapes (Brush *et al.*, 2000).

Debates on productivist agricultural production are by no means restricted to the UK or Western European contexts. Many commentators have confirmed that almost all of the above characteristics of productivist agriculture also find parallels in the Australian (Argent, 2002; Holmes, 2002) and New Zealand situation (Jay, 2004). Indeed, both Australia and New Zealand saw dramatic intensification of agriculture during the 1980s in the wake of the deregulation of agriculture that forced many farm businesses to intensify, specialise and concentrate their production with the aim to access new national and overseas markets (Losch, 2004; Robinson, 2004). Similar processes have been identified for other developed countries (e.g. Potter, 1998, for the USA; Wilson and Wilson, 2001, for Germany). Yet, productivist agricultural production techniques have not been restricted to advanced economies. As highlighted above with the example of China, the historical agricultural geography of the non-Western world demonstrates that 'productivist' agricultural production existed long before many 'modern' technologies of production. The Chinese example highlights that many conceptualisations of agricultural change falsely link productivism with modernity and modernism (themselves problematic concepts as discussed in Chapter 4) and modern technology (Wilson and Rigg, 2003). Thus, Bray (1986: 2), in her study of Asian rice economies, noted the "failure (in the main) to recognise the relativity of our conception of technological progress". Indeed, co-ordination of effort to achieve increasing output must be seen as more important than the technologies *per se* that are brought to bear. If this is so, drawing a neat distinction between the modern Green Revolution and more ancient 'green revolutions' is difficult (Wilson and Rigg, 2003). Moreover, even in terms of the technology employed – as shown above with reference to China – the division is not as clear-cut as it might, at first, appear (see also Chapter 7).

The **farming techniques** dimension provides relatively tangible evidence for conceptualisations of both productivist *and* post-productivist agriculture, again partly linked to the relative ease with which data and information can be gathered on such changes (e.g. changes in machinery or agricultural labour force for which long-term data are available). The literature particularly refers to three interconnected changes. First, *increased mechanisation* is seen as closely associated with productivism, not only in the European context (e.g. Ilbery and Bowler, 1998) but also in the context of other advanced economies (e.g. Holmes, 2002, for Australia). This has partly been led by productivist agricultural policies that have provided financial incentives to farmers to mechanise their farms, and partly by increased economic buoyancy of farms during boom periods after the Second World War. The case of former West Germany is particularly instructive where the combination of national and EEC policies, for example, led to a rapid rise in the number of tractors from 70,000 in 1960 to 170,000 in 1990 (Wilson and Wilson, 2001).

Second, the *decline in agricultural labour* has gone hand-in-hand with increased mechanisation of farms as a characteristic of productivism (Lowe *et al.*, 1993; Whitby and Lowe, 1994), and almost all advanced economies have witnessed rapid falls in agricultural workforce since the 1960s (Potter, 1998; Pretty, 1998). In Europe agricultural workforces declined particularly rapidly between the 1950s and 1980s, with the UK, for example, seeing reductions from six full-time workers per 100 ha in 1959 to three workers/100 ha in 1989 (reduction of 50%), with even more dramatic workforce reductions in Denmark from nine to four (- 55%) and West Germany from 27 to only seven (- 73%) during the same time period (Wilson and Wilson, 2001). Productivism is, therefore, associated with substantial loss of people working in rural areas (most former agricultural workers moved to urban areas to find employment) with severe repercussions for the survival of rural communities (Halfacree and Boyle, 1998; Woods, 2005).

Third, above-mentioned productivist increases in the quantity of agricultural commodity production could only be achieved through substantial *increases in the use of biochemical inputs* (artificial fertilisers) to make up for rapid nutrient losses caused by intensive farming (Potter, 1998; Pretty, 1998). This change in farming techniques has had particularly severe implications for environmental conservation. The Netherlands is often mentioned as the most dramatic example of how increased biochemical inputs between the 1950s and 1980s enabled the country to intensify agricultural production to such an extent that by 1990 it was able to produce 9% of the EU's agricultural exports on only 2% of the EU's land area (Robinson, 2004). Similarly, in West Germany the application of nitrates increased from 29 kg/ha in 1950 to nearly 150 kg/ha in 1990 (five-fold increase) and the use of phosphates increased four-fold during the same time period. Many similar quantitative examples illustrating changes in farming techniques can be found in non-European countries (e.g. USA, Australia) and developing countries (e.g. India, China). As will be highlighted in Chapters 6 and 7, the indicator 'changing farming techniques', therefore, emerges as one of the least contested dimensions of the p/pp transition conceptualisation.

There is also little debate surrounding the fact that **environmental impacts** of agriculture have increased through productivist farming practices. For many commentators, therefore, it is the *environmental dimension* that forms one of the crucial pillars for conceptualisation of the p/pp transition (Ward, 1993; Wilson, 2001). As Chapters 6 and 7 will discuss, many see the post-productivist transition as essentially an *environmental transition* that started with productivism characterised as a period of severe environmental imbalance between food production and environmental conservation (Wilson, 2002). Key arguments revolve around the fact that an increasing incompatibility of agriculture with environmental conservation became apparent between the 1960s and 1980s (Knickel, 1990; Potter, 1998). To a large extent, the environmental impacts of productivist agriculture has mirrored those of *industrial productivism* under Fordist modes of production highlighted in Chapter 3 with, at times, similar patterns linked to the industrialisation of agricultural production, mass production and consumption, and the creation of relatively uniform and easily manageable farm landscapes devoid of any 'unnecessary' encumbrances such as 'unproductive' wildlife habitats. As Lipietz (1992: 55) argued, "as productivism spread throughout our planet by imitation or the pressure of foreign debt, it saturated our ecosystem".

Several environmental issues have come to the fore through productivist farming. First, intensification of agriculture is usually associated with increased *biochemical inputs*, in particular fertilisers and pesticides,[17] with harmful polluting effects on water and wildlife (Harvey, 1997). During the early phases of productivism in the USA, Carson's (1962) book *Silent spring* especially highlighted the harmful effects of biochemical compounds in the human food chain, arguing that humans are exposed to unnatural and often toxic substances from intensive agricultural production from human conception to death. As a result, many authors have argued that agriculture has become increasingly detached from the natural environment (Swagemakers, 2003).

Second, the continuing *destruction and removal of remnant wildlife habitats* in the countryside has been a key focus of the literature. Although this has been a process that has accompanied agriculture over the past 10,000 years (Simmons, 1989; Mannion, 1995), advocates supporting the notion of productivism have emphasised the increasing pace and severity of these impacts in the decades after the Second World War (Marsden *et al.*, 1993; Marsden, 2003). In the UK, for example, reference is often made to the rapid disappearance of wildflower-rich meadows (90% lost since 1940), heathland (minus 50%), ancient woodland (30-40% lost since 1940) and hedgerows (minus 50%) through field enlargement and intensification (Pretty, 2002; Robinson, 2004). Similar patterns have been described for other advanced economies, where productivist agricultural practices have led to further decline of biodiversity, such as New Zealand (Wilson and Memon, 2005), Australia (Holmes, 2002; Cocklin and Dibden, 2004) or Germany (Knickel, 1990; Wilson and Wilson, 2001). In Australia, for example, productivist practices after the Second World War have exacerbated the trend of ongoing dramatic flora and fauna decline, leading to overall alteration of over 50% of the floral community, further removal of original tree cover (over two-thirds since European settlement began) and with further pressures on biodiversity (Hobbs and Hopkins, 1990). In the USA, meanwhile, Mitsch and Gosselink (1993) showed how agricultural intensification has led to further wetland reclamation and destruction, with

---

[17] Each year farmers globally apply 5 million t of pesticides to agricultural land (Pretty, 2002).

wetlands declining rapidly between the 1950s and 1980s, and some states (e.g. Arkansas, California) losing 30-50% of wetlands in a few decades.

Third, productivist agriculture has had a substantial *impact on soils* through soil pollution linked to increased use of biochemical inputs, soil compaction through the use of heavy machinery, and soil erosion and salinisation through over-intensive irrigation. Based on farm surveys in Greece, Italy, Spain and Portugal, research has highlighted how productivist agricultural policies since the 1960s have exacerbated the extent and severity of soil erosion, in particular linked to intensification of former environmentally sustainable traditional extensive farming systems (e.g. conversion to highly subsidised durum wheat cultivation in Italy and Portugal), over-irrigation of horticultural crops for the highly lucrative export market (Spain), and overstocking of desertification-prone hillsides based on CAP livestock subsidies (Greece) (Wilson and Juntti, 2005). Globally the rate of soil loss and desertification has reached unsustainable proportions leading to substantial reduction of soil fertility in many formerly highly fertile agricultural districts (UNEP, 1992; UNCCD, 1994).

Reasons for the 'laissez-faire' attitudes towards conservation that have dominated productivist agriculture are linked to dimensions of productivism highlighted above. Key has been the lack of power of stakeholder groups willing to challenge environmentally destructive over-intensive agricultural practices in different parts of the world. In particular, and as Chapter 6 will highlight, conservation organisations only began to assert themselves in close-knit 'productivist' agricultural networks in the 1970s and 1980s in most advanced economies and even later in developing countries (Bryant, 2005). This has meant that destructive agricultural practices have (until recently) not been substantially challenged, agricultural intensification has often not been questioned by a society still remembering the hardships of wartime, and that farmers are still often seen as the best 'stewards of the land' (Ward, 1993). This has given farmers freedom to manage their land as they see fit, resulting in farming techniques progressively incompatible with sustainable environmental management (Mannion, 1995).

5.4 Conclusions

The discussion in this chapter has highlighted that there is some consensus in the academic literature about the definition, delineation and dimensions of 'productivist agriculture'. Key features of productivism can be summarised as a push for *quantity* over *quality* of agricultural production with the aim to achieve national self-sufficiency, supported by *production-oriented agricultural policies* (especially the EU CAP, but also similar productivist policies in other advanced and developing economies) that have facilitated the intensification and specialisation of agricultural production with little regard for environmental conservation. Some have argued that, in the European and US context at least, the emergence of productivism was facilitated by the predominance of the *Atlanticist Food Order* with its orientation towards an Anglo-American axis of agricultural decision-making affecting national agricultural policies underpinned by free market liberalist ideas (Cafruny, 1989; Potter and Burney, 2002). This chapter has also shown that the emergence of productivism (arguably in the 1950s) was only possible because of specific *ideological stances* that have dominated society's view of agriculture in the UK, Europe and other advanced economies, especially the *central hegemonic position of agriculture* in society based on its crucial role of providing self-sufficient supplies of food after the ravages of the Second World War. As highlighted, this has been closely associated with powerful and *close-knit stakeholder groups* linked to productivist agricultural interests who are able to dictate, shape and influence productivist farming policies and agricultural development pathways without much challenge from other groups in society.

These debates show that the conceptualisation of productivism shares many underlying assumptions with other debates on 'isms' (see Chapters 3 and 4). It is no surprise, therefore, that many authors have closely associated agricultural productivism with modernism (and, concurrently, post-productivism with post-modernism; see Chapter 6) (Cloke and Goodwin, 1992; Halfacree, 1997b), in particular concerning productivism as a (possibly) distinctly definable *epoch* with a clear beginning and end. As Chapter 3 highlighted, modernism includes many processes that have also been identified in this chapter as possible components of agricultural productivism, namely rationalisation, functionalism, disembedding mechanisms, a belief in progress and order, individualism, liberalism and consumerism. Further, Bradbury (1976) and Berman (1992) also emphasised processes of industrialisation, scientification, commodification, technologification and globalisation as key components of the modernist project, not dissimilar to many 'dimensions' of productivism discussed above. We also find close associations between conceptualisations of agricultural productivism and notions of Fordist production (Goodwin and Painter, 1996; Potter and Burney, 2002), as many above-mentioned processes and farming techniques show elements of Fordism, in particular through the production of high quantities of relatively uniform foods for mass consumption. It is important to reiterate that, as Chapter 3 highlighted, the notion of 'productivism' is by no means restricted to the agricultural/rural arena. In the context of debates on the transition to post-Fordism, Lipietz (1992) referred to notions of liberal *productivism* characterised by increases in casual work, subcontracting, the relocation of production into Third World countries, mass production with sharp increases in productivity, and a relative stability in the profitability of firms – elements which also characterise conceptualisations of agricultural productivism. The notion of productivist agriculture can, therefore, not be disassociated from parallel debates and developments in the social sciences that, inevitably, have influenced the way agricultural change has been conceptualised – a conceptualisation that, as Chapter 7 will argue, suffers from the same pitfalls and underlying fallacies as other transitional debates outlined in Part 1.

I also highlighted in this chapter that underlying problems associated with conceptualising agricultural productivism may stem from relatively rigid approaches used to conceptualise productivism – in particular issues about the time scale of the productivist 'era', the UK-centric nature of the concept, the predominance of structuralist approaches, and the problem of presentist interpretations of productivism. In particular, this chapter highlighted that, although some literature is available that has discussed the p/pp transition outside the UK context, few attempts have been made to challenge, and indeed refute, assumptions underlying the conceptualisation of productivism outside the UK. As with conceptualisations of post-productivism (see Chapter 6), the possibility of the transposition of the notion of productivism beyond the situation of advanced economies, therefore, still needs to be discussed – echoing Said's (1983) cautionary note about the 'exportability' of conceptual notions based on the experience of advanced economies. We will return to these issues in Chapter 7, with specific emphasis on the four fallacies inherent in many transitional models outlined in Part 1.

Chapter 6

Post-productivist agriculture

6.1 Introduction

Chapter 6 will discuss conceptualisations of *post-productivist* agriculture which has, it is argued, superseded productivism. Section 6.2 will assess the dimensions of post-productivism and will argue that it has been largely conceptualised as the 'mirror-image' of productivism, with associated conceptual and empirical problems. Section 6.3 will then discuss each of these dimensions in detail, and will highlight debates surrounding their relative importance in conceptualisations of post-productivist agriculture. The issue of *transition* underlying the p/pp model is discussed in Section 6.4, while Section 6.5 provides concluding comments. Both Chapters 5 and 6 form the basis for the deconstruction of binary and linear assumptions underlying the p/pp transition model in Chapter 7.

6.2 Conceptualising post-productivism

Many commentators have suggested that by the mid-1980s, the logic, rationale and morality of productivism were increasingly questioned on the basis of ideological, environmental, economic and structural problems (Whitby and Lowe, 1994), leading some to argue that the productivist ideology was "in disarray" and that *post-productivist*[18] tendencies were increasingly taking hold (Marsden *et al.*, 1993: 68). From a European perspective Potter and Tilzey (2005: 16) defined 'post-productivism' "as an empirical description of a mode of production (and its corresponding policy supports and discourses) which seeks to move away from the traditional model of high input, high output agricultural systems and an agri-centric model of farm households and the rural economy that has dominated the CAP over the last forty years". Lang and Heasman (2004: 21) further argued that "the sustainability and profitability of the productionist [sic] paradigm is now far from certain", while Marsden (2003: 9) suggested that the 1990s have "witnessed the growth of a new 'post-productivist' dynamic which has challenged the relevance of industrial agricultural production". Delgado *et al.* (2003: 22) similarly suggested that "society is currently questioning the rationale of the Fordist model of agriculture, oriented completely around production", while Burton (2004: 211) posited that "at the moment a paradigm change in agriculture is occurring in the so-called move away from 'productivism' towards 'post-productivism'". As a result, Marsden (2003: 100) argued that "the period of progressive productivism has given way to what has been termed a 'post-productivist countryside'", and that "the 'post-productionist' [sic] conditions [are] now prevailing, and, moreover, likely to continue" (Marsden, 2003: 117). Evans *et al.* (2002), thus, suggested that it has become 'fashionable' to conceptualise recent

---

[18] Similar to the notion of 'productivism' that was used in a multitude of different non-agricultural contexts (see Part 1), 'post-productivism' has also not been restricted to agriculture or forestry. Fitzpatrick (2004), for example, in his investigation of a 'post-productivist' future for social democracy, used the term in the context of politics, while Goodin (2001) applied the notion of 'post-productivism' to work and welfare issues.

shifts in agriculture as a 'post-productivist' transition from a previously 'productivist' agriculture, and that post-productivism "appears to have appeal for academics because it encompasses both micro and macro changes and pulls together a wide range of rural issues" (Evans *et al.*, 2002: 314). McCarthy (2005: 773), therefore, emphasised that "the empirical and analytical validity of 'postproductivism' [sic] continues to be a major topic of debate", highlighted by the continuing use of the term in many current publications (e.g. Clark, 2006; Daugstad *et al.*, 2006; Mather *et al.*, 2006).

The notion of post-productivism has now also begun to be used in extra-UK and extra-European contexts. Holmes (2002: 364; emphasis added), for example, although critical of the applicability of the notion of a p/pp transition in an Australian context (see Chapter 7), suggested that in the context of Australian rangelands, "lack of success in pursuit of productivist goals enhances capability in satisfying *post-productivist* values" and that "it is on the most marginal ... pastoral lands that *post-productivist* amenity values can most readily displace a flimsy mode of pastoral occupance". For some of the dimensions of post-productivism (see below), therefore, Holmes (2002: 379) argued that "evidence presented on the Australian rangelands is strongly supportive of the validity and utility of the post-productive transition concept", and that "no other conceptualization can so effectively provide an understanding of current directions in Australia's rangelands, whether scrutinized from an internal perspective or placed in a wider global context".

Yet, contrary to the clearly defined dimensions of productivism, Lowe *et al.*'s (1993) and Ward's (1993) earlier caution that there is a lack of a clear definition of post-productivism still holds true at the beginning of the 21st century. This is because there continues to be a lack of consensus as to whether productivist agriculture has indeed been superseded by post-productivism (see Chapter 7). Thus, one of the reasons for the ongoing interest in 'post-productivism' is that there continues to be fundamental theoretical, conceptual and empirical debate about the nature, pace and even existence of the p/pp transition (Mather *et al.*, 2006). While scholars such as Ilbery and Bowler (1998: 135) argued that "there can be little doubt that agriculture in most developed market economies has entered a post-productivist period", or Evans *et al.* (2002: 314) suggesting that "some might argue that the post-productivist descriptor is increasingly matched by empirical reality" (see also Ward, 1993) and Marsden (1998a: 28) referring to "post-productivist conditions now prevailing", other authors are more cautious. For example, I suggested that notions of productivism and post-productivism have been useful in highlighting existing spatial differences in contemporary agricultural landscapes (Wilson, 2001, 2002), but I also argued that the notion of post-productivism is still highly questionable due to its implied temporal linearity and binary assumptions (see also Argent, 2002; Wilson and Rigg, 2003). Other critics (e.g. Evans *et al.*, 2002) go even further and largely reject the notion of post-productivism as a 'myth', arguing that there is very little evidence of post-productivism anywhere in the world.

In the absence of a commonly agreed definition, previous work has suggested that post-productivism can be conceptualised as the 'mirror image' of the seven inter-related dimensions of productivism discussed in Chapter 5. As Argent (2002: 99) highlighted, "as with all post-prefixed terms [see Part 1], post-productivism is a relational concept, drawing much of its meaning from its 'constitutive outside, productivism' ... the two become polar opposites in order for both to survive as meaningfully contrasting higher-order concepts". Table 6.1 highlights that discussions on the transition towards post-productivism have focused largely on the policy dimension, changing ideologies in society, the emergence of post-productivist rural governance, changes to food regimes and agro-commodity chains, agricultural production, farming techniques and changing environmental impacts (Wilson, 2001; but see also Mather *et al.*, 2006). As with our discussion of productivism, this highlights that a multitude of different characteristics need to be considered to understand post-productivism, and that a focus on just one dimension would be insufficient to highlight the multi-faceted processes underlying the postulated post-productivist transition.

Table 6.1. Dimensions of post-productivism (Source: after Wilson, 2001; Evans *et al.*, 2002; Mather *et al.*, 2006)

Agricultural policies:
- Reduced financial state support; move away from state-sustained production model (Marsden, 1999b, 2003)
- Demise of state-supported model of agricultural development which placed overriding priority on production of food (Lowe *et al.*, 1993)
- New forms of rural governance (Marsden *et al.*, 1993; Mather *et al.*, 2006)
- Enhancement of local planning controls (Munton, 1995; Halfacree and Boyle, 1998)
- Encouragement of environmentally friendly farming; greening of agricultural policy (Potter, 1998; Marsden, 2003)
- Increased regulation of agricultural practices through voluntary agri-environmental policies (Ward, 1993; Hart and Wilson, 1998)
- Move away from price guarantees; decoupling (Pretty, 1998; Potter and Burney, 2002)
- Increasing planning regulations for agriculture (Lowe *et al.*, 1993; Marsden *et al.*, 1993)
- Loss of security of property rights (Whatmore *et al.*, 1990; Marsden *et al.*, 1993)

Ideology:
- Loss of central position of agriculture in society (Ward, 1993; Burton and Wilson, 2006)
- Move away from agricultural fundamentalism and agricultural exceptionalism (Marsden *et al.*, 1993; Winter, 1996)
- Loss of ideological and economic sense of security; farmers branded as destroyers of countryside (Body, 1982; Potter, 1998)
- Changing attitudes of public towards agriculture; agriculture as villain (Harper, 1993; Marsden *et al.*, 1993)
- Changing social/media representations of the rural (McHenry, 1996; Winter, 1996)
- Changing notions of countryside idyll; contested countrysides (Hoggart *et al.*, 1995; Halfacree, 1997b)
- Main threats to countryside perceived to be agriculture itself (Pratt, 1996; Marsden, 1999b)
- Loss of security of property rights (Marsden *et al.*, 1993)
- 'Rural' increasingly separated from agriculture; new social representations of the rural (Cloke and Goodwin, 1992; Marsden, 2003)

Governance of rural spaces:
- Agricultural policy community widened; inclusion of formerly marginal actors at the core of policy-making process (Cox *et al.*, 1988; Hart and Wilson, 1998)
- Weakening of corporate relationship between agriculture ministries and farming lobby (Lowe *et al.*, 1993; Winter, 1996)
- Changing power structures in agricultural lobby (Winter, 1996)
- Counterurbanisation and social and economic restructuring of the countryside (Halfacree, 1997b; Halfacree and Boyle, 1998)
- Increasing demands on rural spaces by reconstituted 'urban' capitals through new manufacturing and service industries (Murdoch and Marsden, 1994; Marsden, 2003)

Food regimes and agro-commodity chains:
- Challenge to the Atlanticist Food Order (Goodman and Watts, 1997; Goodman, 2004)
- Post-Fordist agricultural regime; non-standardised demand for goods and services; vertically disaggregated production (Lowe *et al.*, 1993; Potter and Tilzey, 2005)
- Critique of protectionism; free market liberalisation; free trade (Potter and Burney, 2002; Potter and Tilzey, 2005)
- Increased market uncertainty (Marsden *et al.*, 1993)
- Changing consumer behaviour (Winter, 1996; Lang and Heasman, 2004)

Agricultural production:
- Critique of industrialisation, commercialisation and commoditisation of agriculture; critique of corporate involvement (Lowe *et al.*, 1993; Lang and Heasman, 2004)
- Less emphasis on securing national self-sufficiency for agricultural commodities (Potter, 1998)
- Extensification (Ilbery and Bowler, 1998)

- Dispersion (Ilbery and Bowler, 1998)
- Diversification; pluriactivity (Evans and Ilbery, 1993; Shucksmith, 1993)
- Farmers wishing to leave agricultural 'treadmill' (Ward, 1993)
- Move from agricultural production to consumption of countryside (Marsden *et al.*, 1993; Marsden, 2003)

Farming techniques:
- Reduced intensity of farming (Munton *et al.*, 1990; Potter, 1998)
- Reduced use or total abandonment of biochemical inputs (Ward, 1995; Pretty, 2002)
- Shift towards sustainable agriculture (Pretty, 1995, 1998; Marsden, 2003)
- Replacing physical inputs on farms with knowledge inputs (Winter, 1997; Ward *et al.*, 1998)

Environmental impacts:
- Move towards environmental conservation on farms; critique of notion of production maximisation (Wilson, 1996; Morris and Winter, 1999)
- Re-establishment of lost or damaged habitats (Mannion, 1995; Wilson and Memon, 2005)

Although at first glance the use of mirror image dimensions may neatly support conceptualisations of a transition from productivism to post-productivism, this approach masks the continuing underlying problems linked to specific approaches used to conceptualise productivism outlined in Chapter 5. As Mather *et al.* (2006) rightly criticised, 'importing' the seven dimensions highlighted in Chapter 5 to understand post-productivism also means continuing with the argument that both productivism and post-productivism are *clearly definable* 'epochs' with clear start and end points.[19] Indeed, if post-productivism is the mirror image of productivism, then this implies an underlying assumption of *temporal linearity* in the transition towards post-productivism. Thus, if the mid-1980s is seen by many as the end point of productivism, it is similarly seen to mark the beginning of post-productivism (Marsden *et al.*, 1993; Ilbery and Bowler, 1998). It also means that the UK-centrism underpinning the conceptualisation of productivism will continue to be a problem when attempting to broaden discussions of post-productivism beyond the UK framework. Nonetheless, most research that has discussed the applicability of the notion of productivism outside the UK has also discussed the applicability of the concept of *post-productivist* agriculture. However, and as Chapter 7 will discuss, most commentators concede that not all dimensions of post-productivism resonate well with patterns and processes of agricultural change observed outside the UK (e.g. Argent, 2002; Jay, 2004). Beyond this, and as with conceptualisations of productivism, structuralist approaches continue to underpin our understanding of post-productivism. As Mather *et al.* (2006) reiterated, problems associated with political economy approaches that underlie most of the concepts of productivism (see Chapter 5) will, therefore, also be apparent in our discussion of the different dimensions of post-productivism below (see also Chapter 7).

Arguably the most severe problem with the conceptualisation of post-productivism as the mirror image of productivism is that 'productivism' has only been conceptualised from a presentist 'post-productivist' vantage point. Indeed, only the concept of 'post-productivism' enabled commentators to *retrospectively* define 'productivism'. As with other conceptualisations of transitions from 'isms' to 'post-isms' discussed in Part 1, the prefix 'post' in post-productivism may, therefore, merely denote something which comes after another thing, and that does *not necessarily mean its opposite* – in other words post-productivism has only been defined in the 'negative' as *what it is not*, rather than as *what it may be*. As a result, Part 3 of this book will argue that a *normative* notion of

---

[19] Based on evidence from Australia which has witnessed different pathways in the transition to post-productivism than Europe, Holmes (2002) suggested an eighth dimension of 'changing amenity values' of the countryside. I argue that for ease of conceptualisation amenity-related changes are contained within both the 'ideology' and 'governance' dimensions (see below).

'multifunctionality' rooted in a revised productivist/non-productivist model better encapsulates processes of agricultural change and contemporary agricultural practices and norms.

6.3 The seven dimensions of post-productivist agriculture

This section will investigate the individual components of what the literature has characterised as 'post-productivist agriculture'.[20] The aim here is to provide insights into the debates on what the literature suggests are the 'ingredients' of post-productivism, as a basis for critical analysis of the implied linearity underlying the p/pp transition discussed in Chapter 7.

As with conceptualisations of productivism, changes in **policies** are one of the most commonly mentioned dimensions in conceptualisations of post-productivism, mainly because policy documents provide one of the more 'tangible' and easily accessible sets of information necessary for analysis (Baldock and Lowe, 1996; Whitby, 1996). As a result, policy change during the mid-1980s (in a UK and European context) is usually seen as a key ingredient of the shift towards post-productivism (Marsden *et al.*, 1993; Ilbery and Bowler, 1998). Marsden (1995: 289) argued that post-productivism implies a policy shift "from encouragement of food and farm production to one that also attempts to deliver other environmental and consumer-based benefits". Evans *et al.* (2002: 314) similarly argued that "post-productivism implies that agricultural policies have moved beyond a principal emphasis upon sustaining and increasing levels of production", and that "evidence for the emergence of post-productivism might be reasonably anticipated to include a strong shift in agricultural policy away from production support towards restraints on productivism" (Evans *et al.*, 2002: 323; see also Marsden, 2003). Burton (2004: 211) also suggested that "there can be little doubt that policy changes are directing farmers towards … new 'post-productivist' roles … and away from the role of intensive agricultural producers". This was reiterated by Mather *et al.* (2006) who, in their defence of the notion of 'post-productivism', suggested that changes in policy climate occurred in the second half of the 1980s and early 1990s that could be interpreted as an *abrupt* and *radical* transition towards post-productivism.

Post-productivist environmental discourses are seen to permeate policy documents from the mid-1980s, best highlighted in EU and national documents on the future of agriculture and rural society (e.g. CEC, 1988, 1996; House of Lords, 1990). Such discourses have particularly emphasised notions of *extensification, environmental protection* and gradual *withdrawal of state support* for farming (Wilson, 2001). In the UK, these changes have been particularly highlighted through government documentation such as the 1990 White Paper *This Common Inheritance* (DoE *et al.*, 1990), the 1994 publication *Sustainable Development: the UK strategy* (DoE *et al.*, 1994), the 1995 White Paper *Rural England* (DoE, 1995), or the *Sustainable Development* report (House of Lords, 1995). Similar policy milestones have been identified in the context of other advanced economies, for example the 1974 'Green Paper on Rural Policy' in Australia (Argent, 2002). Compared to the 'productivist' 1970s and 1980s, these changes that emphasise the importance of non-agricultural activities have been interpreted as epitomising a 'post-productivist' sustainable development strategy that also includes environmental conservation in the countryside as a crucial component (Marsden, 2003; Mather *et al.*, 2006).

In policy terms, therefore, post-productivism is generally characterised by *reduced state subsidies*, indicative of a move away from state-sustained production models, and signalling a gradual loss of faith in the ability of the state to influence agricultural regeneration (Marsden, 1999a, 2003). This is further seen as closely associated with the demise of the

---

[20] As in Chapter 5, I will avoid referring to the 'post-productivist era', as this would imply that a transition towards post-productivism *has* taken place – a proposition I will contest in Chapter 7.

productivist state-supported model of agricultural development which, as Chapter 5 highlighted, has placed overriding priority on the production of food (Lowe *et al.*, 1993). Further, post-productivist agriculture has been characterised by the loss of security of property rights (e.g. with reference to recent debates about access to the countryside in the UK or the tightening of on-farm pollution regulations across the EU), increasing planning regulations for agriculture (Lowe *et al.*, 1993; Marsden *et al.*, 1993) and the further blurring of the divide between 'public' and 'private' use of rural resources (Whatmore *et al.*, 1990; Marsden *et al.*, 1993).

Yet, many authors have also highlighted that for most advanced economies state retreat from financial regulation of agriculture has, in turn, been accompanied by *increased regulation* of agricultural practices through voluntary AEPs, encouraging farmers to farm in environmentally friendly ways (Wilson, 1997a; Buller *et al.*, 2000), and the enhancement of local planning controls (Munton, 1995; Halfacree and Boyle, 1998). Recent research has, for example, highlighted that nearly 30% of all farmland in the EU is now entered into some form of agri-environmental scheme (shallow or deep), with some European regions showing high agri-environmental scheme uptakes (90-100%) (Buller *et al.*, 2000; Wilson and Hart, 2000). As Munton (1995) and Potter (1998) highlighted, it is one of the paradoxes of contemporary rural change in North America, Australia, New Zealand and most of Europe that just when commodity markets are extended and liberated, the processes of production and consumption are increasingly regulated. However, "it has been questioned whether these new types of policies should be seen as true indicators of a shift towards post-productivism" (Wilson, 2001: 84). As Chapter 7 will highlight, what some have described as the 'greening of agricultural policies' (e.g. Harper, 1993) has been criticised by others as mere 'incrementalism' instead of 'reform', indicative of policies aimed at farm income support rather than environmental conservation (e.g. Baldock *et al.*, 1990; Pretty, 1998).

In a European context, many have argued that policy changes to the CAP – such as the introduction of milk quotas (Ward, 1993), set-aside policies (Cloke and Goodwin, 1992), the 'Alternative Land Use and Rural Economy' package of the late 1980s (Cloke and Little 1990), the decoupling of agricultural from 'green' subsidies (Potter, 1998; Potter and Burney, 2002), the LEADER and former Objective 5b programmes (Ray, 1998, 2000), or the Rural Development Regulation (RDR) as part of Agenda 2000 (Marsden, 2003; Lang and Heasman, 2004) – show signs of a shift towards post-productivism (e.g. Buckwell *et al.*, 1998; Buller *et al.*, 2000). Thus, Clark (2003: 18) argued that "the ratification of the RDR … marked … Member State's acceptance that agriculture's exclusive emphasis on commodity production was no longer tenable". In particular, the move away from price guarantees, the 'decoupling' of payments for environmental protection, and notions of 'cross-compliance' embedded in recent CAP reforms, forcing farmers to provide minimum environmental standards before being eligible to receive production subsidies (e.g. through the newly introduced Single Farm Payments in the EU), have all been seen to embody post-productivism (Pretty, 2002; Marsden, 2003). Shucksmith (1993: 466), therefore, argued that "the [EU] Commission and member states have [developed] post-productivist agricultural policy instruments", while Clark (2003: 28) argued that in the EU "the argument for … post-productivist agriculture is buttressed by incremental change in the overarching policy framework of the CAP, away from a purely production-based focus to address wider social, environmental and rural development objectives".

The importance of policy changes in conceptualisations of post-productivism highlights why many have seen the post-productivist transition as largely *orchestrated* by policy-makers and the policy environment (Marsden *et al.*, 1993; Ilbery and Bowler, 1998). Similar to other societal transitions that have been heavily influenced and shaped by policy (in particular the demographic and post-colonial transitions and to some extent technological transition; see Chapters 3 and 4), there is no doubt that policy, as an often exogenous and top-down process influencing rural and agricultural communities, has had a large influence in shaping the nature and processes associated with a possible transition towards post-productivism. However, the overemphasis in much of the literature on policy indicators of change can, yet again, be interpreted as a legacy of the predominance of structuralist interpretations of agricultural transitions that have dominated Anglo-American rural research over the past three decades (Wilson, 2001; Evans *et al.*, 2002). As Holmes (2002: 380) highlighted for Australia, "too much emphasis is placed on agricultural policies and particularly on their assumed role as a driving force towards a transition … certainly, the CAP has been highly influential in shaping … agricultural directions and rural fortunes in Western Europe, but to an extent not matched elsewhere, not even in the US nor in Japan, where any post-productivist transitions appear to be less strongly influenced by changes in agricultural policies". As a result, Chapter 7 will show that the policy dimension of the postulated p/pp transition has been particularly heavily criticised for its simplistic assumptions of unilinear change.

With regard to *ideology*, post-productivism is often characterised by the loss of the central position of agriculture in society (Ward, 1993), strongly associated with the loss of the privileged political place of agriculture in the (mainly European) liberal 'new left' governments such as the UK's New Labour Party (Wilson and Wilson, 2001; Marsden, 2003). As Hoggart and Paniagua (2001: 50) argued, "there is little doubt that rural economies have become less dependent on farming and associated service activities", indicating that many authors see changes in ideology and civil society as some of the most tangible agricultural and rural changes that have taken place over the past few decades. In these views, post-productivism is seen as a move away from agricultural fundamentalism and exceptionalism that characterised productivism (Winter, 1996; Argent, 2002), with the associated loss of the ideological and economic sense of security for farmers (Potter, 1998). Indeed, in post-productivist rhetoric farmers are branded as 'destroyers' of the countryside rather than as 'stewards of the land' (Marsden, 1999a, 2003). In the UK, this has been closely linked to changing public attitudes portraying agriculture as a 'villain' (mainly in environmental and health terms) (Marsden *et al.*, 1993; Marsden, 2003), accompanied by changing media representations of the 'rural' (Harrison *et al.*, 1986; McHenry, 1996), and changes to the notion of the countryside idyll through new 'contested countrysides' (Marsden *et al.*, 1993; Hoggart *et al.*, 1995). Marsden (2003: 93), therefore, argued that "the farm becomes a criminalised space, a place where the 'dirty business' of intensive agriculture occurs". The establishment of the UK Department for the Environment, Food and Rural Affairs (DEFRA) that replaced the former, arguably more productivist, Ministry of Agriculture, Fisheries and Food (MAFF), is often seen as a relatively clear indicator of a shift towards post-productivism (Clark, 2006; Mather *et al.*, 2006). This reform has meant that, for the first time, there is now no UK central government department with the word 'agriculture' in its title (Lowe *et al.*, 2002).

Evidence on changing ideologies is also seen as coming from increasing urban-rural migration with resulting challenges to conservative and traditional rural values, and increasing criticism of agricultural impacts on the landscape and on public health. The latter has been particularly obvious in the wake of the recent bovine spongiform encephalopathy (BSE) and foot-and-mouth epidemics (Hinchcliffe, 2001; Scott *et al.*, 2004). Marsden (2003) particularly highlighted the importance of the post-productivist 'consumption' of the countryside by non-rural and non-agricultural stakeholder groups for recreation and leisure purposes, leading to conflicts over the most 'appropriate' use of the countryside between farming and non-farming groups (Winter, 1996) – conflicts exacerbated in the UK by the above-mentioned loss of security of property rights that have reduced farmers' opportunities to prevent public access to their land. Indeed, some argue that a clear tendency "in the post-productivist paradigm is public external access to the farm resource, particularly the aesthetic character of the agricultural landscape, and the regulation and restrictive planning of this for the broader symbolic good" (Marsden, 2003: 93). Thus, in the post-productivist countryside nature is seen to become *commodified* as a consumption good, to be exploited not by agricultural pursuits but by urban populations in search of the rural idyll. In these views, the countryside is increasingly treated as the 'edge' of the suburban frontier, with the welfare and degree of sustainability of the countryside increasingly tied to the 'outside' urban world of work and play, with Marsden (2003) estimating that in the UK 60% of the population visit the countryside at least once a year (1.2 billion day visits/year). Such figures emphasise the scale of non-agricultural activities now taking place in the countryside of most advanced economies (Pretty, 2005). In the post-productivist view, therefore, the main threats to the countryside are perceived to be agriculture itself, and less 'other' non-agricultural activities (Pratt, 1996; Marsden, 2003). The result has been that – at least in the UK context – conceptualisations of the 'rural' and the 'countryside' are becoming increasingly separated from conceptualisations of 'agriculture' and 'farming'[21] (Murdoch and Pratt, 1993; Pratt, 1996). Yet, such debates have not been restricted to the situation in the UK. For the Australian context, for example, Holmes (2002: 378) argued that "the entry of a diverse array of amenity-oriented modes of land use is dissolving the productivist agricultural/pastoral hegemony, requiring continuing resource revaluations through differentiated land markets, driven by urban-based interests with markedly different value-orientations, goals and priorities".

Debates on post-productivist ideological positions are arguably less clear-cut than the above discussion of the policy dimension, although it is evident that changing ideologies have been an important argument in conceptualisations of post-productivism. In Chapter 7 we will see that the question of whether ideologies have become more 'post-productivist' continues to be the focus of heated debates. As with conceptualisations of productivism, the *UK-centric approach* underpinning the p/pp transition on the basis of ideological change is obvious and, as a result, transposition of post-productivist ideological patterns and processes to other geographical contexts is difficult.

In contrast, little work is available on the reconfiguration of actor spaces and ***governance*** as a possible indicator of the transition towards post-productivism (Little, 2001; Wilson, 2004). Nonetheless, an argument can be made that post-productivism is characterised by the gradual breaking down of the traditional 'corporate relationship' between agriculture ministries and powerful farmers' unions to allow formerly politically marginal actors (such as environmental groups or local grassroots organisations) into the decision-making and policy formulation networks (Cox and Winter, 1987; Hart and Wilson, 1998). The recognition that the security of the national food supply has come at considerable environmental and social cost led to a "public challenge of the hegemony of 'agricultural corporatism' by a vocal coalition of interest groups" (Argent, 2002: 100). This suggests weakening power structures within the agricultural lobby (Winter, 1996), partly linked to the introduction over the past decades of neo-liberal principles in public administration in many developed countries that have resulted in a large increase in the range of actors, organisations and institutions involved in public service delivery *outside* of the core executive of state-related actors and agencies. Thus, Halfacree (1997b: 72) argued that "post-productivism may signal a search for a new way of understanding and structuring the countryside. A space in the imagination is opening, whereby non-agricultural interests and actors are given an opening to strive to create a rurality in their image". Such changes in governance have involved partial diminution of 'pure' state activity through the rise of public-private partnerships, privatisation processes, changes in modes of regulation, emergence of new organisations or altered roles for established ones, or new access routes for stakeholder groups into policy-making decision-making frameworks (Hoggart and Paniagua, 2001). This, in turn, has facilitated the injection of 'green' ideas into the agricultural policy-making process by newly empowered actors such as environmental NGOs (Lowe *et al.*, 1986; Hart and Wilson, 1998). In the UK, this has meant the inclusion of *new* actors and stakeholder groups into policy-making processes that also include statutory bodies, reflecting the newly

---

[21] This is maybe best highlighted by a statement of UK Prime Minister Tony Blair in February 2000 that "farming in the UK may be in crisis, but the countryside is not".

found political power of several environmental NGOs with large memberships (e.g. in the UK the National Trust, the Royal Society for the Protection of Birds, Friends of the Earth) (Wilson and Hart, 2001). This shift can be associated with changing trajectories of governance with empowerment of local stakeholders, and, ultimately, the erosion of the state as the sole deviser and shaper of policies and decisions affecting rural communities (Bryant and Wilson, 1998; Jessop, 2002).

Of equal importance, it is argued, has been the social and economic restructuring of the countryside and the reconstitution of actor spaces through urban-rural migration (counterubanisation) in many advanced economies (and to some extent also in the developing world; see Wilson and Rigg, 2003) that has brought mainly middle class and conservative migrants into rural areas for lifestyle, environmental and security reasons (Cloke and Goodwin, 1992; Lowe *et al.*, 1993). These changes to the 'traditional' countryside since the 1980s have led some to argue that "migration of people to the more rural areas of the developed world … forms perhaps the central dynamic in the creation of any post-productivist countryside" (Halfacree and Boyle, 1998: 9), and that "farmers become the bystanders to the powerful coalitions of middle-class fractions and non-agricultural entrepreneurs who have gained more than simply a physical foothold in rural space" (Marsden, 2003: 110-111). In these debates it is, therefore, argued that counterurbanisation brings with it social and economic restructuring of rural governance (Cloke and Goodwin, 1992; Halfacree, 1997b), resulting in new interests and actors coming on the scene in an attempt to create a rurality in their (usually urban) image of the rural that now also permeates the 'deep' countryside in many advanced economies (Wilson, 2001). Halfacree (1999) argued that the result of this process is that traditional productivism is increasingly moulded into middle class post-productivist space underlain by the rural idyll, further exacerbated by increasing demands placed on rural spaces by reconstituted urban capitals through new manufacturing and service industries (Murdoch and Marsden, 1994). Hoggart and Paniagua (2001: 53), therefore, suggested that "in-migrants are seen to introduce an idealised rural lifestyle that penetrates their actions on arrival". Farmers, therefore, face new challenges to their authority by these new neighbours over such matters as on-farm pollution, access disputes and environmental management practices (Ward *et al.*, 1995).

Such changing post-productivist governance structures are seen to be closely associated with changing attitudes of newly empowered stakeholders vis-à-vis their own position in rural communities (Jessop, 1998; MacKinnon, 2000), and with changing attitudes towards destructive environmental management practices on farmland at grassroots level (Wilson, 2002). As an extension of Offe's (1985) concept of 'new social movements' challenging the boundaries of institutional politics, this conceptualisation can also be set in the wider context of new 'post-productivist' social movements in post-industrial societies (e.g. Habermas, 1981; Beck, 1992), or, in Williamson's (1996) terms, 'horizontal non-market co-ordination' between various state and non-state actor groups. Post-productivist 'indicators' of such movements would include a mission towards environmental sustainability of rural systems (Cocklin, 1995; Bowler *et al.*, 2002), with emergence of new 'innovative' institutions and practices within civil society, as well as expanding opportunities for interaction between stakeholder groups associated with further democratisation of civil society and with increasing capacity to hold the state accountable for its actions (Jessop, 1998, 2002). Critically, these expanding opportunities for stakeholder interaction should also facilitate participation of hitherto marginalised groups in rural society, in particular women, immigrants and native people (Cloke and Little, 1997; Rhodes, 1997). It would also mean full access for grassroots actors to knowledge networks available at state and intermediate levels and, in turn, acknowledgement of local grassroots knowledge by state-level actors (Jordan, 1999). Such governance would also have to entail evidence of the state's retreat from the position of provider of support and sole policy-maker to one of 'coordinator', 'manager' and 'facilitator' of the many stakeholders embedded in new forms of rural governance (Little, 2001). Marsden (2003: 73), therefore, described such new structures as "a consequential search and contestation for new forms of governance which do not rely upon heavy state intervention and rule by a bureaucratic elite, but rather seek (selectively) more public participation, often mobilised through local and regional (rather than national) systems of governance, partnerships between levels and interests, and greater reliance on private and public 'entrepreneurship' and innovation".

These debates are closely associated with suggestions that a *re-regionalisation* of governance of rural areas may be occurring (e.g. Ray, 2000; Marsden, 2003), filling a political vacuum in many advanced economies left after the gradual retreat of the state from local/regional agricultural governance, and characterised by deregulation dynamics, re-regulation through private sector initiatives, and increased complexity and divergence within parastatal agricultural institutions (Halfacree and Boyle, 1998). In other words, post-productivist governance could be seen to be contributing towards reduction of local stakeholder alienation, conflict avoidance, support building for public-private policy-making structures, tapping local knowledge, contributing to local community education, general enhancement of democratic processes by empowering local stakeholders, and increasing government accountability (Offe, 1985; Rhodes, 1997). Yet, the literature highlights that caution is necessary, as it is difficult to automatically equate local community action with 'post-productivism' and state-led action with 'productivism', and past research has shown that often a more nuanced analysis is necessary (Wilson, 2001, 2004). In line with Jordan's (1999) recent conceptualisation of 'multilevel environmental governance', there is a need to acknowledge that many state policies and actions could also be classified as 'post-productivist' (see above), while policy influence emanating from the grassroots level may also, at times, be highly 'productivist' (Holmes, 2002; Wilson, 2002) – issues I will return to in Chapter 7.

Forces related to the possible composition of post-productivist *food regimes and agro-commodity chains* have been more difficult to conceptualise. The problem lies in the question of scale, as no consensus has yet been reached as to whether 'local' is necessarily a virtue in post-productivist agriculture (Wilson, 2001; see also Chapters 7 and 10). While some stress the importance of the re-regionalisation of agro-food chains and associated processes of *vente directe* (e.g. Pretty, 1998), some argue for 'post-Fordist' agricultural regimes that emphasise vertically disaggregated food production on the basis of non-standardised demand for high quality goods and services (Lowe *et al.*, 1993; Lang and Heasman, 2004). Marsden (2003: 197), for example, suggested that "through developing new quality definitions associated with locality/region or speciality and nature, new associational networks can be built ... which involve radically different types of supply chain". Others, meanwhile, stress the new consumption-oriented roles of agriculture operating at various scales (recreation, leisure, environmental conservation) (Marsden *et al.*, 1993). Interestingly, Argent (2002) argued from an Australian vantage point critical of the p/pp transition that the food regimes literature has provided the *key pillar* for conceptualisations of the transition – an assertion that, in my view, is not fully supported by the evidence available in the literature.

What is less contested is that post-productivist actors challenge the Atlanticist Food Order from the 1950s to the 1970s (Goodman and Watts, 1997; Argent, 2002), and call for free market liberalisation and the rapid dismantling of protectionist nation state (and EU) policies (Tangermann, 1996; Potter and Burney, 2002) to create a 'level playing field' for farmers in the global capitalist economy. This highlights that any conceptualisation of post-productivism cannot be divorced from broader debates on globalisation, with its associated positive and negative effects on global agriculture (Pretty, 2002). Potter and Burney (2002), in particular, argued that the parameters of long-term agricultural policy reform were changed fundamentally by the Uruguay Round Agreement on Agricultural Trade (URAA) in 1986 that brought agriculture more fully into the GATT negotiations. This established a clear agenda for the progressive liberalisation of agricultural markets through the WTO negotiation rounds via improved market access, the elimination of export subsidies, and the decoupling of domestic support (Lowe *et al.*, 2002; Marsden, 2003). While the liberalisation of agricultural markets can be interpreted as strengthening productivist tendencies by giving new agricultural producers access to the global market, the (planned) elimination of export subsidies resulting from the URAA and the decoupling of domestic support can be interpreted as engendering post-productivist tendencies (Potter and Tilzey, 2005). These developments have led to increased market uncertainty for some farmers, while others have been quick to grasp new opportunities offered through changing consumer behaviour. This has been particularly true in the wake of animal and food health scares, increasing criticisms of genetically modified crops (GM crops), and high levels of toxic pollution in foods (Pretty, 1998; Scott *et al.*, 2004).

What is now increasingly referred to as the 'alternative food economy' (Venn *et al.*, 2006), therefore, emerges as a new component for conceptualisations of post-productivism, where the postulated *food quality transition* is seen to be as important as the p/pp transition itself (Morris and Young, 2000). Thus, Evans *et al.* (2002: 317) argued that "there has been a rise in consumer concerns about the impact of productivist agriculture on ... food safety". Lang (2004) particularly emphasised that tackling obesity and coronary problems linked to poor diet is now emerging as one of the key concerns in advanced economies, with estimates that one-third of all global premature deaths are now linked to poor diets. Murdoch (2002), therefore, referred to the 'post-modern culture of food' that suggests a rejection of 'modern' food production practices characteristic of productivism, and that is based on increasing consumer demand for specialist foods from 'protected designation of origin' (e.g. French wines and cheeses), farmers' markets[22] (i.e. embodying re-regionalisation of agro-commodity chains), and clearer food labelling highlighting to consumers what specific ingredients are contained in food (Jones and Clark, 2001; Dabbert *et al.*, 2004). According to Bell and Valentine (1997), quality foods now provide enhanced opportunity for consumers to differentiate themselves, so that quality foods become a mark of 'cultural capital'. Health and nutrition, therefore, emerge as new and powerful components of post-productivism, which should also be interpreted as a move towards a more diverse diet away from the nutritional (and arguably productivist) dependence on animal fat. These trends have been further exacerbated by the fact the value of food is increasingly constructed in the *post-farm* parts of post-productivist food networks (Marsden, 2003). Yet, Argent (2002) rightly cautioned that conceptualisation of post-productivism through the emergence of different food regimes may be *tautological*. He suggested that to the extent that the food regimes concept rests upon the thesis that the organisation and configuration of production, consumption and accumulation has (arguably) shifted to a post-Fordist mode (see also Potter and Tilzey, 2005), so the p/pp transition concept has been implicitly based on the problematic binary histories of Fordism/post-Fordism (see also Chapters 4 and 7) – issues I will return to in more detail in Chapter 7.

These changes also have important repercussions for the conceptualisation of 'new' forms of post-productivist **agricultural production**. Marsden (2003: 146), for example, argued that "probably most effort has been placed by researchers in specifying the nature of agriculturally related changes in land development in rural areas over recent years". Actors embracing post-productivist action and thought are, for example, seen as critical of industrialisation and commercialisation of agriculture, critical of corporate involvement (as highlighted in the recent criticisms of multinationals [e.g. Monsanto] in GM crop debates), and wish to leave the 'agricultural treadmill' (Lowe *et al.*, 1993; Ward, 1993). Further, post-productivism is also seen to be characterised by stakeholder groups placing less emphasis on securing national self-sufficiency for agricultural commodities, largely due to massive food surpluses in most advanced economies (Potter, 1998). Commentators also refer to the issue of agricultural overcapacity as a key driver for the transition to post-productivism, with

---

[22] In the UK, it is estimated that the number of farmers' markets has risen from only 10 in 1997 to about 400 by 2004 (Robinson, 2004).

Argent (2002: 106) arguing for the Australian context that objectives for farmers "have shifted from the need for a quantitative expansion of the farm base and the production of food and fibre to a more qualitative stance in which the farm base must be … self-adjusting and compatible with the local environment" (see also Holmes, 2002).

Ilbery and Bowler (1998), therefore, suggested that *extensification* of agricultural production should be seen as a key indicator of post-productivism, in particular the setting-aside of agricultural land and the reduction of stocking rates on pastures. Many authors also see *diversification* as an important component of post-productivism, in particular activities linked to the production of non-food commodities, *pluriactive* farm household pathways, and on-farm activities linked to the consumption of the countryside such as horse riding, the establishment of golf courses or farm animal zoos (Ilbery and Bowler, 1998; Wilson, 2001). Burton (2004: 195) suggested that policies played an important role in encouraging farmers to move away "from reliance on traditional agriculture and towards becoming shopkeepers, leisure providers, foresters, nature conservers and public custodians of the countryside". The literature has particularly emphasised the importance of farm tourism (tourist accommodation, farm holidays) as a key farm diversification activity indicating post-productivist trends (Marsden, 2003), especially as tourism is linked to reduction in farming intensity (e.g. through the loss of one or more family members working for tourists instead of agricultural commodity production) and *consumption* of the countryside, with associated marketisation and commodification of countryside 'goods' in a 'post-modern' society (Evans and Ilbery, 1993; Macnaughten and Urry, 1998). More recently, interesting new components of post-productivist diversification have been suggested including, for example, the establishment of wind farms which could be seen as a new type of 'farming', as farmers now 'harvest' wind not dissimilar to harvesting a field of wheat (Littlefair, 2005). Similarly, suggestions have been made that the increasing trend towards the planting of biofuel or energy crops (e.g. oilseed rape for energy production) could be seen as a post-productivist diversification activity (Wilson, 2002), although Chapter 7 will highlight that the productivist agenda behind the planting of these crops makes it difficult to argue that these often intensively planted crops lead to a dramatic change in agricultural production practices.

As Chapter 7 will discuss, Ilbery and Bowler's (1998) suggestion that *dispersion* of farm production should also be seen as an indicator of post-productivism has been more contentious (Wilson, 2001; Evans *et al.*, 2002). Ilbery and Bowler suggested that a de-concentration of farm production has taken place in some farming areas in advanced economies, similar to a move back towards environmentally more sustainable mixed farming systems. As Chapter 7 will highlight, evidence for such processes is largely lacking, thereby questioning the utility of this post-productivist 'indicator'. Less debatable has been the importance of *commoditisation* of former agricultural resources (e.g. land, wildlife habitats, barns, cottages) by urban migrants to rural areas (Marsden, 2003). Indeed, first conceptualisations of 'post-productivism' placed great emphasis on this specific indicator (e.g. Kneale *et al.*, 1992; Murdoch and Marsden, 1994), and current research (in the UK at least) continues to stress the importance of commoditisation and 'rural fetishism' of on-farm resources as an important ingredient of post-productivism (Halfacree, 1999; Marsden, 2003). These changes in approaches to farm production have led some commentators to argue that traditional, arguably productivist, agricultural education institutions also had to adapt to the new rhetoric of post-productivist extensification, diversification and 'de-agrarianisation'. Lang (2004), for example, argued that the recent closure of agricultural colleges in the UK, such as Seale Hayne (Devon) and Wye College (Kent), could be seen as a clear indication that there is now less need for such institutions still firmly embedded in the productivist paradigm and based on outdated 'productivist' teaching structures geared towards training farmers to intensify and increase agricultural production.

As Part 3 will discuss in relation to notions of multifunctionality, there is also a deeper conceptual problem linked to conceptualisations of post-productivist *agricultural* production. I will argue that the terms 'agricultural' production and 'post-productivism' are, to some extent at least, anathemas, especially as agricultural production implies some form of food and fibre production. In other words, can an activity that involves *production*, even in its most modest and extensive form, ever be labelled *post-productivist*, i.e. 'beyond production'? This highlights some of the inconsistencies and conceptual problems linked to the p/pp transition model discussed in more detail in Chapter 7.

Conceptualisations of post-productivism have also been closely linked to new types of *farming techniques* (Wilson, 2001). The adoption of new techniques is particularly seen as a reaction to the technological and input-driven 'treadmill' of productivism (Ward, 1993), and is usually characterised by reduced intensity of farming, reduced use or total abandonment of biochemical inputs (Morris and Winter, 1999), and a critique of the notion of production maximisation and its harmful effects on the environment (Potter, 1998). Many commentators argue, therefore, that there is a strong conceptual link between post-productivism and the shift towards environmentally sustainable agricultural practices and *environmentalism* (Altieri and Rosset, 1996; Pretty, 1998), also characterised by the gradual replacement of physical inputs on farms with knowledge inputs (Wilson, 1997c; Winter, 1997). Here, debates on post-productivism intertwine with the long-standing discussions on sustainability, with many commentators suggesting that the shift towards 'sustainable agriculture' should be seen as part-and-parcel of the post-productivist project (Pretty, 1995; Wilson, 2001) – thereby re-emphasising the above mentioned importance of environmental parameters in conceptualisations of post-productivist agriculture.

Examples of post-productivist farming practices include, for example, the shift from conventional to organic farming (Clunies-Ross and Cox 1994; Tovey, 1997). In most EU countries, organic farming now forms an important new environmentally friendly farming technique where even countries such as the UK – initially slow to adopt organic agriculture – now have over 500,000 ha under organic production, with 20% annual increase in organic food consumption (Goodman, 2004), while in countries such as Denmark over one-third of all milk is now organically produced (Marsden, 2003). Changing farming techniques are also embedded in new notions of 'best' agricultural practice that emphasise environmentally friendly forms of production (Wilson and Wilson, 2001), integrated production (Edwards *et al.*, 1993; Morris and Winter 1999), precision farming, and new sustainable management practices implemented by farmer self-help groups such as Landcare Australia (Campbell, 1994; Wilson, 2004). Debates on post-productivism have increasingly suggested that farm practices in advanced economies have *moved away from farming* – in other words that the notion of 'farming' and 'agriculture' (or 'agri-culture' in Pretty's, 2002, words) itself is changing to include more non-farming and non-agricultural activities.

As highlighted repeatedly in this chapter, the ***environmental component*** forms an important aspect of post-productivist conceptualisations. The literature has especially focused on the notion of *sustainable agricultural practices*, compatible with environmental protection, as a key element of post-productivism (Wilson, 2001; Marsden, 2003). While this inclusion of environmental issues is less contested in the literature, there continue to be disagreements (as with debates on 'sustainability' itself) about what 'sustainable agriculture' means (Cocklin and Dibden, 2004). Thus, as part of the conceptualisation of post-productivism as the mirror image of productivism (see above), most commentators suggest that the transition to post-productivism implies a move towards *environmental conservation* on farms, closely associated with a critique of the notion of productivist production maximisation (Wilson, 1996; Morris and Winter, 1999). Some have focused on the re-establishment of lost or damaged habitats (Adams *et al.*, 1994; Mannion, 1995) as a crucial environmental component of post-productivism, while others have placed more emphasis on pollution issues related to agriculture, characterising the post-productivist countryside as one where external inputs are substantially reduced and where more care is given to protection of water resources (surface water and ground water) and soil protection (Pretty, 2002; Wilson and Juntti, 2005). Recent debates on the use of GM crops have added an additional dimension to discussions on post-productivism. On the one hand, GM crops can be interpreted as a truly post-productivist innovation, in particular reducing the need for high applications of external inputs such as pesticides and herbicides (Pretty, 2002). On the other hand, the GM debate has also been used to highlight the continuation of productivist modes of production, as GM crops help farmers to further maximise production in an increasingly globalised farming environment driven by multinational corporations dictating the GM agenda (Marsden, 2003; Herrick, 2005).

Discussions on the environmental transition towards post-productivism have to be embedded with wider transitions in society that also carry strong environmental undertones, in particular the transitions towards *post-Fordism* (Lipietz, 1992; Lowe, 1992) and *environmentalism* (O'Riordan, 1995; Wilson and Bryant, 1997) discussed in Part 1. The environmental component of conceptualisations is also firmly embedded in other 'dimensions' of post-productivism, in particular the above-mentioned greening of agricultural policies and discourses (Harper, 1993), changing agrarian ideologies (Hoggart *et al.*, 1995) and changing agricultural production practices (Potter, 1998; Wilson and Hart, 2001). Chapter 7 will discuss in detail whether and how the environmental dimension of the p/pp transition model stands up to closer critical scrutiny.

6.4 Conceptualising the transition towards post-productivism

Having outlined the dimensions of post-productivism as the mirror image of productivism, I wish to briefly focus attention on conceptualisations of the post-productivist *transition* in this section. That both productivism and post-productivism have been treated in the literature as relatively separate entities with different (and indeed often opposite) characteristics highlights the underlying assumption that a transition from one type of agricultural regime to another *has* taken place (e.g. Mather *et al.*, 2006). Nonetheless, the literature has partly shied away from labelling current agricultural systems as *solely* embedded within post-productivist paradigm (see, in particular, Wilson, 2001; Marsden, 2003; but see also Ilbery and Bowler, 1998, and MacFarlane, 2000), and has instead often referred to the 'post-productivist *transition*' – emphasising the uncertainty associated with the question whether a new agricultural regime has emerged that fully replaces productivism (see Chapter 7). Yet, the explicit focus on transition highlights an underlying assumption of *linearity* as a structured move away from the productivist paradigm, at least in the UK but also with the assumption that these transitional processes apply to other geographical contexts (Wilson and Rigg, 2003). As a result, the p/pp transition model in its most basic form can be shown as a relatively *straight linear* transitional pathway from productivism to post-productivism over a specific time period (Fig. 6.1). Whether this transition takes the form of a straight line, a stepped transitional pathway, or the shape of an s-curve (see Chapter 2), is still a matter of debate and interpretation of what different proponents of the transition have suggested (e.g. Holmes, 2002; Mather *et al.*, 2006). For the sake of simplicity, Figure 6.1 depicts this transition as a straight linear line, although I acknowledge that most of the literature has proposed a more nuanced transitional process (see, in particular, Marsden *et al.*, 1993; Ilbery and Bowler, 1998; Marsden, 2003).

Figure 6.1. shows two key assumptions underlying the conceptualisation of the p/pp transition. First, it suggests the assumption of *directionality*, i.e. that agricultural regimes have moved from one state of organisation to another with little chance of either a move backwards or, indeed, sideways. Second, the figure suggests inherent *value judgements* underlying post-productivist transitional debates, i.e. that the transitional direction is 'positive' or 'pointing upwards' and suggesting an 'improvement' in agriculture. As the discussion in Part 1 highlighted, this shows interlinkages between debates on agricultural transition and other transitional processes that are also often based on binary and directional assumptions about the 'positive' nature of change. In other words, conceptualisations of the transition towards post-productivism have, seemingly, fallen into the same 'trap' as many other transitional debates based on *Cartesian dualistic and binary interpretations* of how the world around us changes (cf. Hallinan, 1997; Altvater, 1998) – a trap facilitated by the relatively simplistic approaches underlying conceptualisations of the p/pp transition linked to the predominance of structuralist and political economy approaches, as well as presentist and UK-centric interpretations of this transition (see Chapter 5).

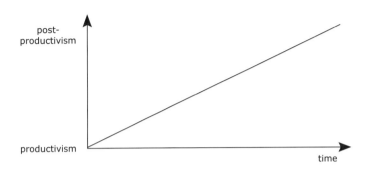

Fig. 6.1. The productivism/post-productivism transition model

As a result, we can see elements of all four transitional *fallacies* highlighted in Part 1 in the dimensions of productivism and post-productivism. Thus, assumptions about temporal linearity, spatial homogeneity, global universality and structural causality underpin most of the debates on the p/pp transition (see Chapter 7). We see particular parallels with debates on *post-Fordism* and its associated linear transitional assumptions, as post-productivist agriculture is seen to have followed similar pathways towards rejection of Fordist modes of production (Potter and Burney, 2002). Further, we also see parallels with debates on *post-modernism*, as post-productivist ideology is seen to include many 'post-modern' elements that range from changing consumer behaviour, critiques of the structured role of agriculture in society, and increasing diversity, flexibility and unpredictability in agricultural decision-making pathways (Halfacree, 1997b). In addition, Chapter 6 has highlighted the close associations of conceptualisations of post-productivism with the *transition towards environmentalism* where green ideas and environmentally sustainable farm development pathways are accorded great importance (Evans *et al.*, 2002). *Technological transition* has also underpinned many arguments in the post-productivist transition, again with similar problems surrounding assumptions of linearity, homogeneity and universality in the acceptance of 'new' post-productivist ideas. The large influence of agricultural extension officials in innovation diffusion, as well as the wealth of literature generated on the back of Green Revolution technology diffusion (e.g. Pretty, 1995; Parayil, 1999), need to be particularly emphasised in this context. Indeed, the post-productivist project is essentially about the *adoption of innovation*, not only concerning adoption of new post-productivist ideas/ideologies, but also about acceptance of more tangible and practical ideas such as new technologies or new farming practices. The conceptual closeness between the post-productivist and technological transition debates explains why so many examples attempting to illustrate technological transitional pathways have come from the agricultural arena (e.g. Rogers, 1995).

It is these explicit linkages between debates on the p/pp transition and the discussion of 'transition theory' presented in Part 1 that will form the basis for our analysis and eventual deconstruction of the p/pp model in Chapter 7. I will argue that the assumption of linear and spatially homogenous patterns and processes underlying the postulated transition towards post-productivism is indeed a *fallacy*, and that, instead, much more temporally, spatially and structurally complex patterns are evident that lead us to question whether a full transition towards post-productivism has taken place.

6.5 Conclusions

Both Chapters 5 and 6 have discussed debates surrounding conceptualisation of productivist and post-productivist agriculture. It has been highlighted that there is a growing body of literature that suggests that a transition towards post-productivism *has taken place*, at least in the context of most advanced economies, although both chapters already hinted at underlying problems surrounding the assertion that a *full* transition has occurred. The chapters also discussed that conceptualisations of the p/pp transition have been based on specific, arguably relatively narrow, approaches based largely on *political economy and structuralist* interpretations of agricultural change that can be interpreted as a legacy of the dominant scientific discourses that permeated agricultural and rural studies during the 1980s, in particular, and the 1990s to a lesser extent. It was highlighted that post-modern and cultural geography perspectives have, so far at least, been largely lacking in discussions of the p/pp transition, and I will argue that this one of the reasons why notions of transition surrounding *agency* have largely been neglected in conceptualisations of post-productivism (see Chapter 7). The chapters also highlighted problems associated with *presentist interpretations* of the p/pp transition, especially regarding the retrospective conceptualisation of productivism from an, arguably, post-productivist vantage point. Similarly, the *UK-centrism* of the p/pp transition debate has been emphasised, with associated problems linked to the 'exportability' of the post-productivist concept to other geographical arenas – the main reason why most of the examples discussed above in the context of the different dimensions of the p/pp transition have come largely from a UK context. The post-productivist project is, therefore, largely a *UK-based concept* that has still not been fully accepted outside of the relatively narrow realm of UK agricultural history and experience. As a result, current debates have only provided partial answers.

Mather *et al.* (2006) emphasised that these issues are partly linked to the relatively *weak conceptualisations* of both productivism and post-productivism where, as Chapter 6 highlighted, post-productivism has been largely seen as the 'mirror-image' of productivism. This, in itself, suggests an inherent weakness in the assumed p/pp transition and lends credence to Cloke and Goodwin's (1992: 324) criticism that, in their eagerness to join in with new developments in theories of rural change, rural researchers during the 1980s and early 1990s may have borrowed inappropriate ideas and may have used "somewhat overarching concepts in a rather cavalier fashion". In Chapter 7, I will, therefore, argue that the *binary* and *dualistic* assumptions underlying the p/pp model have been an 'easy way out' for many rural researchers struggling with explaining the complex changes witnessed in almost all agricultural regimes around the world. I will also argue that rural researchers have fallen into the same trap as many other commentators on transitional processes – be they social or natural scientists – in attempting to provide too simplistic linear and seemingly homogenous solutions to complex transitional systems.

Let us, therefore, turn to a *detailed critique* of the underlying assumptions of the p/pp transition model in Chapter 7, and then develop a new framework for understanding productivism and post-productivism that can act as a *conceptual anchor* for understanding multifunctionality in agricultural transitional pathways and processes (Part 3).

Chapter 7

'Post-productivism' or 'non-productivism'?

7.1 Introduction

Chapters 5 and 6 have analysed current conceptualisations of productivist and post-productivist agriculture. It was shown that a large section of the contemporary literature on agricultural change has assumed, and continues to assume, that a transition from productivism to post-productivism has taken place (the p/pp transition), and that the 'era' of productivism, thought to have lasted from about 1950 to 1985 in most advanced economies, has been superseded by the 'post-productivist era'. However, Chapters 5 and 6 also highlighted that the notion of a transition towards post-productivism has been heavily contested. The aim of Chapter 7 is to look in more detail at these critiques. I will suggest that post-productivism has *not* replaced productivism and that evidence shows that both *co-exist* temporally, spatially and structurally. I will also argue that, when shorn of its association with the notion of transition, the concept of agricultural productivism continues to be useful, but that continuation of the use of the term 'post-productivism' is *untenable*. Instead, I will suggest that *non-productivism* provides a better conceptual term that describes the *true opposite* of 'productivism' in a temporally non-linear, spatially heterogeneous and globally complex way, and that also acknowledges structure-agency inconsistencies in stakeholder adoption/rejection of productivism. I will argue that 'productivism' and 'non-productivism' constitute extreme ends *of a spectrum* of agricultural and rural decision-making pathways within which the concept of 'multifunctionality' can be theoretically anchored.

Section 7.2 will, first, discuss scientific critiques of the p/pp transition model and will show that, from the beginning, the implied temporal linearity and spatial homogeneity underlying the p/pp model have been questioned by many commentators. Based on the discussion in Part 1, Section 7.3 will then deconstruct the notion of a transition towards post-productivism by linking these critiques to the concept of transition. I will particularly focus on the four transitional 'fallacies' that emerged from discussions of other transitions in Chapter 4, with specific reference towards evidence that suggests that the postulated p/pp transition has been spatially heterogeneous, globally complex, characterised by structure-agency inconsistency and, as a result, temporally non-linear. Building on the discussion in Section 7.3, Section 7.4 presents a *non-linear* model of 'productivism' and 'non-productivism' that will form the basis for conceptualisation of the notion of 'multifunctionality' discussed in Part 3. Concluding remarks will be given in Section 7.5.

7.2 Scientific critiques of the productivism/post-productivism transition model

Chapter 6 showed that most of the literature has suggested that by the mid-1980s, the logic, rationale and morality of the productivist era were increasingly questioned by various stakeholder groups on the basis of ideological, environmental, economic and structural problems (Whitby and Lowe, 1994), leading some to argue that the productivist ideology was in 'disarray' (Marsden *et al.*, 1993). As a result, many have uncritically argued that agricultural systems (in developed countries at least) have become post-productivist (Ilbery and Bowler, 1998; Mather *et al.*, 2006). Yet, Chapter 6 also highlighted that, contrary to the clearly defined dimensions of productivist agriculture, Lowe *et al.*'s (1993) and Ward's (1993) earlier suggestion that there is a lack of a clear definition of post-productivism still holds true at the beginning of the 21st century. This is because there continues to be a lack of consensus as to whether the productivist era has been superseded by post-productivism, and whether a *transition* to post-productivism has taken place (Wilson, 2001). Mather *et al.* (2006: 451), therefore, argued that "the debate on post-productivism bears many of the typical characteristics of academic debate. A new concept was introduced; it was stretched to unrealistic extents, and then the call went up for its abandonment". This chapter will show that it is far from clear that a new post-productivist era has replaced productivism and whether it has created a 'new period of stability' or a new 'structured coherence'. Contrary to Mather *et al.*'s (2006) suggestion that 'post-productivism' continues to be a viable concept, I will propose that we should do away with this post-ism and replace it with a conceptually more robust true opposite of 'productivism' (see Section 7.4).

Problems have particularly emerged with regard to the implied linearity of the p/pp transition model. It has been argued that this bipolar assumption does not fully encapsulate the diversity and heterogeneity that can be observed in modern agricultural systems (Wilson, 2001; Evans *et al.*, 2002). The discussion of transition theory in Part 1 highlighted that debates on various transitions have not occurred in a vacuum, and that several important synergies exist between conceptualisations of social transitions and debates on the p/pp transition. This was particularly obvious in the context of post-modernism, and the preoccupation with post-productivism can be seen to parallel the fact that "postmodernism has risen to prominence just at a time of peculiar ferment in the social sciences" (Dear, 1986: 368). As with other transitions to 'post-isms' discussed in Part 1, the question whether both productivism and post-productivism can be seen as clearly defined epochs, problematic and expression, has been particularly challenged. There are also striking parallels between the critiques of post-productivism in agriculture and critiques of regulation theory views of the transition from Fordist to post-Fordist modes of accumulation (Potter and Burney, 2002). Potter and Tilzey (2005: 2), therefore, argued that "the phenomenon of post-productivism ... is only one symptomatic part of the process that is the fracturing of Fordism". Further, as Hoggart and Paniagua (2001: 50) emphasised, "as various commentators have recognised, within rural areas Fordism only had a light touch ..., which questions the capacity for these areas to experience fundamental transformation into post-Fordism". Throughout this chapter, we will see that the discussion about the p/pp transition has to be embedded in parallel debates on the lack of clear evidence of a shift to post-Fordist modes of accumulation.

As Chapter 5 highlighted, problems linked to the over-reliance on political economy theorisations of the transition towards post-productivism were also recognised early in the debates (e.g. Cloke, 1989; Marsden *et al.*, 1996). While Marsden *et al.* (1993: 172) were instrumental in popularising the concept of post-productivist transition, they also criticised the "unreflexive application of structuralist concepts to rural change", and argued that in the early to mid-1980s "the overbearing structuralism of previous work was challenged through attempts to tie structural and local changes together in a non-deterministic fashion" (Marsden *et al.*, 1993: 130). More recently, Marsden (2003: x) suggested that "it is necessary [in agricultural and rural research] to completely break with the (somewhat spurious) coherence of former approaches in political economy", and that we now need "a revised political economy which takes forward a more agro-ecological and socio-spatial approach" (Marsden, 2003: 161) – a point also reiterated in a broader critique of current agri-food studies by Whatmore (2002). Marsden (2003: 100) further suggested that "during the past two decades ... the role of theory for the rural economy has both shifted and been reconstructed by different groups of scholars less tied to, and sometimes highly critical of, the earlier agricultural productionist [sic] paradigm". Marsden and Sonnino (2005: 29) similarly argued that "the profound critical political economy which emerged in the 1980s and 1990s concerning the analysis of the agricultural modernisation process of the late 20th century has not been matched by a parallel critical project concerning the pragmatic state shifts towards … post-productivism". As a result, Hoggart and Paniagua (2001: 45) argued that single perspectives, such as political economy approaches, for the understanding of rural change "are limited, which is worrying … when academic fashion dictates that one tunnel of vision is grasped to the neglect of others (as with political economy two decades ago)" (see also Walford, 2003b).

These critiques already suggested that structure and agency may not move at the same pace towards post-productivism (see Section 7.3.4 below). Similarly, Cloke and Goodwin (1992) stressed that, just as regulation theory was greatly challenged as an explanation of 'post-Fordist' agriculture in the early 1990s, so too unidimensional explanations of post-productivism provided by political economy approaches need to be questioned. They argued that "what appears to be a sea-change to a new epoch [i.e. post-productivism] may well be the latest in a long line of 'constant revolutions', and hence any search for an extensive shift in rural society from Fordism [i.e. productivism] to its successor [i.e. post-productivism] would seem to us to be somewhat premature" (Cloke and Goodwin, 1992: 324). However, as Robinson (2004) emphasised, it is only recently that political economy and regulation theory conceptualisations of post-productivism have been openly challenged. Morris and Evans (1999: 349), for example, argued that rural geography and conceptualisations of recent changes in agriculture contain "greater diversity than the dominant political economy discourse would suggest", and criticised that "political economy has become the dominant discourse to the extent that, for many, it has come to represent agricultural geography" (Morris and Evans, 1999: 350). Burton (2004: 211) also suggested that "much of the work looking at the [post-productivist] transition has failed to take account of the perspectives of individual actors". Clark (2003: 31-32), meanwhile, argued that while political economy approaches have proved successful in identifying broad interrelationships between sectoral and territorial change in rural areas, they have "tended to focus on the importance of factors and processes originating at the national and global scales and, by implication, have omitted detailed theoretical and empirical consideration of the sub-national scale". Yet, the few studies that have used an actor-oriented perspective for theorisations of the p/pp transition have highlighted pronounced differences of adoption of productivist and post-productivist action and thought between different stakeholder groups (Wilson, 2001), as well as across different European regions (Wilson, 2002, 2005). Walford (2003a, 2003b) also critically analysed the policy dimension of the productivist era in the UK and highlighted the doubts that have emerged whether a post-productivist transition has taken place at grassroots level. In particular, he emphasised problems linked to definitions and concepts of the p/pp transition through his study of large Fordist farms in the southeast of the UK. Walford's data extend back to the beginnings of the productivist era in the 1950s and have highlighted that productivist trends continue to dominate on large holdings (see also Section 7.3.3).

As a result, Evans *et al.* (2002) criticised post-productivism as the 'new orthodoxy' in contemporary rural research. They argued that the postulated existence of post-productivism for 'only' 10 to 15 years is insufficient to boldly proclaim that a new agricultural era has emerged. They criticised the p/pp transition model for being a 'theoretical cul-de-sac', and expressed surprise at the 'looseness' of the conceptualisation of post-productivism, the "high degree of consensus over the term post-productivism" among many academics, and "the emergence and widespread uncritical use of such an all-encompassing term" (Evans *et al.*, 2002: 314). They argued that considerable effort has been expended on the rejection of dualistic thinking from various disciplinary angles (see Chapter 4), but that in agricultural and rural studies little engagement has been evident with such wider critiques, as "the active creation and reinforcement of a productivist/post-productivist dualism has emerged as a means of explaining the uneven development of rural areas" (Evans *et al.*, 2002: 314). They, therefore, argued that the concept of post-productivism has been a distraction from developing theoretically informed perspectives on agriculture. As the discussion below will highlight, I agree with their assertion that "there are clear difficulties with the notion of post-productivism", and that "more progress in agricultural (and rural) geography could be achieved by abandoning post-productivism" (Evans *et al.*, 2002: 325). Yet, at the same time, these authors fail to provide a *new alternative* to the notion of post-productivism, although they suggest that more use should be made in conceptualisations of agricultural change of regulation theory, actor network theory and more culturally informed approaches (see also Morris and Evans, 1999). As a result of these critiques, Mather *et al.* (2006: 441) highlighted that "post-productivism is a contested concept, and some argue that it should be abandoned".

It is also interesting to analyse changes in academic thinking by those who initially suggested the p/pp transition model. For example, Terry Marsden – one of the most instrumental academics for popularising the notion of post-productivism – has moved away from the relatively unilinear assumptions inherent in the original conceptualisation (e.g. Marsden, 1999b). Indeed, Marsden (2003: 241) argued that our 'new' knowledge about agricultural and rural change processes has allowed us now to "build upon the somewhat false, and dare I say now, somewhat futile and oversimplified dualisms between structures and actor strategies and networks, social constructivism and realism, globalisation and localism, economistically determined productionism [sic] and culturally confined consumptionism". Although productivist modes of production continue to feature strongly in one of Marsden's (2003) three heuristic dimensions of contemporary agricultural change (the *agro-industrial dynamic*), he suggested that post-productivism should only be seen as a *transitory* phase and not as a means to an end. Thus, he argued that "the post-productivist paradigm does not … attempt to radically solve the problem of agri-industrialism", and that there is an increasing realisation that "post-productivism by itself was not the saviour for agriculture" (Marsden, 2003: 160). He further posited that "even when the post-productionist [sic] logic adopts a more environmental protectionist agenda …, the particular ways it does this both in compartmentalising the environmental problems and solutions and in restricting its appreciation of the role of agriculture, tends to continue to fragment the establishment of real and more viable alternatives from taking root in the countryside" (Marsden, 2003: 238). Yet, authors such as Marsden continue to use 'post-productivism' as a heuristic 'opposite' of productivism, particularly evident through his newly conceptualised *post-productivist dynamic* (see also Mather *et al.*, 2006). Yet, as I will argue in Part 3, Marsden has also acknowledged the importance of the conceptual space *between* these two opposing dynamics by conceptualising the new *rural development dynamic* – a dynamic characterised by a new role for agriculture, re-embedded food chains, a revised combination of nature/value/region with co-evolving supply chains, and by a recapturing of lost values of rural space through strategies such as quality production, place differentiation and value-adding. Marsden has, therefore, acknowledged that all three dynamics may occur *together*, and that they are co-evolving and competing at the same time.

Linking such criticisms to the notion of 'agricultural restructuring', Hoggart and Paniagua (2001) provided a good critique of current unilinear assumptions of contemporary agricultural change. They concluded that the empirical evidence cautions "against readily accepting that current changes are as profound as the literature suggests", and that "the pace of most agricultural change is slower than the literature suggests …, while qualitative dimensions of change are more muted than academic and populist accounts propound" (Hoggart and Paniagua, 2001: 50). In particular, they argued that there is little that is 'new' about current debates on agricultural change, that farmers have been buffeted by external forces for decades, that rural economies have long been driven by demands (e.g. consumption) of non-markets, and that pressures on farmers to change production practices to meet the requirements of non-local populations have also been long felt. Their study, therefore, provided yet another crucial building block in the deconstruction of simplistic unilinear conceptualisations of agricultural change.

Approaching the problematic transition towards post-productivism from a different angle, Burton (2004) convincingly highlighted that the p/pp transition needs to be questioned not only at the level of economic and political changes, but also at the cultural identity-based level (see also Burton and Wilson, 2006). Burton argued that perceptions of the 'good farmer' are still closely associated with productivist production goals (e.g. maximisation of production), as well as with 'tidy' productivist farm landscapes through which farmers can demonstrate their skills to their farming neighbours and rural community. He argued that "entire cultural and symbolic systems can, and have, been constructed based on the productivist role of the farmer. While farmers may be encouraged to move away from productivism through financial incentives, unless there is a corresponding compensation for the social loss a farmer may experience, there may be a strong reluctance to change roles" (Burton, 2004: 211). As I will discuss below, Burton rightly argued that this productivist mindset among most farmers in the developed world should lead us to question the implied directionality in the transition to post-productivism.

Interesting critiques of post-productivism have also come from an Antipodean perspective (e.g. Argent, 2002; Jay, 2004; Holmes, 2006). Commentators have argued that the Antipodes provide "a challenging test of the applicability and utility of the post-productivist concept, by scrutinizing its relevance in a context far removed from its Western European provenance" (Holmes, 2002: 379). Critiques have been linked to the partial inappropriateness of the p/pp transition model to the agricultural situation in Australia and New Zealand. Holmes (2002: 362), therefore, argued for a "post-productivist transition with a difference" in the Australian context. He particularly criticised the UK- and Euro-centric discourses of post-productivism, and suggested substantial divergences in the Australian context concerning 'post-productivist' impulses, actors, processes and outcomes. He criticised the all-embracing notion of a post-productivist transition and, instead, suggested pronounced geographical territorialisation of productivist space (which he termed 'post-productivist occupance'; see also Section 7.3.4 below). Holmes argued that UK-typical post-productivist dimensions such as the 'contested countryside' (see Chapter 6) are absent in the Australian context, in particular in the outback where most of the dimensions of post-productivism highlighted in Chapter 6 "have little relevance to Australia's rangelands, which provide few opportunities for new modes of agricultural production" (Holmes, 2002: 364). Holmes argued that this is particularly true for 'post-productivist' notions of the consumption countryside linked to middle-class in-migrants into rural areas (see Halfacree and Boyle, 1998) – processes that are absent from large areas of Australian 'countryside'. In particular, Holmes argued, the transition to post-productivism in Australia has not been characterised by major redirections in policies as in Europe, but that *territorialisation* of productivist and post-productivist spaces has been a distinctive attribute.

These arguments have been echoed by Argent (2002: 97) who, again in the Australian context, argued that "while there is some evidence of a productivist regime operating in Australia from 1945 to the early 1980s, and some more recent incipient trends consistent with a transition to a post-productivist countryside, there is much stronger evidence that the Australian farm sector and rural landscapes are shaped by the complex interactions between the 'productivist' ideals held by farmers and key policy makers alike, and the growing environmental regulation of farming". Like Holmes, therefore, Argent (2002: 97) suggested that no post-productivist transition has taken place in Australia, and that "while the concept of 'post-productivism' is superficially appealing, it has little practical or conceptual application to Australian conditions". Argent concluded that, based on the problematic binary concept of post-productivism, the p/pp transition may be merely a triumph of *narrative* over analysis, as little empirical evidence can be found in the Australian context to support the assertion that such a transition has taken place. This has also been reiterated by Roche (2005: 302) who argued that "from a vantage point in New Zealand, much of the discussion about a postproductivist [sic] transition in agriculture over the last several years appears of more significance to the UK than elsewhere" (see also Jay, 2004).

These debates highlight that the notion of a post-productivist transition is increasingly questioned among agricultural and rural researchers from different disciplinary and geographical vantage points, even by those who originally proposed the concept. Let us now investigate in more detail the different components of these debates by linking these critiques to the concept of transition in general. The following section will, therefore, use the framework of the four transitional 'fallacies' that emerged from the discussion of other transitions from 'isms' to 'post-isms' (Chapter 4) to investigate temporal, spatial and structural inconsistencies underlying the p/pp transition model.

7.3 Transition theory and the four fallacies of the productivism/post-productivism transition model

In this section I wish to link critiques of the p/pp transition model to our discussion of transition theory in Part 1. As highlighted in Chapter 2, transition theory forms the ideal approach to analyse and dissect specific patterns and processes identified as crucial indicators of the transition towards post-productivism. To this end, Section 7.3.1 will remind us of the basic results of our analysis in Chapters 2-4 and will discuss how this can be applied to the p/pp model. Section 7.3.2 will then briefly investigate the notion of discursive barriers, and will highlight how UK-specific developments may have hindered the potential for 'exporting' the post-productivist concept to the wider world. Sections 7.3.3-7.3.6 will then analyse the four fallacies inherent in the p/pp transition concept, starting with the question of temporal linearity, and then analysing whether the transition has been spatially homogenous, globally universal and structurally causal. This discussion will form the basis for a modified p/pp transition model discussed in Section 7.4 which, in turn, will be used as a platform for conceptualising multifunctional agriculture in Part 3.

7.3.1 Transition theory and the post-productivist transition

Part 1 showed that the issue of 'transition' has been critically analysed from different disciplinary vantage points and that both conceptualisations and critiques of any transitional processes – be they social or natural transitions – share many similarities. In particular, the discussion highlighted that substantial criticism has been voiced against the often *unilinear* and *binary* assumptions inherent in many transition models and that transitions rarely follow smooth linear pathways. As a result, I argued that four 'fallacies' can be identified that underpin virtually all debates on transition: the fallacies of temporal linearity, spatial homogeneity, global universality and structural causality. These four fallacies will provide the conceptual framework for our critical analysis of the p/pp transition model in this section, highlighting that any discussions of a transition towards post-productivist (and multifunctional) agriculture need to be *embedded* in the wider context of 'parallel' debates on transitions. We can not understand the problems inherent in the assumed directionality of the p/pp transition model without, at the same time, acknowledging similar unease and criticism about *other* conceptualisations of transitional processes. Although individual transitions may have their own temporal and spatial dynamics, it is important to recognise the *intertwined* nature of the various transitions (Rotmans *et al.*, 2002). Indeed, the discussion in Chapters 5 and 6 has shown that many historical and disciplinary biases and erroneous evolutionary transitional assumptions also find parallels in our discussion of the transition towards post-productivist agriculture.

Chapter 4 highlighted that these parallel debates have important repercussions for conceptualisations of the p/pp transition, leading us to ask several questions. First, can we assume that post-productivist agriculture will necessarily supersede productivism – i.e. that the transition to post-productivism follows a linear temporal pattern (**fallacy of temporal linearity**)? Why should the assumed direction of change – from productivism to post-productivism – not be reversed in some circumstances? Second, is it meaningful to suggest that the p/pp transition is spatially homogenous – i.e. that different agricultural regions are affected in similar ways and are simultaneously taking part in the transition (**fallacy of spatial homogeneity**)? Third, can we assume that the p/pp transition can be universally applied to agricultural systems in developing countries – i.e. that the transition can be observed in rural areas in advanced and less developed economies at the same time (**fallacy of global universality**)? Fourth, is it possible to assert that the p/pp transition has affected all stakeholders in similar ways and that all actors are moving along transitional pathways towards post-productivism at the same pace (**fallacy of structural causality**)? Inevitably, these four fallacies are interlinked, and I will show that evidence for spatial heterogeneity, global complexity and structure-agency inconsistency in contemporary agricultural change also intertwine with questions about the temporal non-linearity of agricultural trajectories.

7.3.2 Discursive barriers, UK-centrism and 'exporting' the post-productivist concept

Before discussing whether and how the p/pp transition model stands up to scrutiny in light of the four transitional fallacies, we briefly need to discuss issues surrounding the problem of 'exporting' notions of post-productivism beyond the UK. Chapters 5 and 6 have highlighted that conceptualisations of the p/pp transition have largely been developed by English-speaking academics, and that these debates have been heavily biased towards the UK experience and history of agricultural development. Indeed, Chapter 5 highlighted that it was UK researchers who developed and popularised the concept of post-productivism during the early 1990s. Thus, "discussions about the possibility of a post-productivist agricultural regime have almost exclusively taken place in the UK, and the notion of post-productivism has not been taken over (yet) by European … rural theorists" (Wilson, 2005: 111). There is, therefore, a need to discuss both to what extent this may simply constitute *discursive barriers* between British and non-British contexts (including the USA), rather than tangible differences in the nature of agricultural trajectories in non-British countries. How and to what extent may UK-specific dimensions of the p/pp model hinder the 'exportability' of the concept beyond the UK? As Marsden (1999b: 242; emphasis added) argued, although there is "recognition amongst the main literatures about the integrative and holistic nature of the new processes of rural change, the approaches thus far have yet to theoretically develop an approach which begins to guide a clearer understanding of the processes which are making things different in the *European* post-productivist countryside".

Discussing the 'exportability' of the concept of post-productivism beyond the UK needs to address the problem of conflicting terminology (Wilson, 2001). Because of the heavy focus on the UK, conceptualisations of post-productivism have inevitably been strongly biased towards UK terminology of the 'rural' and the 'countryside' (cf. Hoggart, 1990). In Part 2, I deliberately used the term post-productivist *agriculture* – rather than post-productivist *rural* areas or the post-productivist *countryside* – precisely because in an international context the English notions of 'rural' and the 'countryside' are strongly contested, often non-existent, or ambiguous (Wilson, 2005). There is a wealth of literature on conflicting definitions of 'rural' across Europe (e.g. Hoggart, 1990; Hoggart *et al.*, 1995). In Germanic countries, for example, translations can be found for 'agriculture' (*Landwirtschaft*) or 'rural areas' (*ländlicher Raum*), but direct translations of 'rural' (*ländlich?*) or 'countryside' (no equivalent word) are problematic. As a result, it is problematic to refer to the *post-productivist German countryside*, at least from the viewpoint of German agricultural and rural researchers (Wilson, 2002). UK-centric terminologies, language and discursive barriers may have, therefore, hindered, rather than advanced, the applicability of the concept of post-productivism beyond the UK (this is also relevant for the fallacy of global universality discussed below). Further, differences continue to exist between agricultural/rural trajectories in the UK and other countries, which may suggest difficulties for the applicability of some of the dimensions of the p/pp transition model beyond the UK. In the following, I wish to highlight debates surrounding *UK-specific indicators* of the p/pp transition model that may not find applicability elsewhere.

A good starting point is UK land ownership rights. The UK is a country that, unlike many of its European counterparts (e.g. France), never experienced a revolution that led to the breaking up of large land-owning estates (Hoskins, 1955; Wilson, 2002) – one of the reasons for Marsden's (2003) conceptualisation of the 'paternalistic' countryside as one of the main UK countryside types. As a result, the UK still has one of the highest percentages of very large landowners (i.e. >1000 ha) in Europe with 20% of landholders owning 80% of the countryside, current yearly receipts of CAP subsidies of large estates often exceed €500,000, and access to the countryside continues to be highly restricted despite recent drastic (post-productivist?) countryside access regulations (Mandler, 1997; Woods, 1997). This highlights why one of the indicators of the post-productivist countryside from a British perspective has been the loss of legal security of British farmers with regard to *land access rights* (see Table 6.1. in Chapter 6). Hoggart *et al.* (1995) found little evidence of the paternalistic countryside elsewhere in Europe, suggesting that changing land ownership rights may be an indicator with little resonance in non-UK geographical contexts (Wilson, 2002; see also Holmes, 2006, for the Australian context).

Applying notions of the 'consumption countryside' as a possible indicator for the transition towards post-productivism beyond the UK may also be problematic. As highlighted in Chapter 6, hitherto neglected 'resources' in the British countryside, such as old farm buildings, abandoned cottages and former farm workers' cottages, as well as the wealth of environmental and landscape-based resources, have gained new importance in an arguably post-modern and post-industrial society (Macnaughten and Urry, 1998). Although similar patterns can be identified in other European countries (but less so in North America, Australia and New Zealand), the UK has the highest rate of conversions of former farm buildings into private accommodation (e.g. second homes in countryside settings) or offices (Robinson, 2004; Woods, 2005). It is, therefore, important to point out the different historical trajectories between the UK and most other European countries. The UK was the first country to industrialise, with concurrent large-scale urbanisation and, most importantly, a relatively early separation between urban and rural areas (i.e. since the mid-18th century). Some have argued that this urban/rural divide continues to be more pronounced in the UK than in many other advanced economies (Halfacree, 1994, 1997b) – one of the reasons for the focus of UK debates on the 'rural' (Hoggart, 1990). As Hoggart *et al.* (1995) highlighted, such pronounced urban/rural dichotomies – both physical and mental – can not be found to the same extent in the rest of Europe. France is a case in point, where most French people still have a strong attachment to the countryside and often have family connections in 'rural' areas (Buller, 2004). This may be one of the reasons why translations of the term 'rural' are more difficult in other European contexts that have not witnessed the historical development of separate dualistic and binary territories of urban and rural to the same extent as the UK (Wilson, 2002, 2005). This, of course, begs the question whether the dualistic and binary notion of the p/pp transition has, in turn, been strongly influenced by the existence of seemingly distinctive urban and rural territories? For conceptualisations of the post-productivist transition, this urban/rural divide in the UK has reinforced the demand by wealthy urbanites for 'rural' resources (including buildings) or, in a broader sense, the urban quest for the rural idyll (Yarwood, 2005). This suggests that post-productivist indicators linked to the commoditisation of rural resources may not find direct applicability in non-UK contexts, and that assessing notions of the spatial homogeneity of the p/pp transition may have to exclude this specific indicator of the p/pp transition (Wilson, 2002).

The issue of *governance* of rural areas emerges as another contested arena. Indeed, "the re-regionalisation of governance (especially in rural areas) has become an important goal of recent politics (especially as part of New Labour's 'Third Way') and, consequently, has become a key indicator of post-productivism in the UK" (Wilson, 2005: 116). Yet, UK-specific developments in recent political history may make it difficult to directly transpose notions of a UK-based post-productivism to other countries (see also discussion below on the influence of national political history for the analysis of spatial homogeneity of the p/pp model). Many argue that during and after the Thatcher era in the UK (since 1979) a gradual withdrawal of the state has taken place, away from a state-centric model towards empowerment of localities and the regions (Goodwin and Painter, 1996; Jordan, 1999), characterised by the devolution of powers from the centre (i.e. Whitehall in London) to the periphery (i.e. Cardiff, Edinburgh and Belfast as the centres of newly devolved structures). Local communities have, therefore, gained substantial decision-making powers, while central authorities may have lost direct power (Gibbs *et al.*, 1998). Although there continue to be heated debates as to whether these processes have weakened the powers of the nation state (Giddens, 1998), this discussion is important as re-regionalisation/re-localisation of power is one of the key indicators for the post-productivist agricultural era. This post-productivist indicator will, therefore, only be applicable in advanced economies outside the UK that have seen similarly rapid processes of decentralisation. In relatively centralised countries such as France, this indicator will, therefore, only find limited applicability, as political and decision-making structures have not substantially changed since the 1970s (at least not to the same extent as in the UK) (Lowe *et al.*, 2002). On the contrary, it could be argued that in most Western European countries the empowerment of regional and local decision-making structures already occurred during the 'productivist era' (e.g. Germany, Spain, Italy) and that a reversal of this indicator (i.e. *weakening* of regional powers) could be seen as an indicator of post-productivism (Wilson, 2002).

Finally, post-productivist indicators based on the breakdown of the Atlanticist food order may not be easily transferable to advanced economies beyond the UK and USA. Questions have arisen whether other European countries were ever part of this food order (Cafruny, 1989; Goodman and Watts, 1997) – a food order that particularly emphasised the close agricultural policy ties between the USA and the UK prior to British accession to the EEC in 1973 (Goodman and Redclift, 1991). It has been argued that in countries of the EEC, the CAP played a much more important role (since the 1960s), and that only the UK was strongly influenced by bilateral agricultural policy agreements with the USA. As many commentators have argued, since the 1970s the continuing closer integration (ideological and economic) of the UK with the USA may mean that pressures to liberalise agricultural markets and to dismantle protectionist subsidy policies linked to the CAP may be more pronounced in the UK than in other EU countries (Potter and Burney, 2002; Potter, 2004). This suggests that the 'post-productivist' characteristics of free market trade may be overemphasised from a UK perspective.

It is important to bear these caveats in mind in the following discussion of the four transitional fallacies in the p/pp transition debate. In particular, these caveats emphasise problems associated with both exporting theory based largely on the experience of one country and with transitional terminology that may not find universal applicability (or even linguistic comprehension) elsewhere.

7.3.3 The fallacy of temporal linearity

With these caveats in mind, we can now analyse the p/pp transition through the lens of transition theory. I will begin with the fallacy of temporal linearity because, as Section 7.3.2 highlighted, it is temporal issues surrounding the p/pp transition that have generated the most vociferous debates. Linked to this is the problem of the binary nature of the p/pp model, in which productivism and post-productivism are seen as two clearly definable 'eras' or 'epochs'. Underlying these assumptions is the belief that post-productivist agriculture will necessarily supersede productivism – i.e. that the transition to post-productivist agriculture follows a *linear temporal* pattern. In other words, it is generally assumed that action, thought and agricultural processes all move along the post-productivist transitional trajectory *at the same pace*. Yet, as highlighted above, the assumed directionality of the p/pp transition has been strongly contested, and commentators have rightly begun to ask why the assumed transition can not be *reversed* in some instances. The following discussion will highlight that for all the dimensions seemingly characterising the p/pp transition, evidence can be found that no full transition to post-productivist action and thought has occurred yet and that, instead, productivist and post-productivist processes occur simultaneously with some aspects of the transition operating at different time scales to others. Akin to similar debates in transition theory, the notion of 'pre-productivism' will also be considered (see debates on 'pre-modernism', 'pre-colonialism' or 'pre-Fordism' in Part 1). The discussion will largely focus on patterns and processes in the developed world, while Section 7.3.5 will investigate in detail the applicability of the p/pp transition concept to the developing world.

Agricultural policies: incrementalism or reform?

Let us start with agricultural policies as one of the most commonly used indicators of the transition to post-productivism (see Chapter 6). Few commentators would doubt that substantial changes have taken place in agricultural policy since the 1980s. This includes, in particular, the so-called 'greening' of agricultural policy (Harper, 1993). Yet, many would also argue that, in a European context for example, policy change has not been as dramatic as many suggest. Some have stressed that, based on Hall's (1993) three orders of policy change, most agricultural policy changes in the EU are at the second order at most (new policy mechanisms to accommodate *non-radical* change), and not *radical* third-order changes (changes made to a policy's guiding principles) (Clark *et al.*, 1997; Mather *et al.*, 2006). Potter (1998) and Wilson (2005) re-emphasised this by suggesting that most agricultural policies in the EU and USA are currently either at the 'discourse' or 'argument' stage, but rarely at the 'persuasion' stage.

A heated debate has particularly revolved around the notion of AEP as a post-productivist indicator (Wilson and Buller, 2001; Evans *et al.*, 2002). Although many have suggested that new policies that encourage sustainable environmental management on farms indicate a shift towards post-productivism (Whitby and Lowe, 1994; Burton, 2004), AEPs in the EU possess three features which question their eligibility as indicators of post-productivism: farmer participation in higher tier options is often voluntary (leaving many non-participants who may continue to farm in productivist ways) or based on relatively shallow tiers linked to new cross-compliance requirements; land may only be enrolled for a temporary period, especially for higher entry level prescriptions (usually 5-10 years after which farmers can often revert back to productivist production); and different stakeholder groups have different agendas about the purpose and goals of AEP (e.g. EU Commission, agriculture ministries, NGOs). The latter point is particularly important as it often means that, although policy prescriptions appear relatively clear at the point of formulation, during implementation different interpretations and national agendas dilute the original policy purpose (Hanley *et al.*, 1999; Wilson and Juntti, 2005). This has been particularly obvious during implementation of AEPs in Mediterranean countries in the late 1990s, leading to inconsistencies in implementation and confusion about the 'real' goals of these policies (Buller *et al.*, 2000). This has led to discrepancies between policy formulation at EU level that may show post-productivist tendencies, and interpretation and implementation of these policies on the ground in often productivist ways (Wilson and Hart, 2000; Wilson and Juntti, 2005; see also discussion of the fallacy of structural causality below). Further, while some critics argue that recent policy initiatives such as the 1992 CAP reforms or Agenda 2000 may well be categorised as post-productivist in their basic philosophy, recent moves towards trade liberalisation of global markets through recent WTO agreements, together with the watering down of Agenda 2000 as part of national policy bargaining, may have led to a re-emphasis of productivism (Potter and Burney, 2002; Ward and Lowe, 2004). This has led Delgado *et al.* (2003: 31) to argue that "many observers expected Agenda 2000 to mark the beginning of the roll back of the emphasis on the productivist model of the traditional CAP... [but] this turned out not to be the case." Similarly, Marsden (2003: 6) suggested that "the current policy reforms under Agenda 2000 ... expose a policy framework which will do little to shift the basic philosophy beyond its bias towards the industrial [productivist] model".

Evidence, thus, shows that AEP is still used by most participating landowners as an economic subsidy. Although part of the 'green box'[23] agreement as part of the WTO URAA, the use of AEP money can be seen as a continued 'blue box'[24] subsidy. As a result, "for the majority of farmers ... participation in an agri-environmental scheme does little or nothing to challenge the nature of conventional (productivist) food production practices" (Evans *et al.*, 2002: 321). Indeed, critics argue that AEPs are just one way with which the EU (and countries such as France in particular) continue to subsidise their farmers through hidden or green subsidies (Potter and Tilzey, 2005), with Winter and Gaskell (1998) even suggesting that AEPs *legitimise* productivism by providing a source of finance for investment in productivist food and fibre production. In addition, it remains questionable whether AEPs can be categorised as post-productivist on the basis of their contributions towards sustainable environmental management. Most agri-environmental schemes in the EU (and beyond) continue to be about *maintenance* and not *change* (including new schemes such as the basic entry-level tier of the UK Environmental Stewardship Scheme[25]), which suggests that there is little potential yet for bringing about substantial environmental recovery on farms (Buller *et al.*, 2000; Burton, 2004). Thus, Potter and Burney (2002: 45) argued that "many critics of the EU's agri-environmental policy now accept that schemes are often poorly designed, insufficiently linked to environmental outcomes (or, more feasibly, outputs), counter-intuitive in paying for income foregone rather than environmental gain and rewarding most those who threaten damage rather than those with a consistent record of long-term stewardship".

AEP, therefore, can be interpreted as only a 'thin veneer' of policies with post-productivist tendencies over a staunchly productivist CAP – or, in Hoggart and Paniagua's (2001: 53) words, a "changing of the colour of the paint, rather than the construction of a new building". This led Evans *et al.* (2002: 321) to argue that "it is well known that the emphasis on agri-environmental policy in research and political rhetoric has far outweighed its significance on the ground" (see also Hanley *et al.*, 1999). Thus, "it is ... dangerous to equate agri-environmental policy ... with post-productivism, as agri-environmental policy is often used as an economic rather than environmental tool" (Wilson, 2005: 113), while Walford (2003b: 491) argued that "policy reform measures have been characterised as contributing to a structural transition from a 'productivist' to 'post-productivist' era in agriculture ... [but] empirical evidence for such a reorientation at the farm level is less than conclusive". Although agricultural subsidy expenditure as part of the overall CAP budget has been gradually reduced to about 40% of the overall expenditure (by 2006), 2nd pillar expenses are still minimal (currently less than 10% of the CAP) (Delgado *et al.*, 2003). Indeed, agricultural subsidies in the CAP still amount to about £45 billion/year, most of which is paid on production subsidies (Mather *et al.*, 2006). Similar patterns can be observed in the USA where the subsidisation of cotton farmers forms a direct continuation of productivist policies aimed at shielding US farmers from global markets and competition

---

[23] According to the traffic light classification of domestic subsidies agreed under the WTO URAA, 'green box' payments are related to agri-environmental and rural development purposes and are regarded as *least trade distorting* and, thus, notionally immune from WTO challenges on trade grounds (Potter and Tilzey, 2005).

[24] Support measures that are *trade distorting* because they are tied to the land or given on a headage basis, but which enjoy temporary shelter under the terms of the 'Peace Clause' agreed between the EU and USA at the conclusion of the URAA.

[25] At the time of writing (2006), it is still too early to gauge the environmental success of this new scheme. By the end of 2005, over 6000 farmers had joined the scheme, with nearly one million hectares of land entered.

(Freshwater, 2002). Thus, for AEP as a possible indicator of the p/pp transition, there is still little evidence at the beginning of the 21st century of reduced financial state support for agriculture (at least in the EU), and only fledgling attempts have been made at moving away from a productivist state-sustained production model.

This suggests that we continue to see the prioritisation of policies on the production of food, despite attempts at encouraging more sustainable environmental management practices (see also below). Although decoupling and cross-compliance have been included in recent rhetoric of EU policy-making, empirical evidence on the ground highlights that European agriculture is still reliant on protectionist price guarantees. Evans *et al.* (2002: 324), therefore, suggested that "it would be hard to characterise either the 1992 reforms or the Agenda 2000 reforms as anything other than a means of limiting CAP-induced surpluses". Winter (1996) further argued that recent CAP policy changes do not represent a 'new-look' agriculture and that much of the rhetoric surrounding recent changes has been concerned with making European agriculture *more* competitive, not less (see also Delgado *et al.*, 2003). Similarly, Marsden (2003: 2) suggested that "despite over fifteen years of debate and policy crisis concerning the negative effects of productivist support mechanisms within the CAP ... it is still the case that the main pillar of the CAP remains in this area in terms of funding. Moreover, it still tends to reinforce the logic of agricultural productivist scale economies". Recent debates in the WTO and at G8 summits continue to highlight the defensive position of EU leaders about dismantling the productivist subsidy regime. Potter and Burney (2002), in particular, argued that support measures in the URAA blue box have offered the EU an important safety valve in its pursuit of CAP reform, allowing reductions in price support to be offset by livestock headage or arable area payments. To many, the use of blue box support measures within the EU suggests a perpetuation of staunchly productivist policies aimed at encouraging farmers to reap maximum subsidies through continuation of intensive production. Thus, Potter and Burney (2002: 38) argued that "it might be argued that packages of measures, such as arable area payment tied to a set-aside requirement, can be designed to be production neutral, but such payments still require farmers to maintain land in arable production or, in the case of headage payments, to keep cattle or sheep, in order to qualify for the subsidy".

Such policy developments led Evans *et al.* (2002: 316) to question policy change as an indicator for the transition to post-productivism, as the "political emphasis on the need for farmers to be able to compete in a liberalized global market seems to place greater emphasis worldwide on the continuation of productivist principles". Indeed, "the slow and contested progress ... under the Agenda 2000 reforms to the CAP testifies to the tenacity of productivist thinking in the agricultural policy community" (Evans *et al.*, 2002: 320). Similarly, Potter and Burney (2002: 40-41) argued that "the reality is that the CAP has operated in different directions, with both positive and negative environmental consequences; price support interacting with market forces and technological change in complex and paradoxical ways to encourage more intensive use of land". Thus, Baldock *et al.*'s (1990) earlier suggestion that policy change in the European context has been characterised by *incrementalism* instead of *reform* still holds true today. The 'greening of agricultural policy' (Harper, 1993) continues to be an elusive goal rather than a tangible reality. As a result, it is difficult to argue that agricultural policies (at least in the EU) can be classified as post-productivist. Instead, productivist policies continue, occasionally *punctuated* by attempts at introducing policies with post-productivist agendas. This means that productivist and post-productivist policies occur *simultaneously* and that policy change is not characterised by temporal linearity but by *temporal non-linearity*.

Changing ideologies?

In contrast to debates on policy change, the claim that productivist ideologies have changed towards post-productivist ways of thinking is more difficult to dispel, partly because of lack of empirical evidence and tangible 'data'. Yet, few would doubt the claim of those who conceptualised the p/pp transition that in many developed countries agriculture has lost its central position in society (e.g. Lowe *et al.*, 1993). Rural society has, therefore, witnessed definite changes, and it is possibly with regard to ideology that the most pronounced move towards post-productivism has occurred (Wilson, 2002). Indeed, Holmes (2002) saw changing ideologies as one of the three key driving forces for the post-productivist transition in an Australian context.

In Europe, the loss of the powerful hegemonic position of agriculture accelerated particularly in the 1980s (Ward, 1993). This was exacerbated by counter-urbanisation trends that have changed the nature of rural society in the UK and elsewhere profoundly (especially near urban centres), and through the purchase of second homes, especially in areas of outstanding natural beauty such as the south-west of the UK or the Dordogne in France (Hoggart *et al.*, 1995). Bearing in mind the above-mentioned caveat on 'exporting' notions of post-productivist consumption beyond the UK context, evidence is emerging of increasing consumption of the countryside across Europe (e.g. Schmid and Lehmann, 2000, for Switzerland; Wilson and Wilson, 2001, for Germany) and other advanced economies (e.g. Holmes, 2002, for Australia). Yet, and as will be discussed in detail below for the fallacy of spatial homogeneity, the influence of counter-urbanisation and associated consumption of the countryside as indicators of post-productivism should not be overemphasised, due to different historical developments in the UK and its, arguably, more pronounced rural/urban divide (Newby *et al.*, 1978; Newby, 1985). Indeed, within Europe, counter-urbanisation trends may be most pronounced in the UK, especially after more than a decade of unprecedented economic growth and wealth creation among the middle classes hungry to live the dream of the 'countryside idyll' (see also recent critique by Holmes, 2002, of the use of 'counter-urbanisation' as a post-productivist indicator in a non-European context).

There is also no doubt that recent catastrophes such as BSE and foot-and-mouth in the UK and beyond have led to further societal criticism of how agriculture is run and regulated (Hinchcliffe, 2001; Scott *et al.*, 2004). This has led to a further marginalisation of farmers, a move away from mass consumption and production of foodstuffs, and an alienation of farming and rural communities who often bear the brunt of societal critiques of how food is produced, transported and processed (Lang and Heasman, 2004). In the UK and beyond, this has been closely associated with changing media representations of the rural (Harrison *et al.*, 1986; McHenry, 1996), portrayals of agriculture as the main threat to the countryside (Pratt, 1996; Marsden, 1999a) and, arguably, the increasing separation of the 'rural' from 'agriculture' linked to new social representations of the rural (Cloke and Goodwin, 1992). However, some of the indicators of the move towards post-productivist ideologies remain highly questionable. The productivist rhetoric of self-sufficiency in agricultural production is still evident in political discourse, coupled with notions of a 'strong' and 'assertive' agricultural sector (Robinson, 2004). Although the number of farms in all developed countries continues to dwindle rapidly (e.g. in Germany from 1.7 million in 1949 to 0.4 million in 2005) agriculture continues to be the dominant global land use, occupying up to 80% of the land area in some countries (e.g. Ireland, Denmark). Indeed, more food is produced today than ever before because of increasing *productivity* on many farm holdings, highlighting that agriculture continues to be a dominant process despite reduced farm numbers and the loss of political power of agricultural actors (see below). This may be the reason why at the institutional level not much ideological change is evident for institutions and state departments controlling and regulating agriculture. Indeed, Marsden (2003) highlighted that continuing productivist pressures have provided a new raison d'être for productivist-driven agricultural ministries in many advanced economies who now espouse consumer-driven needs through a highly bureaucratic and rationalistic approach.

The cases of both the UK and Germany are illustrative in this context. In both countries the 'traditional' ministries of agriculture (MAFF in the UK; the Bundesministerium für Ernährung Landwirtschaft und Forsten [BML] in Germany) were replaced in 2001 by new ministries with arguably more post-productivist agendas (the Department for Environment, Food and Rural Affairs [DEFRA] in the UK; the Bundesministerium für Verbraucherschutz, Ernährung und Landwirtschaft [BVEL] in Germany). However, despite inclusion of the word 'environment' in DEFRA, and although the German BVEL was headed by a minister from the Green Party until 2005, there is little evidence that both DEFRA and the BVEL have embarked on a post-productivist ideological trajectory. Instead, and as will be highlighted in more detail below, both ministries continue to be hijacked by corporatist interests of powerful lobby groups that have often managed to impose productivist agendas at times of political crises and, in particular, before national elections (see also Argent, 2002, for parallel developments in Australia). The continuing powerful role of the productivist German farmers' union (the Deutscher Bauernverband) has been particularly highlighted, as it continues to be able to impose a highly influential productivist policy agenda during national elections, despite rapidly dwindling membership numbers (Wilson and Wilson, 2001). Similar lobbying processes with productivist undertones have been described by Bell (2004) concerning how US environmental authorities are lobbied by productivist-oriented groups.

While there is some evidence that ideologies in advanced economies have shifted towards post-productivist action and thought, we simultaneously witness the continuation of productivist ideologies at various stakeholder levels. In other words, there is insufficient evidence that the postulated shift towards post-productivism has embodied major qualitative, and not just quantitative, change in agricultural action and thought. Productivist and post-productivist ideological processes, therefore, occur side-by-side and are not necessarily characterised by temporally linear transitional patterns.

Towards post-productivist rural governance?

As Chapter 6 highlighted, the key argument underpinning notions of post-productivist rural governance are based around the notion that in the so-called 'post-productivist era' the traditional 'corporate relationship' between agriculture ministries and powerful farmers' unions is gradually broken down to allow formerly politically marginal actors (such as environmental groups or local grassroots organisations) into the decision-making and policy formulation networks. Marsden (2003: 128) argued that post-productivism is characterised by "the steady marginalisation of the farmers unions as a powerful broker in agricultural based clientelism". But is there evidence to suggest a weakening of power structures within agricultural lobbies in advanced economies, leading to a widening of the agricultural policy community with inclusion of formerly marginal actors into the core of the policy-making process?

Bearing important caveats in mind about differential governance pathways between the UK and other countries highlighted above, for advanced economies empirical evidence suggests that recent 'inclusive processes' *have* facilitated the injection of 'green' ideas into the agricultural policy-making process by newly empowered actors such as environmental NGOs and other stakeholder groups in rural society (Marsden, 2003; Wilson and Juntti, 2005). In the UK, this has meant the inclusion of new stakeholder groups into policy-making processes such as the National Trust, the powerful RSPB, Friends of the Earth, or the Countryside Alliance, reflecting the newly found political power of such interest groups with large memberships (Lowe *et al.*, 1997; Wilson and Hart, 2001). Similar processes have been described elsewhere (e.g. Buller and Brives, 2000, for France; Andersen *et al.*, 2000, for Denmark; see also reference to Green agricultural minister in Germany above). These processes have been reinforced by the social and economic restructuring of the countryside and the reconstitution of actor spaces through urban-rural migration (counter-urbanisation) that has enabled middle class and non-farming stakeholders access to decision-making processes within their rural constituencies (e.g. through political representation in parish or district councils). There is little doubt that these new interests and actors are influencing decision-making processes, especially through attempts by newcomers to create a rurality in their usually urban image of the countryside idyll. In the UK and elsewhere, this has shaped planning regulations and restrictions (arguably by loosening regulations and enabling further housing developments in greenfield sites), and the pressure on conversions of farm buildings into luxury private accommodation has, concurrently, increased (Marsden, 2003).

Yet, parallel to these processes, underlying productivist governance structures continue to be visible. Despite the partial loss of the ideological position of agriculture highlighted above, and although farm numbers are dwindling, farmers and their political representatives at the local, regional and national levels continue to be a powerful political force (Winter, 1996). For the UK, Hoggart and Paniagua (2001: 52) argued that "the vision that a key element of state restructuring in the countryside is the overthrow of a farmer-dominated local hegemony is a dangerous image to cling to … we still lack a solid empirical base from which to derive clear judgements about change in state action in rural areas". Although the agricultural policy community has widened to include non-farming stakeholder groups, in most rural areas of the developed world this community continues to be relatively tight-knit and to maintain some degree of internal strength (Winter, 1996; Little 2001). Thus, many studies suggest that little is changing despite new programmes, organisations, policies and modes of stakeholder empowerment (Clark et al., 1997). This is particularly true for the continuing powerful position of farmers' unions in many countries, with the Deutscher Bauernverband in Germany (Wilson and Wilson, 2001), the National Farmers' Union in the UK (Winter, 1996) or the Greek Farmers' Union (Beopoulos and Vlahos, 2005) as obvious examples of groups with powerful influence in local/regional governance structures. Although the rhetoric within many farmers' unions has become more accomodationist due to external ideological pressures, on the whole they continue to advocate productivist ideologies. This has hindered the establishment of post-productivist governance structures (see Povellato and Ferraretto, 2005, for the Italian case) which has often led to a watering down of post-productivist governance systems (Jordan, 1999; Bowler et al., 2002).

Traditional productivist governance structures also continue to prevent participation of hitherto marginalised groups in rural society, in particular women, immigrants and native people (Cloke and Little, 1997). For Australia, Liepins (1995, 1998), for example, highlighted how governance patterns continue to prevent full access for some grassroots actors to knowledge networks available at state and intermediate levels. Warren (2006) also showed how continued exclusion of certain groups in rural society from internet-based communication technologies can lead to political disenfranchising (e.g. no possibility to vote online). Although some indications exist that changing governance structures can help conflict avoidance in rural areas and may help support public-private policy-making structures, there is little evidence so far for the emergence of post-productivist governance that contributes towards reduction of local stakeholder alienation, that extensively taps local knowledge, or that contributes greatly towards improved local community education. This suggests that in many rural areas the 'corporate' relationship between powerful political agricultural actors and the farming lobby still continues (Winter, 1996). Further, even where local grassroots actors may be empowered through changing governance structures, this does not necessarily mean that they will promote post-productivist action and thought. Indeed, much evidence from Europe suggests that it is often local rural grassroots actors who may be more firmly embedded in productivist ideologies, by emphasising rural development trajectories aimed at increasing production and productivity, profit maximisation and protectionist farm subsidies (Wilson and Juntti, 2005).

A recently undertaken analysis of post-productivist rural governance patterns in Australia illustrates many of these issues (Wilson, 2004). The study analysed whether the Australian Landcare movement complies with notions of 'post-productivist rural governance'. Landcare was selected as it has been hailed as one of the most successful rural-based grassroots-led movements in advanced economies (Campbell, 1994; Carr, 2002) and, therefore, provided an ideal testing ground for assumptions about changing rural governance structures as an indicator for post-productivism. Results showed that Landcare has been a vast improvement on previous governance structures for the management of the countryside in Australia, and that it has managed to mobilise a large cross-section of stakeholders. However, the study also showed that the Landcare movement only depicts certain characteristics of post-productivist rural governance. Although Landcare has some elements that fit in with theorisations of post-productivist governance structures, it still depicts many characteristics that show its close affiliation with the state and its agencies, in particular through budgetary shackles that mean that Landcare can not be seen as entirely acting independently from the state. Thus, Landcare can not be conceptualised as a fully inclusive movement, and there is little evidence that it has been able to actively shape government policy (Curtis and Lockwood, 2000; Carr, 2002). However, Landcare has also contributed towards changing environmental attitudes, which can be seen as a key precondition for the successful implementation of post-productivist rural governance structures (Wilson, 2004). In particular, Landcare's innovative approach of mutual farm visits, and its emphasis on the demonstration of 'best practice', has led to both an increased awareness of land degradation problems and the creation of grassroots 'information networks' (Martin and Halpin, 1998). There has also been some success regarding Landcare's ability to change attitudes of the wider Australian public. Two important lessons for conceptualisations of post-productivist rural governance emerged from this study. First, individual components of post-productivist rural governance may change at different times, with the *attitudinal* level most influenced by Landcare, while underlying socio-political productivist structures will take much longer to change, thereby suggesting a temporally non-linear pattern of governance change. Second, the problem in labelling Landcare – arguably the most innovative rural programme in advanced economies – as an expression of post-productivist rural governance shows how far away most other rural programmes in advanced economies still are from such new forms of governance. The results, therefore, support those advocating that post-productivism may only be a theoretical construct in the minds of academics, rather than an expression of reality on the ground (Wilson, 2004).

The Landcare example from Australia highlights that some attitudinal changes, as well as policy-related changes, can be induced through inclusion of grassroots voices in rural governance structures. However, it is also important to note that despite the inclusion of hitherto silent voices in agricultural policy-making processes, this empowerment can also be a problem as it has, arguably, deradicalised certain groups who now feel that they may have less power than before. This was recently highlighted by a representative of the RSPB, one of the largest environmental NGOs in the UK: "You could say that [the Ministry of Agriculture] has been very clever, whether they've done it by accident or on purpose or not I don't know, but things like the Agri-environment Forum has brought people together, which is exactly what the NGOs were asking for, this consultation. But perversely, by bringing everyone together, it's also allowed [the Ministry of Agriculture] to say 'OK, you do this, you do that and we'll have what we all want won't we', and meanwhile [the Ministry of Agriculture] gets away with doing bugger all. But it's very difficult to say, if I want to change things, do I withdraw from that process, take our organisation out of it? Of course, no individual organisation is going to do that, they want to be involved. So it's stalemate really. Where do you go next, what do you do?" (Hart and Wilson, 1998: 267). Thus, while many NGOs may have lost their position at the fringe of policy-making they have also, at the same time, lost the opportunity of being able to aggressively lobby against state regulations (Bryant, 2005; Wilson, 2005). They now have to steer an accommodationist and 'diplomatic' approach, often leading to more incremental policy change rather than reform that, at its core, continues to be largely dominated by powerful agricultural interests and, therefore, often oriented towards productivism (Winter, 1996; Hart and Wilson, 1998). It is here, in particular, that parallels can be drawn with debates on the transition towards environmentalism highlighted in Chapter 4. While a segment of society represented by organisations such as the RSPB shows growing disenchantment with the environmental degradation associated with productivist farming (e.g. RSPB, 2001), a large proportion of contemporary society still continues to be staunchly utilitarian (Wilson and Bryant, 1997).

The evidence suggests, therefore, that although some changes to governance structures have occurred, on the whole there is still little evidence of empowerment of local stakeholders and, concurrently, little evidence of the erosion of the power of the state and state-related agencies and institutions as the main devisers and shapers of policies and decisions affecting rural communities in advanced economies (Rhodes, 1997; Jessop, 2002). The general enhancement of democratic processes through the empowerment of local stakeholders and increasing government accountability continues to be a vision for the future rather than reality on the ground (Habermas, 1981; Dean, 1999), especially in cultural contexts where powerful agricultural elites continue to control agricultural policy-making trajectories (Oñate and Peco, 2005; Povellato and Ferraretto, 2005). In most European countries there is also limited evidence of the state's retreat from the position of provider of support and main policy-maker to one of 'coordinator', 'manager' and 'facilitator' – concepts that, in theory, are key to post-productivist rural governance (Imrie and Raco, 1999; Little, 2001). In particular, there are few indications of a *re-regionalisation* of governance of rural areas (Pretty, 1998; Ray, 2000), although the above evidence (e.g. the Australian Landcare example) suggested that some deregulation dynamics are beginning to take shape. The weight of evidence, therefore, suggests that although some post-productivist tendencies are emerging for the reconstitution of rural governance structures, productivist tendencies continue to exist simultaneously and, at times, dominate stakeholder discourses. As Marsden (2003: 13) emphasised, "the productivist and post-productivist system of governance, while coming from different origins, have found that they can peacefully co-exist – at least for the time being". The proposed transition towards post-productivist governance structures is, therefore, not yet a real transition but a temporally non-linear process in which productivist and post-productivist governance structures exist side-by-side.

Emergent post-productivist food regimes and agro-commodity chains?

Chapter 6 highlighted that the breakdown of the Atlanticist food order forms a key indicator of post-productivism. Yet, evidence for the complete dismantling of this food order is not conclusive. Although there is some evidence in the UK that the Atlanticist regime has been largely replaced by the 'Europeanisation' of agricultural policies linked to the CAP and the UK's closer ties with Europe since EEC accession in 1973, other European countries have been, arguably, more weakly embedded in, or affected by, the Atlanticist food order (Goodman and Redclift, 1991; Goodman and Watts, 1997). For the UK, the continuing close integration (ideological and economic) with the USA has undoubtedly influenced the free marketeering ideology that underpins the British approach to the CAP, in particular linked to calls to liberalise agricultural markets and to dismantle protectionist subsidy policies (Potter and Burney, 2002; Potter and Tilzey, 2005). This may suggest that the 'post-productivist' characteristics of free market trade may be *overemphasised* from a UK perspective. Conversely, it may also highlight that Atlanticist rhetoric and ideology still underpin many of the UK's agricultural policies and that no substantial transition away from ideologies associated with the Atlanticist food order has yet taken place.

Evidence on changing agro-commodity chains is, arguably, more clear-cut. Reference needs to be made to four intertwined issues: the move away from the Fordist regime of agricultural production and associated critique of the industrial agriculture model; the push for food quality instead of quantity; debates on the re-localisation of agro-commodity chains and associated moves towards 'vente directe'; and the increasing importance of regional food brands akin to the French system of 'appélation d'origine controlée'. First, linked to above-mentioned debates on ideological changes, there is much evidence of an increasing critique of Fordist regimes of agricultural production and, concurrently, rejection of the industrial Fordist agricultural model. Partly spurred by food-related catastrophes such as the foot-and-mouth and BSE crises in the UK (and beyond), the ever-recurring epidemics of swine fever in European countries, and increasing scares about the possibility of a 'bird flu' pandemic, many stakeholder groups in society (including consumers) have increasingly voiced their concerns about Fordist mass agriculture (Scott *et al.*, 2004). Yet, the reaction against industrial agriculture has been temporally (as well as spatially; see below) uneven. While many stakeholder groups in countries such as the USA, Australia, the UK or Germany have made concerted efforts to challenge productivist industrial agriculture processes (Potter, 1998; Wilson and Wilson, 2001), many continue to purchase, consume and, indeed, condone mass-produced industrial (and heavily processed) foods. Thus, Hoggart and Paniagua (2001: 49) suggested that "calls for improved food quality do not primarily result from heightened consumption emphases but owe more to recent food scares... [as a result] we have to question how far food (quality) issues provide good evidence of a shift towards a consumption-oriented countryside". Although empirical evidence is sketchy, Lang and Heasman (2004) highlighted that health problems in developed countries are often due to consumption of mass-produced processed foods with little nutritional value. While some consumers may have adopted post-productivist attitudes regarding Fordist production, the majority continue to support, encourage and condone industrially produced commodities through their food purchase and consumption habits. These problems do not only occur in the developed world but also increasingly in developing countries (Lang and Heasman, 2004). Marsden (2003: 42), for example, highlighted how in South-east Asia "as incomes rise, pre-packed products ... and products which have undergone further processing begin to replace some staple foods", leading to growing nutrition-related health problems. We, therefore, witness what may be termed a *temporally non-linear territorialisation of food consumption patterns* with productivist (or even super-productivist) tendencies evident through increasing obesity problems (see below) while, simultaneously, post-productivist food consumption pathways are evident through a growing trend towards healthier eating habits.

Second, debates on the push for quality instead of quantity are closely linked to the critique of industrial agricultural production processes. As Marsden (2003: 26) emphasised, "we are moving towards food supply chains which are qualitatively regulated rather than quantitatively". Undoubtedly, many consumers in advanced economies have given more attention to food quality, since evidence emerged showing links between food quality and human health in the 1960s (e.g. Carson, 1962; Goodman, 1999). Debates on the consumption of organically produced food are particularly interesting in this respect (Tovey, 1997). While the purchase of organic products continues to increase, organic food currently only comprises about 1-2% of the diet of Europeans, with most organic food currently sold via large supermarkets that offer little opportunity for knowledge transfer between producers and consumers (Marsden, 2003; Goodman, 2004). Yet, while some consumers have chosen the arguably post-productivist (but see also below and Chapter 9) pathway of organic high quality consumption, many consumers have not – often for financial reasons as organic produce tend to be 10-20% more expensive than non-organic (Dabbert *et al.*, 2004). In particular, there have been pronounced temporal differences across different cultural contexts. While in Germany 'biologischer Anbau' formed an important part of changing consumer demands for food quality as early as the 1970s (Wilson and Wilson, 2001), in the UK the emphasis on organic produce is a relatively recent phenomenon, although demand has increased dramatically in the wake of the BSE and foot-and-mouth crises (Hinchcliffe, 2001; Marsden, 2003).

Of great concern has been the issue of imported organic produce, mainly from developing areas. Although discussions about 'food miles' are more complex than simple arithmetic calculations of distance transported (Lang and Heasman, 2004), that a large proportion of the organic food sold in European and North American supermarkets travels long distances from its source of origin can hardly be seen as a 'post-productivist' indicator (Friedberg, 2003). Dabbert *et al.* (2004) highlighted further differences in the temporal adoption of the idea and philosophy behind the organic farming movement across Europe, with countries such as Sweden, Switzerland and Austria at the 'enthusiastic' end of the spectrum, but with countries such as the UK initially characterised by less enthusiastic adoption of the idea. That no uniform temporal transition towards organic consumption has occurred within advanced economies is even more evident if we scrutinise 'typical' food consumption habits of middle-class consumers where the cost of produce is not the key determinant for purchasing decisions. For many, the shopping trolley will be comprised of *both* organic and non-organic products, suggesting that productivist and post-productivist action and thought continue to occur side-by-side even within the arena of personal food purchase decision-making (and who is to say that all non-organic products should be categorised as productivist?). Marsden (2003: 213), therefore, argued that "as organics become subsumed into mainstream food supply networks several facets or elements of the more radical philosophy, particularly associated with local economic development and the support for small-scale agriculture, potentially become side-lined or ignored" (i.e. notion of the 'mass niche' or 'quality treadmill'). At a broader level, evidence shows that consumer behaviour about food quality purchases varies considerably between different cultural contexts and between different stakeholder groups. It is evident that among some groups in society no transition towards healthy food consumption has taken place, and that the 'junk-food-dependent' segment of society may, indeed, be gradually increasing. Hawkins (2002), for example, highlighted that only 11% of UK consumers were interested in knowing more about food issues and changing food habits. Such non-linear pathways in the development of *multiple food cultures* within individual countries are particularly evident through increased levels of obesity. Lang and Heasman (2004) highlighted that the USA, Australia and the UK are now among the most obese societies on Earth (although many other countries are quickly catching up), indicating that while one, arguably small, segment of society consumes healthy (and possibly largely organic) food, large sections of society do not. Marsden (2003: 8) further argued that "the provision of 'new' quality lines in the superstores (e.g. 'freedom foods'; welfare lines; free range) really only represent new innovations in supermarket category management principles ... rather than clearly defined alternative food supply networks" (see also Hollander, 2003). If we accept the push for food quality production and consumption as an indicator of the p/pp transition model, then evidence suggests that productivist and post-productivist food consumption behaviour occur side-by-side in a temporally non-linear way.

Third, the re-localisation of agro-commodity chains and associated moves towards 'vente directe' (or 'alternative agro-food networks' in Goodman's, 2004, terminology) as possible indicators of a post-productivist transition are equally problematic. Such vertically disaggregated production is seen as key to post-productivism and has become more evident in many developed countries (e.g. Ermann, 2005, for southern Germany). Yet, the UK-centrism of the p/pp transition debate may, again, be evident, as the de-localisation of agro-commodity chains – i.e. the gradual dismantling of local processing and increasing agglomeration of the food industry away from places of agricultural production – has, arguably, been more pronounced in the UK than in many other developed countries (Pretty, 1998). Many local abattoirs in the UK were closed, so that today only a quarter of those that existed in 1970 have survived. In most cases, remaining abattoirs are located long distances away from centres of livestock production – for some commentators a key reason for the severity of both the recent BSE and foot-and-mouth crises in the UK (Scott *et al.*, 2004). Further, it is now estimated that one quarter of all lorry journeys in the UK are associated with transporting food long distance, with an annual cost of £9 billion and an additional estimated cost of £2 million in atmospheric pollution damage (Pretty, 2005).

It is little wonder, then, that *relocalisation* of such food processing structures is seen as a key ingredient of the p/pp transition in a UK context. Yet again, processes have been temporally uneven. While the UK may have lost most of its local food processing industries to centrifugal agglomeration tendencies, other countries such as France arguably never delocalised food production processes to the same extent. Indeed, most rural areas in France have maintained a 'traditional' locality-based food production and processing infrastructure that has meant much shorter 'food miles' and, arguably, a closer connection of food consumers with places of food production (Jones and Clark, 2001; see also Guthey *et al.*, 2003, for California, and Ermann, 2005, for southern Germany). It is no coincidence, therefore, that UK food analysts have looked with envy at the 'vente directe' networks in France – processes of direct or more localised food sale structures that had to be *reinvented* in the UK through a revaluation of farm-gate sales and, in particular, 'new' farmers' markets (Kneafsey, 2002). Yet, by the early 2000s only 6% of UK farms were involved in direct selling activities, only 400 farmers' markets operated in the UK, and only 6000 farms were engaged in direct sales to supermarkets (Kirwan, 2004; Robinson, 2004) – emphasising the general persistence of non-localised agro-commodity chains. Roche (2005: 301), therefore, argued that "advocates of localised food … may have missed the point that [this] may conceal a range of agricultural forms, consumer motivations and politics". This was reiterated by Winter (2003), who emphasised that the ideology of localism does not necessarily ensure 'post-productivist' food safety or environmental sustainability, and by Hinrichs (2003) and Allen *et al.* (2003) who, through studies of relocalisation of food systems in Iowa and California respectively, suggested that such relocalisation processes do not necessarily lead to progressive societal changes akin to the post-productivist transition model. Thus, productivist and post-productivist processes related to both de- and relocalisation of agro-commodity chains become increasingly blurred, and temporally non-linear transitional patterns tend to predominate.

Fourth, emphasis has been placed on the increasing importance of regional food brands akin to the French system of 'appélation d'origine controlée'. For some this highlights a post-productivist move towards a new valuation of regionally or locally produced food (see Chapter 6), while for others this may signal a continuation of patterns and processes of food production that had never substantially changed in the first place. The importance of food branding and the (re)discovery of the importance of the *geography of food production* may highlight the detachment of consumers that, as some would argue, has occurred through productivism, and that has now been 'rediscovered' through post-productivist processes. However, such notions of temporal linearity are again problematic, as many areas never lost, nor had the need to reinvent, the link between locality and food. The French example is, again, revealing, where the notion of 'appélation d'origine controlée' always constituted an important aspect of a regional French food culture, in particular related to wines, cheeses and other regional speciality products (Jones and Clark, 2001; Pretty, 2002).

Overall, it is, therefore, easy to concur with Evans *et al.*'s (2002: 318) critique of the notion of a post-productivist transition towards high quality foods, as "it is the tendency uncritically to assume a relationship between quality and post-productivism to which objection can be raised". Similarly, Marsden (2003) concluded that food issues related to nature, region and quality are configured in ways which *suppress variation* in the standardised lines of food supply, especially as innovation and capital is increasingly located towards the globalised retail end of supply chains (see also Goodman, 1999). This evidence, therefore, questions the implied temporally linear transition towards post-productivist agro-commodity chains.

Changing agricultural production strategies and farming techniques?

As Chapters 5 and 6 highlighted, debates as to whether agricultural production strategies and farming techniques have moved towards post-productivist approaches are closely intertwined with discussions on changing agro-commodity chains, policies and ideologies. Linked to the above-mentioned critique of Fordist agriculture has been a critique of 'productivist' agricultural practices, often associated with environmental degradation. Debates have focused on the three issues of extensification, diversification and 'alternative' production strategies, such as organic, integrated or precision farming, as key opportunities for 'new' farm development pathways in the post-productivist countryside (Ilbery and Bowler, 1998; Robinson, 2004).

The notion of *extensification* provides one of the most interesting debates on the temporal linearity of the p/pp transition model. While many policies have encouraged EU farmers to extensify farm production in line with the greening of agricultural policy arguably moving away from a productivist production-maximisation model (see above), processes of intensification are continuing simultaneously. While EU upland farming areas currently tend to *extensify* production encouraged by the 2nd pillar of the CAP, in much of the more fertile and infrastructurally better embedded lowlands evidence suggests continuing *intensification* of production (Walford, 2003a, b). Evans *et al.* (2002: 320), for example, argued that in many areas of Europe "actual stocking densities have remained high and have had much scope to increase in real terms" (see also Wilson and Juntti, 2005, for Mediterranean countries). In Europe, intensification trends are particularly evident in the Paris Basin (France), East Anglia (UK), Emilia Romagna (Italy) and central Germany, while in the USA, Australia and the Ukraine further intensification is particularly apparent in arable areas. It is in this context that Halfacree (1999, 2004) coined the term 'super-productivism' – a process of agricultural intensification that goes even beyond the most intensive production strategies during the so-called 'productivist era'. Here, debates on the post-productivist transition most closely intertwine with discussion of sustainable agriculture, as it is in these intensively farmed arable areas that some of the most severe environmental problems linked to agricultural production can be identified (Cocklin and Dibden, 2004). Although there is much discussion of the use of various 'indicators' to measure agricultural outputs and policy effects (Brouwer and Crabtree, 1999), key indicators of increasing farming intensity include rising levels of productivity and increasing use of biochemical inputs (see below). The German case is, again, illustrative. While yields of selected crops increased gradually during the so-called productivist era (e.g. wheat from 2.5 t/ha in 1950 to 4 t/ha in 1980; barley from 2.4 t/ha to 3.5 t/ha; or rapeseed oil from 1.6 t/ha to 2.3 t/ha), yields increased at a *much faster* rate during the so-called post-productivist period (e.g. wheat from 4 t/ha in 1980 to 7 t/ha in 2000; barley from 3.5 t/ha to 5.5 t/ha; and rapeseed oil from 2.3 t/ha to 3 t/ha) (Wilson and Wilson, 2001), and similar trends can be observed in most other advanced economies (e.g. Potter, 1998, for the USA; Robinson, 2004, for other European countries). Thus, productivist (and even 'super-productivist') and post-productivist strategies occur side-by-side, and no temporally linear p/pp transition can be identified for intensification or extensification processes in most agricultural regions.[26]

The notion of *diversification* as a possible indicator for the p/pp transition provides insights into an equally challenging debate (Evans *et al.*, 2002). As highlighted in Chapter 6, many authors see diversification as an important component of post-productivism, as processes of diversification are explicitly linked to both a reduction in core agricultural production (not dissimilar to notions of 'deagrarianisation' discussed below) and to post-productivist countryside consumption activities (Walford, 2003a, b). Debates have particularly focused on the production of non-food commodities, pluriactive farm household pathways, and on-farm activities linked to the consumption of the countryside, such as horse riding, the establishment of golf courses or farm animal zoos, many of which are closely associated with tourist activities in rural areas (Wilson, 2001, 2002; Marsden, 2003). Yet, diversification emerges as a particularly problematic indicator for the transition towards post-productivism, leading Evans *et al.* (2002: 319) to argue that "caution must be exercised in the use of diversification as a descriptor and theorization of post-productivism". While some diversification activities are more clearly associated with a shift away from agricultural

---

[26] Even more contentious is the empirical evidence for farm dispersion as a key indicator of the transition towards post-productivism (Ilbery and Bowler, 1998). As yet, no studies (in a European context at least) have shown that the trend towards concentration of specific types of production has been halted or, indeed, reversed.

production (e.g. farm holidays, conversion of former farm buildings to non-agricultural activities), others can be associated with the *continuation* of productivist farm development pathways (see also Chapter 10). This is particularly true for new pathways associated with energy production. For energy crops, for example, although the output may no longer be associated with 'traditional' production of food or fibre, cultivation (e.g. oilseed rape) is often associated with high-intensity, and often environmentally harmful (e.g. high biochemical input), productivist cultivation strategies. This is also true for 'alternative' production strategies involving the breeding of rare or exotic breeds, as activities such as deer or llama farming may lead to increased agricultural production, may be environmentally harmful, and could, therefore, be classified as 'productivist' (Evans *et al.*, 2002). The most problematic issue relates to the fact that diversification on farms often occurs *simultaneously* with intensification of agricultural activities on non-diversified parts of the farm (see also discussion below on on-farm territorialisation).

Further, although on-farm tourist accommodation has now become an important part of diversified farm holdings in many developed countries, Morris and Evans (1999) showed for the UK that, even during boom times for diversification in the late 1980s, only about 6% of farms had on-farm accommodation (the most popular type of diversification activity). Similarly, Marsden (2003) suggested that by the early 2000s, only 7% of UK farms had diversified into agri-tourism ventures, and only 20% were involved in nature and landscape management schemes. Evans *et al.* (2002), therefore, argued that even if farmers were willing to embark on pathways of diversification, they are often unable to do so due to 'lock-in' mechanisms into quota systems (especially in the EU) or arable area payment schemes which, according to Winter and Gaskell (1998), have hardened structural (and often productivist) rigidity in farming. Hoggart and Paniagua (2001: 48), thus, concluded that "when the diversification of farm activities is put under the microscope, surprise, surprise, we find farmers are more prone to diversify into production activities than into consumer services". This was also emphasised by Argent (2002: 108) from an Australian perspective, where "pluriactivity is not an unequivocally reliable measure [for the transition to post-productivism] because many diversification strategies – particularly those that are land based (e.g. niche livestock or crop enterprises such as emu farming and Asian vegetable growing) are still dependent upon the agronomic qualities of the farm to produce economically viable quantities and qualities of food and fibre … Conceptually, therefore, only those diversification strategies that are based totally off-farm, or are non-farm oriented but create market values out of the farm's amenity or other use values, can properly be said to represent a shift to post-productivism". Productivist non-diversified activities, therefore, continue to exist alongside – and indeed continue to dominate – post-productivist farm diversification processes in temporally non-linear ways. For most farmers in the so-called post-productivist era it appears to be 'business as usual' (Marsden, 2003).

The importance of acknowledging temporal non-linearity in agricultural production has also been highlighted in debates on the introduction of GM crops (Lang and Heasman, 2004). Most of the criticisms refer to the fact that there has been no *gradual* transition in the introduction of GM organisms, and that effects of the mixing of plant and animal genes within one organism have unforeseen impacts. While 'traditional' selective breeding and associated genetic modification over time can be associated with a long transition period (e.g. basic cereal crops have been modified over 10,000 years) and, therefore, allowed for the parallel existence of various levels of selectivity among species, the sudden introduction of GM crops breaks this transitional cycle. This *accelerated transition* can be seen as one of the key reasons for the reluctant adoption of GM crops by most people (Levidow, 2005). Criticism exists, thus, not necessarily because these organisms have been created *per se*, but because there has been a limited transition phase that has not permitted gradual social and mental readjustment.

The cluster of organic, integrated and precision agricultural production strategies also warrants closer attention. The above discussion already highlighted that the labelling of organic *food* consumption as a post-productivist activity may be problematic. Similarly, organic *agricultural* production is equally ambiguous as a possible indicator for the shift towards post-productivism (Clunies-Ross and Cox, 1994). Although organic farming contains a host of approaches that may safely fall into the category of post-productivism (e.g. abandonment of biochemical inputs) it can, at the same time, also be interpreted as a highly productivist activity. Indeed, most organic farmers seek, like their non-organic counterparts, to maximise farm income and production (Marsden, 2003). Organic farming can be a very intensive agricultural production strategy, especially if it involves horticultural products (a large part of organic production). Organic farming strategies can, therefore, be seen as a *pragmatic productivist strategy* adopted by farmers for financial reasons to satisfy rapidly growing consumer demand. In addition, environmental benefits of organic farming remain doubtful, especially if organic farms are located near high-traffic road arteries or if they border on non-organic farms whose owners continue to use high biochemical inputs (Tovey, 1997; Dabbert *et al.*, 2004; see also Chapter 9).

Similar criticisms apply to integrated farming.[27] Morris and Winter (1999) acknowledged that while integrated farming undoubtedly has some environmental benefits over 'conventional' (i.e. productivist high external input) farming due to reduced use of biochemical inputs, they also highlighted that it can entail productivist farm development strategies. As a result, they described integrated farming as a 'third way' between conventional and organic farming, epitomising an agricultural production strategy that sits relatively comfortably *between* productivist and post-productivist farm production pathways (see also Schmid and Lehmann, 2000). Similar arguments have been made about 'precision farming' – an approach that also aims at reducing over-use of external inputs (Edwards *et al.*, 1993). Using a high-tech approach involving, for example, geographical positioning systems and remotely sensed field-level information about external input needs of a given terrain, precision farming reduces the need to apply biochemical inputs (especially fertilisers) evenly across a field, thereby reducing environmental impacts. This approach has gained ground in the UK, but has been particularly widely adopted in countries such as Switzerland and Austria (Schmid and Lehmann, 2000). However, as with integrated farming, there is also a productivist strand to precision farming, as it enables farmers both to save money (e.g. through reduced use of fertilisers) and to maximise agricultural commodity production in the most efficient way. Thus, both integrated and precision farming offer insights into *multiple pathways of farm decision-making* bounded by both productivist and post-productivist action and thought.

As highlighted in Chapters 5 and 6, the literature has focused on three further farming technique 'indicators' in conceptualisations of the p/pp transition: the reduced emphasis on mechanisation and associated reduced farming intensity, the reduction in farm labour inputs, and the reduction or total abandonment of biochemical inputs and associated replacement of physical inputs with knowledge inputs. In line with the lack of evidence of a transition to post-Fordist agricultural production, there is little evidence in many farming areas of reduced mechanisation and reduction in farming intensity. While some de-mechanisation has occurred in areas encouraged by agri-environmental policy to extensify agriculture (especially in uplands; see discussion below), intensively farmed areas have often witnessed increases in levels of mechanisation in the last decades. Thus, Marsden (2003: 5) argued that "it is clear that the realities of the current agro-food conditions are still largely pulling it along an agro-industrial [productivist] path which is based upon a neo-classical 'virtual' logic of scale and specialisation using industrial technologies". Germany, for example, continues to have one of the highest levels of farm mechanisation in Europe, and although tractor numbers increased substantially during the 'productivist' era, the so called 'post-

---

[27] Integrated farming is associated with farming techniques that do not go as far as organic farming in renunciation of biochemical inputs, but that see the farm as an 'integrated' unit in which productivist farming techniques can be used *side-by-side* with environmental conservation. The timing of external inputs, as well as selective and reduced application, are key to a successful integrated farming approach (Morris and Winter, 1999).

productivist era' has seen further rises (see Chapter 5). Indeed, farm machinery has become relatively cheaper over the past few decades, and while ownership of a tractor would have been unthinkable for many 'productivist' small family farms in the 1960s, it has become commonplace on most 'post-productivist' European farms at the beginning of the 21st century (Robinson, 2004). These changes have had severe repercussions for manual farm labour inputs. There is less debate about the 'post-productivist' transitional tendency of reduced agricultural workforce, including reductions in employed agricultural contractors and use of farm family labour inputs (Robinson, 2004; Woods, 2005). Indeed, data across Europe show rapid reductions in agricultural employment, with the UK witnessing reductions from 6 full-time workers/100 ha agricultural land in the 1950s to less than 3/100 ha in the early 21st century, Germany (territory of former West Germany) from 27 to 6 over the same time span, Italy from 31 to 12 and the Netherlands from 20 to less than 10 (Wilson and Wilson, 2001). This has gone hand-in-hand with rural depopulation trends in many developed countries, with some of the most pronounced effects occurring in Southern European countries (Oñate and Peco, 2005; Povellato and Ferraretto, 2005). However, for the hypothesised reduction or total abandonment of biochemical inputs and associated replacement of physical inputs with knowledge inputs, the picture is more varied. While some farmers have chosen farm development pathways based on total or partial abandonment of biochemical inputs (e.g. organic or integrated farming; see above), the majority continue to rely on application of fertilisers, pesticides and herbicides. This is best exemplified by the fact that Austria, currently the country with the largest share of organic agricultural land in the world, still has 87% of agricultural land farmed by non-organic practices.[28] Although awareness has undoubtedly increased about harmful environmental and health effects related to biochemical inputs (Carson, 1962; Ward et al., 1998), the postulated post-productivist transition towards a cleaner and greener agriculture has, therefore, only partly taken place.

Overall, the evidence suggests that a mixed picture emerges on the postulated transition towards post-productivist agricultural production strategies and farming techniques. While there are some tendencies towards post-productivist farming strategies (e.g. reduction in farm labour; extensification), productivist farming techniques continue alongside especially in the lowlands. Thus, Walford (2003b) suggested that in many advanced economies farm labour is now used much more intensively that in the early years of the productivist era. Further, Burton (2004: 208) emphasised that "it becomes relatively simple to understand why farmers are resistant to many of the suggested changes to the industry encouraged as part of the post-productivist modernisation of agriculture. While productivism and its consequences for the landscape may represent to us the excesses of an over-subsidised agricultural industry, for many farmers it represents a picture of good farming practice". Farmers use a *mixture* of productivist and post-productivist coping strategies to ensure farm survival by, for example, adopting arguably post-productivist agri-environmental policies to reduce farming intensity on parts of their farm, while continuing with productivist farming techniques on others. Cynics, therefore, argue that farmers continue with productivist farming techniques while *exploiting* the post-productivist policy environment for economic gains (Wilson, 2001). In the developed world, farming techniques have, therefore, not moved along clear temporally linear pathways, but depict complex patterns of temporally non-linear parallel strategies.

---

[28] It should be noted here that Austria leads the rank of countries with 'official' organic production (i.e. based on organic accreditation schemes linked to the CAP). Although no official statistics exist, many developing countries have shares of organic production (i.e. without artificial inputs) much higher than those of Austria (Pretty, 2002). However, as many reports suggest (e.g. Trippel and Davenport, 2003), the use of chemical inputs (especially pesticides) is increasing rapidly in many developing countries, and of the 1100 t of pesticides exported daily by the USA between 1997 and 2000, 75% were destined for developing countries.

Towards environmentally sustainable farming?

Debates on increased *environmental sustainability* form one of the key pillars of the conceptualisation of the p/pp transition. Indeed, some continue to argue that the environmental component forms one of the key reasons why debates on the p/pp transition have become so prominent in the literature, and that the only 'real' measure of the postulated transition can be found in farmers' contribution towards environmental sustainability (Argent, 2002; Walford, 2003b). Although debates are complex and could cover a large part of this book, I will restrict the discussion to key debates over the past 10 to 15 years on shifts towards environmentally sustainable farming. First, key components of sustainable farming in terms of soil, water and habitat conservation need to be considered. Second, debates have focused on whether 'post-productivist' AEP has contributed towards environmental sustainability.[29]

There is a wide range of literature that has discussed the characteristics of 'environmentally sustainable farming'. There is no room to delve into these debates here and good summaries can be found in several key texts (e.g. Pretty, 2002; Cocklin and Dibden, 2004). The general essence is that landholders who farm in environmentally sustainable ways guarantee the survival of remnant habitats and the cleanliness of water resources on the farm as well as the maintenance of soil quality. Sustainable farming is, therefore, about striking the right 'balance' between inputs and outputs and is generally seen as an approach to farming that is in tune with environmental limits. Yet, there is increasing evidence that, at the global level, farmers are moving *away* from environmental sustainability and that some farm regions are at the brink of environmental collapse due to overuse of external inputs, the indiscriminate destruction of remnant wildlife habitats, and the pollution of surface and groundwater resources (Pretty, 1995, 2002). This is particularly the case in areas such as the American Midwest, the Australian wheat belt, the Paris Basin, Emilia Romagna (northern Italy) and most of the Netherlands where 'super-productivist' agricultural practices have led to irreversible environmental damage and degradation (Wilson and Bryant, 1997; Robinson, 2004). Environmental problems are often exacerbated in these areas due to the time lag of many pollution problems, and it is only now that productivist practices of high external input applications of the 1970s and 1980s are coming to the fore.

Evidence from a study in Southern Europe suggests that many intensive agricultural systems (e.g. durum wheat areas in Italy and Portugal or intensively irrigated areas in southern Spain) have exacerbated desertification processes and, in some cases, have led to irreversible environmental damage to soils, original plant cover and water resources (Wilson and Juntti, 2005). Despite the 'greening' of agricultural policy in recent decades highlighted above, in some of these areas there is an impression that many environmentally worst affected areas have been abandoned by policy-makers and that, instead, the focus of policy has been on areas that already have high conservation values (e.g. in or near national parks; areas with high tourism potential), suggesting a rather pragmatic and compartmentalised policy approach. The continuation of incompatibility of intensive agriculture with environmental conservation in many productivist farming areas, thus, continues to be evident (Knickel, 1990; Potter, 1998). It is in this context of worsening environmental conditions that debates have focused on the potential environmental benefits of AEPs. As highlighted in Chapter 6, AEP now forms a key policy approach in many developed countries, with about one quarter of all farmers and land currently enrolled into voluntary agri-environmental schemes (or higher scheme tiers) in the EU (Buller *et al.*, 2000; Wilson and Juntti, 2005). Yet, debates continue as to whether AEP is leading to substantial environmental improvements on farms. Increasing evidence shows that AEP continues to be used by most landowners as an *economic subsidy* rather than a means for improved environmental

---

[29] The question whether farmers' attitudes and identities have changed from productivist to post-productivist will be discussed in detail in Section 7.3.6 on the fallacy of structural causality.

sustainability (Wilson and Hart, 2000; see also above). This suggests that there is little potential yet for bringing about substantial environmental recovery on farms (Buller *et al.*, 2000). In particular, little emphasis continues to be placed on re-creation of destroyed habitats (Fish *et al.*, 2003). A similar criticism relates to the concern that AEP generally continues to focus on areas that already have high conservation value (especially uplands), while neglecting intensively used (and at time super-productivist) lowlands (see below). Further, until recently many AEP contracts were on a 5-year renewable basis, and farmers could decide to opt out after 5 years and, in theory, destroy all environmental gains accrued over the 5-year agreement phase (this may be different from 2006 onwards with new cross-compliance measures coming into force across the EU). The key concern has, therefore, been with the lack of permanence of changes induced by AEP (Wilson, 1997b; Fish *et al.*, 2003). The potentially positive environmental effects of AEP can also partly be destroyed by accidental or conscious productivist misinterpretation of policies at regional and local level by stakeholder groups holding specifically powerful positions in policy implementation networks (Wilson and Juntti, 2005). Thus, well intended policies do not always trickle down to grassroots level and, although in theory a crucial component of WTO 'green box' agreements, the way some farmers are allowed to use AEP moneys can be equated with a continuation of productivist 'blue box' subsidies (Potter and Burney, 2002). This has led Evans *et al.* (2002: 316) to argue that the quest for agricultural sustainability can only be seen as a 'societal wish' "that as yet can only begin to be approached through a rather preliminary and incremental set of agri-environmental policy measures".

As a result of these controversies, the notion of 'best agricultural practice' – a notion that now underpins cross-compliance as part of Agenda 2000 – has been very difficult to define. Different countries in the EU, for example, define 'best agricultural practice' in different ways, some emphasising environmental conservation, others focusing on soil protection, while many also focus on identification of agricultural practices that ensure the survival of farming and rural communities (Wilson and Buller, 2001; Lowe *et al.*, 2002). The notion of environmental sustainability in the countryside, therefore, varies considerably, depending on cultural, political and farming interests. Nonetheless, there have also been some success stories of improved environmental sustainability. Evidence collected by European environmental NGOs suggests that some of the wildlife threatened by agricultural intensification is beginning to recover (RSPB, 2001), while nitrate pollution levels in some farming areas targeted by specific incentives to reduce nitrate applications are improving. It is particularly in the uplands and farming areas where 'traditional' low-intensity production systems have survived where most of the recent environmental successes can be witnessed. Yet, in these areas farming may have never had a substantial impact on the environment (i.e. low external inputs, low intensity farming with high levels of biodiversity; Bignal and McCracken, 1996), and it remains questionable whether the targeting of these already biodiversity-rich areas for AEP is the right approach for encouraging sustainable management (Wilson, 1997b; Potter, 1998).

Evidence, therefore, suggests parallel temporally non-linear pathways of environmental trajectories on most farms in the developed world. While many farms caught in the 'treadmill' of production maximisation (Ward, 1993) continue with relatively environmentally unsustainable farm trajectories, we also witness a move towards more environmental sustainability on other farms – partly spurred by policy, partly, as we will see below, by farmers' own initiatives and changing attitudes, perceptions and identities (Burton and Wilson, 2006). Modern farming has not unequivocally moved towards sustainable environmental management as the p/pp transition model implies. Instead, multiple pathways of temporally simultaneous processes are apparent, some of which resonate more with the productivist end of the spectrum while others follow post-productivist trajectories.

Temporal non-linearity and the p/pp transition model

The above discussion highlights that a critical analysis of the dimensions of the p/pp transition model reveals that *temporal non-linearity* characterises productivist and post-productivist transitions. Based on evidence from the developed world it is, therefore, questionable whether a *full* transition towards post-productivism has taken place – an issue also highlighted by Evans *et al.* (2002: 313) who were surprised at the continuing and "widespread and uncritical use of such an all-encompassing term [post-productivism] … given debates elsewhere … on the rejection of dualistic thinking". This has particular repercussions for conceptualisations of the timing of the post-productivist transition highlighted in Chapter 5. First, productivist and post-productivist action and thought most often occur *simultaneously*. In other words, most agricultural practices contain some productivist elements while, at the same time, also showing evidence of post-productivism. This means that productivist and post-productivist action and thought can be conceptualised as a spectrum of *parallel temporal pathways* akin to the Deleuzian transition model (see Fig. 2.1) that has characterised many other societal transitions outlined in Part 1. There are also striking parallels here with debates on how globalisation processes fragment transitional pathways. Nederveen Pieterse (2004: 67), for example, argued that "globalisation … increases the range of organisational options, all of which are in operation simultaneously. Each or a combination of these may be relevant in specific social, institutional, legal, political, economic, or cultural spheres. What matters is that no single mode has a necessary overall priority or monopoly". In contrast to the p/pp transition model shown in Figure 6.1, Figure 7.1 shows temporally simultaneous productivist and post-productivist Deleuzian transitional pathways.

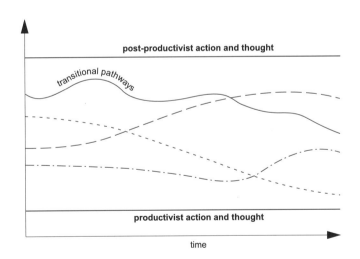

Fig. 7.1. Temporally simultaneous productivist and post-productivist transitional pathways

Second, and bearing in mind this Deleuzian nature of the p/pp transition, different dimensions of the p/pp transition exist along different temporal horizons. While productivist tendencies continue to be apparent in all the seven dimensions discussed above, some dimensions – in particular agricultural policies, food regimes, governance structures and agricultural production strategies – contain elements with post-productivist characteristics. On the other hand, ideologies (and farmer identities; see discussion below), farming techniques and the questionable move towards more environmentally sustainable farming systems may be seen as dimensions with a stronger, and continuing, legacy of productivist action and thought (but with large variations, depending on specific regional and cultural contexts; see below). Thus, Marsden (2003: 7) emphasised the problem of "the longevity of the transitions … in creating radical dissonance in the structure and functioning of the local and regional agricultural production systems".

This suggests that – as with most other transitions from 'isms' to 'post-isms' discussed in Part 1 – the delineation of both productivism and post-productivism as specific *eras* or *epochs* in the history of agricultural change is problematic. The era of productivism was fractured and complex – or, in Hoggart and Paniagua's (2001: 51) words "in many arenas, we find heavy traces of the past in current changes, rather than fundamental transfiguration". Mather *et al.* (2006) similarly argued that "the temporal dimension of post-productivism is contested as much as the concept itself, and the search for the existence and definition of a precise start date may be unhelpful". Elements of the productivist Atlanticist food order may have begun to break down at the end of the 1970s (Marsden *et al.*, 1993), and some aspects of agricultural (and environmental) policy-making can be seen to have moved since the 1980s towards the post-productivist end of the p/pp spectrum (see Fig. 7.1 above) (Wilson, 2001). Similarly, some productivist governance structures may have begun to shift towards post-productivism during the 1990s and 2000s (Wilson, 2004), and some agricultural production strategies may have shifted away from entrenched productivist action and thought over the past few decades (Evans *et al.*, 2002; Holmes, 2002). However, many other processes linked to agriculture have *not* changed substantially from the productivist paradigm.[30] This non-linear nature of the p/pp transition lends credence to Sayer's (1989: 269-270) critique that most transitional analyses are based on "a linear, sequential form which inevitably favours the expression of the episodic over the configurational. Despite the fact that experience commonly registers many things happening at once, the representation of that experience has difficulty in reflecting this. Grasping the whole is more difficult than grasping what happens next in the story".

Deconstructing the fallacy of temporal linearity in the p/pp transition model has severe repercussions for conceptualisations of agricultural change. If productivism as an epoch becomes a problematic concept, this leads us to ask whether identifying a starting point for the era of 'productivism' is a meaningful objective. As with other transitional debates, is it also meaningful to conceptualise an agricultural period pre-dating productivism – i.e. 'pre-productivism'? Many agricultural historians have grappled with the issue of finding meaningful temporal boundary layers between individual 'agricultural periods'. In some textbooks, the notion of the 'imperial agricultural regime' (about 1850-1940) is seen to pre-date the productivist era (McMichael, 1995; Woods, 2005) – a concept based on equally questionable indicators, dimensions and temporal horizons of change. If we deconstruct the temporal linearity of the p/pp transition model (as implied in Fig. 7.1 above), then it could be argued that the notion of a pre-productivist agricultural era also no longer makes much sense. Indeed, conceptual notions of pre- and post-productivism merge into one, meaning the same thing, namely *non-productivist* processes (see below).

Evidence also suggests that 'traditional' (i.e. low-input extensive) farming is not necessarily always environmentally sustainable in the post-productivist sense, thereby further challenging notions of a unilinear p/pp transition. Similarly, Marsden (2003) argued that the 'post-productivist dynamic' is not necessarily synonymous with sustainable environmental management. This suggests that *post-productivist* action and thought may not be as environmentally sustainable as *pre-productivist* agricultural systems. In other words, the legacy of the productivist era is still apparent on many 'post-productivist' farms, for example through high nitrate levels in soils, or because a productivist neighbour may adversely

---

[30] Our discussion below will also highlight that the least evidence of a shift towards post-productivism can be found with regard to farmers' attitudes and identities (see Section 7.3.6)

influence a farmer's attempts at sustainable environmental management. It is in this context that Marsden has increasingly moved away from the original concept of post-productivism as an epoch, reconceptualising post-productivism, in his view, as a *transitory* phase that is not a means to an end towards sustainable rural systems as suggested earlier (Marsden, 2003). This echoes debates on other 'post-isms', in particular the transition towards post-modernism. Latour (1993), for example, mentioned the problem of conceptualising a 'pre-modern' era, and King (1995) also challenged the implied temporal linearity of the post-modern transition by arguing that 'relabelling' was required: the era of modernism should be renamed 'pre-modern', while the 'modern' should become what was termed the 'post-modern'. Acknowledging temporal discontinuities in the transition towards post-modernism, King (1995: 121), therefore, argued for a 'respatialization of the modern' as a way to overcome the problem "that what some have labelled 'postmodern' culture pre-dated what they have labelled 'modern' culture". For the demographic transition model, Cleland (2001: 64) similarly suggested that "any theory of profound social change must start with a detailed analysis of the relevant system prior to the change … Misspecification of the system prior to the point of departure inevitably results in misspecification of the forces of change".

Part of the problem with the p/pp transition model may, therefore, be that the notion of 'productivism' was conceptualised without considering its possible continuity (or discontinuity) with a previous, but hitherto weakly defined, pre-productivist agricultural era. Further, and similar to other debates on transitions from 'isms' to 'post-isms', the evolution of the term 'post-productivism' only brought into being the term 'productivism'. We have, therefore, witnessed the *retrospective definition* of the 'productivist' era from a 'post-productivist' vantage point, with the result that *during* the productivist (or indeed pre-productivist) era no theoretically grounded reference can be found that acknowledged the existence of that specific regime. This bias is important, as Chapter 5 highlighted that 'productivism' has been conceptualised from a *presentist* vantage point – in other words, from a perspective that benefits from the knowledge that agriculture may have 'moved on' from the productivist era – thereby influencing an 'objective' assessment of what made up the so-called 'productivist era'. This dilemma is nothing new, as virtually all conceptualisations of 'isms' (e.g. Fordism, modernism) have suffered from presentist interpretations of the past (see Chapter 3). For post-modernism, for example, Dear (1986: 373) argued that the central issue in assessing 'isms' and 'post-isms' "is the problem of theorizing contemporaneity". The presentist bias in conceptualisations of the p/pp transition is, therefore, one of the reasons why assumptions that advanced economies are now in a post-productivist era can not be substantiated.

It is interesting to speculate whether the so-called productivist era (if it exists) could be seen as a brief 'blip' (the 'productivist trough') in what has been, for thousands of years, a pre- or indeed post-productivist era characterised by high environmental sustainability, low intensity and productivity, localised production and food consumption, and close-knit farming communities (Wilson, 2001). As Figure 7.2 highlights, such a concept would imply, that if we take longer time horizons into account, then the past 50 years of relatively intensive agricultural production may pale into conceptual insignificance (see also Fig. 10.11). As Pretty (2005) highlighted, if we were to accept the notion that a productivist era did exist, then this era comprised, at most, two to three farm generations out of about 400 farm generations since the Neolithic period (in Europe). The 'productivist era' may, therefore, be an irrelevant time period in the grander scheme of agricultural evolution.[31] The problematic issue of temporal linearity embedded in the p/pp transition model becomes particularly evident if we ask whether pre-productivism automatically implies that *all* societies are 'aiming' towards productivism (i.e. implied directionality) (Wilson, 2001)? Can

---

[31] Note parallel criticisms regarding debates on both post-colonialism and post-socialism (see Part 1), where recent contributions to the debates suggest that both 'colonialism' and 'socialism' can be interpreted as Euro-centric discourses of a phase in history that was relatively brief (i.e. decades or a few centuries at most) and that neglects historical pathways before these specific 'eras' (Simon, 2006).

post-productivism, therefore, only occur in rural areas that have 'gone through' the productivist phase, or is it possible to conceive of rural areas that move directly from pre-productivism to post-productivism – in other words, can societies *leapfrog* one stage of agricultural transitional development to another?[32]

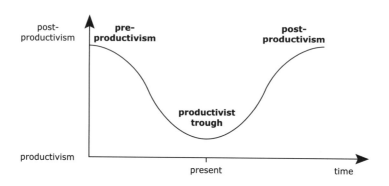

Fig. 7.2. The productivist trough

Acknowledging the non-linearity of the p/pp transition enables us to better understand the issue of 'discontinuance' of post-productivist activities (i.e. farmers reverting back to productivist activities after a brief spell of 'post-productivist' farming) – issues that are also at the heart of debates on technological and evolutionary transition (Dennett, 1995; Rogers, 1995; see Chapter 4). Most evidence comes from research on agri-environmental policy and farmers' diversification strategies, where studies have shown that farmers can quickly *revert back* to more productivist farming techniques, for example, after deciding not to renew agri-environmental contracts or to withdraw from diversification activities (e.g. Wilson and Hart, 2000, 2001; Fish *et al.*, 2003). Rantamäki-Lahtinen (2004), for example, showed that some Finnish farmers intended to revert back to productivist farming activities once their agri-environmental contracts expire, while one-fifth of farmers in her survey also argued that, once they were legally allowed to leave schemes, they would re-focus back on productivist agriculture and withdraw from diversification activities altogether. Seeing productivism and post-productivism as activities in a wider *spectrum* of farmer decision-making opportunities, and not as two processes that follow each other in a temporal linear way, allows us to better understand such temporally discontinuous farmer decisions. As Chapter 9 will highlight, this is the essence of what I will term 'multifunctional' farm decision-making pathways – pathways that, at times, may straddle the post-productivist (better: 'non-productivist'; see below) end of a wide decision-making spectrum but that, at other times, may veer towards the productivist end of the spectrum.

7.3.4 The fallacy of spatial homogeneity

As Chapters 5 and 6 highlighted, assumptions about spatial homogeneity underpin conceptualisations of the p/pp transition. It is generally assumed that different agricultural regions are affected in similar ways by post-productivist transitional processes, and that they

---

[32] These questions will have particular relevance in the discussion of the fallacy of global universality in Section 7.3.5.

are simultaneously partaking in the transition. Parallel to conceptualisations of other transitions discussed in Part 1, early conceptualisations were based on the assumption that the p/pp transition model can be applied to other advanced economies beyond the UK, and that most countries and regions would go through a productivist phase before moving onto post-productivism. Although criticisms of such simplistic assumptions emerged relatively early (see Section 7.2) – and despite the issue of discursive barriers and the UK-centricity of the p/pp transition highlighted above – none of these critiques explicitly highlighted that the transition towards post-productivism may be characterised by *spatial heterogeneity*. This section will explore the 'exportability' of the p/pp transition model into the *developed world*, while Section 7.3.5 will explore the p/pp transition model with particular reference to the *developing world* (fallacy of global universality).

A review of the international literature reveals surprisingly little reference to 'post-productivism' (and indeed 'productivism') in advanced economies beyond the UK (e.g. Wilson, 2002, 2005), although Australian and New Zealand researchers have recently begun to critically analyse the applicability of the p/pp transition to their national contexts (e.g. Holmes, 2002; Jay, 2005). This suggests either that there is little faith among the academic community that the concept can be applied in other advanced economies, or that advanced economies other than the UK have not yet moved beyond productivism. Despite UK-specific indicators of post-productivism, both these explanations are highly unlikely. Much work has highlighted, for example, that many similarities exist between agricultural and rural developments in the UK and other Western European countries (e.g. Hoggart *et al.*, 1995; Buller *et al.*, 2000), that similar processes of rural diversity are occurring in different parts of Europe (Van der Ploeg, 1997; Wilson, 2002), and that external policy pressures under the umbrella of the WTO are affecting most advanced economies in similar ways (Potter and Burney, 2002; Potter and Tilzey, 2005). Most of the post-productivist dimensions should, therefore, also be applicable in a wider European context.

As highlighted in Chapter 6, it is particularly regarding development of CAP policies that researchers have ventured beyond the UK framework in their conceptualisations of post-productivism (e.g. Potter, 1998; Marsden, 2003) – arguably one reason why so much emphasis has been placed on the *policy dimension*, as policy change has been seen as the only truly 'common' indicator to assess post-productivism across the EU. Similarities are, however, also emerging for common EU member state responses to changes in the organisation of food regimes under the WTO URAA (Potter and Burney, 2002), in responses by member states to CAP policy frameworks (e.g. similar adoption of AEP structures; Buller *et al.*, 2000), in efforts to tackle desertification problems in agricultural areas of Southern Europe (Wilson and Juntti, 2005), or growing consumer demand for organic products across the EU (Dabbert *et al.*, 2004), to name but a few. Further, several authors have highlighted that ideological changes towards the role of agriculture in society have been similar to the UK situation in many advanced economies, especially concerning rejection of increased industrialisation of agriculture, criticisms of the commercialisation of agricultural production, and criticisms of the increasing power of multinational agro-businesses (e.g. Wilson and Wilson, 2001, for Germany; Bell, 2004, for the USA; Argent, 2002, for Australia). In particular, similarities in most European countries can be found for the relative loss of the central hegemonic position of agriculture in society and the associated ideological and economic security of farmers – a process particularly evident in Germany (Wilson and Wilson, 2001), but also in Denmark (Andersen *et al.*, 2000), Greece (Beopoulos and Vlahos, 2005) or Portugal (Vieira and Eden, 2005). Further, in many European countries and in North America farmers have also increasingly been labelled as 'destroyers of the environment', possibly best highlighted through changing rhetoric concerning the definition of what is meant by 'good agricultural practice', which increasingly incorporates criticisms of conventional agricultural practices (McCarthy, 2005). Similarities can also be found across Europe for counter-urbanisation trends and their effects on the countryside (Wilson, 2005), as well as with farmers' AEP policy behaviour (Buller *et al.*, 2000; Wilson, 2002).

Finally, similarities have also emerged concerning consumer behaviour, especially in the context of improved food quality and healthier food production (Lang and Heasman, 2004).

This suggests that, at the national level at least, sufficient similarities exist that would suggest that the p/pp transition model *should* be applicable across most of Europe and, indeed, the developed world. Yet, evidence suggests that there are large *spatial differences* in the transition to post-productivism within advanced economies. I have termed this the *territorialisation* of productivist and post-productivist actor spaces – a notion which acknowledges that agricultural territories that show productivist and post-productivist characteristics may lie side-by-side, often in close vicinity (Wilson, 2001; see also Buller, 2005). While at the broader geographical level some commonalities appear to be identifiable that may suggest a common transition towards post-productivism in some spheres (but bearing in mind the temporal non-linearity of the p/pp transition highlighted above), closer scrutiny reveals fundamental differentiation into agricultural spaces ranging from extreme productivism to extreme post-productivism. Such spatialisation was already recognised early on in debates by those who conceptualised the notion of the post-productivist transition, with Marsden *et al.* (1993: 13), for example, arguing that "there has emerged a fragmentation of localistic orientations as individual rural communities and areas express their specific consumption and rural development needs".

Several patterns of p/pp territorialisation can be identified. Although common patterns exist between countries that have shared common approaches under the umbrella of the CAP, different political, ideological and economic developments in European countries until recently not part of the EU and the CAP make it more difficult to argue for common p/pp transitional patterns (with exception of Switzerland; see Wilson, 2002). It is here that discussions on the p/pp transition most evidently overlap with debates on the post-socialist transition, especially as trajectories of agricultural change in Eastern European countries (e.g. many new EU member states such as Poland, Hungary, Slovenia) have varied substantially from developments in the West, particularly between 1945 and 1990 (Stiglitz, 2002). Under socialism, agriculture in these countries operated under different political (i.e. non-democratic), ideological (productivist tendencies based on the need for self-sufficiency) and economic (i.e. non-capitalist) conditions that have led to different farm development pathways compared to the West. The former GDR is a particular case in point, as agriculture was largely characterised by large state-owned production cooperatives that worked hand-in hand with state officials and state-developed production goals that often showed no sense of agricultural reality (Wilson and Wilson, 2001). In most Eastern European countries, therefore, agricultural productivism was *artificially orchestrated* and shaped by the state, with little input by farming or rural populations who most often were mere pawns in a wider political game. It would, therefore, be futile to attempt to apply some of the dimensions of the p/pp debate (e.g. ideology, changing governance spaces) to these areas, as agricultural structures were constructed and maintained under artificial conditions (the same is true for non-democratic countries in the developing world; see Section 7.3.5 and Chapter 10). Nonetheless, *productivist* characteristics can be identified for farming techniques and agricultural production patterns that showed very little regard for sustainable environmental management (Pickles and Unwin, 2004). This *productivist artificiality* became particularly evident after the fall of socialism in 1990, when countries such as the former GDR suddenly found themselves exposed to 'post-productivist' pressures (e.g. AEP; but see also above). The slowness of the adaptation of the GDR agricultural systems to these changes is a particular case-in-point (Wilson and Wilson, 2001). Different political, social and economic trajectories will, therefore, inevitably create spatially different pathways in the transition towards post-productivism, emphasising the *spatial heterogeneity* of the processes (Potter and Tilzey, 2005). This was also emphasised by Argent (2002: 111) for the Australian context where the claim for a p/pp transition "has largely been met with growing incredulity by their Antipodean counterparts who see all around them the evidence of the continuing strong grip of productivism as financially stressed farm families strive to attain higher levels of productivity to survive" (see also Holmes, 2006).

Spatial heterogeneity in the p/pp transition can not only be identified in an east-west context within Europe, but also in inter-European north-south differentiation of productivist and post-productivist actor spaces. While most Northern European countries may be more firmly embedded in the post-productivist transition (or even in 'strongly' multifunctional agricultural pathways; see Part 3) (Andersen *et al.*, 2000), Mediterranean countries may not even have fully entered the productivist phase yet (as productivism itself can be seen as a Northern European and North American project; see Chapter 5), let alone moved towards post-productivist modes of thinking (Oñate and Peco, 2005; Povellato and Ferraretto, 2005). Some regions of the Mediterranean are still characterised by pre-productivist patterns of agricultural production (but see Section 7.3.3), with post-productivist policies 'imposed' onto these countries through the CAP framework (Wilson and Juntti, 2005). Thus, "we are witnessing pronounced temporal differences in the transition to post-productivism within the EU. While many Northern and Western EU countries may be on their way towards post-productivist agricultural regimes, Mediterranean countries are still largely caught in the productivist treadmill" (Wilson, 2005: 116).

There may, therefore, be a large gap in Mediterranean countries between post-productivist policies imposed by Brussels and farmers with staunchly productivist attitudes – a major explanation for the often low uptake of agri-environmental schemes in these countries (Buller *et al.*, 2000; Wilson and Juntti, 2005). This has been reiterated by Louloudis *et al.* (2000: 104) who argued for Greece that "productivist thinking has been favoured over post-productivism, and … environmental considerations have been subordinate to commodity production". Similarly, Peco *et al.* (2000: 167) argued that "Spain had only little experience with implementation of AEPs, and the persistent productivist ethos that had marred implementation of agri-environmental schemes under Regulation 797/85 has continued to strongly influence the implementation of Regulation 2078". As a result, Spain, Portugal and Greece have criticised the EU for imposing policies that aim at *extensification* of agriculture (a key indicator for post-productivism; see Chapter 6), at a time when they are still mostly concerned with 'catching up' with their Northern European counterparts through *intensification* (Vieira and Eden, 2005; Wilson and Juntti, 2005). These *imposed post-productivist discourses* have potentially devastating effects for extensification processes in the Mediterranean, especially by encouraging land abandonment – a rapidly accelerating depopulation process leading to severe environmental degradation through lack of maintenance of agricultural infrastructure (Oñate and Peco, 2005; Vieira and Eden, 2005). Thus, "while extensification can be seen as a key post-productivist indicator in Northern and Western Europe …, this indicator may not be applicable in Mediterranean countries where intensification is still a key political goal and where extensification often goes hand-in-hand with land abandonment and desertification" (Wilson, 2005: 116). It is little wonder, therefore, that Mediterranean policy-makers resist the imposition of post-productivist policies that are seen as insufficiently tailored towards their specific situation (Wilson and Juntti, 2005). It is also difficult to equate the extensification witnessed in the Mediterranean with 'post-productivist' processes, in particular as Mediterranean extensification is closely coupled with land abandonment and *not* with a perceived need to de-intensify agriculture based on 'post-productivist' ideologies. The transition towards post-productivism may, therefore, only be a Northern European project. Both the north-south and west-east territorialisation of European productivist and post-productivist spaces is, therefore, akin to a *cultural territorialisation* of post-productivism, in which cultural factors influence the interpretation of post-productivist policies imposed by Northern European countries. Indeed, evidence suggests that some stakeholder groups in the Mediterranean use the post-productivist policy agenda imposed by Brussels to further *productivist* goals at the national level (Povellato and Ferraretto, 2005). This was particularly evident in a recent study on the interlinkages between policy and desertification in Southern Europe, where weak actor-networks in some Mediterranean rural regions have enabled powerful stakeholder groups to use and misuse existing agricultural and environmental policy networks to further their own productivist – and often environmentally degrading – goals (Wilson and Juntti, 2005).

Such cultural territorialisation of post-productivism is also apparent in the Australian context where, due to different environmental conditions and policy frameworks, there is little evidence that post-productivist agriculture is beginning to emerge. Holmes (2002) suggested that spatial factors should be seen as the key component of the conceptualisation of the post-productivist transition in the Antipodes and, therefore, argued for a new concept of post-productivist *occupance*. In Australia, while certain post-productivist practices may be recognisable in specific localities (e.g. in designated national parks or areas with little agricultural land use), productivist land use continues to dominate in more intensively used agricultural areas in the east and south-east. Holmes (2002: 364), therefore, argued that in the Australian context "post-productivist dimensions are being super-imposed on pre-existing dimensions, with marked regional variability in the pace of change ... this is consistent with Wilson's interpretation of territorialisation, non-linearity and spatial heterogeneity". He further suggested that the sharp regional delineations evident in Australian rangelands "lend themselves to a provisional multi-attribute regionalization of Australia's rangelands, recognizing spatial heterogeneity and diversity" (Holmes, 2002: 378; see also Holmes, 2006). Similar patterns have also been identified in New Zealand, where a clear territorialisation into productivist agricultural land and post-productivist protected indigenous forest territories (with virtually no timber production) can be identified (Jay, 2004; Wilson and Memon, 2005).

Much reference has also been made to the difference between *uplands* and *lowlands* (Buller *et al.*, 2000; Marsden, 2003). Commentators have suggested that the European lowlands tend to be characterised by continued productivist farming practices predicated on highly intensive and often environmentally damaging agricultural practices (Walford, 2003a, b). Brassley (2005) argued that such territorialisation is also accompanied by 'professional' farming in the lowlands and by farmers with often less 'formal' education in the uplands. Halfacree (1999) suggested that some of the current lowland productivist holdings may even have become 'super-productivist' – i.e. more productivist than during the productivist era (e.g. East Anglia, Paris Basin or parts of the Netherlands). Although the territorialisation of farm holdings is most easily (but not necessarily most usefully) conceptualised on the basis of the dualistic poles of productivism/post-productivism – also referred to as the 'two-track countryside' by Ward (1993) – territorialisation can also mean a broader *spectrum* of responses encompassing a wide variety of adjustments in farming, whether linked to uplands or lowlands or other spatial disparities in adoption of post-productivism. As Lowe *et al.* (1993: 206) argued, "the retreat from agricultural productivism has been varied. Some farming areas continue to experience intensification of production; others face new types of productivism linked to other external capitals ... while others are experiencing a partial decoupling from the high-tech model through various forms of extensification and diversification and reintegration into local and regional economies".[33]

As a result, Marsden *et al.* (1993) suggested four ideal types of 'countryside', termed the preserved countryside (established preservationist and anti-development interests), the contested countryside (rural spaces outside core commuter catchment areas), the paternalistic countryside (owned by large private estates) and the clientelist countryside (remote uplands where agricultural interests still hold sway), without claiming that these necessarily needed to be reflected in neat geographical entities. Although these categories have been criticised for their applicability beyond the UK context (e.g. Hoggart *et al.*, 1995), they nevertheless usefully highlight the spatial territorialisation of countryside concerns in most advanced

---

[33] See also Cloke and Goodwin (1992) who argued that post-productivist territorialisation can be conceptualised through 'rural structured coherences' that comprise arenas of contestation such as those embedded in the new hi-tech service sector economy, rural areas as places of commodification, or areas of slow and steady decline.

economies. Marsden (2003: 106) argued that "such emerging ideal types begin to capture the processes of differentiation and dispersal occurring in the post-productivist countryside" – processes that Marsden terms the 'new post-productivist rural diversity'. Focusing on the equally differentiated spaces of the consumption countryside, Marsden (2003: 99) further argued that "while there are national and indeed international trends which suggest the development of the consumption countryside, its shape and effects vary considerably concerning the particular social and regional configuration of that rural space". Holmes (2002) added an interesting new dimension to these debates by arguing that in the Australian context a particular territorialisation is occurring whereby the 'outback' (the remote interior and very dry parts of Australia) is increasingly depicting post-productivist characteristics (e.g. retreat or pronounced de-intensification of farming), while the intensively farmed eastern and south-eastern parts of the continent continue to be staunchly productivist or even super-productivist. Thus, most actors and rural/agricultural spaces are located 'in between' productivist and post-productivist action and thought, and it is this territory that is the most likely arena of conflict for financial resources, actor spaces and ideologies (Murdoch and Marsden, 1995). This leads one to question the implied assumption of spatial homogeneity embedded in conceptualisations of the p/pp transition model.

In this context, it is interesting to revisit Marsden's (2003) recent thoughts on the spatiality of the p/pp transition. As one of the key proponents of the p/pp model in the early 1990s, Marsden suggested a more nuanced approach that acknowledges the spatial heterogeneity inherent in recent agricultural change trajectories. He now identifies three broader categories of territorialisation that may also be more applicable outside the UK (Marsden, 2003). First, the productivist *agro-industrial dynamic* (synonymous with productivist action and thought) is characterised by the squeezing out of nature in the agro-food system, standardised products, capital-intensity, sophisticated food-supply chains, high concentration of farms, and large farm units. In line with the productivist ideology, rural space is seen essentially as agricultural space. Second, the *post-productivist dynamic* (the other extreme of territorialisation in my terminology) sees rural space as consumption space, rural land as development space, the use of the natural as attractor in the counter-urbanisation process, and is characterised by commodified nature. The third dimension is the *rural development dynamic* (i.e. the 'bit in the middle' in my territorialisation concept), characterised by a new role for agriculture, re-embedded food chains, a revised combination of nature/value/region with co-evolving supply chains, and by a recapturing of lost values of rural space. In this third territorial space, the emphasis is on rural livelihoods and new associational designs and networks – in the European context a 'Europe of rural development regions' (Ray, 2000), where rural development acts as a counter-movement and activates the potential of rural resources. All three dynamics or 'territories' may occur together, re-emphasising that the territorialisation of productivist and post-productivist action and thought is spatially heterogeneous. Thus, Marsden (2003: 192) argued that "the agro-industrial and the post-productivist development dynamics are juxtaposed in space". As Section 7.4 will highlight, Marsden's latest p/pp concept comes closer to what I will term the 'modified p/pp transition model' (see also Chapter 9).

Yet, territorialisation is not necessarily restricted to larger spatial entities such as countries or regions. Indeed, there is increasing evidence of what I termed *on-farm territorialisation* of productivist and post-productivist spaces, where action of one agricultural actor may be located at different places within the productivist/post-productivist spectrum (Wilson, 2001). A farmer may adopt (arguably) post-productivist agri-environmental schemes, while at the same time continuing to adhere to productivist farming ideologies (Wilson and Hart, 2001). Further, some parts of a holding may be farmed in post-productivist ways (e.g. conservation of biodiversity through extensification), while other parts may be farmed for maximum production. Thus, Evans *et al.* (2002: 321) argued that "percentage reductions in output do not match the percentage of land set aside due to the combined effects of farmers intensifying production on the remainder of their land and retiring the least productive land first" (see also Wilson, 1997b). The discussion below on the fallacy of structural causality will also highlight that productivist/post-productivist territorialisation can also occur within the *farmer's mind*, where at certain times a farmer's identity may depict post-productivist characteristics while at others productivist counter-identities may come to the surface (Burton and Wilson, 2006). On-farm territorialisation particularly emphasises the highly complex nature of identifying 'clear-cut' productivist and post-productivist spaces at any geographical scale and in any cultural context.

The evidence presented in this section suggests that there are large *spatial differences* in the transition to post-productivism within most advanced economies. Lowe *et al.* (1993: 221), therefore, argued that it is "hardly surprising that no coherence can be identified in the post-productivist phase of rural development. Local unevenness is its quintessential and necessary feature". Similarly, Boyle and Halfacree (1998: 7-8) referred to a "much more heterogeneous countryside" in which "agriculturally-based structured coherences will be more spatially selective". Further, the spatially varied nature of the p/pp transition also suggests that some indicators are too UK-specific to be applied to non-UK contexts – referred to as the 'scale insensitivity' of the p/pp transition model by Argent (2002). Such territorialisation processes are particularly linked to specific historical agricultural developments that differ between countries, especially concerning land ownership rights (very specific history in the UK, for example), the emergence of the 'rural idyll' and the commoditisation of the countryside based on urbanisation of society, or changing governance structures (see Section 7.3.2). I, therefore, agree with both Holmes' (2002: 379) recent assertion that assumptions about post-productivist spatial generalisations based on Western European experience "may have encumbered and biased conceptualizations of the rural transition", and with Argent (2002: 106) who argued that "as a macro-structural concept, post-productivism fails to account accurately for the complex nature of regional- and farm-level actions". This means that, similar to most other transitional debates, *spatial heterogeneity* of productivist and post-productivist actor spaces is the norm, suggesting that, from a geographical context, there is *no evidence of a transition* towards post-productivism. Instead, productivist and post-productivist spatial territories lie side-by-side, often in close geographical vicinity (e.g. boundary uplands/lowlands; neighbouring farms), or even within the territory of a single farm holding. As theorists of post-modernity reminded us, "the time-space landscape is more likely to consist of a melange of the obsolete, current, and newborn artefacts commingling anachronistically in each region" (Dear, 1986: 374). McCarthy (2005: 774), thus, emphasised in his critique of post-productivism that "the areal differentiation at multiple scales ... would seem to be glossed over in a theory framed in terms of an epochal [unilinear] shift".

7.3.5 The fallacy of global universality

The discussion in the previous section largely focused on debates within the context of advanced economies. As an extension to this debate, and linked to the transitional fallacy of *global universality*, this section will investigate to what extent the p/pp transition model can be applied to the situation in the developing world.[34] As Chapters 5 and 6 highlighted, little discussion has taken place on the p/pp transition in a developing world context. The reason for this may be that the theoretical debate on post-productivism (and the supporting empirical data) has been based on various assumptions that may only be found in advanced economies,

---

[34] While I refer to the 'developing world' or 'South' (as well as 'developed countries' or 'North'), I acknowledge that there is enormous diversity within these categories (Hettne, 1990; Berger, 1994). Some of the statements will only apply to certain *parts* or *regions* of the developing world, and I recognise that the multiple experiences and uneven development of agricultural systems in developing countries should not be over-generalised. By looking to the South, I am not suggesting that there is a North-South divide *per se*, but wish to highlight the way in which research in different regions of the South sheds new and instructive light on the p/pp transition model that has its roots in research based in the North.

and that these assumptions can not easily be transferred to the South – in other words, that the theory of post-productivism can not easily be 'exported' into developing world socio-cultural, political and economic contexts (Wilson and Rigg, 2003). Can we, therefore, assume that the p/pp transition model can be universally applied to agricultural systems in developing countries – i.e. that the transition can be observed in advanced and less developed economies at the same time? It could be argued that if the notion of post-productivist agriculture does not find applicability at a global scale, 'post-productivism' may be an insufficiently robust theoretical framework to explain agricultural change in any locality.

There is an inherently evolutionary and unilinear undercurrent to the work on rural change in the developed world and its assumed application to developing countries. Bryceson (2000), for example, worried about the tendency for scholars to apply theories developed in the North without much thought to the South. Said (1993) also questioned the notion that Europe has found the 'royal road to wisdom'. The linearity of change embedded in our views of rural change and the transition from a peasant to an industrial society, makes it difficult to imagine anything more nuanced. As Byres (1995, 1996) showed, the historical narrative is most often one of multiple, not single, paths of agrarian transition, and the same can be said of the pre-productivist, productivist and post-productivist transition when applied to a developing world context (Wilson and Rigg, 2003). As the discussion in this section will show, the multiple experiences of countries in the developing world indicate that there may be *many paths* of rural change, some of which resonate with the work on post-productivism in the North, but many more that do not – echoing criticisms about the exportability of the p/pp transition model beyond the UK as discussed above.

I will pay particular attention to possible 'discordant concepts' beyond theoretical assumptions of agricultural change between advanced economies and the developing world. Do different socio-cultural, economic and political parameters that distinguish agricultural change in advanced economies from that of many developing world regions mean that Northern theoretical notions of post-productivism can never be applied elsewhere? In line with similar discussions about the applicability of the exportability (or lack thereof) of Northern-centric theories discussed in Part 1, the discussion will also address the gap that continues to exist between often 'parallel' theoretical frameworks of agricultural change between advanced economies and the developing world, by arguing that holistic theoretical concepts need to be developed that cross the divide between North and South (Pretty, 1995). Such a link may be possible by combining theoretical elements of debates on 'post-productivism' centred on the North, debates on 'deagrarianisation' focused on the South (Wilson and Rigg, 2003), as well as new conceptualisations of multifunctionality discussed in Part 3. Only once we acknowledge that there is as much (or as little) 'productivist/post-productivist' diversity in advanced economies as in the developing world, can we use this knowledge to develop a new understanding based on theoretical and conceptual models that will help us better understand global agricultural structural change.

Northern-centric assumptions underpinning the p/pp transition model

As highlighted above, conceptualisations of post-productivist agriculture have largely been based on the experience of the UK, and, to a lesser extent, on the situation in the EU, Australia and New Zealand. Inevitably, this has led to a theoretical framework of post-productivism predicated on the specific historic and contemporary socio-economic and political situation of the UK (see Chapter 5). The result has been the development of a set of both theoretical and practical assumptions about the shift to post-productivism that may not be easily transferable to Europe (Wilson, 2002), the Antipodes (Holmes, 2002; Jay, 2004), and let alone the South. In the following, I will analyse these assumptions as a basis for understanding the difficulty of 'exporting' the theory of post-productivism into the developing world context.

We first need to address the 'problem' of the implied directionality inherent in the p/pp transition model and repercussions this may have for its applicability to developing countries. Not only has the notion of directionality and temporal linearity been questioned in the context of debates on agricultural change in both the UK and other advanced economies (see above), but this assumption also does not sufficiently take agricultural development trajectories into account that differ from the 'traditional', seemingly linear, intensification/extensification model based on the UK/EU experience – trajectories in developing countries exposed to historical, political and socio-cultural factors that have often differed markedly from the UK or European experiences (Francks *et al.*, 1999; Rigg, 2001). As Chapters 5 and 6 illustrated, the p/pp transition model is, for example, based on the notion of a 'strong' dirigiste (but democratic) governance structure that often actively 'guides' agricultural change through top-down policies – exemplified by the importance of state policies as a crucial dimension in conceptualisations of the transition. It is tempting to contrast this with a developing world 'weak' state model in which the local grassroots/village community level of governance is often crucial in directing agricultural change. This links with a rich literature on 'autonomous' communities, where local social and political forms and structures from so-styled redistributive and consumption-smoothing 'moral' economies (Scott, 1976) to indigenous production systems based on local/indigenous technologies and social and agro-ecological diversity shape activities (Chambers, 1983; Pretty, 1995). However, any fine-grained analysis of the South quickly demonstrates that the state has, at times, played a defining role in guiding agricultural change. Further, the vision of 'peasant' communities as "in an important sense, still a world to themselves" (Elson, 1997: 33) is historically problematic – at least as an all-inclusive generalisation. As Bowie (1992: 819) argued for 19th century northern Thailand, "this examination ... reveals a society with a complex division of labor, serious class stratification, dire poverty, a wide-ranging trade network, and an unanticipated dynamism". There are also developing countries that have gone through an agrarian transition where the shadow of the state is both long and dark, and yet they have emerged looking far from 'post-productivist' in the classic UK-centric sense (Wilson and Rigg, 2003). The cases in point here are Japan, South Korea and Taiwan (Freeman, 1987; Francks *et al.*, 1999). In the case of Japan, for example, a unique set of historical circumstances and a state that massively supported agriculture has permitted an anachronistic 'productivist' small-farm sector to persist (Francks, 1995, 2006).

As highlighted above for the fallacy of temporal linearity, the p/pp model also does not sufficiently conceptualise the state of agriculture *before* productivism. We have already seen that the evolution of the term 'post-productivism' only brought into being the term 'productivist', linked to the retrospective definition of the 'productivist era' from a 'post-productivist' vantage point (Wilson, 2001), with little theoretical consideration of the possible existence of other regimes pre-dating 'productivism'. Yet, as highlighted above, no theoretical work has so far been conducted on whether the notion of 'pre-productivist' agriculture may apply to the agricultural situation that existed 'before productivism' – whether in advanced economies or the poorer world. It is likely, however that the notion of pre-productivism is particularly applicable to the historical (and, in some cases, contemporary) agricultural situation in much of the South, where hundreds of millions of small-scale subsistence farmers still continue to manage their holdings in what could only be described as 'pre-productivist' ways (cf. Losch, 2004).

If this is the case, and if we acknowledge the (problematic) linearity of the shift towards post-productivism (as highlighted above), does pre-productivism then imply a 'direction' of agriculturally based societies towards productivism? Can post-productivist agriculture, therefore, only occur in rural areas that have 'gone through' the productivist era? In other words, can the notion of post-productivism *not* be applied to most countries in the South because they have never been 'productivist' in the first place? Indeed, would it be possible to conceive of developing world rural areas that have moved directly from pre-productivism to post-productivism without ever being 'productivist'? As Part 1 highlighted, there are interesting parallels here with other shifts from 'isms' to 'post-isms', notably debates surrounding Fordism and post-Fordism where some rural areas in both the developed and developing world do not appear to have 'gone through' the Fordist stage, but have jumped straight from pre-Fordism to post-Fordism (Lipietz, 1992; Rogers, 1995). Literature on technological leapfrogging (e.g. Perkins, 2003) also highlights that many developing countries may completely by-pass 'intermediate' stages of development, thereby creating *ruptures* in seemingly linear development pathways. Further, as Chapter 5 highlighted, the historical agricultural geography of the developing world demonstrates that 'productivist' systems existed long before many modern technologies of production (Wilson and Rigg, 2003). The example of China was highlighted above as the most persuasive, where the imperial state embarked upon a focused and well-supported attempt to boost production from AD 1000, and possibly much earlier. Other arguably 'productivist' systems where central structures of authority directed agriculture include Central and South America (the Maya, Aztec and Inca civilisations), Southeast Asia (Java, Burma, Cambodia and Thailand) or the Baliem Valley in New Guinea. Each, in its different way, demonstrates clear evidence of an early 'productivist' ethos.

It would seem, therefore, that the p/pp conceptualisation of agricultural change falsely links productivism with modernity (itself a problematic concept) and modern technology. Bray (1986: 2) in her study of Asian rice economies noted the "failure (in the main) to recognise the relativity of our conception of technological progress". The co-ordination of effort to achieve increasing output rather than the technologies *per se* should be brought to bear, which means that drawing a neat distinction between the modern Green Revolution and more ancient green revolutions is difficult (Parayil, 1999). Moreover, even in terms of the technology employed (e.g. China) the division is not as clear-cut as it might, at first, appear. So, if we assume that the application of technology, co-ordinated and driven by some sort of state apparatus, is an essential element in the shift from pre-productivism to productivism, then examples from developing countries suggest caution. There were, quite clearly, 'productivist' agricultural systems in what might be considered the 'pre-productivist' period. Therefore, "the image of farmers living immutable existences in subsistence communities overlooks the dynamism, degree of market integration, and level of state co-ordination that existed in more than a few areas of the world in the pre-modern period" (Wilson and Rigg, 2003: 692). It appears, therefore, that countries in the South have not made clear and complete transitions from pre-productivism, to productivism, to post-productivism.

Further, evidence shows that individual households in the developing world may also *at the same time* embody elements of all three. For example, a rural household may continue to cultivate rice for home consumption, possibly using techniques that we would describe as organic; another portion of their land might be allocated to cash cropping driven and organised by purely commercial considerations (it may even be contract-farmed for a global agro-industrial conglomerate; see Rigg and Nattapoolwat, 2001); they might have a son or daughter living in the city, possibly as a student or maybe as a factory worker; and they may have sold off some land to a housing developer permitting non-rural classes to infiltrate the countryside (see also Chapter 10). From a household perspective, such a *multi-stranded* and *spatially fragmented* (indeed 'multifunctional') approach to building a livelihood is entirely sensible. But when, in conceptual terms, it comes to placing such a household in one of three boxes neatly labelled 'pre-productivist', 'productivist' and 'post-productivist' their *hybridity* presents a considerable challenge (Wilson and Rigg, 2003). This echoes debates surrounding the *territorialisation* of productivist and post-productivist actor spaces in advanced economies highlighted above, where European researchers have highlighted the simultaneous existence of productivist (or even 'super-productivist') and post-productivist action and thought across farming regions and even within individual farm enterprises. As a result, a more nuanced and refined conceptualisation of agricultural transition in the developing world is necessary, where local histories and particularities are brought to bear and that, therefore, challenges the directionality and global universality underpinning the p/pp transition model.

Northern-centric indicators of post-productivism and the developing world

There are similar problems in terms of the specific 'indicators' used to conceptualise both productivism and post-productivism from an advanced economies perspective. Many examples could be mentioned here, but I will restrict myself to six key debates surrounding the use of 'fashionable' indicators of a shift towards post-productivism (see also Wilson and Rigg, 2003). These six indicators have also been 'tested' to a certain extent beyond the UK framework of post-productivist conceptualisations, in particular in recent work on the exportability of the post-productivist concept as either a 'myth' or 'reality' in the European context (Wilson, 2002), and in Holmes' (2002, 2006) analysis whether recent agricultural and non-agricultural changes in the Australian outback confirmed a shift to post-productivism or not (see above).

First, an obvious starting point is the investigation of possible policy change as a key indicator of post-productivism in the South, as one of the most 'tangible' indicators to assess 'real' shifts in agriculture (see Chapters 5 and 6). Further, broader political economy considerations and policies linked to changing global agro-commodity chains need to be investigated (Friedmann and McMichael, 1989; Goodman and Watts, 1997). Here, the argument may be that the globalised food economy with its 'global' policies associated with the WTO framework may act as a 'great leveller' for productivist and post-productivist policies in *both* developed and developing countries (Potter and Burney, 2002). Thus, we also need to consider to what extent a global capitalist regulatory structure in the agricultural sector may water down possible theoretical and material differences between North and South for post-productivist policy rhetoric and discourse. For policy change, there is some evidence that pressures emanating from supranational bodies such as the WTO and the EU have influenced agricultural policies in the South and propelled them in a post-productivist direction (Potter and Burney, 2002). In 2002, for example, the EU began to test imports of shrimps and poultry meat from Thailand for the presence of Nitrofurans, veterinary drugs now banned in the EU. Following 'useful and fruitful discussions' between the EU and the Thai authorities, the Thai government agreed to ban the use of 16 chemicals, to test for their presence in export consignments, and to apply stringent penalties for those who infringe export conditions (Wilson and Rigg, 2003). The result of this scare was that exports of shrimps from Thailand declined by more than 50% in 2002.

There is also a body of evidence which suggests that rather than national governments changing policies in response to global capitalist regulatory structures, what we see in the South is a sidelining of national governments (and policy making) with the rise of global agro-food systems. The incorporation of poorer countries into such systems has seen their inclusion within broadly productivist rather than post-productivist agriculture. Thus, we can identify across the South from India (Singh, 2002), Mexico (Sanderson, 1986; Barkin, 1990), the Dominican Republic (Reynolds, 2002) and Thailand (Goss and Burch, 2001) the rising cultivation of non-traditional crops, often under systems of contract farming, using new technologies and high levels of chemical inputs. For Barkin (1990: 3) this has 'wrenched' farmers from their local contexts and "increasingly subjugated [them] to the designs of an international market". Others likewise see integration into the global food economy as having the effect of marginalising national governments (Goss and Burch, 2001; Reynolds, 2002). Ponte's (2002) global commodity chain analysis of the coffee industry, for example, demonstrated how real power has not shifted away from national governments to supranational organisations, but rather from national and local governments and farmers towards 'buyers' and, more generally, from producers to consumers. Indeed, "as governments retreat from the regulation of domestic coffee markets ... the weakness and inherent instability of the institutional framework falls straight on the shoulders of coffee farmers in developing countries" (Ponte, 2002: 1116). Consumers could, of course, make a difference – and do. There has been a growth of fairly traded coffee since the 1990s, where producers receive a 'fair' return for their labour based on a minimum floor price. In addition, consumer consciousness is driving the demand for organic, shade-grown and bird-friendly coffees (Ponte, 2002).[35] Such developments could be seen as reflecting the intrusion of post-productivist ideals into developing world production systems. But the motivation, in this instance, comes not from changes in government policy, nor from the roles played by the EU and WTO, but from consumer consciousness in the rich world, often promoted by NGOs like the Fairtrade organisation.

Second, debates about 'organic farming' as a possible indicator of post-productivism and its applicability to the South are also informative for testing the assumption of global universality of the p/pp model. Although there are explicit organic farming debates in all the major regions of the developing world (see Sanders, 2000a, b, for work on China), the organic farming concept is less relevant in the South and, indeed, in some regions of advanced economies such as Greece or Portugal that have, arguably, always farmed organically, "where agricultural production has, often by economic necessity, been 'organic' for thousands of years" (Wilson, 2001: 92). Should an 'organic' subsistence farmer in India, therefore, be classified as 'pre-productivist' – i.e. at an agricultural stage that pre-dates possible intensification – or 'post-productivist' – i.e. at an agricultural stage that is already environmentally sustainable and extensive and often accompanied by well-developed horizontally integrated local governance structures (Wilson and Rigg, 2003)? As highlighted above, even in advanced economies there is debate over whether organic farming is post- or super-productivist (Kaltoft, 2001; Dabbert *et al.*, 2004), in particular as organic farming is, in many instances, driven by the same market forces that created intensive farming. Indeed, if intensity is taken to mean more than chemical inputs, then organic farming can be characterised as sometimes even more intensive – in terms of labour inputs, the application of science, and the assiduousness with which the land and the crop is maintained. Take the case of Laos, a so-styled 'least' developed country. Here, organic production remains very much part of the agricultural landscape, and levels of use of chemical inputs are very low in regional terms. Yet, organic farming has also been identified as one of Laos' (few) comparative advantages as it attempts to carve out a development niche for itself. Commercial organic farming and the sale of organic produce (particularly to neighbouring Thailand) are, therefore, seen as a means to raise incomes in poor rural areas (Wilson and Rigg, 2003). In such a context, organic farming should not be seen as a profound shift in farmers' strategies, but as another attempt by farmers to maximise their returns in a changing market with shifting consumer preferences. The work by Sanders (2000a, b) on 'eco-villages' in China also shows that the emergence of organic farming is not 'new', but rather the restructuring of a traditional practice – very similar to patterns apparent in Mediterranean farming areas (Wilson and Juntti, 2005). Nor is it a case of China leapfrogging from pre-productivism to post-productivism. China has been highly productivist (as argued above and in Chapter 5) – and sometimes within a broadly 'organic' system.

Third, 'counter-urbanisation' is another useful example, especially with its associated changes in attitudes of formerly traditional rural communities now influenced by 'urban' and more 'progressive' middle-class values and environmental attitudes which, it is suggested, may have led to changes in farming practices and a questioning by the farming community of 'traditional' and often environmentally destructive countryside management behaviour (see Chapter 6). Counter-urbanisation is a feature of rural-urban relations in some, albeit a few, areas in the developing world. Yet, counter-urbanisation may be interpreted as a developed world-centric concept and, so far, rarely applicable – at least directly – in a developing world context. Thus, 'traditional' values of developing world rural communities (whether they are pre-productivist or productivist) may not (yet) be challenged by counter-urbanising 'middle class newcomers' (Wilson and Rigg, 2003). The lack of this key ingredient for

---

[35] For example, since 2002 the UK-based Co-op chain of more than 2000 supermarkets only sells their own brand chocolate produced from fairly traded Ghanaian cocoa.

conceptualisations of post-productivism should, therefore, lead to a weakening of the post-productivist concept in a Southern context.

Fourth, conceptualisations of post-productivism have also been based on the assumption that formerly marginal actors, such as NGOs, have been increasingly included in the 'core' of the policy-making process. This assumption is based on the model of a democratic and pluralist society that enables (most) non-state actors to air their views by organising themselves into lobby groups that often challenge and change state policies (Bryant and Wilson, 1998; Bryant, 2005). Yet, such enabling democratic conditions are sometimes not in place in the developing world (e.g. North Korea, Myanmar) (Princen and Finger, 1994; Bryant, 1996, 1997), which means that a possible shift towards post-productivism can not necessarily be gauged in such countries by analysing the role, or even existence, of non-state actors in the agricultural and environmental policy-making process. This is not to say that NGOs are not important actors in developing countries. They are increasingly so, and in some instances have been actively courted by government and by multilateral organisations such as the World Bank. Nonetheless, in most cases they continue to play an *oppositionist* role where they challenge, rather than work with, governments (Bryant, 2001, 2005).

Fifth, and as highlighted in Chapter 6, Marsden *et al.* (1993) saw the move away from agricultural *production* towards *consumption* of the countryside as a key ingredient of post-productivism. This 'indicator' implies that there is a willingness and an ability among society as a whole to 'consume' the 'new' goods produced by actors in the countryside (e.g. golf courses, walking tracks, farm tourism), and assumes a relatively wealthy, mobile and essentially urban society that has an 'urge' to pay regular visits to the countryside for intangible benefits such as solitude, aesthetic enjoyment or relaxation. Yet, again, the example of the South does not mimic the experience of the North. In Thailand, for example, there are strong notions of the rural idyll – usually articulated by the urban-based, middle class – that draw on imagery that can be dated back to the late 13th century (Rigg and Ritchie, 2002). Housing estates, populated by urbanites keen to consume rural areas, are collectively called *mubaan* or 'villages'. Hotels draw on the past in their promotional literature and in their architecture and, in some cases, twist agricultural production into a tourist consumption practice (Wilson and Rigg, 2003). Further, Singhanetra-Renard's (1999) research highlighted that in the mid-1970s a northern Thai case study village was more pre-productivist than productivist, but by 1993 the last village rice field had been sold, and by the end of the decade within a 15-km radius of the village there was a cavalcade of "golf courses, reservoirs, and elephant shows; orchid, butterfly and snake farms; restaurants, five-star hotels, karaoke bars, brothels, massage parlours and resorts" (Singhanetra-Renard, 1999: 77). Not only have consumption practices left their mark in rural areas, but the knock-on effects for traditional production practices are also clear. In the village of Ban Lek in northern Thailand, for example, land prices rose by 2300% as the area was increasingly drawn within the orbit of the regional centre of Chiang Mai and the buying power of the middle classes (Ritchie, 1996a, b). Poorer 'pre-productivist' households are simply rendered unable to compete in such circumstances. The environmental impacts of land speculation (leaving idle land where farm pests can multiply) and housing estate construction (disrupting traditional water management systems) can also have the effect of undermining pre-productivist and productivist systems (see Kelly, 1999, on the Philippines; and Kirkby and Xiaobin, 1999, on China), thereby echoing debates on the undermining of productivist ideologies in advanced economies. In Malaysia, meanwhile, scholars have described a similar dual process of de-village-isation and de-agriculturalisation (Courtenay, 1988; Kato, 1994), where villages have become little more than retirement settlements for the parents of young men and women attracted to the city (see also discussion of 'deagrarianisation' below). Yet, although wealth and mobility in societies in the South have undoubtedly increased, and there is also evidence of increasing consumption of countryside settings in urban fringe areas for relaxation and aesthetic enjoyment (Rigg and Ritchie, 2002), mobility continues to be often driven by economic necessity underpinned by social and cultural change, as rural people access non-farm employment, whether through commuting, circular migration or more permanent spatial transitions (Rigg 1998; Singhanetra-Renard, 1999). In particular, the work of Rawski and Mead (1998) on China suggested that there may be as many as 100 million 'phantom farmers' – farmers who have left the countryside, most probably for urban or peri-urban areas. Exceedingly few of these, we can surmise, have moved for any post-productivist 'consumption' purpose. Thus, although elements of the consumption debate of the p/pp transition model find parallels in the developing world, it remains questionable whether this post-productivist indicator can be easily applied in most developing world contexts.

A final factor relates to on-farm diversification as a key indicator of the p/pp transition, especially as the limitations of the conceptual sequence and the changes that are seen to underpin it may be clearest of all. In developing countries, there has always been a good degree of diversity in livelihood terms, both in agriculture and non-agriculture. Pluriactivity[36] may be the term most commonly used to describe economic farm diversification in the North, but something (apparently) very similar has gone by a range of alternative terms in the South such as 'occupational multiplicity', 'multi-stranded livelihoods' and 'diverse portfolios of activities' (Grandstaff, 1988). Indeed, "the logic of embracing diversity in the pre-productivist era was clear: it was a means of smoothing consumption in a context where environmental threats (particularly) to livelihoods were real and common" (Wilson and Rigg, 2003: 695). Indeed, farmers would cultivate a range of agro-ecological niches, plant complex assemblages of cultivars, and supplement farm production with the collection of non-timber forest products, all with an eye to spreading risk through time and space. Today, diversity remains attractive to many inhabitants in the rural South. However, the avenues by which diversity can be secured have changed. So too have sources of risk in rural areas. Thus, in the 'pre-productivist era' threats to livelihood were largely environmental, while today, with thorough-going market integration and an increasingly global capitalist regulatory structure in agro-food chains, risk is more likely to be embedded in fluctuations in the market economy (Pretty, 2002). However, the need for diversity has remained – for the time being at least – a constant. A second important change concerns the means by which rural people can achieve diversity in their livelihoods. In the past, activities were largely (though not entirely) embedded in the farm economy and often in agriculture. Over time, these traditional opportunities for diversity have often been compromised as land has become a scarce resource. But, at the same time, new avenues have opened up, many non-farm in orientation and *ex situ* in location. While diversity traditionally may have relied on ingenious combinations of agricultural systems (as well as diversity within agricultural systems; Rigg, 1993), today it is more likely to be founded on new activities such as factory work or various forms of informal sector employment (Eder, 1999; Hayami and Kikuchi, 2000). There is a spatial as well as a sectoral component to these changes, as formerly spatially rooted livelihoods have become increasingly delocalised as household members have searched beyond the village for work (Wilson and Rigg, 2003).

Looking at the issue of diversification from a Southern perspective, therefore, raises some significant questions about the applicability, and conceptual resilience of this indicator of post-productivism. First, diversity and diversification, whether on- or off-farm, could be seen as something of a red herring when it comes to the debate over post-productivism. It is the *form* that diversification takes, and the *way it is embedded* in the unique histories of rural places that are key in determining its significance in the South. Rural histories and their development trajectories create various opportunities for diversification, and some may resonate with post-productivism while others will not. Second, a Southern perspective highlights the limitations imposed – at least in terms of understanding livelihoods – by an on-farm/off-farm distinction. Households are socially constructed, and overly rigid spatially determined delineations of diversification will obscure important aspects of people's

---

[36] Pluriactivity entails a combination of agricultural and non-agricultural activities performed by the farmer or members of the farm household (non-agricultural sources of income).

livelihoods. It is again questionable whether diversification, as a possible indicator of the p/pp transition, can easily be transferred to the developing world situation. To be sure, diversity is a feature of many rural lives in the South, but the experience of the South also amply demonstrates the difficulties of assuming that indicators informed by, and developed from, one historical experience can be applied to another context (Ellis, 2000).

It is highly questionable, therefore, whether these indicators of post-productivism (or other indicators not mentioned here) can be easily transferred to the South. The transposability of Northern-centric assumptions underlying the p/pp transition as well as the implied linearity of the p/pp model are, therefore, deeply problematic when viewed from a Southern perspective. Indeed, "the difficulty with seeing the presence of such debates (and processes) as evidence of nascent post-productivism is that it assumes that their 'meanings' are the same as those deduced from the UK experience" (Wilson and Rigg, 2003: 693). Thus, on the basis of the experience of the South, a central weakness in the p/pp model is that it may confuse *appearances* with *driving forces*.

'Deagrarianisation' and 'post-productivism' from a developing countries' perspective

In the previous discussion we have seen that it is problematic to transpose Northern-centric indicators of the shift from productivism to post-productivism to the situation in the South. Yet, there exist debates about agricultural and rural change in developing countries that, at times, show striking parallels with developments in the North. For example, in her work on Africa, Bryceson (1996, 1997a, b), and Bryceson and Jamal (1997), argued that a widespread process of *deagrarianisation*[37] is a feature of the continent – thereby echoing debates on both post-productivist reduced intensity of agricultural production and the withdrawal of 'classic' agricultural production towards other 'alternative' uses of land highlighted in Chapter 6.[38] Scholars working on Asia (Rigg, 1998) and Latin America (Zoomers and Kleinpenning, 1996; Lanjouw, 1999) have also marked out a parallel process of rural change. Rigg (2001) further argued that for Asia 'spatial interpenetration' is increasingly becoming a characteristic of rural change, with evidence of a shift from farm to non-farm activities (but usually without the complete abandonment of the rural base), both in terms of employment and income; a series of cultural and social changes that have served to undermine the social utility of traditional farming, making it in many areas a low status occupation; and a series of important spatial changes that are largely a product of these first two changes. These debates have strong parallels with some regions in advanced economies, notably Mediterranean countries where similar patterns of deagrarianisation can be observed (Garrido and Moyano, 1996; Peco *et al.*, 2000), and where the applicability of the UK-based model of post-productivism has been equally challenged (e.g. Louloudis *et al.*, 2000; Wilson and Juntti, 2005). Deagrarianisation can, therefore, be viewed as both a product of *post-productivist* tendencies as well as creating the economic and physical space within which *pre-productivist* systems can make the transition to productivism, and productivist to post-productivist ones. To take just one example: the shift to non-farm work increases the cash available for farm investments, while also rendering households short of labour, so encouraging mechanisation of production (Hayami and Kikuchi, 2000; Rigg, 2001). The theorisation of 'deagrarianisation' in the South can, therefore, be usefully set alongside discussions on 'post-productivism' in the North, and components, ingredients and indicators of these two theoretical concepts may, indeed, be similar (Wilson and Rigg, 2003). It is here that,

---

[37] Bryceson (2002: 726) defined deagrarianisation "as a long-term process of occupational adjustment, income-earning reorientation, social identification and spatial relocation of rural dwellers away from strictly agricultural-based modes of livelihood".

[38] Debates on 'deagrarianisation' are not restricted to the developing world. In his recent critique of post-productivism from an Australian perspective, for example, Holmes (2002) highlighted how large parts of the Australian outback are suffering from the retreat of pastoralism from marginal lands, but with little 'post-productivist' activities to replace this loss.

arguably, some of the most interesting and interweaving parallels can be found in the debates surrounding the transposability of the p/pp transition model to the South. Yet, how far the mosaic of social, economic and spatial changes evident in developing countries are way stations between different 'eras' is not at all clear yet.

Some scholars (e.g. Grandstaff 1988, 1992) have made much play of interlocking livelihoods, occupational multiplicity, and the existence of 'shadow' households as inherent features – rather than short-term transitory stages. Rural people's primordial links with the land, and the social resilience of the family and the household are all seen as factors why we should expect an equal level of resilience in the hybrid states we see. Yet, Bryceson (2002: 727) links deagrarianisation in Africa with de-peasantisation as another key process, as after a century of colonial and post-colonial peasant formation "depeasantization has now begun, representing a specific form of deagrarianization in which peasantries lose their economic capacity and social coherence, and shrink in demographic size relative to nonpeasant populations".[39] Given debates on the possible role of the EU and WTO in changing the character of agricultural production in the South (although in what direction is not clear; cf. Potter and Burney, 2002), it is significant that Bryceson identified structural adjustment policies (SAPs) and market liberalisation as instrumental in propelling de-peasantisation. Drawing on a study of six African countries (Ethiopia, Malawi, Nigeria, Tanzania, South Africa and Zimbabwe), these policies, although to varying degrees and in different ways, have undermined productivism by unpicking the state-orchestrated structures of support and subsidy. Bryceson (2002: 728) argued that "SAP policies largely dismantled African marketing boards and parastatals that had serviced peasants' input requirements, enforced commodity standards, and provided single-channel marketing facilities and controlled prices. … Farmers were faced with a more uncertain market environment, producer prices were subject to wide fluctuations, input prices skyrocketed and supply became tenuous as most [private] traders did not have the rural outreach of the parastatals they replaced". These policies led to two important changes relevant to the discussion here. First, a shift towards the production of 'fast crops' such as tomatoes and potatoes where farmers could recoup their costs quickly, and away from crops requiring large capital inputs. Second, they led to a progressive move out of agriculture – de-peasantisation – reflecting, in part, the steep decline in returns to farming. Thus, parallels to Northern-centric debates over post-productivism can be identified in Southern-focused discussions of deagrarianisation, in particular as the literature is geographically richer and, therefore, socially and historically more differentiated while, simultaneously, there is not such a concern with mapping out a particular trajectory of change and embedding this within a certain conceptual framework (Pretty, 1995, 2002).

Different 'takes' on pluriactivity in rural households, therefore, provide an illustration of the richness – and diversity – of the empirical evidence from the South, and the contrast that this provides with the rather a-historical, a-geographical linear and binary Northern-centric p/pp transition model. In Europe, the growth of pluriactive farm households has been taken to indicate a *re-peasantisation* of agriculture (Van der Ploeg *et al.*, 2000), while in the developing world it has been interpreted as indicative of a process of *de-peasantisation*, and the same may be true of the situation in Southern Europe (Buller *et al.*, 2000; Wilson and Juntti, 2005). In Ireland, half of farm households are engaged in off-farm activities (Kinsella *et al.*, 2000), for Japan the figure is 90% (Motoki, 2002), and in many other countries in the developing world (e.g. Germany) figures around 50% are common (Wilson and Wilson, 2001). What are we to make of this growth in pluriactivity? Van der Ploeg *et al.* (2000), writing from a European perspective, see pluriactivity as a new pillar supporting farming. In contrast, in the developing world pluriactivity may be a point on a transition towards the thorough-going restructuring of farming and rural areas. In Japan, meanwhile, pluriactive households may represent – and this in spite of the massive subsidisation of agriculture – the death throes of the Japanese small farm sector (Pretty, 1995; Francks, 2006).

---

[39] Kato (1994) used the same term to describe agrarian transitions in Malaysia.

Post-productivism and the South: discordant concepts?

The lens of deagrarianisation discussed above may offer a different vantage point from which to view notions of post-productivism in the South. Is it, therefore, possible to apply the basic underlying principles of the p/pp transition model to processes of agricultural change in the South, despite the difficulty of using Northern-centric 'indicators' of the p/pp transition? The Southern context reveals that reality may be more complex than is assumed by many authors outlining the theoretical framework of post-productivism, and that the transferability – indeed 'exportability' – of complex theoretical concepts such as 'post-productivism' to the South is highly problematic.

The analysis of key indicators of post-productivism has particularly highlighted that the successful 'exporting' of the p/pp transition model to the South relies on *shared definitions, meanings* and – ultimately – *discourses* of post-productivism (Wilson and Rigg, 2003; see also Section 7.3.2). Although similar patterns that may be vital ingredients of post-productivist agriculture can be observed in some developing world contexts, there is confusion about the exact meaning of these complex activities (e.g. 'green policies'; 'organic farming'; 'counter-urbanisation'; 'countryside consumption'; 'diversification'; etc.). Transferring the notion of post-productivism to the South, therefore, does not only face the problem of whether such indicators find expression in a *material sense* (i.e. through practical agricultural and rural changes), but particularly whether the imbued meanings of post-productivist indicators couched in a framework of agricultural/rural change in advanced economies (and there mainly from a UK perspective) can be transferred in a *theoretical/conceptual sense* to the situation in the South – not dissimilar to issues of transferring the p/pp transition model to the agricultural situation outside of the UK discussed above. Thus, "by attempting to apply notions of post-productivist agricultural regimes that have been developed in the North to the South, we witness, therefore, a perpetuation of the problem that our ideas of agrarian transitions and the direction and forms that they take continue to be based largely on *models* developed in the North and, perhaps more importantly, based on the *experiences* of the North" (Wilson and Rigg, 2003: 699; original emphases). To take a term from historical studies, it may be academically beneficial to 'permit' the countries of the South to construct their own autonomous conceptualisations of agrarian change (e.g. deagrarianisation), rather than uncomfortably shoe-horn their varied experiences into the existing conceptual sequence. Indeed, it may be the conceptualisation of this historical sequence based on the p/pp transition model that may be flawed and that, as Part 3 will argue, a new conceptualisation of 'multifunctionality' along a *spectrum* may better encapsulate the different historical and contemporary agricultural change trajectories that can be observed in both the North and South. Moreover, it may be that work on agrarian change in the South has important lessons for the North. As Chapter 4 highlighted, a similarly cautious tone has been adopted by those who have attempted to link theoretical debates surrounding the notion of a shift from Fordist to post-Fordist modes of production to the South (Potter and Tilzey, 2005). Lipietz (1992), Amin (1994) and Byres (1996), thus, argued that post-Fordist modes of production may take a very different trajectory in developing countries or those making the transition to the market (e.g. China) to that indicated by Northern-based models.

The analysis of various indicators of post-productivism in the context of the South provides further ammunition to deconstruct the implied linearity of the p/pp transition model. Above examples show that the rigid sequence of historical steps implied in the p/pp transition model needs to be questioned. The discussion has shown that both productivist and post-productivist activities can occur *simultaneously* in many rural areas of the developing world, thereby mirroring debates about a possible 'territorialisation' of productivist/post-productivist actor spaces in advanced economies discussed in Section 7.3.4 above. This was also reiterated by Marsden (2003) who suggested that different agricultural 'dynamics' can be identified to occur simultaneously across the globe, including, for example, an agro-industrial dynamic visible in some of the intensively farmed and globally inter-linked agricultural districts of Brazil (see also Roux *et al.*, 2004), territories of quality consumption and retail spaces in most of Northern Europe, and areas of 'agricultural retreat and local vulnerability spaces' in the Caribbean. Thus, Marsden (2003: 46) argued that we are witnessing at the global level the creation of "uneven agrarian spaces which express the inherent contradictory nature of recent agrarian development". Such complexity is further encouraged through the globalised food economy with its 'global' policies associated with the WTO framework. On the one hand, the WTO has acted as a 'great leveller' for productivist and post-productivist policies in *both* advanced economies and the South (Potter and Burney, 2002). Indeed, as examples above suggest, the global capitalist regulatory structure in the agricultural sector has indeed watered down some of the theoretical and material differences between the North and South. On the other hand, although Potter and Tilzey (2005) have emphasised that at the broader supra-national level there is a clear convergence of policies in both North and South addressing problems associated with free trade (or the lack thereof), the outcome has been a set of policies with *both* productivist (e.g. aiming at commodity production maximisation and intensification) and post-productivist tendencies (e.g. ensuring better food quality and aiming to reduce negative environmental impacts) that apply to *both* the North and South simultaneously (see also Goodman and Watts, 1997).

The experience of the rural South, therefore, demonstrates the need to avoid being seduced by mere appearance. Indeed, in the developing world it is possible to find the same agricultural systems side-by-side, one embedded in a pre-productivist landscape and the other in a post-productivist one. It is only by highlighting the underpinning *meaning* of the agricultural form that the difference makes 'sense'. Again, similarities are apparent with the discussions surrounding the applicability of the Fordist/post-Fordist transition model to the developing world. In particular, some have suggested that some rural areas in the South may have moved directly from the pre-Fordist to the post-Fordist stage (e.g. Webber and Rigby, 1996). This suggests that if the formative conceptual stage is missing (i.e. productivism or Fordism) then the use of the 'pre-' and 'post-' concepts may be questionable, except in comparison with spaces where these concepts do, to a certain extent, apply (e.g. debates on the 'productivist' era in the EU; see below).

Overall, this section has highlighted that it is difficult to apply the p/pp transition model in the context of the South, as post-productivism may have been defined against a set of criteria that may not be applicable beyond certain advanced economies. The model can certainly not be imported wholesale and would need to be substantially adapted and developed to address conditions outside of the developed world. As McCarthy (2005: 774) emphasised, "the [post-productivist] thesis fits countries outside Western Europe poorly if at all". The transition to post-productivism (if it exists) can not be observed in developed and developing countries *at the same time*. Agricultural development trajectories in the developing world are characterised by a *temporally simultaneous juxtaposition of several pathways*, some of which show tendencies towards productivism while others share characteristics of post-productivism. Instead of *global universality*, we witness *global complexity*. As I will argue in Part 3, one possible way out of this conceptual impasse will be to link developments in the South with a newly conceptualised understanding of what 'multifunctional' agriculture may mean, in particular by embedding it within theoretical discussions surrounding the above-mentioned Southern-based concept of 'deagrarianisation'. Indeed, multifunctionality may be a more robust *normative* term to consider *relative* changes in action and thought of rural societies (i.e. changing perceptions of agriculture relative to earlier societal conceptions; willingness of grassroots actors to adopt new forms of environmentally friendly farming techniques; etc.) than the attempts to establish *absolute* indicators of change based on the UK-centric p/pp transition model.

7.3.6 The fallacy of structural causality

As Chapters 5 and 6 highlighted, conceptualisations of the p/pp transition have been based on the assumption that all stakeholders in rural areas (and beyond) have been affected in similar ways by the transition, and that all actors are moving along similar transitional pathways towards post-productivism in the same manner and at the same pace (fallacy of structural causality). If the implied linearity in the p/pp transition model was true, it could be hypothesised that both structure (e.g. agricultural policies, rural political economy) and agency (farmers' identities and rural culture) would 'move along' transitional pathways at the same pace and in the same manner. Chapter 5 highlighted that this assumption has been largely based on the over-emphasis of structuralist approaches, in particular political economy-based theorisations (Evans *et al.*, 2002). Despite the multitude of dimensions of the post-productivist transition discussed in Chapters 5 and 6, the dominant political economy discourse has inevitably led to a strong focus on the importance of macro-economic factors in actor decision-making. This has led to a p/pp transition concept largely defined through structural *exogenous* 'indicators' of agricultural change (in particular policy changes; the political economy framework; farmers' economic adjustment strategies to external forces; etc.; see Chapters 5 and 6), rather than through agency-related *endogenous* characteristics that may accompany this change (e.g. attitudes, perceptions, behaviour and identities of specific agricultural and rural actors) (Wilson, 2001; Burton and Wilson, 2006). Burton (2004: 197), therefore, argued that "there has been very little work conducted on the culture of 'productivism' or 'post-productivism' from the actor perspective ... we still know very little about the farmers' perspective of 'production' and 'productivism'". As a consequence, many traditional features of post-productivist enquiry have focused on specific actor groups (e.g. policy-makers) or larger structural entities (e.g. 'the state') to the neglect of individuals and their actions (Evans *et al.*, 2002). Thus, "the dominant political economy discourse has ... inevitably led to a heavy emphasis on the importance of the state and policies, a strong focus on the importance of macro-economic factors in actor decision-making ... and a heavy emphasis on food production and global market regimes. ... As a result, the farming community has often been viewed as responding almost entirely to *outside* forces, with little acknowledgement of possible changes from *within*" (Wilson, 2001: 85-86; original emphasis).

Philo (1992) was among the first to criticise rural research by arguing that rural populations and farmers have often been depicted as 'homogenous entities', and that the diversity of individual opinion has been neglected by political economists[40] (see also Ward and Munton, 1992, who argued for combining political economy and socio-cultural approaches to understanding pesticide pollution regulation on farms). Ward (1993: 362) further noted that "to understand how new sets of regulatory, market and social pressures impact upon farm businesses and households, models will need to be more sensitive to the actions and values of individual actors involved". Similarly, Marsden *et al.* (1993) advocated a better understanding of rural actors through analysis of 'action-in-context', akin to emphases on the interface between 'local' and 'extra-local' (Lowe *et al.*, 1995; Marsden, 1999b), further echoed by Lowe *et al.* (1993: 210) who stressed that "a top-down causal argument, which portrays local areas as merely the passive recipients of general movements of capital ... is inadequate". Marsden *et al.* (1993: 172) also criticised the "unreflexive application of structuralist concepts to rural change", and that "current notions within the literature, emerging as they have largely from a political economy perspective, tend to retain an excessive economism and a set of 'top-down', structuralist assumptions about the nature of change" (Marsden *et al.*, 1993: 20).

---

[40] It should be noted that there is a wealth of literature in rural sociology on understanding the behaviour of farmers from within the farming community (e.g. MacKinnon *et al.*, 1991; Tovey, 1997), but this research has not often been linked to conceptualisations of the p/pp transition model (see Chapters 5 and 6).

I, therefore, agree with Morris and Evans (1999: 350) that "consistently approaching an analysis of agricultural change from one theoretical position has tended to eclipse the rich variety of work on agricultural change which exists alongside that adopting a political economy perspective". Yet, although adding a crucial dimension to existing debates, revised conceptualisations of post-productivism – including, for example, Halfacree's (1997a, b, 1999) discussion of counterurbanisation and post-productivism (see Chapter 6); work on changes in the governance of rural areas (Ray, 2000; Little, 2001); and investigations into different actor-networks affecting post-productivist rural spaces (e.g. Murdoch and Marsden, 1995; Marsden, 1999a) – have remained largely within the boundaries of structuralist arguments. Halfacree (1997a, b), for example, still saw migration as an *external* impact on the farming community. Even where attempts have been made to critically assess the role of agency in the post-productivist transition, there has been a tendency within typologies to classify actors as motivated by single parameters along simplistic spectra (e.g. 'conservationists'/'non-conservationists'). As a result, *identities* of specific stakeholder groups – especially farmers – have, until recently, remained relatively unexplored. Many have, therefore, called for the injection of more agency-based approaches into conceptualisations of the p/pp transition (Burton, 2004). In particular, conceptualisations of post-productivism would benefit from actor-oriented and behaviourally grounded approaches that would also consider changing *endogenous* perceptions and attitudes of rural actors (Wilson, 2001). Many studies have taken up the challenge with the purpose of exploring whether, and to what extent, specific segments of society have moved towards post-productivism (e.g. Jay, 2004; Burton and Wilson, 2006). In line with Long and Van der Ploeg (1995) who argued for an injection of a more thorough-going and better theorised actor-oriented approach in research on rural change, I argue that treating the farming community as a homogenous entity – merely *reacting* to external change – has neglected one of the most important dimensions of the p/pp transition: whether there has been a shift in grassroots actors' attitudes concurrent with the postulated p/pp transition, and whether such a shift has been reflected in changing farming techniques and behaviour. Thus, has there been a shift in grassroots actors' *attitudes* and, most importantly, *self-concepts* and *identities* concurrent with the suggested shift towards post-productivism?

Much work in social psychology and sociology (e.g. Stryker, 1994) has suggested that a much more complex and multi-dimensional self-structure exists (Burton and Wilson, 2006). This section will explore to what extent findings from these studies support the notion of 'structural causality' in the adoption of post-productivism or whether, instead, 'structure-agency inconsistency'[41] is a better descriptor. Only if agricultural/rural grassroots actors' attitudes and self-concepts show signs of post-productivist characteristics, can we begin to assume that modern agriculture and rural society are moving towards post-productivism. For farmers, as one of the key rural actors, we should argue that only if farmers' attitudes (and eventual changes in their farm management behaviour) indicate substantial shifts towards post-productivist thinking (i.e. concern for environment; adoption of environmentally friendly farming practices; acceptance of new forms of policy regulation; changing perceptions of role of farmers and agriculture; acknowledgement of multiple actor spaces in the countryside), can we fully acknowledge that a transition towards post-productivism has taken place. Based on social psychology theory, this section will, therefore, investigate whether *farming identities have moved towards 'post-productivism'*, in order to test the assumption that rural agency is moving according to the same patterns and pace as agricultural/rural structure. Based largely on Stryker's (1968, 1994) 'identity theory', I will discuss various 'identity groups' (based on Burton and Wilson, 2006) and will investigate if, and to what extent, these groups share characteristics with the various dimensions of post-

---

[41] The notion of 'structure-agency' invokes both neo-Gramscian views of how societal segments are ordered and vie for power (cf. Gramsci, 1971; Bieler and Morton, 2001), as well as Giddens' (1979, 1984, 1991) conceptualisations of the role of 'structure' and 'agency' in defining contemporary societal evolutionary pathways.

productivist agriculture suggested in the p/pp transition model. Farmer identities will often be linked to specific geographical, cultural, farm structural or economic factors in a given locality, and generalising patterns and processes of identity may be difficult beyond a limited geographical context. Although studies on farmer identities are still scarce, and almost entirely restricted to case examples from advanced economies, the analysis below will, nonetheless, draw on a wide variety of studies across the EU and beyond based on differing agricultural and farming cultures, structures and styles (but admittedly to the neglect of developing countries due to lack of studies). What should emerge is an illustrative picture of farmer self-concepts that can be used to challenge existing notions of structural causality in the p/pp transition model.

Actor-oriented and behaviourally grounded approaches: towards post-productivist thinking?

Until recently, there was little evidence that the wealth of actor-oriented and behavioural literature on rural/agricultural change had been incorporated into conceptualisations of the p/pp transition model. However, many actor-oriented studies investigated a wide variety of issues *implicitly* related to the p/pp transition, but little was said about *explicit* links of this body of research to conceptualisations of post-productivism. Shucksmith's (1993) investigation of understanding motivations and behaviours of Scottish farmers in a post-productivist context, and our own study of farmer identities and post-productivist action and thought (Burton and Wilson, 2006), are some of the few exceptions. Studies that help answer whether farmers' identities have become more post-productivist include behavioural approaches to understanding farmers' land use decision-making processes (e.g. Ward and Lowe, 1994; Wilson, 1997c), work on reactions of grassroots actors to agricultural policies and AEPs (e.g. Froud, 1994; Buller *et al.*, 2000), research on farmers' attitudes towards farming and the environment (e.g. Morris and Potter, 1995; Lobley and Potter, 1998), analyses of interactions between pollution officials and farmers (e.g. Ward *et al.*, 1995, 1998), studies of the roles and attitudes of agricultural extension services and officials in policy implementation (e.g. Winter, 1996; Lowe *et al.*, 1997), as well as investigations of actor perceptions and motivations at the macro level (e.g. Wilson *et al.*, 1999; Clark, 2006). Results from these studies have highlighted that conceptualisations of the p/pp transition need to go beyond analysis of broader ideological changes mentioned in Chapters 5 and 6, and that we should also consider whether values of actors *directly* involved in processes of agricultural/rural change (e.g. farmers, agricultural extension services, agri-business managers, policy officials, etc.) reflect the postulated shift towards post-productivism.

First, research has shown that grassroots actors hold a *plurality* of often highly disparate opinions on issues surrounding environmental, agricultural and rural change, and that they are diverse actor communities that can often neither be branded 'productivist' or 'post-productivist' (Morris and Potter, 1995; Ward *et al.*, 1998). Second, studies have shown that many agricultural actors continue to adopt 'productivist' action and thought. A few examples relating to farmer environmental attitudes and adoption or non-adoption of AEPs may illustrate the point. Studies have suggested that most farmers interviewed in regional, national and international surveys continue to depict productivist attitudes, best expressed through persisting perceptions of farmers as the best 'stewards of the land', emphases on production maximisation as the ultimate goal of farming, and critical views, if not outright rejection, of new types of 'green' policies (e.g. Wilson, 1996; Lobley and Potter, 1998). This has been echoed by: Morris and Evans (1999: 352) who argued that "for most farmers it is 'business as usual' in meeting food output goals"; Shucksmith (1993: 467) who found that most farmers were unwilling "to adapt their businesses to, or engage with, the new [post-productivist] imperatives"; or Potter (1998: 88) who suggested that existing evidence "does little to support the hypothesis that farmers generally are becoming more conservation minded".

According to Shucksmith (1993) and Burton (2004), such behaviour can largely be explained through the fact that farmer's attitudes and behaviour derive in large part from the subconscious and cumulative assimilation of an established ethos of being a 'farmer', and that *farmer identity* and rootedness are often situated in traditional conceptions of the role of agriculture as a producer of food and fibre. Thus, Shucksmith (1993: 468) argued that many options potentially open to farmers, such as 'post-productivist' forms of diversification, "may never seriously be considered because they are literally 'unthinkable'". Similarly, Burton (2004: 196) suggested that "it is becoming increasingly evident that farmers may also resist change on the basis of an anticipated loss of identity or social/cultural rewards traditionally conferred through existing commercial agricultural behaviour". This was supported by Morris and Potter's (1995) 'participation spectrum' for farmers' adoption of AEPs that suggests a plurality of responses ranging from 'resistant non-adopters' and 'conditional non-adopters' to 'passive' and 'active adopters' – similar to Shucksmith's (1993) farmer attitude classification into 'accumulators' (productivist), 'conservatives' (productivist) and 'disengagers' (post-productivist) – with most farmers in Morris and Potter's survey falling into the 'conditional non-adopter' and 'passive adopter' categories. Morris and Potter, as well as subsequent research conducted in the UK and Europe (e.g. Wilson and Hart, 2000; Wilson and Juntti, 2005), found that of those farmers participating in new agri-environmental schemes, the majority joined schemes for financial reasons and/or because schemes fitted well with current farm management practices. Other work has suggested that many farmers only enter a limited quantity of eligible land into schemes, while intensifying production on the rest of the farm (Whitby, 1996; Lobley and Potter, 1998), highlighting that many farmers continue to farm in productivist ways while adopting 'post-productivist' policies. Indeed, that most AEPs and higher scheme tiers throughout the EU are voluntary, and not regulatory, should be seen as a recognition by policy-makers that large parts of the European farming community continue to be sceptical about changes to productivist ways of farming. That most agri-environmental schemes in the EU (and beyond) continue to be about *maintenance* and not *change* also suggests that there is little potential yet for bringing about substantial shifts in farmers' attitudes (Wilson and Juntti, 2005). Marsden *et al.* (1993: 65), therefore, argued that "adjustments are not easy … and the effort is handicapped by the legacy of the productivist ideology which is deeply engrained in the outlook and behaviour of many farmers, landowners and agricultural officials".

There are, therefore, continuing discrepancies between formulation of (arguably) post-productivist policies 'at the top' (e.g. AEPs and CAP 2nd pillar incentives) and productivist interpretation of these policies by grassroots actors (Buller *et al.*, 2000). Even if farmers cannot 'think' in post-productivist terms, it could be argued that they are more or less forced into moving towards post-productivist action, not least via other intermediary (e.g. NGOs) or state-level actors (e.g. farm extension services as implementers of state policies). It would be wrong, however, to argue that policy-makers (or other state actors) are necessarily post-productivist in thought and action, as there is much evidence to suggest an equally varied set of attitudes among officials and government extension services (Wilson *et al.*, 1999; Povellato and Ferraretto, 2005). The different discourses that permeate the 'implementation gap' highlight that different actors are situated at different points of the productivist/post-productivist spectrum (Burton and Wilson, 2006). Most studies, therefore, continue to emphasise the productivist orientation of farmers. Halfacree (1997b), for example, suggested that some farmers in the most intensively used European farming regions have even become 'super-productivist' (see above). Similar findings also come from rural areas outside the EU. In Australia, for example, studies have suggested that farmers are still largely caught up in a productivist profit-maximising mindset that focuses on further intensification of farming and increased use of biochemical inputs, with little evidence of a shift towards post-productivism (e.g. Martin and Halpin, 1998; Holmes, 2002). These and other findings emphasise that the persistence of farmers' productivist identities has encouraged a conservative view towards policies perceived to be imposed by 'others' (e.g. politicians, planners, conservationists) –

especially policies requiring substantial shifts in farm management practices and those with environmental components perceived as 'threatening' to traditional pro-production farming roles (Curry and Winter, 2000).

Research on farmers' environmental attitudes and decision-making also largely supports the persistence of productivist attitudes (e.g. Wilson, 1996; Fish *et al.*, 2003), as does Ward *et al.*'s (1995, 1998) work on interactions between pollution officials and farmers, work investigating farmers' land use decision-making processes (e.g. Ward and Lowe, 1994; Wilson, 1997c), and studies on farmers' reactions towards agricultural extension services and officials in policy implementation (e.g. Lowe *et al.*, 1997). Many studies investigating farmers' 'world views' also point towards a persistence of productivist interpretations and a lack of an 'ecocentric' world view by many rural actors (e.g. Selby and Petajistö, 1995; van der Meulen *et al.*, 1996). Thus, "where farmers choose to undertake alternative activities that are distant from productivist agriculture, such as tourism, there is still an emphasis on maintaining production simultaneous to the new development, rather than on shifting the emphasis of the unit completely away from agricultural objectives" (Burton and Wilson, 2006: 106; see also Hjalager, 1996).

A final arena of concern relates to farmers' aesthetic appreciation of the landscape and its role as a symbol of farming ability (Burton, 2004). Most evidence suggests that an ecologically rich and diverse landscape with high biodiversity value, theoretically the objective of the 'post-productivist' farmer, is still strongly associated with *poor* farming practices. For example, Egoz *et al.* (2001) suggested that in New Zealand controlled (tidy and without weeds) mixed farming landscapes represent hard working farmers, while a landscape that appears uncontrolled represents 'laziness'. Similarly, in the USA, Nassauer (1997: 68) observed that "cultivated fields are expected to have straight rows and no weeds. Farmers who allow weeds in their fields risk being seen as lazy or poor managers". Fish *et al.* (2003: 25) also highlighted that in the UK "productive concerns of farming [are] often positively linked to notions of an ordered landscape aesthetic". Thus, failing to observe the productivist ideal can have a significant and direct impact on the social position of a farmer within the local agricultural community and, therefore, a strong influence on their ability to perceive themselves as 'good' or 'real' farmers. Indeed, "within the farming community, this role-related productivist farmer identity has been suggested as providing members with a strong sense of well-being, particularly amongst older farmers and small-scale farmers, and, as such, is vital in many instances for the maintenance of farmers' self-esteem and perceptions of being a 'successful' farmer" (Burton and Wilson, 2006: 107).

Although most European farmers appear to be situated towards the productivist end of the spectrum, it also needs to be acknowledged that an increasing proportion of the farming community is beginning to adopt post-productivist ways of thinking. In an international survey, for example, 54% of farmers stated that they were participating in agri-environmental schemes because they wished to promote environmental conservation (Wilson and Hart, 2000), highlighting that productivist farmers' responses to policies are not incompatible with post-productivist concerns. Similarly, Fish *et al.* (2003) found that 90% of respondents responded positively to the question whether they were 'in sympathy with the conservation goals of agri-environmental schemes'. Yet, other studies have suggested that the dominance of productivist thinking may increase with the degree of proximity of survey respondents to the farming community (e.g. Wilson *et al.*, 1999, for Spain). This would suggest that actors closer to the policy core (i.e. those driving the policy agenda) may be more post-productivist than those at the periphery (Hart and Wilson, 1998). As Lowe *et al.* (1993) suggested, this highlights possible tensions in actor spaces where post-productivist differentiation (both in thought and action) may be contested through various actors attempting to impose their respective moral representations of the 'rural' and 'farming' over others (see also Lowe *et al.*, 1997; Wilson and Juntti, 2005). Further differentiation in post-productivist thinking is becoming evident within farming communities themselves. Various studies looking at the importance of farmer age for environmental thinking and responses to policies, for example, suggest that older farmers tend to have more traditional notions of farming and may, therefore, be more productivist than their younger counterparts (Wilson, 1997a; Potter, 1998). The latter have benefited from more modern education systems and better access to information, and more often show greater interest in conservation-oriented innovative farming practices. That most farmers in the EU are older than 50 years may be one of the explanations why productivist thinking currently tends to predominate. There is little evidence, therefore, that most European farmers (as well as US, Australian and New Zealand farmers; Potter, 1998; Holmes, 2002, 2006) have whole-heartedly engaged in new forms of post-productivism. Mental landscapes of European agriculture appear to remain relatively unchanged and embedded in productivist modes of thinking (Wilson, 2001).

This, together with a similar lack of evidence of a shift towards post-productivist thinking among many 'intermediate' and 'upper' level actors such as street-level bureaucrats, pollution officials and national and EU policy-makers (Clark *et al.*, 1997; Lowe *et al.*, 1997), begs important questions. First, is the notion of post-productivism only a conceptual construct that describes patterns at the macro-structural level and that has not yet permeated to the grassroots level? Are we, therefore, facing a 'post-productivist myth' as suggested by Morris and Evans (1999)? Second, considering that grassroots actors' attitudes may not have yet shifted towards post-productivism, is it then possible to pinpoint exactly, as some of the literature would suggest, *when* the transition towards the post-productivism occurred? The following sections will turn to these issues, with a specific focus on identity theory and whether farmer identities have become post-productivist.

Identity theory and farmers' post-productivist action and thought

Application of social psychology theory (especially identity theory) has shed interesting new light on debates about the structural causality of the p/pp transition model (Burton and Wilson, 2006). Based on this theory, identity salience is determined by the strength of the relationship between the individual and the social group as measured by the individual's commitment, i.e. the degree to which an individual's relationship to a specified group of people or social network depends on him/her being a particular kind of person and observing a particular set of social rules (Stryker and Serpe, 1982). Levels of affective and interactive attachment to the group determine how committed an individual will be to maintaining this identity and, therefore, in combination with situational demands, the likelihood of the identity being salient in the decision-making process. In return for commitment to the group, the individual may expect to receive through both social affirmation and self-assessment the security offered by a sense of group belonging and self-esteem – both of which provide substantial motivation for future compliance with group norms (Cast and Burke, 2002). The strengths of adopting the identity theory perspective are, therefore, three-fold. First, unlike the 'behavioural approach', identity theory takes account of social influences on decision-making through its representation of the relationship between self and society. Second, its focus on actor-oriented decision-making negates the criticisms often applied to the top-down structuralist approaches predominant in conceptualisations of the p/pp model, and, third, it provides a defined and theoretically based notion of identity linked to a body of literature that uses the self-concept as its central research theme (Deschamps and Devos, 1998).

While identity theory suggests that an individual can maintain as many identities as roles played in distinct sets of social relationships (Stryker, 1994), spatial and temporal variations in the opportunity to express identity inevitably result in some categories of identity becoming more important for self-construction than others. It has long been recognised that occupational, gender, family, ethnic and national identities are fundamental in this respect (Callero, 1985) and that *occupational identities* are, along with gender identities, seen as "amongst the most important in the individual's pantheon of idealised role identities" (Gordon, 1976: 407). An individual may maintain a number of distinct work-related identities that are situationally likely to influence behaviour. In the context of farmers' identities, this would be, for example, exemplified in hobby farming where the land manager may be more committed to a professional occupation as measured through time spent in that specific role, and yet the 'farmer' identity may be more salient through the affective commitment to agricultural production and the sense of place the farm environment provides (see also Chapter 9). Indeed, many studies suggest that small/part-time farmers often identify strongly with their consumptive farming occupation, even where their main productive occupation lies elsewhere (Coughenour, 1995). Farmers' occupational identities, therefore, derive in large part from work roles associated with agricultural production, with farmers subconsciously accumulating the ethos of being a farmer and learning, through socialisation, roles and meanings that are symbolic of the 'real' or 'good farmer' amongst the peer group (Burton and Wilson, 2006; see also Shucksmith, 1993). That occupational identities are among the most important has particularly strong implications for *agricultural identities*, as the proportion of time spent by farmers on their farms or within their own communities performing production roles highlights the potential for intense production-oriented identities. Yet, social psychology research highlights that people engaged in occupations such as 'farming' rarely depict a *single* identity. Instead *multiple* identities appear to be the norm, with some identities coming to the surface at certain times while others predominate occupational action and thought at others (Burton and Wilson, 2006).

There is, therefore, a strong argument for conceptualising the farmer 'self-concept' as comprising multiple identities (e.g. 'agricultural producer' *and* 'conservationist'), each with different notions of what comprises good farming practice, and each capable of becoming the focus for action. In addition, "there may be important non-farming identities that may guide farming behaviour, in particular family oriented identities which may determine how the farmer follows a specific economic development path (e.g. business expansion) for the successor/s, even where his/her personal agricultural preference lies elsewhere" (Burton and Wilson, 2006: 100). Also important is the notion of 'other' as expressed through counter-identity or counter-role, and there is evidence that notions of 'other' are important in farming identities. For example, McEachern (1992) noted that farmers used stereotypes of urban dwellers in order to discredit conservationists. More importantly, however, are studies in which farmers identify certain types of farmers as 'other' on the basis of their specific role behaviours. For example, 'traditional' farmers interviewed by Seabrook and Higgins (1988) perceived farmers following 'progressive' approaches to agriculture negatively. Bell and Newby (1974: 99) similarly observed that "many traditional farmers oppose the modern large-scale agribusiness techniques of their neighbours, while the latter may be equally scornful of the lack of innovatory zeal of their farming peers", while Gasson (1974: 134) argued that "many smaller farmers oppose the image of modern large scale farming, and the 'barley barons' and 'broiler kings' may receive little esteem from their more traditional neighbours". Farmers' identities can, therefore, only be understood in a *relational context*, i.e. how they view themselves within their wider farming community.

In a recent study, we used a hypothetical conceptualisation of farmer-selves that focused on four recurring typological groups identified in previous studies as key farming typological categories that represent specific role strategies towards agricultural production (Burton and Wilson, 2006). These farmer types broadly comprise: (i) 'traditional' – a conservative productivist farmer who maintains cultural notions of stewardship; (ii) 'agri-businessperson' – a farmer who concentrates agricultural production to the extent that the profit motive dominates and stewardship concerns are lessened; (iii) 'conservationist' – a farmer who focuses on environmental and lifestyle concerns; and (iv) 'diversifier' – a farmer who is shifting away from standard agriculture towards non-agricultural sources of income. We then linked these conceptualisations of identity to the different dimensions of productivism and post-productivism highlighted in Chapters 5 and 6. Productivist farmer identity roles can be conceptualised as pro-development, stressing increasing reliance on, and intensification of, agriculture, and a degree of faith in technocentric approaches to land management. In contrast, post-productivist identities would stress a lesser role for agriculture (yet perhaps not peripheral), an increased consideration of environmental factors, and a diversity of approaches to land management. The next step was to compare this conceptualisation with the main farmer groups identified, and to arrive at hypothetical models of productivist and post-productivist farmer-selves. The assumption was that a clear division emerges between 'traditional' and 'agribusiness' farmers who can be classified as 'productivist', and 'conservationist' and 'diversifier/entrepreneurial' farmers falling into the 'post-productivist' dimension (Burton and Wilson, 2006). We hypothesised, therefore, that a productivist farming self will be defined not only by her/his own 'agricultural producer' and 'businessperson' identities, but by a rejection of the role-behaviours of farmers from the 'conservationist' and 'diversifier' groups.

However, hypothesising the *post-productivist* farming self proved more difficult. Despite the lack of emphasis on the more commercial aspects of agriculture, 'conservationist' and 'diversifying' farmers continue to be farmers and, therefore, it was not possible to conceptualise these identities through the rejection of farming roles (Burton and Wilson, 2006). Nevertheless, it can also be argued that the salience of the 'agricultural producer' identity is likely to be lessened, and one or both of the post-productivist identities will correspondingly increase in importance as the farmer takes on new roles and forges new social contacts. However, the 'agribusiness' approach to agriculture emphasising intensification, specialisation and commercialisation may still be regarded as a counter-identity, particularly amongst those with a high salience of the 'conservationist' identity. By applying an identity theory approach, therefore, we arrived at a hypothetical conceptualisation of productivist and post-productivist farmer-selves that may better reflect the complexities of self-identity construction. Rather than, for example, simply labelling farmers as 'conservationists' or 'diversifiers' as previous studies often did, our hypothetical conceptualisation recognised that farmers may maintain all identities *simultaneously* and, where the situation arises, appropriate the most suitable identity with its expression of beliefs, roles and attitudes. Therefore, "this conceptualisation of a structured, hierarchical and situationally dependent self may aid greatly in explaining the inconsistencies many researchers ... have found between farmer's *expressed attitudes* towards conservation and their *actions towards the environment*, in that the 'conservationist' identity may be made temporarily salient through the social interaction with the researcher ... thereby seemingly overshadowing underlying productivist attitudes towards farming" (Burton and Wilson, 2006: 103; original emphasis).

Applying this social psychology approach to a case study area in the UK in which afforestation of farmland was one of the policy goals (Marston Vale, Bedfordshire) suggested that farmers remained strongly productivist in their self-concepts (Burton and Wilson, 2006). Identities constructed of productivist roles ('agricultural producer' and 'agribusiness') were salient in most cases compared to only few with salient 'conservationist' and 'diversifier' identities. Of particular importance, unsurprisingly, was the 'agricultural producer' identity which was, on average, almost twice as salient as the next closest 'agribusiness' identity. Not only has the farming culture developed to value the role of 'agricultural producer' above all other alternatives in most advanced economies, but, in addition, by simple virtue of their permanent role of managing the vast majority of rural land, farmers have evolved a strong self-image as 'stewards of the countryside', such that the image of the farmer steward appears at the centre of the farming culture, despite evidence that intensive farming leads to considerable environmental damage (Wilson, 1996). Thus, in general, the 'good' or 'real' farmer has strongly incorporated within her/his identity the productivist ideals of the era in which s/he emerged and conforms – a pattern not necessarily in accordance with expectations of what 'good farming' practice means for other stakeholders such as agricultural policy-makers or academics. As a result, the 'conservationist' identity was the second lowest in the Marston Vale, while the 'diversifier' identity was the least salient. Farmer's self-perspective as an 'agricultural producer' is closely tied with the notion of 'agribusiness' and, therefore, some elements of the role of the 'agribusiness' identity (in particular business expansion and an emphasis on profit maximisation) also contribute to the 'agricultural producer' identities. Further, our findings showed that the 'diversifier' identity had the highest dissimilarity with the productivist identities. This discrepancy cannot simply be attributed to the lack of diversification among farmers in the survey, as respondents showed high levels of pluriactivity. We argued that this discrepancy should be expected, as the 'diversified farmer' has been seen in the past as representing a 'failed' farmer, and farmers have shown a general unwillingness to move away from agricultural production towards diversification (Burton and Wilson, 2006). Our results also showed that the 'diversifier' identity is negatively correlated with all other identities, which suggests that it represents a counter-identity ('other') to the agricultural identities. However, a lower correlation with the 'agribusiness' identity suggested that there is more commonality between 'diversifier' and 'agribusiness' roles than between 'diversifier' and the 'conservationist' or 'agricultural producer' roles. This definition of diversifier as 'other' was further supported by in-depth interviews, and when asked to define what a 'good farmer' is, mention of conservation was made occasionally under the broader heading of 'stewardship', as well as to business roles in order to keep the name on the land. However, no mention was made of diversification. Rather, a 'good farmer' was defined largely by productivist roles, such as 'the chap who can up his output by a ton an acre or whatever – and continue to do so', or as the one who 'tries to get three heads of corn where there used to be two or three blades of grass where there used to be two' (Burton and Wilson, 2006).

An increasing body of work has also begun to argue that farmers in advanced economies are becoming more 'conservation-oriented' (see above). This would imply that there should be a perceptible shift in the salience of the conservationist identity within existing farmer self-structures. However, our findings from the Marston Vale did not support any radical shift towards new conservationist attitudes among farmers (including younger farmers). Indeed, an interesting feature of the data was the apparent incorporation of the 'conservationist' identity within the existing production-oriented self-concept (Burton and Wilson, 2006). Younger farmers had a significantly lower salience of the conservationist identity than older farmers, and the majority of conservation-oriented farmers were older farmers for whom financial security provided the opportunity to manage land more extensively. Younger farmers – often at a stage of building up a business and/or supporting a young family – were either oriented towards agribusiness ideals or reliant on diversification, with conservation not seen as a priority in both cases. This finding is not unique to the Marsden Vale case study. For example, Short (1997: 49; emphasis added) observed from a case study in Malta that "farmers under 35 were *by far* the most sceptical about traditional practices and enthusiastic about the use of chemicals." However, as high conservation value is more often associated with traditional agriculture, loss of traditional social norms can lead to a more utilitarian rather than conservationist approach. What we may be detecting in younger farmers is greater environmental awareness – i.e. their understanding of environmental issues is increased through the education received and increased societal awareness. However, this is not necessarily reflected in environmental concern nor prioritisation over other production-oriented roles/identities. Neither does correlative evidence of a relationship between age and the presence of a conservation scheme necessarily indicate environmental concern. Although some researchers have assumed that engaging in a conservation scheme qualifies a farmer to be classified as a 'conservationist' (e.g. Battershill and Gilg, 1996), this should be regarded with scepticism as in many situations the motivations for participation in conservation schemes can be aligned with productivism (Wilson, 1997a; Wilson and Hart, 2001).

As a result, "the development of the post-productivist 'conservationist' role and its absorption within the 'good farmer' identity may not be as linear as reported in other studies" (Burton and Wilson, 2006: 108). One possible alternative to a linear model is that although farmers have integrated the conservationist identity within the productivist self-concept, the conservationist identity is applied *situationally* around the farm, particularly as identity is situational – i.e. the conservationist identity is suitable for expression only to land deemed 'appropriate' for conservation. Other studies provide evidence to suggest this may be the case. Wilson and Hart (2001: 268), for example, suggested that farmers may hold different attitudes with respect to different parts of their farms, and participants in the Devon Countryside Stewardship Scheme (which targets biodiversity-rich areas of 'culm' grassland in the south-west UK) "largely expressed their [conservationist] attitudes with regard to the area of culm grassland entered into the scheme, [while] a few interviewees revealed less conservation-oriented views with regard to farming practices on the rest of the farm". Wilson and Hart (2001: 269) concluded that it was, therefore, "questionable whether the conservation concern is actually transposed beyond the small area of culm grassland and implemented on the rest of the farm". This, and other research (e.g. Morris and Potter, 1995; Lobley and Potter, 1998) has suggested that some farmers may, therefore, view their participation in so-called post-productivist policies as separate from the rest of the farm which is, arguably, still managed in productivist ways.

While until recently studies were arguing that some sections of the (European) farming community had become more 'conservationist' on the basis of their investigation of unidimensional attitudinal attributes, the new evidence highlighted above suggests that these studies may have only scratched the surface, and that they may have neglected that these seemingly predominant attitude traits are in reality embedded in a wider pantheon of farmer self-identities that, in their entirety, remain largely focused on the productivist 'agricultural producer' and 'agribusiness person' roles. What exists then – observed by researchers as a discrepancy between conservation attitudes and behaviour – is, in fact, evidence that the 'conservationist' identity is present within the farmer self-concept, but that it is frequently suppressed by the dominance of production-oriented identities. In other words, a good farmer is a conservationist, but s/he is an *agricultural producer* first. If this is true, this throws further doubt on the p/pp transition model, as "transition is clearly not a simple case of converting a production-oriented farmer to a conservation-oriented farmer, but rather involves making existing but latent role beliefs more salient" (Burton and Wilson, 2006: 108). Similarly, a 'post-productivist' identity can be identified, apparent through a strong negative correlation between the 'agribusiness' and 'conservationist' identities. However, when viewed at an individual level, there does not appear to be any conflict between these two identities, as those who hold post-productivist identities as salient may still stress 'agribusiness' as more important than 'diversification'/'conservation'. In other words, business roles associated with the 'agribusiness' identity permeate across the spectrum of farming self-concepts – be they 'productivist' or 'post-productivist'. This continued position of 'agribusiness' as an important component of the farmer self-concept and the position of 'diversifier' as a counter-identity suggest that farmers are still a long way from fully adopting 'post-productivist' identities into their self-concepts. Recent research, therefore, has found a relative absence of clear post-productivist farmer identities, confirming that the notion of new 'conservationist' farmer identities can not be seen yet as a new orthodoxy taking hold in European farming (and beyond) (see also Cayre *et al.*, 2004, for similar debates from a French point-of-view).

Farmer identities and the fallacy of structural causality

What implications do these findings have for debates on the p/pp transition model and the fallacy of structural causality? At the beginning of this section I asked whether it is possible to assert that the p/pp transition has affected all stakeholders in rural areas (and beyond) in similar ways and that all actors are moving along transitional pathways towards post-productivism at the same pace. As highlighted, 'traditional' behavioural studies that have focused on single attitudinal and behavioural attributes of farmers have lent some credence to this suggested transition, by implying that actors (i.e. farmers) at the grassroots level are beginning to change their outlook towards 'post-productivist' and 'conservation-oriented' thinking – with associated 'positive' environmental management practices. Yet, recent studies suggest that when farmers' self-concepts are viewed as multiple, hierarchical and situational, following the dominant conceptualisation within the social psychology literature, a much more complex picture emerges (Burton and Wilson, 2006). Thus, farmers' expressions of pro-conservationist or diversification attitudes may be made within the context of individual identities, and thus measurements of general attitudes may bear little relationship to actual behavioural choices. This is given added weight by a body of evidence that is emerging in social psychology which argues that, while attitudes are stable across time and context, individuals can nonetheless hold *multiple attitudes* towards a given attitude-object, thus creating a variety of behavioural choices that may be expressed depending on situational factors (Wood, 2000). Thus, "although farmers' environmental management practices and expressed attitudes seemingly show signs of what could be interpreted as post-productivism, we argue that these must be viewed in the context of the continued domination of the self-concept by identities based on production-oriented roles" (Burton and Wilson, 2006: 110). This was reiterated by Hoggart and Paniagua (2001: 48) who argued that "the much-touted shift towards a more ... environmentally friendly farm sector is contradicted by evidence that farmers are not passive agents, but resist change or creatively recreate themselves in a manner of their own choosing" (see also Cayre *et al.*, 2004). It can, therefore, be hypothesised that most 'farmers' in the context of advanced economies still perceive themselves as actors who, first and foremost, produce food and fibre with the aim to maximise food production and to pass on an economically viable farm business to the next generation. As a UK farmer recently commented, "maintenance [engendered by many agri-environmental schemes] is not as satisfactory for farmers as production" (Mercer, 2005), while Allison (1996: 142) observed that "farmers want to farm. It gives them their identity and their sense of achievement".

    This interpretation of entrenched productivist farmer selves has obvious repercussions for conceptualisations of the p/pp transition model. First, it supports criticisms made above regarding the fallacies of temporal linearity, spatial homogeneity and global universality in that the transition from productivism to post-productivism is far from linear in a temporal sense, and that different stakeholder groups in rural society are moving towards post-productivist action and thought at a *different* pace. This was supported by Argent (2002: 107) who, from an Australian perspective, argued that "in the case of post-productivism ... its central tenets are weakly supported by what actually happens at the 'grass roots' level", and that "post-productivism as a macro-structural concept is largely unable to account for ... local-scale events and processes, mainly because farm-level dynamics do not fit neatly into any productivist/post-productivist divide" (Argent, 2002: 111). This is also emphasised by the fact that beyond the grassroots level, there is some evidence that not all actors at the macro-structural level (e.g. in Brussels or national policy-makers) are post-productivist in their outlook, and "there is much evidence to suggest an equally varied kaleidoscope of attitudes among officials and government extension services [as there is among farmers]" (Wilson, 2001: 88). At the farmer level, any move towards post-productivist behaviour is likely to be underlain, at least in the short and medium term, by strong productivist identities that could rapidly become dominant again were the pressures of 'post-productivist' policies to be relaxed. While some aspects of macro-structural agricultural policy may be moving towards post-productivism, many studies suggest that the same is not occurring at the grassroots level. This, therefore, contests the notion that both structure and agency – in Giddens' structuration theory terms – 'move' along the p/pp transition at the same pace.

    This *structure-agency inconsistency* partly explains agricultural policy failures described in many studies, especially for 'post-productivist' agri-environmental policies and some of the new policy elements contained under the 2nd pillar of the CAP (see Section 7.3.3). The lack of evidence suggesting that farmers have adopted a 'post-productivist' ethos explains why schemes designed to move farmers away from productivism have met with limited success. There is little wonder, therefore, that when farmers' *self-concepts* are still firmly entrenched in productivism, their acceptance of policies with a post-productivist tendency does not represent any real change in their beliefs in the role of the 'farmer', and that, as a result, agri-environmental schemes are often used as an income support measure (Wilson and Hart, 2000). Similarly, the dominance of productivist occupational identities also explains why so-called 'post-productivist' farmers still have problems adapting to new consumer demands for organic produce (Tovey, 1997). Indeed, "by simply adapting the 'conservationist' identity … farmers can build in both agri-environmental policy and organic farming production into their self-concept, thereby strengthening, rather than weakening, their productivist outlook" (Burton and Wilson, 2006: 112). This structure-agency inconsistency, therefore, leads one to question many of the implied assumptions in the p/pp transition model. Linked to Section 7.3.3 (fallacy of temporal linearity), it challenges the assumption that contemporary agricultural systems in advanced economies have moved towards post-productivism, and also leads us to question whether a specific 'starting point' for the 'era of post-productivism' can be identified as suggested by various commentators (see Chapter 5). Instead, multiple pathways of structure-agency interactions towards post-productivism appear much more likely.

7.4 The productivist/non-productivist spectrum of decision-making

7.4.1 Deconstructing post-productivism

The discussion in Section 7.3 has highlighted that the p/pp model is not robust enough to explain agricultural change – be it from the perspective of developed or developing countries. Echoing scientific critiques of the p/pp transition model (Section 7.2), the discussion has highlighted that the so-called 'transition' towards post-productivism is characterised by temporal non-linearity, spatially heterogeneity, global complexity and structure-agency inconsistency. This makes the notion of a shift towards post-productivism difficult to substantiate. Marred by issues of UK-centrism, presentism, an overemphasis of political economy and confusion about temporal linearity and spatial homogeneity, the p/pp transition model does no longer stand up to critical scrutiny as a model adequately explaining contemporary *transitional* processes in agricultural change. Evans *et al.* (2002: 324; emphasis added), therefore, argued that "overall, some commentators declare that post-productive conditions now prevail … If these conditions are founded on the theorization that productivist processes are being progressively reversed, then current evidence shows them to be *untenable*". This was also echoed by Marsden and Sonnino (2005: 28) who argued that many British social scientists are advising politicians with "the ideas of post-productionism [sic] influencing policy debates in Whitehall. There is now a real need for some serious critical reflection on this process, however" (see also Holmes, 2006). Similarly, Potter and Tilzey (2005: 13; emphasis added) suggested that primary production should be seen as "an increasingly *bifurcated* agricultural industry, comprising *both* productivist and post-productivist sectors", where "productivist and post-productivist functions are delivered side by side" (Potter and Tilzey, 2005: 15). From a rural sociology perspective, Goodman (2004: 11, emphasis added) further argued that "several recent contributions to the productivism/post-productivism debate have articulated strong critiques of this representation of farm output diversification and its empirical validity… and have questioned whether the values, attitudes and behaviour of rural actors actually correspond to this *uncompromising binary definition* of their alternatives". Robinson (2004), therefore, suggested that the term 'post-productivism' was adopted too readily and uncritically.

The problem of linearity of the p/pp concept has partly been linked to the fact that the conceptual literature on post-productivism largely failed to take into account the wealth of actor-oriented and behaviourally grounded research (Wilson, 2001). As Section 7.3.6 showed, an inclusion of such research points towards considerable differences in productivist and post-productivist action and thought at different levels of actor spaces, that grassroots actors have remained largely productivist in their agricultural occupational identities, and that we are dealing with a *spectrum* of different views rather than two easily definable and 'separate' conceptual entities. As Knickel and Renting (2000: 523) emphasised, "agricultural change in post-productivist societies is characterized by complex differentiation processes". Acknowledging findings from actor-oriented and behaviourally grounded studies suggests that the breakdown of productivist attitudes and ideologies has not occurred yet. This highlights that "political economy approaches have traditionally led us to search for a post-productivism, while an integration of agency, and the acknowledgement of the complex adoption and non-adoption of post-productivist thought by various actors, allows us to conceptualise both multiple post-productivisms and different mental stages of post-productivism along a wide-ranging spectrum. Different arenas of agriculture have adopted post-productivist thinking at different times ... The transition from productivism to post-productivism may, thus, be made more difficult because of the perceptual gap between grassroots and higher level actors, and will require not only time and effort with regard to practical considerations, but also negotiation in balancing the numerous and complex goals and ideologies of often competing agricultural/rural interests" (Wilson, 2001: 90). Throughout our discussion, it has also been evident that the problem also partly lies in the fact that the p/pp transition model has been conceptualised from a UK-centric perspective that has largely failed to discuss whether, and to what extent, the concept has wider applicability within Europe and beyond. Section 7.3, therefore, also highlighted the time-lag and *spatial* inconsistencies in the adoption of post-productivist action and thought, and emphasised that different localities are positioned at different points in the p/pp transition. Implications of these critiques for the understanding of rural spaces 'beyond agriculture' are fundamental, and the notion of the 'territorialisation' of productivist and post-productivist actor spaces (see Section 7.3.4) highlighted the wide-ranging diversity that exists within the p/pp spectrum. Marsden (2003: 151), therefore, argued that different rural spaces have "their own temporal and regulatory dynamics, involving different networks of actors, agencies and relationships with local, national and global markets". That such productivist and post-productivist action and thought occur in *multidimensional coexistence* leads one to question the implied directionality of the productivist/post-productivist debate.

Our evidence presented in Section 7.3, therefore, suggests that post-productivism has *not* replaced productivism. Indeed, it appears that both processes *co-exist* temporally, spatially and structurally. We witness, therefore, temporally non-linear, spatially heterogeneous and structurally inconsistent parallel pathways of productivist and post-productivist action and thought. As Goodman (2004: 3) emphasised, there are "doubts about the conceptualization of contemporary European rural development in the radical terms of paradigm shift rather than continuities in change". These 'continuities of change', therefore, highlight that post-productivist agriculture runs *concurrent* with (rather than 'counter' to) productivism – an idea also supported by Halfacree and Boyle (1998: 7) who emphasised that "the idea of a post-productivist countryside does not mean a countryside in which agriculture is either no longer present or in which it has been eclipsed in significance by other land uses". Similarly, Marsden (2003: 11) suggested that the productivist (agro-industrial) and the post-productivist dynamics should be seen as co-existing heuristic entities (rather than as temporally linear transitional pathways), thereby highlighting "the embodiment of conflict between productivist and post-productivist models being played out amongst the farm and wider rural population ... These conflicts occupy one contested territory concerning rural nature; a territory which brings together productivist with consumption concerns".

The so-called transition to post-productivism, therefore, should not imply that productivist institutional forms, networks, ideologies and norms have been superseded. Potter and Tilzey (2005: 1), thus, argued that "rather than witnessing the shift towards a post-productivist agriculture anticipated by some recent commentators, we argue that the dominant framing is in favour of a neoliberal regime of market productivism", and that we see "a bi-modal policy strategy emerging which is premised on the assumption that so-called post-productivist consumption spaces and policy strategies can co-exist alongside 'market' productivism" (Potter and Tilzey, 2005: 4). 'Post-productivism' has not been radical, but rather incremental and accommodationist to productivist action and thought. The assumption highlighted in Chapters 5 and 6 that, after a brief period of crisis and restructuring, a new, qualitatively different, post-productivism has risen from the ashes of productivism, therefore, needs to be questioned. This means that questions relating to specific dates *when* productivism was replaced by post-productivism (see Chapters 5 and 6) are meaningless. Again, we see parallels with debates on other transitions to 'post-isms'. Cloke and Goodwin (1992) and Goodwin and Painter (1996) also questioned that new regulation has formed a dynamic element of a new 'post-Fordist' mode of regulation, thereby also allowing for Fordist and post-Fordist modes of regulation to occur simultaneously in spatial, temporal and conceptual terms. Similarly, Hoggart and Paniagua (2001: 56) questioned the notion of 'rural restructuring' by arguing that "a transition has taken place but as a package not something that is sufficiently different from past processes to be seen as a pivotal, 'radical break'".

Some commentators (e.g. Wilson, 2001; Evans *et al.*, 2002) have, therefore, rightly asked whether the notion of post-productivism may only be a conceptual construct that describes patterns at the macro-structural level. Morris and Evans (1999) even suggested that we may be dealing with a 'post-productivist myth', while I argued that "post-productivism may only … be a theoretical construct in the minds of academics, rather than an expression of reality on the ground" (Wilson, 2004: 481). Like any terminology beginning with 'post', the notion of 'post-productivism' may also indicate a hesitance by those who propelled the term into the academic terminology to create a *new* terminology. This has been echoed by those who developed the term 'post-productivism', themselves conceding that "the spatial and sectoral unevenness of the substitution of the new for the old shows how ambiguous such terms as 'post-Fordism' and 'post-modernism' are, and points up the weakness of unilinear arguments" (Marsden *et al.*, 1993: 19). Marsden (2003: x) further argued that conceptualisations of agricultural and rural change should take us "beyond the modernist and post-modernist critiques of agrarian political economy, which have been so dominant in the past twenty years". From a theoretical perspective, these debates warn against the widespread inclination to posit clear distinctions between productivist and post-productivist agricultural systems, and encourage the search for new tools to explore the nature and dynamics of the conceptual ground 'in between' the extreme pathways of productivism and post-productivism. Yet, the main conceptual problem rests less with the notion of 'productivism', but rather with the concept of 'post-productivism' (especially with the prefix 'post') and the suggestion of an *irreversible transition* from one organisational state to another. Walford (2003b: 500) rightly argued based on evidence from the UK that "far from the demise of productivism, it is alive and well on most large-scale farms" (see also Wilson, 2001; Lang and Heasman, 2004). This suggests that a *reconceptualisation of the p/pp transition model* is necessary if this model (or parts thereof) are to have wider scientific credibility in the future. In the following section, I will investigate to what extent we can salvage positive aspects of the p/pp debate and discuss how a *revised transition model* may form a useful basis for reconceptualising multifunctional agriculture (Part 3).

7.4.2 Towards a revised transition model: the productivism/*non*-productivism spectrum

The failure of the reductionist p/pp transition model to explain contemporary agricultural change, identified through the lens of transition theory, means that the continued use of the term 'post-productivist' is untenable. I, therefore, disagree with Mather *et al.*'s (2006) recent assertion that "post-productivism is a reality, and should not be abandoned [as a term]". In my view, and as transition theory has amply demonstrated, 'post' implies something following after something else, and the evidence does not support the suggestion that a coherent and conceptually well defined era has 'followed on' from the productivist era. I, therefore, concur with both Evans *et al.* (2002: 313), who suggested that "future progress in agricultural research will only be made if [the notion of] post-productivism is abandoned", and Argent (2002: 97) who argued that "'post-productivism' is fundamentally misconceived, largely owing to its inherent binary narrative form and logic". Hoggart and Paniagua (2001: 42) similarly argued that "restructuring does not involve the complete transition from one of these isms to its associated post-ism", while Clark (2003: 28-29) emphasised that "productivism and post-productivism [should] be seen as a 'spectrum' rather than a duality, with post-productivist forms of agriculture often coexisting in rural regions alongside productivist patterns of farming".

Yet, this is not to say that the p/pp transition debate has been futile. On the contrary, as with many theoretical propositions it has usefully forced social scientists to re-evaluate conceptions of agricultural change. As both Evans *et al.* (2002) and Mather *et al.* (2006) argued, the p/pp model has had real heuristic value as a descriptor of agricultural change, and, as a consequence, I do not wish to do away entirely with useful terms and terminologies that have evolved from these debates. In line with Marsden (2003) who has equated his new concept of the *agro-industrial dynamic* with the concept of 'productivism', I wish to argue that the term 'productivism' continues to be relatively robust, as all indications are that productivist tendencies *can* be identified in contemporary action and thought affecting agriculture and rural areas. This is particularly the case as the indicators of 'productivism' outlined in Chapter 5 have, to some extent at least, withstood the test of time and are based on tangible, at times measurable, evidence. Indeed, many authors continue to refer to 'productivism', while increasingly shying away from using 'post-productivism' (e.g. Evans *et al.*, 2002). This has also been confirmed for advanced economies outside of Europe where the utility of the concept of 'productivism' has been much less questioned than 'post-productivism' (Holmes, 2002, 2006; Wilson and Memon, 2005). In the Australian context, for example, Argent (2002: 107) argued that "a strong case can be made that agricultural 'productivism' was pursued by national and State governments during the post-war period". My main criticism is, therefore, less with the notion of 'productivism' but with the *implied transition* away from productivism to a so-called post-productivist era. When shorn of its association with the notion of a *transition*, the concept of agricultural 'productivism' continues to be useful.

Instead, agricultural action and thought should be viewed as occurring along a temporally non-linear and spatially heterogeneous **spectrum of decision-making** possibilities, akin to what Duram (1997) termed the 'continuum' of farmer decision-making, and similar to Morris and Potter's (1995) farmer policy participation spectrum (see also Meert *et al.*, 2005). The problem with the p/pp model was, therefore, that it treated both 'productivism' and 'post-productivism' as *distinctive entities* with seemingly relatively clear and separable temporal boundaries. Viewing agricultural transition through a spectrum, meanwhile, negates the need to define such distinctive boundaries and acknowledges the fluidity and flexibility of agricultural action and thought. I argue that such a spectrum shares many similarities with the *Deleuzian transitional model* (see Chapter 2) in which multiple development pathways are possible. If we acknowledge that such a spectrum of decision-making exists that continues to include productivist elements at one end, we, therefore, need to find a term that better encapsulates the 'opposite end' of productivism than the current term 'post-productivism'. I suggest here that we should use **non-productivism** as the opposite of productivist action and thought, suggesting that this is a *true* opposite (which 'post-productivism' never was) that allows for the juxtaposition of temporal, spatial and structure/agency-related pathways of agricultural decision-making[42] (Fig. 7.3.). I will refer to

---

[42] The notion of 'non-productivism' has been used by others in the past. For example, Tovey (2000) suggested in the context of Irish agri-environmental policy that such policies contain 'non-productivist' elements, while Potter and

this as the *revised productivist/non-productivist transition model* (hereafter the p/np transition model). Non-productivist action and thought will contain many 'indicators' used to conceptualise post-productivism in the past, but without the underlying assumption that there is necessarily a temporally linear movement *towards* such action and thought over time. Non-productivist action and thought will be linked to high environmental sustainability, a tendency for local embeddedness, short food chains, low farming intensity and productivity, weak integration into the global capitalist market, a high degree of diversification, open-minded farming and rural populations who see 'farming' and 'agriculture' as processes that go beyond productivism, and open-minded societies who accept that the nature of 'farming' and 'agriculture' (or 'agri-culture' in Pretty's, 2002, words) is in the process of change (see also Chapter 9).

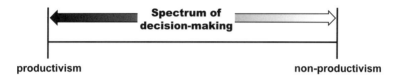

Fig. 7.3. The productivism/non-productivism spectrum of decision-making

One of the key advantages of 'non-productivism' is that, as opposed to 'post-productivism', it is a *neutral* term, simply denoting action and thought that is *not productivist*. This may help us get away from the danger of *subjectivity* that has hitherto branded 'productivism' as the 'bad' era and 'post-productivism' as the 'good' era that we should all be striving for. There is, therefore, no longer any intrinsic reason why the assumed direction of change – from productivist to non-productivist – should not be reversed in some circumstances, and, as this chapter highlighted, there is plenty of evidence to suggest that this is often the case. Further, embedding both productivism and non-productivism as part of a spectrum allows us to recognise that there may be virtually no action and thought affecting agricultural and rural areas that is *entirely* productivist or non-productivist, but that each action and thought may contain elements of both along Deleuzian development pathways – an issue I will discuss in detail in Part 3 in the context of multifunctionality. It is this implied *fuzziness, complexity* and *absence of distinctive boundaries* between productivism and non-productivism that makes this model more robust than a unilinear transition model. This enables us to envisage productivist and non-productivist spaces of decision-making along a spectrum of *Deleuzian decision-making pathways* that take into account the above identified temporal non-linearity, spatial heterogeneity, global complexity and structure-agency inconsistency that characterise contemporary agricultural change (Fig. 7.4). Further, and linked to our discussion of the fallacy of global universality in Section 7.3.5, this model also allows us to do away with the notion of 'pre-productivism' (see also Wilson and Rigg, 2003). Indeed, 'pre-productivist' action and thought can now easily be conceptualised as a set of approaches that often include non-productivist elements, but that may also, at times, contain elements of productivism. 'Pre-productivism', therefore, covers an equally varied range of

---

Tilzey (2005: 25) referred to "a non-productivist form of agrarianism" in which farming incomes derive primarily from the sale of farm products valued on the basis of an environmental and social tariff. Similarly, Delgado *et al.* (2003) and Gallardo *et al.* (2003) referred to 'non-productive' functions of agriculture not valued by the market (see also Buller, 2005), while Rapey *et al.* (2004a) used the notion of 'activité agricole non productive' in their analysis of multifunctional land use in France. In their recent contribution advocating the continuation of the use of the term 'post-productivism', Mather *et al.* (2006) also referred to 'non-productive' objectives of UK forestry policy.

farmer decision-making in temporally non-linear and spatially heterogeneous ways as does modern agriculture, all bounded by the extreme pathways of productivism and non-productivism.

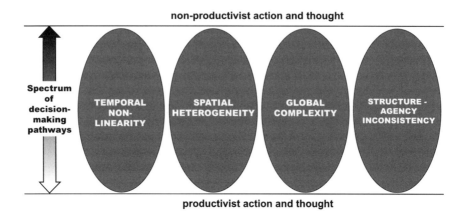

Fig. 7.4. Transitional fallacies and the productivism/non-productivism spectrum

This revised model shares many similarities with other models of decision-making pathways bounded by two extreme boundaries, such as 'nature-nurture' debates (i.e. in most cases 'reality' lies somewhere in between these two extremes) (Braun and Castree, 1998), or 'structure-agency' (Giddens, 1979, 1984, 1991) to name but a few. In particular, and referring to the discussion in Part 1, it begs the question whether other debates on transitions from 'isms' to 'post-isms' would also benefit from revised concepts based on a *spectrum* of decision-making that allows for the mutual coexistence of 'isms' and 'non-isms' (as opposed to 'post-isms'; see also Section 4.6). Thus, while I agree with Potter and Tilzey's (2005: 16) recent suggestion that "post-productivism therefore captures only part of the dynamic of transition", I disagree with their suggestion that "post-Fordism ... describes the totality of the processes and the juxtaposition of productivism and post-productivist tendencies under a new regime of accumulation/mode of regulation" (Potter and Tilzey, 2005: 16). In my view, and as Part 1 demonstrated, the implied linearity in the notion of 'post-Fordism' needs to be equally questioned. Instead of a transition towards *post*-Fordism one could envisage a 'Fordist/*non*-Fordist' spectrum that does not necessarily imply that a full transition has taken place, but that also leaves room for non-linear and heterogeneous parallel transitional processes. The same could apply to spectra ranging from 'modernism' to 'non-modernism' or 'environmentalism' to 'non-environmentalism' (with the latter acknowledging the mutual coexistence of several 'shades of green'). There are clearly interesting philosophical questions emerging from such possible reconceptualisations, and I hope that the discussion in this chapter will also encourage others interested in reconceptualising *societal change* from a transition theory perspective to critically scrutinise possible repercussions of a reframing of 'post-isms' based on temporally non-linear and spatially heterogeneous concepts.

7.5 Conclusions

While Chapters 5 and 6 provided insights into the different components that make up the so-called transition to post-productivism, Chapter 7 provided a critique of this transition and argued, instead, that productivist and post-productivist action and thought continue to occur simultaneously. As no clearly defined 'post-productivist' territory can be defined, I suggested in Section 7.4 that we should do away with both the notions of a p/pp transition and with the term 'post-productivism'. Instead, the notion of *non-productivism* should be used to describe the opposite end of a spectrum ranging from productivist to non-productivist action and thought within which all agricultural/rural decision-making can be explained and conceptualised. This echoes debates surrounding other transitions from 'isms' to 'post-isms' discussed in Part 1. Indeed, Section 7.3 highlighted that many debates surrounding 'post-isms' and other transitions have found resonance in our discussion of productivist and post-productivist agricultural systems. This discussion has, therefore, confirmed that the *intellectual situatedness* of the issue of agricultural transitions takes on a different, and arguably more significant, dimension if we recognise that parallel pathways are a result of social processes that are interrelated in complex ways and that go well beyond discussions of 'agricultural change' alone.

Why has this discussion of transition theory and criticism of the p/pp transition model been necessary? As outlined in Chapter 1, transition theory was used to highlight that four key fallacies have underpinned most conceptualisations of transitions in the social (and to some extent natural) sciences. Identification of these four fallacies has formed a crucial argument in our deconstruction of the notion of an implied transition towards post-productivism. The application of transition theory to the p/pp transition model confirms that debates on agricultural change share more than just parallel critiques with other debates on 'post-isms', but that they should, indeed, also be seen as reflections of *mutually constituted processes*. Parallels in intellectual debates surrounding transitions may, therefore, simply be a *mirror* of social processes that are interrelated in complex ways. In particular, *Deleuzian transitional pathways* (see Chapter 2) appear to provide the best explanation for contemporary societal and agricultural transitions, rather than rigid linear and bi-polar attempts at explaining transition. It is this Deleuzian nature of transitional agricultural and rural pathways, bounded by productivist and non-productivist decision-making pathways, that will form the conceptual framework for our discussion in Part 3. This will be used for *theoretically anchoring* the notion of 'multifunctionality' in transition theory, and to argue that 'multifunctionality' can be conceptualised as embedded between the two extreme pathways of productivism and non-productivism. Thus, in Part 3, I will discuss how and to what extent this reconceptualisation of multifunctionality may help us better understand trajectories of agricultural change, and I will argue that the notion of *multifunctional agriculture* better encapsulates the diversity, non-linearity and spatial heterogeneity that can currently be observed in modern agriculture and rural society.

Part 3

Conceptualising multifunctional agricultural transitions

Part 2 of this book deconstructed the notion of a transition from productivism to post-productivism and suggested that, instead, we should interpret agricultural change based on a spectrum of decision-making pathways bounded by productivist and non-productivist action and thought (the p/np model). Building on this concept, Part 3 will theoretically anchor the notion of multifunctionality in this revised p/np model and will discuss the spatial, temporal and socio-economic processes underlying such a new understanding of multifunctionality.

Chapter 8 will first discuss contemporary conceptualisations of multifunctionality and will highlight that, despite crucial advances in current debates on multifunctionality from different disciplinary vantage points, the concept still suffers from vagueness, poor conceptualisation, and weak interlinkages with wider theoretical frameworks of agricultural change. **Chapter 9** will then conceptualise multifunctionality based on the p/np model highlighted in Chapter 7. Emphasis will be placed on how this new understanding of multifunctionality influences our current concepts of what 'agriculture' means, and a new holistic framework based on notions of 'weak', 'moderate' and 'strong' multifunctionality will be suggested. The chapter will also discuss the spatial context of multifunctionality. Building on this discussion, **Chapter 10** will analyse multifunctional agricultural transitions and will take us back to theoretical and conceptual issues explored in Part 1. It will focus specifically on the interlinkages between multifunctionality and decision-making pathways at different levels. The chapter will explore decision-making constraints and opportunities in complex multifunctional decision-making environments ranging from the farm level to the national and international levels. This will help us understand transitional potential available to actors at different levels. The chapter will then investigate multifunctional transitions at farm level over time and discuss how the contemporary era of multifunctional agriculture is situated within much longer and complex historical and future pathways of agricultural change that may help us learn from past and present lessons of (mis)management of multifunctional agricultural systems. The chapter concludes by investigating how multifunctional agricultural transitions can best be managed, by whom this transition should be managed, and what challenges different stakeholders face in the implementation of 'strong' multifunctionality pathways. Conclusions for the book will be presented in **Chapter 11**, where specific focus will be placed on revisiting transition theory and its wider implications for future agricultural and rural research, as well as pointing towards future avenues for empirical research on multifunctionality.

Chapter 8

Contemporary conceptualisations of multifunctionality

8.1 Introduction

Part 2 of this book highlighted that the argument for a fully fledged transition to post-productivist agricultural systems is no longer tenable. Such concern has been echoed by many circles – whether scientific or applied – and the need for a new conceptualisation of contemporary agricultural change has been emphasised. In response to this growing criticism, many authors are beginning to argue that the notion of *multifunctional agriculture* better encapsulates current trends in agricultural and rural processes (e.g. Potter and Burney, 2002; Knickel *et al.*, 2004). In this chapter, I wish to review current conceptualisations of multifunctionality. First, I will discuss the genesis of the notion of 'multifunctionality' and show that the term has been applied in many different arenas from different disciplinary vantage points. This will emphasise that multifunctionality has only relatively recently been appropriated by agricultural and rural researchers and decision-makers as the 'buzzword' to describe the complexity of contemporary agricultural change (Section 8.2). Second, I will argue that contemporary understandings of multifunctionality have been relatively *structuralist* and *reductionist* and based largely on narrow economic and policy-based approaches, although more 'holistic' interpretations of multifunctionality are coming to the fore (Section 8.3). As a result, I will argue that multifunctionality is currently understood more as a *process describing* current agricultural trends, rather than as a *concept explaining* agricultural change. Further, debates on multifunctionality have suffered from relative *discursive insularity* that, in my view, has confused rather than clarified what multifunctionality could be about. Third, the chapter will highlight that 'cultural' and 'territorial' aspects are important for understanding current approaches to multifunctionality, in particular in relation to the Euro-centric notion of multifunctionality used as a smokescreen to defend the continuing productivist subsidy culture of European agriculture (Section 8.4). This will include a discussion of the issue that multifunctionality is perceived by many as a 'European project' with little relevance to non-EU agricultural regions. Fourth, the chapter will highlight that notions of multifunctionality were (mis)appropriated early by policy-makers as a tool for (often) shallow policy decision-making, thereby giving seemingly rapid 'legitimacy' to the term – policy-based decision-making characterised by short-termism, weak theorisation and pragmatism. Section 8.5 will argue that this may have lessened the need for thorough conceptual and theoretical investigations of what 'multifunctional agriculture' means and that, as a result, we are currently left with an understanding of multifunctionality that has not yet gelled into a coherent workable framework for fully *understanding* contemporary agricultural and rural change. In the conclusions (Section 8.6), I will highlight that this currently weakly theorised notion of multifunctionality forms the rationale for the revised model of multifunctionality discussed in Chapters 9 and 10.

8.2 Multifunctionality

8.2.1 The emergence of 'multifunctionality'

The notion of 'multifunctionality' – i.e. something having 'multiple functions' – has been part and parcel of human evolution since the dawn of time. Early Palaeolithic tools, for example, often served multiple functions (e.g. flint scrapers as tools for cutting *and* scraping). This notion of the multifunctionality of tools or machines is still widely used today, as evidenced through the use of 'multifunctionality' to describe the multiple uses of laboratory instruments (Duxbury, 2005) or even 'multifunctional' robots (Hulse *et al.*, 2005). More important for this book is the notion of multifunctional processes, land uses and human actions, as it is here that crucial antecedents to the notion of 'multifunctional agriculture' can be found. The use of the term 'multifunctionality' in the *forestry sector* is of particular importance (Brouwer, 2004b), especially through the notion of 'multiple use forestry' (whereby multiple use = multiple functions) (Mather, 2001; Buttoud and Yunusova, 2002). Debates about multiple uses of forests date back to the emergence of 'scientific forestry' in Germany in the late 18th century, which emphasised that forests have multiple functions that go beyond timber production and that may also include, for example, avalanche protection or shelter for farm animals in winter (Edlin, 1970). Throughout the 1900s, the notion of multiple use developed as a forest management practice as demand patterns shifted in ways that made outputs other than timber more significant (Freshwater, 2002). Dietrich (1953) already emphasised the importance of 'multifunctionality' in forestry policy, although the *scientific* paradigm shift towards multifunctional forests only occurred in the 1980s when productivist concerns about afforestation and destructive logging were replaced by concerns for sustainability, nature conservation and soil protection (Mather, 1991; Larsen, 2005).

In the USA, policies of the National Forest Service changed from a relatively 'monofunctional' agenda (forests for protection *or* production) to an agenda that involved a spatially variable mix of protection *and* production (Loomis, 1993). As early as 1960, the US Forest Service was formally required by the Multiple Use Sustained Yield Act to consider competing and complementary uses of forests (Bowes and Krutilla, 1989; Freshwater, 2002). In the UK, meanwhile, a new concern for multifunctional use of forests was expressed from the 1980s onwards through implementation of specific schemes (e.g. Farm Woodland Scheme) which offered annual payments for the planting, management and maintenance of woodlands for multifunctional purposes (Winter, 1996; Mather, 2001). As a result, Marsden (2003: 242) suggested that "we see the traditional, narrow definition of forestry as productivist rural space beginning to be challenged". Mather (2001: 249) similarly emphasised the multifunctional character of forestry by arguing that in the 'post-industrial' forest "the emphasis placed on timber production is reduced relative to that placed on environmental services (such as biodiversity and recreation)". Mather *et al.* (2006) added that attitudes towards forests have increasingly emphasised *non-productivist* objectives such as landscape protection, while the 'classic' productivist objective of producing timber has decreased in importance. By the early 1990s, forestry in many countries had become inextricably linked with environmental concerns (Winter, 1996). This exemplified a move towards multifunctional forestry ideals, best expressed through a statement by the UK Forestry Commission (1991: 31; emphasis added) that recognised "advantages of basing [forestry] policy on the realisation of *multiple* objectives". By the mid-1990s, these multifunctional ideals had become well embedded in forestry policy of many developed and some developing countries, exemplified in commitments towards multifunctional forestry led by social and biodiversity requirements (e.g. Rangan and Lane, 2001, for India; Wilson and Memon, 2005, for New Zealand). Larsen (2005: 101) referred to this as the shift from monofunctional 'segregated' forest landscapes (i.e. separation of natural forest from intensive land uses such as agriculture) towards 'integrated' landscapes with a multifunctional forestry character, and suggested that "forests and trees play an important role in increasing the multifunctionality of landscapes".

Notions of multifunctionality have not been restricted to forestry and agriculture. A fruitful debate has also emerged in multifunctional *urban planning*, and although the linkages between this body of literature and multifunctional agriculture are not explicit, debates about the changing functions of urban spaces have also influenced debates on multifunctional agriculture. Of particular relevance have been debates on multifunctional urban land use that emerged in the late 1990s, with a recent issue of the journal *Built Environment*, for example, entirely dedicated to the subject (Priemus *et al.*, 2004). Rodenburg and Nijkamp (2004: 287) suggested that "the concept of multifunctional land use has turned out to be a very interesting one in urban planning". In these studies, multifunctionality is defined as a combination of different socio-economic functions in the same area (Priemus *et al.*, 2004). They suggest that urban functions and spaces have gradually become more 'monofunctional', with an increasing separation between workplace and residential functions, and that there is a need to redress this by promoting 'multifunctional' or mixed use (see parallels to our discussion of 'monofunctional' agriculture in Chapters 5 and 6). Multifunctional urban land use is believed to consume less land and generate less traffic than monofunctional use (i.e. less need for commuting and travel). Urban spaces are said to become more multifunctional when the number of functions, the degree of interweaving, or the spatial heterogeneity increase in a given area under consideration (Rodenburg and Nijkamp, 2004). An increased degree of multifunctionality may result from the addition of functions to the area (notion of 'multifunctionality by diversity'), from an increase in dispersion of the number of functions ('multifunctionality by interweaving'), or from an increase in spatial functions ('multifunctionality by spatial heterogeneity').[43]

Beyond such relatively well delineated debates, it is also interesting to note the *silences* surrounding multifunctionality. The absence of multifunctionality discourses in *extractive industries* is particularly prominent. One may understand the lack of debates on multifunctionality regarding mining or quarrying, for example, although the post-mining and post-quarrying use of territories – for example as landscaped areas for amenity – offers some interesting, yet unexplored, questions about the multifunctionality, or change of function, of specific localities. However, the relative silences surrounding 'multifunctional' fishing or fisheries are more surprising. Although fishing, other than most agricultural practices, is an *extractive*, and therefore *super-productivist*, economic sector based on exploitation of common property resources in open systems (in particular open ocean fishing), the recent downturn in EU fishing due to imposed quotas has forced many fisherpeople to either abandon or *diversify* use of their fishing boats for non-fishing activities (in particular fishing for tourists, whale watching, coastal scenic trips, etc.) (Gray and Hatchard, 2003). There would, therefore, be much scope for conceptualisation of *multifunctional* activities linked to fishing. Yet, although implicit allusions are made to the potential for multifunctional use of fishing boats and gear (e.g. Gray and Hatchard, 2003; Hogarth, 2005), few *explicit* references are made to conceptual notions of multifunctional fishing. One of the few attempts has been the International Symposium on Multiple Roles and Functions of Fisheries and Fishing Communities held in 2003 in Japan. The notion of 'multiple roles' in this context referred largely to the increasing recognition of social aspects of fisheries away from monofunctional emphasis on fishing yields alone. Schmidt (2003: 2) asked "whether the concept of multifunctionality, which has been intensely discussed in the agricultural sector, would be a useful analytical framework [for] fisheries". Using the example of Japanese fisheries, Schmidt suggested that the coastal fishing sector can have multiple roles including social, economic and environmental multifunctionality – albeit couched in discourses emphasising maintenance of the sustainability of productivist fisheries rather than diversification activities

---

[43] Note parallels to debates on the deconstruction of post-productivism in Chapter 7.

away from productivist fishing. Schmidt's paper is one of the few that acknowledges the possible conceptual synergies between multifunctional fisheries and multifunctional agriculture, although he says little about how multifunctional pathways could be implemented among fishing communities. It appears that multifunctionality is seen in the fisheries debate relatively narrowly as largely synonymous with environmental sustainability aimed at guaranteeing long-term survival of fish stocks, rather than as a potentially holistic socially based concept that has, at least in part, been characteristic of multifunctional agriculture debates (see Section 8.3).

Multifunctionality has, thus, been discussed in a variety of contexts, although no overall coherent picture emerges. Garzon (2005) argued that there may be a 'natural' linkage between primary production activities (such as agriculture) and the notion of multifunctionality, as it is in such systems that natural resources and land can also be seen as public goods (see below) and often involve the jointness of production of non-commodity goods (which may not be the case for fisheries, for example). Similarly, Luttik and Van der Ploeg (2004) argued that farmers tend to be *generalists* (compared to fisherpeople or miners, for example) and, therefore, may be more suited to 'multifunctionality' than many other professions. In the following, I wish to investigate in more detail issues surrounding the emergence of the notion of 'multifunctional agriculture' and highlight both the relative richness, but also theoretical paucity, of debates surrounding the term.

8.2.2 The genesis of the notion of 'multifunctional agriculture'

Compared with forestry, for example, agricultural and rural research has been relatively late in embracing notions of 'multifunctionality'. What is interesting to note here is that both academic and policy-making discourses embraced the notion of 'multifunctional agriculture'[44] at about *the same time* in the early 1990s. This suggests that – contrary to the notion of 'sustainable development', for example, which was first conceptualised theoretically by academics (1980s) and only later (1990s) embraced by political and policy-making discourses (cf. Redclift, 1987; O'Riordan, 2004) – the notion of multifunctional agriculture was embraced, if not appropriated, relatively early by policy-makers (Brouwer, 2004b). As Section 8.5 will discuss, this 'parallel' development of multifunctional agriculture as an academic *and* policy-based discourse may be the reason for weak theorisations of the concept. In other words, policy-makers propelled the concept into wider debates early during evolution of the term (early 1990s), possibly reducing the need for in-depth conceptualisation of what became a seemingly 'established' – and, therefore, apparently well understood – notion with 'direct' policy relevance.

As Section 8.4 will highlight, there is also an important spatiality in the use of multifunctionality by policy-makers. It is almost entirely within the 'European model of multifunctionality' that policy-makers were quick to embrace the term in the early 1990s, at a time when the CAP was under severe pressure for change (Potter and Burney, 2002; Potter, 2004). In other parts of the world, meanwhile, policy-makers have been much slower to adopt discourses of multifunctionality due largely to perceived associations of the term with a 'European way' of addressing agricultural issues. In the USA, for example, multifunctionality was only enshrined in the 'Conservation Security Program' in 2002 (Dobbs and Pretty, 2004). Most of the strongest initial advocates of agricultural multifunctionality have, therefore, been countries with high cost agricultural regimes and high levels of subsidies (Freshwater, 2002). This spatiality of multifunctional policy discourses is mirrored by a Euro-centric *academic debate* on multifunctionality. As will be highlighted below, the 'European model' of multifunctionality has been partly construed and

---

[44] Throughout this book I refer mainly to the notion of 'multifunctional agriculture'. However, it should be noted that the notion of 'agricultural multifunctionality' is used interchangeably by many authors to describe the same processes and there does not appear to be any conceptual difference between the two terms.

conceptualised by academic discourses, especially with a focus on multifunctional policy evolution and agricultural economic development (Potter and Burney, 2002; Knickel *et al.*, 2004). Nonetheless, due to the interconnected nature of multifunctionality issues – particularly at the policy and global trade level (Potter and Burney, 2002; see below) – commentators have also realised the important implications of European notions of multifunctionality in a North American context (e.g. Freshwater, 2002; Bills and Gross, 2005). Recently (i.e. early 2000s) – and mirroring the gradual 'globalisation' of p/pp transition discourses discussed in Part 2 – Antipodean researchers have also actively engaged with the question whether and how multifunctionality may apply to Australia (e.g. Cocklin *et al.*, 2006; Holmes, 2006) and, to a lesser extent, New Zealand (Jay, 2004; Wilson and Memon, 2005). However, as with the relative lack of debates on the p/pp transition, commentators on, or in, *developing countries* have been much slower to engage in debates on multifunctional agriculture (Wilson and Rigg, 2003; McCarthy, 2005). Although much implicit work exists in which parallels to Northern-centric conceptualisations of multifunctionality can be drawn from developing countries' perspectives (e.g. Rigg, 2001; Pretty, 2002), there is little explicit work that has used the terminology 'multifunctional agriculture' in the South.[45] I will investigate these spatial discrepancies in the use and conceptualisation of multifunctional agriculture in more detail in Section 8.4.

As Garzon (2005) emphasised, the emergence of the notion of 'multifunctional agriculture' can most easily be traced through analysis of policy and political debates. The notion of multifunctional agriculture was already *implicitly* apparent in official policy documents of the late 1980s. As the remainder of our discussion in this book will highlight, this reflected a range of issues from political legitimation of the EU farm subsidy model, changes in the development pathways of agriculture and rural areas, and changes in societal views about what 'agriculture' should or should not be about (see Chapter 7). Thus, in 1988 the Commission of the European Communities in its landmark publication *The future of rural society* highlighted the *multiple contributions* that the EU agricultural sector could make to territorial economic development, environmental management, and the viability of rural communities (CEC, 1988; Clark, 2003). One of the first *uses* of the notion of multifunctional agriculture, meanwhile, can be traced back to the Earth Summit in Rio de Janeiro in 1992, where Article 14 of Agenda 21 stated that agricultural policies should take into account the 'multifunctional' character of agriculture (Delgado *et al.*, 2003).

In Europe, academic allusions were made to multifunctional issues linked to 'non-economic objectives' of agriculture, pluriactivity and diversification in the early 1990s. Ronningen's (1994) publication entitled *Multifunctional agriculture in Europe's playground?* is particularly noteworthy, as it is one of the first to explicitly link agri-environmental policy, integration of cultural landscape considerations, and rural tourism to agricultural multifunctionality (see also Fuller, 1990; MacKinnon *et al.*, 1991). The term 'multifunctional agriculture' was 'officially' used for the first time in 1993 by the European Council for Agricultural Law in an effort to harmonise agricultural legislation across Europe and to provide a legal basis for sustainable agriculture (Losch, 2004; Garzon, 2005) – emphasising the EU-centrism of early multifunctionality debates. As Brouwer (2004b) suggested, it is likely that the appropriation of multifunctionality by policy-makers was influenced by earlier developments linked to notions of multifunctional forestry (see above), although Mather *et al.* (2006) argued that the triggers for the emergence of multifunctionality between agriculture and forestry differed, with overproduction of food and associated budgetary problems in agriculture as key trigger for the former, and uncompetitive costs of timber production and issues of social justice for the latter. Yet, an exact date for the *tangible* incorporation of multifunctional rhetoric in EU policy documents is difficult to find, as

---

[45] However, see Becu *et al.* (2004) for a notable exception on the application multifunctionality in Thailand. However, their main focus was on developing a methodology for the assessment of multifunctionality in rice-based agrarian systems, rather than about conceptual issues linked to transposition of the term to the developing world.

indirect expression of multifunctionality goals can be traced back to the failure of the 1968 Mansholt Plan leading to the EU strategy of 'farm survival policy' (Potter, 1990), to the Less Favoured Areas Directive in 1975 (subsidies for agriculturally disadvantaged areas; Caraveli, 2000), the first European Structural Funds of the late 1980s (diversification and environmental conservation), or the EU LEADER programme since 1991 (rural development).

The notion of multifunctional agriculture has also been used by various bodies and institutions outside the EU. Losch (2004) and McCarthy (2005) suggested that the term was popularised in international discourses during the 1994 URAA, although above discussion highlights that there were important European antecedents pre-dating this 'globalisation' of the term. Delgado *et al.* (2003) and Garzon (2005) emphasised that the Food and Agriculture Organisation (FAO) was instrumental in early popularisation of the term, in particular through its 'Quebec Declaration' (1995), its 'Rome Declaration and Action Plan' (1996), and its conference 'Multifunctional character of agriculture and the land' (1999). More recently, the OECD, as a supra-national policy-making body, highlighted the urgency to acknowledge the multifunctional character of farming (OECD, 2001; Durand and Van Huylenbroek, 2003). Delgado *et al.* (2003) saw this as the 'definitive recognition' of the multifunctionality concept at international level, although the OECD document showed the continuing difficulties in reaching agreement on the *exact* meaning of multifunctional agriculture.

In the EU, meanwhile, the commitment of the European Commission to multifunctionality was formally articulated in the Cork Declaration in 1996 (European Commission, 1996; Potter and Tilzey, 2005). This Declaration recognised the declining economic role of conventional agriculture in marginal rural areas and the need to find other rationales for public subvention (Lowe *et al.*, 2002). It also emphasised that agriculture should be seen as a major interface between people and the environment, and that farmers have a responsibility as 'stewards of the countryside' (Gorman *et al.*, 2001; Losch, 2004). The Cork Declaration suggested that "integrated rural policy must be ... multifunctional in effect, with a clear territorial dimension. It must apply to all rural areas in the Union ... It must be based on an integrated approach, encompassing within the same legal and policy framework: agricultural adjustment and development, economic diversification ... the management of natural resources, the enhancement of environmental functions, and the promotion of culture, tourism and recreation" (CEC, 1996: 2). This formed the basis for the establishment of the 2nd pillar of the CAP (Lowe *et al.*, 2002). However, there are continuing debates about the introduction of the notion of multifunctionality at the Cork meeting, in particular linked to criticisms of Commissioner Fischler's personal interests based on his Austrian background – a country in which implementation of multifunctional farm development pathways may be easier than in others (see Chapter 10). Nonetheless, many have described the Cork Declaration as marking "a new and decisive stage in European rural policy" (Delgado *et al.*, 2003: 29).

The notion of multifunctionality enshrined in the Cork Declaration was *agreed* by the European Council in December 1997, while the European Commission officially *introduced* the notion of multifunctionality at the OECD Agricultural Commission Meeting in March 1998 (the same month in which Agenda 2000 proposals were presented) (Gallardo *et al.*, 2003), although the official *approval* of multifunctionality by the EU Commission did not occur until the Berlin Council in 1999 (Potter and Burney, 2002). As a result, multifunctionality was reiterated in many national agricultural policies of the late 1990s, the best examples of which are reference to multifunctional agriculture in the preamble to the French 'Loi d'Orientation Agricole' of 1999 (Durand, 2003; Lardon *et al.*, 2004). However, it was not until 1999 that a formal attempt to *define* the concept of multifunctionality was made by the European Commission (Garzon, 2005; see below). Policy discourses enshrined in Agenda 2000 have further upheld the 'European model of multifunctional agriculture', in particular through the focus on the role of farmers in maintaining the countryside (Potter and Burney, 2002). Thus, "multifunctionality appears as a central element of the European agricultural model put forward by the European Council in Agenda 2000" (Gallardo *et al.*, 2003: 169), while Marsden *et al.* (2004) emphasised that it was only once EU policy-makers were willing to better integrate agricultural and rural policy by the late 1990s that it became possible to use 'multifunctionality' as a possible conceptual springboard for directing EU agrarian *and* rural change. In other words, the agrarian-oriented (productivist) mindset that has dominated policy discourses in the EU (see Chapter 5) meant that so-called multifunctional policies could only be implemented once *agricultural* development pathways were conceptually embedded in notions of wider *rural* development.[46] As a result, Marsden and Sonnino (2005: 11) argued that EU agriculture "is now generally perceived as a multifunctional activity" (see also Ward *et al.*, 2003).

The WTO has also played a crucial role in forcing nation states and stakeholder groups to refine the notion of 'multifunctional agriculture'. Although policy documents of the WTO rarely *explicitly* refer to multifunctionality, their policy rhetoric has emphasised that agriculture produces non-commodities and that these non-trade concerns need to be addressed through national and supra-national policy mechanisms (Potter and Burney, 2002; McCarthy, 2005). Potter and Tilzey (2005) argued that through the WTO, neo-liberal interests have been offered an opportunity to recast agricultural support in a much more market-oriented form. Potter (2004) further argued that the debate on multifunctionality within WTO negotiations has been centred on the extent to which trading partners would be willing to accept something short of a full decoupling of domestic subsidies in exchange for other concessions on issues such as export subsidies and market access. In particular, continued pressure exerted on the CAP during the late 1990s by global trade blocks through the WTO, coupled with increasing recognition that eastward enlargement of the EU has necessitated a new policy approach that integrates rural and agricultural concerns more directly, has kept up the impetus for implementation of multifunctionality principles at a broader level (Clark, 2003; see also Section 8.5).

Despite the various discourses on multifunctionality, definitions of 'multifunctional agriculture' continue to be wide and varied, reflecting specific approaches, philosophies and disciplinary vantage points of those attempting to explain what multifunctional agriculture could or should be about. Marsden and Sonnino (2005: 1-2), therefore, emphasised the 'scarcity of analyses' of the potentially contradictory nature of multifunctional policy developments which has significantly limited research on multifunctional agriculture, highlighting that "'multifunctional agriculture' is by no means clearly and uniformly conceptualised and understood".

8.3 Current conceptualisations of multifunctional agriculture

The notion of 'multifunctional agriculture' has become an almost ubiquitous term used by different actors ranging from academics, policy-makers to grassroots-level stakeholders (Knickel and Renting, 2000; McCarthy, 2005). Marsden and Sonnino (2005: 1) argued that "in the last decade the expression 'multifunctional agriculture' has steadily entered the political and scholarly debate about the role of farming for the economy and the society as a whole", while Van Huylenbroek and Durand (2003) referred to multifunctionality as a 'new paradigm' for European agriculture and rural development. In this sense, multifunctionality has suffered a similar fate to the notion of 'sustainability' whose meaning has become increasingly watered down, fuzzy and partly *misappropriated* over time (Redclift, 1987; O'Riordan, 2004). Just as Chapter 7 highlighted the vagueness (and inappropriateness) of '*post*-productivism', I will show in the following that 'multifunctional agriculture' has

---

[46] Freshwater (2002) argued that, as society has become more wealthy and food outlays have proportionally declined in household expenditures over the past few decades, positive 'non-market' values of agriculture have become relatively more 'valuable' over time, therefore facilitating the *social acceptance* of the term multifunctionality.

suffered the same fate and that analysts have not yet probed this concept sufficiently conceptually and theoretically.

In particular, I will argue that the theorisation of multifunctionality – especially its theoretical anchoring within wider debates of social change – has, so far, been relatively weak. There are three reasons for this. First, multifunctional agriculture has been conceptualised from various disciplinary and ideological vantage points, leading to a wide array of different emphases and definitions. McCarthy (2005) highlighted that the term has been approached from a wide array of angles, including, for example, rural sociology, agricultural economics, environmental economics, policy studies and geography. Second, there is little evidence that commentators on multifunctionality are aware of 'parallel' debates on multifunctionality in other fields of enquiry. We saw in Part 1 that this was also true for discussions of transitions from 'isms' to 'post-isms', and that the key problem has been the lack of recognition by many commentators that there are fruitful debates on specific concepts *outside* their specific realm of expertise. Third, and as the previous section has highlighted, the notion of multifunctional agriculture was rapidly appropriated by policy-makers and politicians. Indeed, the term was used as a *tool* for guiding policy-making long before it had been thoroughly defined, tested and debated at a more *conceptual* level. Inevitably, there are severe dangers associated with such *political appropriation* of complex terms (and associated processes set in motion through political tools), and the ultimate result has been that many have shied away from thorough theorisations of a term that is already 'firmly' embedded in the policy-making jargon of many states and supra-national organisations. It is little wonder that we find a wide array of different, at times contradictory, definitions of what multifunctional agriculture is about – an issue I turn to in the following.

8.3.1 Definitions

Without claiming to be exhaustive, let us briefly investigate 'definitions' of multifunctional agriculture in contemporary literature (academic and non-academic). I will begin with the most basic understandings of multifunctionality and gradually reveal increasingly multi-faceted definitions that will come closer to the (re)conceptualisation of multifunctionality proposed in Chapter 9. As Garzon (2005: 3) emphasised, "it is difficult to draw a clear definition of multifunctionality". In its broadest sense, multifunctional agriculture is seen by many to imply simply an agriculture or way of farming that serves *multiple functions* and that reduces the emphasis on food and fibre production (e.g. Potter and Burney, 2002; Clark, 2003). As the following will show, these multiple functions are usually seen to include many processes already discussed in the context of the p/pp transition in Part 2: production of environmental goods on farmland, diversification, pluriactivity and, in a broader rural sense, relocalisation of food chains and empowerment of the regions in an attempt to improve agriculture-rural relationships by strengthening social capital.[47] Where definitions vary is in the *relative weighting* of these individual components and, more crucially, the frequent *neglect* of a holistic vision of the interconnectedness between these components.

Probably the most widely used definition of multifunctional agriculture comes from the OECD (2001) that argued from a (relatively narrow) neo-classical perspective that the term should be interpreted as a characteristic of an *economic* activity based on the agricultural production process and its outputs which produces multiple and interconnected results and effects. According to the OECD (2001: 11), "multifunctionality refers to the fact that an economic activity may have multiple outputs and, by virtue of this, may contribute to several societal objectives at once. Multifunctionality is thus an activity oriented concept that refers to specific properties of the production process and its multiple outputs". The OECD, thus,

---

[47] In its broadest sense the notion of 'social capital' is linked to relations of trust; reciprocity and exchange; common rules, norms and sanctions; and connectedness, networks and groups (Bourdieu, 1983; Putnam, 1993). Strong social capital implies that people have the confidence to invest in collective activities, knowing that others will do so too.

emphasised that although the primary role of agriculture is to produce food and fibre, many other functions are important such as land conservation, maintenance of landscape structure, sustainable management of natural resources, biodiversity preservation, and contribution to socio-economic viability and economic vibrancy of rural areas. In particular, the OECD linked the notion of multifunctionality to issues of trade and trade distortion associated with farm subsidy regimes present in many countries. A similar definition is proposed by Blandford and Boisvert (2002: 107) who suggested from an agricultural economics perspective that "the concept of 'multifunctionality' refers to agriculture as a multi-output activity involving not only commodities, but also non-commodity outputs, such as environmental benefits, landscape amenities and cultural heritage, which are not traded in organized markets". Similarly, Durand and Van Huylenbroek (2003: 1) used an economistic interpretation to define multifunctional agriculture as "the joint production of commodities and non-commodities by the agricultural sector" (see also Nowicki, 2004). The emphasis in these definitions is on terms such as 'commodities', 'non-commodity outputs (NCOs)', 'joint production' and 'trade', particularly characteristic of economistic definitions of multifunctionality (see Section 8.3.2). Although elements of the notion of *joint production* will find their way into our new conceptualisations of multifunctionality (Chapters 9 and 10), I will argue below that these economistic definitions are too narrow and often neglect wider intangible factors such as cultural, mental and attitudinal changes, as well as complex changes in society-agriculture interactions (see also Buller, 2004). Nonetheless, economists have usefully highlighted that the results and effects of multifunctional agriculture may be *both* positive or negative, intentional or unintentional, synergetic or conflictive, and valued on the market or not (e.g. Van Huylenbroek and Durand, 2003) – issues that will form an important part of the conceptualisation of multifunctionality suggested in Chapter 9.

In a more holistic view of multifunctionality Daugstad *et al.* (2006: 68) argued that "multifunctionality is generally seen as the production of other values beyond food and fibre, including collective goods such as cultural landscapes and heritage, biodiversity, recreational opportunities, rural settlements and food security". In one of the most refined analyses of agricultural multifunctionality, Marsden and Sonnino (2005: 1) built on this by suggesting that "the concept of the multifunctionality of agriculture embraces all goods, products and services created by farming activities". Although still focusing largely on *economic* rural processes, they highlighted the link between multifunctionality and Marsden's (2003) three rural development paradigms of the agro-industrial, post-productivist and rural development dynamics (see Chapter 7). They suggested that the view of multifunctionality in the rural development paradigm, in particular, should be seen to reconnect agricultural production to the wider markets and social possibilities, and that multifunctionality viewed through the rural development lens also emphasises the potential *symbiotic interconnectedness* between farms and their locality. In their opinion, therefore, multifunctional agriculture acquires its most comprehensive meaning and displays its highest integrative development potential through rural development (Marsden and Sonnino, 2005). Clark (2006: 332) similarly argued that "multifunctionality denotes the multiple positive contributions that agriculture can make to economies, environmental management, and the viability of rural communities". Views of French commentators have concurred with this more inclusive view, with authors such as Rapey *et al.* (2004b) and Lardon *et al.* (2004) suggesting that definitions of multifunctionality should also take into account societal *expectations* about how agricultural multifunctional pathways could be conceived (see also Di Iacovo, 2003). In contrast to economistic interpretations, these authors emphasise the importance of *social* and *cultural aspects* of multifunctionality (see also Section 8.3.4).

Losch (2004) partly agreed with these more holistic views by emphasising that multifunctionality has emerged as a serious objection to the largely *monofunctional* productivist model of agriculture and its negative externalities,[48] increased environmental awareness, demand for food security and quality, and criticisms of the ineffective regulatory environment in advanced economies (see also Pretty, 2002, who referred to 'monoscapes' created by monofunctional agriculture). Similarly, Roux *et al.* (2004) argued that the notion of multifunctionality should also include non-commodity processes that have a 'positive' influence linked to agricultural production or the use of rural areas, while Potter and Burney (2002: 35) suggested that multifunctional agriculture means "producing not only food but also sustaining rural landscapes, protecting biodiversity, generating employment and contributing to the viability of rural areas". Potter and Tilzey (2005), meanwhile, emphasised the need to link notions of multifunctionality both to neo-liberalism and post-Fordist agricultural policy regimes, thereby usefully highlighting the link between multifunctionality and macro-political and macro-economic changes. This was reiterated by Clark (2003: 12) who saw multifunctional agriculture as "an innovative conception of EU agriculture, emphasising the production of a diverse array of economic, social and territorial outputs to complement the sector's traditional emphasis on commodity production". In this view, the point of multifunctionality as a *complementary* process to 'traditional' productivist agriculture is particularly noteworthy (see also Chapter 9). Knickel *et al.* (2004) also highlighted that there are important differences between terms such as 'pluriactivity' that relates to *agricultural processes*, and 'multifunctionality' that refers to the *range of functions* linked to rural space and agricultural activities. These authors, therefore, highlight that multifunctionality has to be seen as a complex and multi-faceted concept that may also include less tangible (and, therefore, less measurable) elements such as environmental conservation, highly variable territorially-based activities (a point particularly emphasised by geographers), and macro-political processes in a globalising world.

An even broader view (and one with which I tend to agree most; see Chapter 9) is provided by human geographer John Holmes (2006: 142), who suggested that "the direction, complexity and pace of rural change in affluent, Western societies can be conceptualized as a multifunctional transition, in which a variable mix of consumption and protection values has emerged, contesting the former dominance of production values, and leading to greater complexity and heterogeneity in rural occupance at all scales". Here, the emphasis is on multifunctionality as a *territorial concept* based on a spectrum of tensions and competing values in the countryside. In a similar vein, Pretty (2002) has possibly taken debates on multifunctionality furthest by adding a temporal dimension to the discussion, suggesting that important transitional forces need to be considered when attempting to understand multifunctionality. Indeed, Pretty is one of the few to acknowledge that "agriculture is … fundamentally multifunctional" – i.e. that the very nature of agricultural production, with its associated diverse and complex actor-environment interactions, is inherently multifunctional. Again, we will look more closely at this latter conceptualisation in the following chapters, especially in the reconceptualisation of multifunctionality suggested in Chapter 9. It is interesting to note that these more 'holistic' interpretations of multifunctionality were recently reiterated by EU Agriculture Commissioner Fischler who suggested that "agriculture as the epicentre of rural development must overcome its purely regional nature and come to be considered a multifunctional activity, since it shapes the rural world thus contributing to the conservation of an intact space of social and economic life, to the protection of an attractive landscape and to the diversification of the activities of rural areas" (cited in Delgado *et al.*, 2003: 28). Fischler (1999), thus, argued (at least rhetorically) that while rural development and agricultural policy were, until the late 1990s, still seen as two separate entities, there was a need for these two arenas to be *intertwined* more effectively around the notion of multifunctionality.

---

[48] The term 'externalities' refers to the side-effects of agricultural activities that are 'external' to markets (e.g. agricultural pollution; habitat destruction) and so their costs to society are not part of the prices paid by producers and consumers.

Despite these more holistic views, the result of this array of different viewpoints has been that conceptualisations of multifunctional agriculture have remained *vague, contradictory* and, at times, *inappropriate* to act as a platform for understanding agricultural transitions. In the following, I will shed critical light on different interpretations of 'multifunctional agriculture' and, as a basis for the discussion in Chapters 9 and 10, highlight inherent shortcomings in the debates.

8.3.2 The economistic view

The 'economistic' view has tended to dominate discourses of multifunctional agriculture since the 'official' emergence of the term in the early 1990s. Garzon (2005: 2) emphasised that "extensive literature, essentially economic, has been devoted to multifunctionality". Most commentators who have attempted to conceptualise multifunctional agriculture have used economic factors or processes as key building blocks for their arguments (e.g. Peterson *et al.*, 2002; Vatn, 2002).

The economistic interpretation of multifunctionality is well highlighted in Gallardo *et al.*'s (2003: 173; emphasis added) suggestion that "the presence of externalities in a productive process prevents an optimum solution for private goods through the existing market. Moreover, the existence of genuine functions of joint production favours the appearance of market errors when changes occur in prices of agricultural products and in the political instruments by which agricultural markets are affected. In the case of positive externalities that feature in the use of *agricultural multifunctionality*, the social ideal will only be reached if the farmer obtains compensation that corresponds to the displacement of his [sic] production to the level of the private ideal. The lack of an accepted price for these externalities and the participation of multiple agents hinders the compensation process and opens the way for public intervention". Similarly, Durand and Van Huylenbroek (2003: 11) argued that "the principal problem of multifunctionality is how to remunerate farmers for providing non-market goods or in other words how the provision can be stimulated in the context of the present market system", and that "multifunctionality is the examination of both the commodities and non-commodities produced by the diverse activities of farmers or of the agricultural sector" (Durand and Van Huylenbroek, 2003: 12; see also Van der Ploeg and Roep, 2003). Romstad *et al.* (2000: 5), meanwhile, suggested that multifunctionality "is a set of interlinked outputs from a productive activity where some goals are private and some are public", while Vatn (2002) saw multifunctionality as implying that several public goods or positive externalities are attached to agricultural production. Similarly, Belletti *et al.* (2003a: 55) argued that "multifunctional agriculture is conceptualized as a process by which NCOs are produced and which maintain or increase local cultural and biological diversity. Many NCOs show non-private characteristics and hence market failure occurs in their provision resulting in sub-optimal levels with regard to social needs". Vanslembrouck and Van Huylenbroek (2003: 83) similarly suggested that "in the overall discussion on multifunctionality of agriculture, both supply and demand aspects need to be considered", while Durand (2003: 129) argued that "the concept of multifunctionality ... above all reflects recognition that the market by itself is incapable of forcing the agricultural sector to address all of society's complex aspirations". In the view of economists, therefore, farming becomes a profession in which natural resources are used and transformed to produce commodities and NCOs in 'multifunctional' ways.

Such economistic conceptualisations permeate almost all multifunctionality discourses, highlighting the powerful positions that agricultural economists hold in many academic and policy-related realms. Even the broad-ranging OECD (2001) definition of multifunctional agriculture (see above) emphasised agriculture as an *economic* activity which produces multiple and interconnected results and effects that may be valued on the market. Even Knickel and Renting's (2000: 522-523) innovative attempt at conceptualising multifunctional farm pathways is dominated by economic indicators,[49] epitomised by their suggestion that "multifunctionality relates to the combination of resources (inputs) or goods such as products, services available both at the farm level and beyond ... The two most important types of relationships are multiple use of the same resources, and complementarity in resource use and production". Similarly, Clark's (2003: 226) suggestion that multifunctional agriculture means "the output by agricultural businesses of a range of public goods and services outside of the conventional commodity production function" is also largely couched in an economistic-oriented view, despite his wide-ranging and incisive analysis of the promotion of multifunctionality at the regional level. Commentators on global multifunctionality such as Losch (2004: 342; emphasis added) have even suggested that "the debate on multifunctionality *must* be put clearly into perspective among the general phenomena that affect the global *economy*" (see also Peterson *et al.*, 2002; Vatn, 2002).

These, in my opinion simplistic, views are predicated on several basic assumptions. First, multifunctionality is seen as a set of multiple contributions that agriculture makes for food production and the production of social and environmental 'public goods'[50] (Freshwater, 2002; Peterson *et al.*, 2002). As McCarthy (2005) criticised, the evaluation of multifunctionality in this context is based on the assumption that economic valuation of *all* these commodity and non-commodity outputs produced by agriculture is possible. Marsden and Sonnino (2005) further emphasised that the economistic view of multifunctionality has been focused too much on the notion of *pluriactivity* based on the combination of agricultural and non-agricultural incomes and activities within the farm household. Second, agricultural economists such as Lifran *et al.* (2004) suggested – in my view erroneously – that through the adoption of 'simple' economistic models the direct translation of policies into multifunctional benefits is possible (see also Durand and Van Huylenbroek, 2003). This view, essentially, sees multifunctionality merely as a *byproduct* of productivist agricultural activity, with authors such as Blandford and Boisvert (2002: 109; emphasis added) arguing that "landscape amenity is an economic *complement* with commodity output". In these views, therefore, policies attempting to place a monetary value on farmers' production of environmental goods are simply branded as 'multifunctional'. This is closely associated with the fact that many economists also tend to place relative faith in the state and public authorities to *orchestrate* agricultural change. Thus, Van Huylenbroek (2003: xiv) argued that "it is often only through the mediation of public authorities that negotiations between demand and supply is possible". In Chapter 10, we will discuss to what extent the state is effectively capable of implementing such mediation processes. Third, economistic views often emphasise the notion of 'joint production'[51] (or 'co-production'; Vatn, 2002; Daugstad *et al.*, 2006) as key to multifunctionality, emphasised by Durand and Van Huylenbroek's (2003: 1) definition of multifunctionality as "the joint production of commodities and non-commodities by the agricultural sector".[52] The OECD (2001) is probably the best example of a powerful supra-national institution emphasising the importance of the joint production nature of multifunctionality, arguing that joint production is based on the assumption of a fixed (or quasi-fixed) relationship between two outputs of an economic activity. It suggested that environmental outputs, cultural heritage values and landscape elements created by agriculture can, at times, be completely separated from farming activities, and that negative externalities caused by agriculture are often linked to one farming activity (e.g. intensification) resulting in another, interdependent, activity (e.g. soil or water pollution).

---

[49] See, in particular, their conceptual model of the structure of multifunctionality (Knickel and Renting, 2000: 517) that focused largely on multifunctionality indicators of distribution of *income* between agricultural and non-agricultural activities.

[50] Public goods usually display low degrees of *excludability* (i.e. access is open) and low degrees of *rivalry* (i.e. consumption by one person does not prevent its consumption by another).

[51] Joint production is usually seen as the production of at least one non-market good through jointness in a production process (cf. Nowicki, 2004; Bryden, 2005).

[52] Adopting a more critical stance, Buller (2005) argued that 'co-production' should only be seen as an important *precursor* to multifunctionality.

Farming is seen here to create a production process "in which nature is converted into goods and services for human consumption" and where it is important for farmers "to reproduce the required natural resources (environmental factors) to safeguard their future source of income" (Swagemakers, 2003: 190). Fourth, economists often emphasise monetaristic approaches to solve conflict in the countryside, best highlighted through the application of methodologies such as the 'consumer pays principle', the 'hedonic pricing approach', 'competitiveness analysis methods' or the 'travel cost method' for addressing the production of positive and negative externalities associated with multifunctionality (e.g. Randall, 2002; Vanslembrouck and Van Huylenbroek, 2003). As Liverman (2004) highlighted, such seemingly 'unproblematic' economic valuations of often highly intangible multifunctional non-commodity goods remain problematic. Fifth, economistic interpretations often assume that humans act in unilinear rational economic fashion based on the notion of 'homo economicus', where economic drivers are thought to explain most human behaviour. For example, Vatn (2002: 324), in a recent economistic analysis of the consequences of multifunctional agriculture for international trade, revealed that "I have assumed farmers are profit-maximisers". Yet, many sociology- and psychology-based studies have shown human behaviour to be much more complex than such mono-dimensional explanations imply (e.g. Stryker, 1994; Burton and Wilson, 2006).

In the view of many economists, therefore, multifunctionality is difficult to implement because the user of public goods created by agriculture is not paying for these goods. In other words, economists argue that not all of society's expectations of agriculture are transformed systematically by market demand (e.g. Peterson *et al.*, 2002; Randall, 2002). This is particularly the case for public goods that are non-tradable (e.g. production of a healthy 'environment') and which, therefore, can not be substituted through imports from outside. Economists, therefore, suggest that non-marketable outputs from agriculture must find other means of evaluation and remuneration modes than the market (Barthélémy *et al.*, 2004). As a result, economists argue, there is growing intolerance among many societies of the 'negative externalities' created by unsustainable agricultural production systems. Van Huylenbroek (2003: xiii), therefore, suggested that "society has the obligation to compensate for this price differential". Engendering multifunctional agriculture in this view assumes that all public goods created by agricultural activities should be adequately priced and paid for by those using these goods (e.g. tourists). Indeed, it is the over-emphasis by economists of interlinkages between 'demand' and 'multifunctionality' that, in my view, leads to erroneous conclusions, epitomised by Brouwer's (2004b: 2) recent suggestion that "the concept of multifunctionality would therefore fail in cases where there is no demand". In Chapters 9 and 10 I will suggest that multifunctionality should instead be seen as a process *inherent* to agriculture, at times also *irrespective of demand*.

More recently, a view of multifunctionality has emerged that attempts to 'scale down' the spatial scale of multifunctionality to the field level. Here, debates intertwine with the above-mentioned economistic notion of 'joint production'. Winter and Morris (2004), for example, argued that multifunctional agriculture can only be fully understood through joint production at field level, and that the differentiated economic use of space within a farm is the only true expression of joint production and multifunctional economic pathways. Although there are some interesting conceptual strands to this debate – in particular regarding the scale at which multifunctionality should apply (see Chapter 9) – this conceptualisation can be criticised for a notion of multifunctionality based entirely on primary production that provides too narrow a conceptual base. Thus, Potter and Burney (2002: 40) argued that for economists "'multifunctionality' is merely another way to describe these long established forms of joint production", while Buller (2004: 104) criticised that "the current use of the multifunctionality concept goes considerably further than the 'positivistic' notion of joint production". Some economists argue that notions of multifunctionality also have to apply beyond the farm level (e.g. Van der Ploeg and Roep, 2003), because multifunctional farm outcomes have economic effects outside the farm as the production of 'joint functions' often requires a collective-systemic scale. Thus, Belletti *et al.* (2003a) argued that multifunctional NCOs will only have real 'value' when these outputs are shared in the local area (e.g. improved environment). Yet, the economic interpretation of how this could be achieved remains narrowly focused on the fact that NCOs produced at the local level show characteristics of "non-complete rivalry and excludability" in order to be successfully implemented at the local or regional scale (Belletti *et al.*, 2003a: 69; see also OECD, 2001).

There is no doubt that economic approaches have been an important component in emerging debates and have forced non-economists to further refine and scrutinise their own views on multifunctional agriculture. Yet, in my view, the economistic interpretation provides a relatively narrow conceptualisation of what Chapters 9 and 10 will highlight is a highly complex transitional process (see also Holmes, 2006). Thus, economistic definitions suggest a relatively narrow concept of multifunctionality centred on agricultural *processes* and based almost entirely on supply and demand issues, the production of commodities and non-commodities, competitive and non-competitive agriculture, and market failure. Economistic approaches that assume that we can identify the most 'efficient' multifunctional policies through mathematical formulas (e.g. Peterson *et al.*, 2002) appear particularly questionable. In addition, the emphasis on production and creation of 'public goods' detracts from key *cultural* and *social* drivers of agricultural change in the form of changing mindsets, identities and attitudes. This is probably best highlighted in Belletti *et al.*'s (2003a: 57; emphases added) suggestion that "rural development, conceived in terms of a *set of techno-economic principles* applied to farm management and policy design, is the *best* approach to strengthen the multifunctional role of agriculture", and is also apparent in Swagemakers' (2003: 204; emphasis added) suggestion that "for a multifunctional farm to be successful it must establish permanent relations with *distributors* and *consumers*. This is the *only* way multifunctional farming will obtain the reward it deserves". As a result, Garzon (2005: 2) criticised narrow economistic views of multifunctionality by arguing that "economists have until recently considered multifunctionality a weak concept that deserved little theoretical significance". I also agree with Di Iacovo (2003: 122; emphasis added) who suggested that the "multifunctionality of agriculture is not yet totally explored. It is clear that it works for marketable products and for semi-public services, but it is *not equally clear* in many other aspects". These critical views emphasise that the notion of multifunctional agriculture should go well beyond unidimensional economic indicators.

8.3.3 The policy-based view

As with conceptualisations of the p/pp transition discussed in Part 2, many commentators have referred to the changing policy environment as a key indicator for multifunctional development pathways (e.g. Abler, 2004; Knickel *et al.*, 2004). Similar to discussions about the 'greening' of society (e.g. Harper, 1993), or debates on the increasing inclusion of sustainability into societal thinking (e.g. O'Riordan and Voisey, 1998), the policy environment has been used both to analyse how ideas about multifunctional agriculture have been gradually *incorporated* into policy thinking (i.e. multifunctionality as an external driver shaping policy) but also how policies themselves can be used to *implement* multifunctional agriculture (i.e. policy as a driver shaping multifunctionality). One of the key strands of the multifunctionality debate, therefore, refers to multifunctionality as a *policy-related discourse* (Garzon, 2005). This was emphasised by Potter and Burney (2002: 39) who highlighted the widespread notion of "multifunctionality as a policy concept", or Bills and Gross (2005: 313) who suggested that "much of the recent economic literature generated in both academic and government circles saves the term 'multifunctionality' for discussions of policies". Authors such as Van Huylenbroek (2003: xiii) suggested that multifunctionality should be "the leading concept for agricultural policies", while Potter (2004: 31; emphasis added) argued that "the concept of multifunctionality appears to be an example of a *policy idea*". From a French perspective, meanwhile, Lardon *et al.* (2004: 6; emphasis added; my translation) similarly posited that "the Loi d'Orientation Agricole of July 1999 gives agricultural multifunctionality an *institutional character* [caractère institutionnel]". Researchers such as Hollander (2004: 300; emphasis added), in her analysis of multifunctionality and the Florida sugar industry, also admitted that "my interest here is with multifunctionality as a *political* strategy and *policy instrument*". Similarly, Garzon (2005: 18) examined the concept of multifunctionality almost entirely by "situating it in the context of policy change" where multifunctionality is seen almost entirely as "a political concept". The policy environment, therefore, emerges as a relatively easy 'inroad' into conceptualisations of multifunctionality.

How have different commentators seen the changing policy environment as an *expression* of increased agricultural multifunctionality? Both Losch (2004) and Marsden and Sonnino (2005), for example, highlighted how powerful policy-making bodies, such as the European Council for Agricultural Law (which used the notion of multifunctional agriculture 'officially' for the first time in 1993; see above) or the OECD (2001), argued that the multifunctional character of farming should be expressed through *policies* that reduce agricultural support and high levels of food production and input use. The OECD has played an important role in 'popularising' the notion of multifunctional agriculture among policy-makers and called for *orchestration* of multifunctional pathways through the policy environment. It argued that if all NCOs were private goods for which markets exist, then private transactions would ensure that resources on the farm are used efficiently and that supply and demand are balanced in all markets. However, as this is not the case in many agricultural systems, most NCOs can be seen as externalities often characterised by market failure – in other words, overproduction of negative externalities occurs by the farmer because they do not represent a cost. It is at this point that the OECD's notion of multifunctionality as a process *orchestrated by policy* takes particular hold: if a policy-driven system could be implemented that would help taking these negative external aspects into account, then the possibility of market failure would be reduced as a decrease in supply (e.g. food production) may be offset by a decrease in negative externalities (e.g. agricultural pollution). Although the OECD acknowledged that policy may not be the best mechanism to orchestrate such re-evaluation of negative externalities, one can see the lure of this policy-driven notion of multifunctionality as a means to offset farmers' losses of CAP 1st pillar subsidies. In other words, enabling a policy system that pays farmers for the reduction of negative externalities (e.g. through AEPs) continues to provide regular payments to farmers (green subsidies) without infringing WTO demands for reduction of 1st pillar subsidies (Potter and Burney, 2002).

This faith in the policy environment is also well highlighted in Durand and Van Huylenbroek's (2003: 11) trust in the policy environment as key to addressing multifunctionality at different spatial scales: "at the international level the question is ... how policies giving incentives to the production of non-commodities can satisfy the conditions of non-trade distortion, while at the regional or territory level the question is how to articulate agricultural policies with rural and/or regional policies. At the farm level questions arise about the practical consequences of these policies". Similarly, Gallardo *et al.* (2003: 169; emphasis added) argued that "however positive the recognition of the multifunctional nature of agriculture may be, the *decisive* factor is to what extent the instruments of agricultural *policy* are in accordance with this assessment" (see also Smith, 2000). In a similar vein, Marsden (2003) highlighted that the rural development paradigm should be seen as a development *tool* or *mechanism* (i.e. policy-led) to promote multifunctional rural development. Further, Durand (2003) argued that only by designing *instruments* and *policies* for regulating the new relations between agriculture and society can multifunctionality be achieved. Durand (2003: 140), therefore, saw policy as a crucial "instrument for managing multifunctionality". Further, Buller (2005: ii) saw policies such as the EU RDR as representing the "primary delivery mechanism for support to agricultural multifunctionality", while Potter and Tilzey (2005) argued that the concept of multifunctionality has its roots in a social welfare justification for state policy assistance. Losch (2004) highlighted that in such structuralist views, multifunctionality is closely associated with the implementation of specific policy tools and instruments that may help *regulate* and *orchestrate* the evolution of multifunctional agriculture. The policy environment is seen here as both the ultimate *driver* for multifunctional pathways as well as a *tool* through which society-agriculture interactions are mediated in multifunctional ways.

In these policy-based conceptualisations several key policy developments have been hailed as harbingers of multifunctional agriculture. In an EU context, Marsden and Sonnino (2005: 9), for example, suggested that Structural Funds (since 1988) played a particularly important role in "redefining the role of agriculture in more multifunctional terms", while Ray (1998, 2000) suggested that initiatives such as the EU LEADER Programme have provided a particularly fruitful platform for the development of local multifunctional initiatives based on (re)construction of local and regional territorial identities. Marsden and Sonnino (2005: 10), therefore, argued that "by reconfiguring local resources, redefining the social role of agriculture and increasing value added on farm products, initiatives of this kind [e.g. LEADER] represent an important step towards multifunctional agriculture". In the context of the UK, they suggested that policies showed an implicit recognition of agriculture's multifunctional character, especially during the 1990s when an economic crisis in the farming sector, combined with the traumatic effects of the BSE and foot-and-mouth crises, determined a shift from a *sectoral* notion of agriculture to a *regional* and *territorial* perspective that helped reintegrate farming into rural development. Yet, Marsden and Sonnino also suggested that the expression 'multifunctional agriculture' has not yet entered the *mainstream* political discourse in countries such as the UK due to the predominance of productivist discourses, although they argued that some empirical evidence is available that suggests that multifunctional agricultural trajectories are, at least partly, implemented at regional level (see also Clark, 2003; Buller, 2005). They particularly highlighted the predominance of neo-liberal approaches to agricultural markets in the UK (see also Potter and Tilzey, 2005), combined with (at times contradictory) efforts for support of CAP reform towards more environmental compliance by farmers. In an EU context, they emphasised the importance of new regionalisation trends that have led governments (e.g. in the UK) to make national commitments to a more rural (rather than agricultural) model of multifunctionality through the adoption of modulation (particularly evident in the way many EU member states have implemented the RDR) (see also Lowe *et al.*, 2002). In a similar vein, Buller (2000: 29) emphasised how the new voluntary farm management contracts in France are "a highly innovative attempt to translate notions of agricultural multifunctionality into the practice of agricultural support" (see also Freshwater, 2002, for discussion of similar policies in the USA). The French policy approach to multifunctionality, therefore, prioritises *equally* multifunctional economic, environmental and social aspects, while UK policy approaches to multifunctionality are seen to focus more on reorientation of farm businesses to support the *rural economy* more generally (Lowe *et al.*, 2002; Clark, 2003, 2005). These discussions highlight fruitful conceptualisations of the role that policy can play to bring about more multifunctional agricultural and rural development. Yet, debate continues about the *relative importance* of such policies in engendering 'multifunctional' behavioural changes at the grassroots level (e.g. Burton and Wilson, 2006).

Some policy arenas have attracted particular attention for their 'multifunctionality effects'. As with discussions of the p/pp transition (see Part 2), much debate has emerged on possible interlinkages between Agenda 2000 reforms and the notion of multifunctional agriculture. Agenda 2000 aims at promoting "a competitive, multifunctional agricultural sector in the context of a comprehensive integrated strategy for rural development" (CEC, 1999: 1; see also Chapter 6), and debates have particularly focused on the RDR as a (possible) vehicle for implementation of multifunctionality at EU level (Clark, 2003). The RDR is the most important feature of Agenda 2000, and is seen by many to epitomise the EU's approach to multifunctionality due to its integrated policy for rural development and because it also extends eligibility for funds to non-farmers and non-agricultural activities

(Clark, 2003). Key features of the RDR that are seen to epitomise multifunctionality include measures for the marketing of quality agricultural products, encouragement of farm diversification activities, encouragement of tourism in rural areas, and landscape conservation policies. The RDR also prioritises the involvement of sub-national actors, organisations and institutions in developing multifunctionality. In particular, the notion of cross-compliance – encouraging minimum environmental standards as a basis for receipt of farm subsidies – has been highlighted as a move towards multifunctionality (Gallardo *et al.*, 2003; Marsden, 2003). As Delgado *et al.* (2003: 32) suggested, the RDR is "outlined as an instrument of great importance for the consolidation of a European model based on multifunctionality". Many researchers have highlighted that, following Agenda 2000 reforms, EU member states now have the discretion to cap direct payments to farms by up to 20% (modulation) and divert the saved expenditure into more 'multifunctional' and 'multi-faceted' rural development measures (Lowe *et al.*, 2002; Potter and Burney, 2002). Marsden and Sonnino (2005: 12) suggested that these reforms "provide an ideal context to further analyse national interpretations of the ideals of multifunctional agriculture", in particular as the CAP 2nd pillar was meant to reflect EU aspirations for a broader rural policy that reduces the role of farming (see also Section 8.4). These authors, therefore, point to the multi-faceted roles of Agenda 2000 as a compromise between market liberalisation (largely due to external pressures; cf. Potter and Burney, 2002) and protectionism. For those emphasising the policy dimension of multifunctionality, this is a key point, as it enabled the emergence of new policy pathways that allow the development of "agricultural and rural policy that recognises and accommodates agriculture's multifunctional roles" (Lowe *et al.*, 2002: 1). Yet, some also argue that the RDR has not fulfilled all expectations regarding enhancement of multifunctional rural development. Commentators particularly criticise *national approaches* in the EU that have tended to focus on new income opportunities for farmers, while little effort has been placed on redefining the agricultural sector and to reconfigure rural resources. In the context of the UK, for example, Marsden and Sonnino (2005: 16) argued that "the UK is still far from having a coherent policy on multifunctional agriculture … By keeping the old 'ruralist' focus … the English RDR in particular makes little or no effort to redefine the role of agriculture in a multifunctional sense".

As I will discuss in Chapter 9, the role of AEPs – as an important component of Agenda 2000 – in promoting multifunctionality has attracted particular attention. Marsden and Sonnino (2005), for example, suggested that AEPs have played a crucial role in the development of a more multifunctional notion of agriculture. However, AEPs are a type of policy that may not be as 'green' or 'non-productivist' as many would assume (see Chapter 7). Thus, AEPs have often been negatively branded as a 'backdoor subsidy' that merely adopts the façade of 'multifunctional' action, while perpetuating farmers' dependence on productivist subsidies for survival and continuation of conventional food and fibre production (Potter and Burney, 2002). Indeed, Barthélémy *et al.* (2004) emphasised that the incremental nature of AEP and other 2nd pillar policies has meant that, by default, these policies had to be gradually adjusted to 'fit' emerging European multifunctionality discourses. That many agri-environmental schemes in the EU merely support *maintenance* of agricultural activities rather than *change* (Buller *et al.*, 2000) lends further credence to the relative lack of enthusiasm for policies encouraging multifunctional farming systems. Indeed, as Buller (2005: 3) emphasised for most AEPs in the EU, "the promotion of agricultural multifunctionality [is not] an explicit and stated objective". Thus, as Falconer and Ward (2000) argued, with all its multifunctional rhetoric the RDR and associated AEPs still adhere to the 'public good' model in which farmers' voluntary provision of agri-environmental goods is key in return for AEP-based compensatory payments for income foregone – a questionable premise regarding the RDR's true multifunctional potential.

Similar issues have emerged for possible linkages between the new Single Farm Payments (SFPs) and multifunctionality as the cornerstone of CAP 1st pillar initiatives. These payments (since January 2005) are independent from production and linked to farmers' compliance with basic environmental, food safety and animal welfare standards through cross-compliance. They signal the first break between the direct linking of subsidy and production in an EU context, although there is much national discretion in their implementation (Winter, 2005). The reason why many see SFPs as multifunctional is particularly linked to the support of various non-agricultural activities such as improving rural services or encouraging tourism. Marsden and Sonnino (2005: 14) argued that different regional and national approaches to the implementation of SFPs "reflect different regional interpretations of the concept of multifunctional agriculture". Further complexity has been added as in many European regions and countries implementation of RDR policies and SFPs have led to varying emphasis of priorities for agriculture and rural development (Lowe *et al.*, 2002). In the UK, in particular, this has fuelled debates about the British notion of multifunctionality, as possible multifunctionality pathways appear to be interpreted differently by individual regions such as Wales, Scotland and England (Marsden and Sonnino, 2005). This means that SFPs can be interpreted as a policy approach that emphasises either more agricultural or rural policy approaches, reflecting different adjustments to the declining importance of agriculture and other primary industries and to new demands of the rural economy (Lowe *et al.*, 2002). Marsden and Sonnino (2005) concluded that the way SFPs have so far been implemented in a UK context lends little credence to the notion of a multifunctional policy transition.

One of the key cornerstones of the policy-based view on multifunctionality relates to macro-political agendas and the predominance of neo-liberal rhetoric and policies at the international level, particularly in the context of multifunctionality and WTO agricultural trade talks. Potter and Burney (2002: 45) emphasised that "multifunctionality has emerged as a key policy concept in the WTO agriculture negotiations", while Mather *et al.* (2006) argued that "multi-functionality [sic] is now probably irretrievably associated with trade negotiations". In particular, Potter and Burney (2002) emphasised the externally driven neo-liberal processes of recent EU policy change, thereby questioning the true nature of multifunctional policy transition claimed by some (see Section 8.4). Similarly, Hollander (2004: 300) suggested that "emerging in the context of agricultural trade liberalization and reform of the CAP, multifunctionality is being promoted as a way to address social and ecological concerns such as farm abandonment and biodiversity loss through domestic agricultural policies that conform to the GATT/WTO". Thus, macro-policy analysts argue that free trade may either jeopardise public good benefits engendered by multifunctional agriculture unless domestic policies are in place, while others believe that such policies create trade distortions (Blandford and Boisvert, 2002). According to WTO discourses (informed by neo-classical economics), there should be no necessary conflict between promoting trade liberalisation and safeguarding a 'contractual type of multifunctionality', provided farmers are paid directly for environmental and other multifunctional services they provide (Potter, 2004). Gallardo *et al.* (2003), therefore, suggested that the *misappropriation* of the notion of multifunctionality as a 'strategic weapon' in WTO negotiations (see Section 8.4) has both weakened the effectiveness of Agenda 2000 and the RDR as well as the utility of the term 'multifunctionality' as a policy-based concept.

Macro-policy theorists have, therefore, usefully highlighted how this notion of multifunctionality has defined the terms of an important debate about non-trade concerns in the liberalisation of agricultural trade. In particular, they have highlighted that a central issue in WTO policy negotiations is the extent to which current policies (especially those linked to the WTO URAA) permit countries to satisfy objectives with respect to a policy-based notion of multifunctionality, and whether such objectives would be undermined by further reductions in tariffs, export subsidies, and domestic support. As Blandford and Boisvert (2002) emphasised, if significant reductions were to be agreed in amber box payments,[53] and

---

[53] 'Amber box' refers to price support and state aids that are directly linked to volumes of output and, therefore, deemed trade distorting (Potter and Tilzey, 2005).

in tariffs and export subsidies, the ability of countries to pursue domestic objectives by maintaining domestic market prices for agricultural commodities above border prices would be substantially reduced. Yet, such authors have also convincingly argued that in these debates the notion of multifunctionality is purely seen as a *smokescreen* put up by EU policy-makers for the continuation of protectionist agricultural policies as part of the 'multifunctional European model of agriculture' (Potter, 2004; see also Sections 8.4 and 8.5). Similarly, Marsden (2003) argued that the WTO has led to a watering down of the notion of multifunctionality by being associated with the logic of industrial commodity markets such as the use of food products for industrial purposes. Thus, Buller (2005: 1) rightly highlighted that many current commentators on multifunctionality "see multifunctionality as essentially a blue box term [see Chapter 7] whose primary function is to provide justification for the continuing use of coupled support to farming". In this sense, the notion of multifunctionality loses its conceptual base and becomes merely a fuzzy concept appropriated by policy-makers to further EU-specific protectionist goals, as well as WTO-related global agro-industrial aims (see Section 8.5). Commentators have rightly questioned how far and in which way the future WTO process is likely to sustain a 'multifunctional agriculture model' under increasingly liberalised and globalised market regimes, with Hollander (2004: 311) arguing that "multifunctionality has become the synecdoche of international agricultural politics to the extent that the acceptance of the term is clearly guided by how a country or region sees its interests in agricultural trade liberalization".

As Chapters 9 and 10 will outline, the policy environment is a vital component of multifunctional agricultural processes, and debates on the interlinkages between policy and multifunctionality have, undoubtedly, provided important insights into recent policy transitions. As Section 8.4 will highlight, understanding policy-based rhetoric on multifunctionality has greatly helped conceptualising the *spatiality* of the concept, especially in the context of a 'European model' of policy rhetoric. Yet, just as agricultural economists may have placed too much emphasis on mono-causal economic processes influencing multifunctionality, in my view the policy-centric multifunctionality debates have also placed too much emphasis on the policy environment as a *singular* force shaping multifunctional decision-making pathways. As highlighted in the social psychology literature, for example, such mono-causal reductionist interpretations often neglect policy-independent decision-making forces that shape human behaviour (e.g. Ajzen, 1991; Burton and Wilson, 2006). Thus, the policy-centred view of multifunctional agriculture provides, in my view, a relatively narrow interpretation of agricultural change. Although it has provided important insights into macro-scalar patterns of multifunctionality, its main shortfall is that it generally tends to place too much emphasis on a structuralist top-down regulatory framework as the solution for guiding agricultural transitional processes, in particular as it accords the state, and associated state policy, the key role in *orchestrating* multifunctional development pathways. Indeed, Clark (2006: 332) highlighted that "to date no attempt has been made ... to assess the challenges inherent in translating multifunctional discourses into concrete policy outcomes". Further, both economistic and policy-based conceptualisations of multifunctionality are based on the need for the development of 'easy' *indicators* to measure the success of implementing multifunctional trajectories (i.e. indicators of economic success or indicators for positive implementation processes) to the neglect of less tangible aspects of what multifunctionality should be about (McCarthy, 2005; see also Wilson and Buller, 2001). As Chapter 10 will discuss, the question of who should orchestrate multifunctionality, and, indeed, whether policy is the right framework for influencing multifunctional agricultural trajectories, warrants closer scrutiny.

8.3.4 Holistic interpretations

Economistic and policy-based interpretations of multifunctionality can be seen as relatively 'narrow' structuralist and reductionist approaches to understand multifunctional agriculture.

In contrast, some commentators (especially rural sociologists and human geographers) have associated multifunctionality with more 'holistic' cultural and social interpretations of agricultural and rural change, linked to broad-based societal changes, rural development issues, the consumption countryside, and grassroots agency-led patterns and processes. Although cultural processes are closely intertwined with economic and policy processes, the difference in these debates is that that they acknowledge human behavioural patterns embedded in social and psychological processes that often go beyond mere structuralist influences (best highlighted in recent 'cultural turn' approaches in human geography).

In her incisive critique of unidimensional explanations of multifunctional agriculture, Garzon (2005: 5) rightly argued that "the fact that the multifunctionality concept is part of a difficult reform process of policy ... is probably not sufficient to explain the intensity of the controversies surrounding it", while Buller (2005: ii) called for "a more inclusive conceptualisation of 'multifunctionality'". From a human geography perspective, Holmes (2006: 145) suggested that concepts of multifunctionality should take into account not only *production* elements, but also *consumption* and *protection* – processes that should be seen as part of "a holistic mode of multifunctional resource use" that take us away from what he identified as 'selective identification' of multifunctionality indicators that characterises economistic and policy-oriented conceptualisations (see also Pretty, 2002). Marsden (2003: 111) further argued that recent patterns and processes of agricultural and rural change "have been far from unilinear or cast simply and strictly along economistic lines". Knickel and Renting (2000: 526) similarly suggested that "farm management economic studies focusing on cost-benefit analyses and the comparative advantages of certain production systems can ... easily miss the point that certain, more traditional production systems have advantages for the farm household that cannot be expressed in monetary terms" (see also Jongeneel and Slangen, 2004). Similarly, Losch (2004: 336; emphasis added) suggested that multifunctionality "offers the possibility of going *beyond* the questions concerning *productivity* and *market competitiveness* towards establishing a debate in terms of strategies for sustainable development". He also argued that the concept needs to go beyond the arena of trade negotiations (in particular the WTO) as advocated by some policy-based concepts.

Hollander (2004) referred to these more holistic interpretations as the 'strong version' of multifunctionality (her 'weak' version refers to the largely policy-based conceptualisation), while Potter and Tilzey (2005) referred to the need for an 'agrarian interpretation' of multifunctionality which sees agricultural/agrarian processes as embedded within social processes of rural change (see also Potter's, 2004, notion of 'narrow' and 'broad' definitions of multifunctionality). Yet, acknowledgements of the interlinkages between multifunctionality, society, rural development and grassroots-led processes have come relatively late in scientific debates on multifunctionality, having only firmly taken hold from the late 1990s/early 2000s. The FAO (1999) was one of the first to acknowledge that multifunctionality and environmental sustainability are closely interlinked (Potter, 2004). It particularly argued that agricultural systems are *intrinsically multifunctional* and that they have always fulfilled more than just the primary aim of food and fibre production. This was echoed by Lowe *et al.* (2002: 15) who suggested that "multifunctionality ... [is] challenging the classic sectoral vision of farming as an exclusively productive enterprise". Despite its strong agricultural economics focus, Van Huylenbroek and Durand's (2003) recent edited book *Multifunctional Agriculture*, meanwhile, is also one of the first to devote a substantial section to the interlinkages between multifunctionality and social processes. Thus, Van der Ploeg and Roep (2003: 37) in the same volume suggested that "multifunctionality and rural development may be the key unifying concept" for a new model of agriculture in Europe. Similar views were espoused in a US context by Boody *et al.* (2005: 36) who argued that "if US farm policy changes to embrace multifunctional agriculture, additional thought ... should be expended ... to include energy production, recreation, education, and other activities that bring income and economic development to rural areas".

Such calls for acknowledging the interlinkages between multifunctionality, society and rural development also echo Marsden's (2003) concept of multifunctionality as the sum total of *exchange* (e.g. food quality, supply chain issues), *production* (e.g. food production, conservation) and *reproduction* (e.g. land, capital) relationships, where a farmer needs to conduct *all* these multidimensional activities simultaneously to be classified as 'multifunctional'. Marsden, therefore, argued that multifunctionality should be seen as a concept that acknowledges that farmers can create *mutually interlinked* products and services which use/exploit the *same* resource base (physical, social or knowledge-based), creating a *synergy* between the creation of different value-added products. This emphasises the notion of *indivisibility* of inputs and outputs rather than the progressive divisibility associated with intensive forms of agricultural production. Durand (2003: 134), therefore, argued that multifunctionality means that "the farmer is recognized not only as a producer of goods and services but also as a manager of the environment and rural space, as well as being one of the principal stakeholders in local development". Similarly, Belletti *et al.* (2003a) suggested that multifunctionality can not be separated from issues related to the 'rediscovery' of the rural development paradigm, while Bills and Gross (2005) asked different stakeholder groups in society (in the UK and USA) about their interpretations of 'multifunctionality', thereby emphasising the importance of including society *as a whole* in understanding multifunctionality. Such authors usefully emphasise the importance of post-structuralist and cultural interpretations of multifunctionality that highlight the role of agriculture as part of a rural culture that is highly valued by an increasingly urbanised society.

Probably the most refined analysis of multifunctionality has come from the EU-funded 'Multagri' project[54] (e.g. Buller, 2005; Marsden and Sonnino, 2005). Building on Marsden (2003), Marsden and Sonnino (2005: executive summary) argued that "in contrast with the 'agro-industrial' paradigm, which restricts the concept of multifunctional agriculture to the notion of farm pluriactivity, and with the 'post-productivist' paradigm, which interprets it as farmland diversification, the emerging 'rural development' paradigm considers multifunctional agriculture as a development tool to promote more sustainable economies of scope and synergy". By acknowledging that multifunctional activities should contribute towards the emergence of a new agricultural sector that takes into account changing needs of wider *society*, thus implying a radical redefinition of rural resources in and beyond the farm enterprise, Marsden and Sonnino (2005) confirmed that any conceptualisation of multifunctionality needs to go beyond economistic (and atomistic) interpretations. Marsden (2003: 179) argued that "it is important to redefine the role of farms as potential multi-functional rural enterprises which serve a variety of markets which contribute to sustainable rural development". Multifunctionality is seen here to intersect with a redefinition of nature by re-emphasising localised food production and by reasserting the socio-environmental role of agriculture as a major agent in sustaining rural economies – with social construction of rural space at the heart these 'new' multifunctionality concepts (see also Potter and Burney, 2002; Delgado *et al.*, 2003).

Marsden and Sonnino (2005) also placed particular emphasis on a conceptualisation of multifunctionality rooted in notions of the *consumption countryside*. Building on Marsden's (2003) post-productivist paradigm (see Chapter 7), they saw multifunctionality as closely linked to the perception of rural areas as spaces of consumption to be exploited not only by industrial capital, but also by urban and ex-urban populations. In this sense, multifunctionality is embedded in cultural processes, where agriculture loses its centrality in society and where 'nature' is conceived mostly in terms of 'landscape value' as a consumption good. However, they also highlighted that in many countries policies (e.g. AEPs) and planning restrictions have reinforced this 'post-productivist' variant of

---

[54] 'Multagri' was a Specific Support Action funded by the EU 6th Framework Research Programme, investigating different aspects of agricultural multifunctionality in Europe with 26 organisations from 15 countries. See also results from the EU-funded projects IMPACT (2001-2004) (Knickel *et al.*, 2004) and TOP-MARD ('Towards a policy model of multifunctionality and rural development') (2004-2007) (Bryden, 2005).

multifunctionality by compartmentalising agricultural and rural land into distinctive multifunctional units. Marsden (2003: 95), therefore, argued that in the UK "through counter-urbanisation and more regular visiting, the countryside has been increasingly seen as a multifaceted consumption space". This was also highlighted by Mather (2001) in the context of 'multifunctional forestry' where forests are increasingly becoming places of consumption rather than places of timber production (see Section 8.2) – emphasising the similarities in debates across different multifunctional land use discourses. In Marsden and Sonnino's (2005: 21) concept of multifunctionality "agriculture thus becomes a multifunctional enterprise that delivers safe and healthy food and non-food products, a visually attractive countryside and distinctive local food products that support tourism and a positive image [of the region]. In other words, agriculture is re-emphasised and repositioned for its contribution to achieving rural sustainability". Contrary to the policy-based and economistic view of multifunctionality, Marsden and Sonnino acknowledge that multifunctionality can only be fully understood by taking into account different options for agricultural pathways linked to stakeholder decision-making processes and rural sustainability. Their emphasis of the interlinkages between multifunctionality and the consumption countryside, therefore, provides an important corrective to the overemphasis of economistic and policy-related indicators.

Other authors suggest an even more socially inclusive view of multifunctional agriculture based on the potential of multifunctional agricultural systems to strengthen *social capital*. Clark (2003, 2005, 2006), in particular, divided multifunctionality into policy goals, outputs and territorial functions. For policy goals he argued that multifunctionality emphasises the production of a diverse array of economic, social and territorial outputs that *complement* the sector's traditional productivist emphasis. Multifunctional outputs and associated territorial functions, in turn, do not only include economic processes, but also important social processes through improved synergies with other economic sectors (e.g. through multiplier effects) and through maintenance of the rural fabric (e.g. by encouraging viable and stable rural communities). Di Iacovo (2003) has been one of the few to highlight that the notion of multifunctionality must also be related to issues of *social inclusion* of groups such as immigrants, the elderly or disabled people in rural areas (which he terms 'the multifunctionality of agriculture in social care') (see also Vatn, 2002; Jongeneel and Slangen, 2004). Using a refreshing approach that shares similarities with the concept of multifunctionality presented in Chapters 9 and 10, Di Iacovo (2003: 108) conceptualised multifunctionality as an "agriculture [that] can also produce different kinds of educational, social, and caring services. They can be designed for particular target groups, as in the case of drug-addicts or psychiatric patients, as well as for the local community, as in the case of [model] farms or for larger groups as is the case in seminar centres and other initiatives". He particularly acknowledged that the inclusion of the social dimension into multifunctional agriculture will also need considerable effort by farmers themselves, in particular through a redesign of production processes to enable participation of marginalised groups such as immigrants, the disabled, the young or the elderly. Here, multifunctionality takes on a role well beyond the farm and is seen as a process that strengthens social capital (see Chapter 9) and that interlinks various functions and stakeholder groups at different scales by improving relationships among locals and local well-being. In particular, multifunctionality is seen here as a *concept* (and not just a process) through which "old social networks and social services in rural areas may be revisited and restyled into a new dimension" (Di Iacovo, 2003: 122) – a concept that may require more educational resources to be made available for the 'multifunctional' re-education of farmers and other rural actors. Di Iacovo (2003: 123) concluded by arguing that "the integration of the social dimension into the concept of multifunctionality may contribute to reshaping the future in rural areas by encouraging a more authentic concept of rural life" (see also Di Iacovo, 2006).

An additional 'holistic' contribution towards agricultural multifunctionality debates has come from French colleagues through the publication 'Les cahiers de la multifonctionnalité'

– arguably the only regularly published journal/series focusing entirely on multifunctionality issues (but only available in French and, therefore, largely inaccessible to an English-speaking audience). Lardon *et al.* (2004), for example, emphasised multifunctional concerns such as the environment, but also the roles and needs of stakeholder groups such as hunters and walkers, while Cayre *et al.* (2004) focused on agro-anthropological concerns (not dissimilar to those discussed in Chapter 10), emphasising the importance of understanding individual farm-level multifunctional trajectories that go well beyond the economic and policy environment. Indeed, Lardon *et al.* (2004: 13; emphasis added; my translation) highlighted that French authors "suggest an evaluation method for assessing decision-making pathways of farmers that is *not restricted to techno-economic considerations*. They, therefore, simultaneously touch upon symbolic, functional and socio-cultural dimensions of multifunctionality". Cayre *et al.* (2004), therefore, suggested a holistic understanding of multifunctionality which urges farmers to redefine the new 'boundaries' of their occupation by recognising the economic, social and environmental repercussions of their agricultural activities. In particular, they suggested that research should focus on understanding the nature of *multifunctional pathways* selected by individual agricultural actors – an approach that will also guide our discussion in Chapter 10. Rapey *et al.* (2004a), meanwhile, addressed the multiple use of rural space *beyond policy and economy* and focused on land use conflicts generated by multifunctional use of rural landscapes, while both Pivot *et al.* (2003) and Roux *et al.* (2004) emphasised the importance of *local embeddedness* of multifunctionality actions – reiterating the importance of social capital in multifunctionality debates. Rapey *et al.* (2004b), in particular, highlighted the importance of spatial and temporal variability in the adoption of multifunctionality pathways by farmers and also emphasised the importance of understanding multifunctionality through the lens of societal demands and perceptions of what 'agriculture' is about. In many of these contributions, the diverse *geography of multifunctionality* is also addressed in more detail than in most Anglo-American contributions (e.g. Guillaumin *et al.*, 2004; Lardon *et al.*, 2004) – an issue I will also focus on in Chapter 9. This is probably best highlighted in Losch *et al.*'s (2004) contribution that asked the interesting question whether *global level multifunctionality* could be a 'viable' concept, and by Roux *et al.* (2004) and Barthélémy *et al.* (2004) who broadened the discussion of multifunctionality to a developing countries context. Overall, a picture emerges here of an intellectual community researching multifunctionality not transfixed by economistic and policy drivers of multifunctionality (like many Anglo-American contributions; see above). Indeed, it may reflect a community rooted in strongly multifunctional agricultural development pathways based on the continuation of, and communication about, relatively strong embeddedness of agricultural actors with their local communities (see Chapter 10).

Finally, although largely focusing on policy dimensions, Potter and Burney (2002: 35; emphasis added) added another dimension to the notion of multifunctionality by arguing that "it is in relation to the perceived threat of extensive agricultural restructuring to *biodiversity* and *landscape values* in the EU that the concept [of multifunctionality] has been most fully realized". This has been reiterated by both Larsen (2005) and De Groot (2006) who see multifunctionality largely as a landscape-related notion defined by both 'functional integration' of landscape elements (including agricultural elements) and by Winter (2001) who focused chiefly on landscape and environmental quality (see also Potter, 2004; Rapey *et al.*, 2004a). Potter and Burney (2002) argued that the underlying argument has been that farming in large parts of the EU was, until recently, compatible with the conservation of biodiversity and other environmental benefits. Indeed, a key argument underpinning this 'environmentally-based concept of multifunctionality' in Europe is that multifunctional agriculture may have also actively moulded the very character of the countryside through a process of 'joint production' (see above) of food and environmental goods. Crucially, they argued that, as far as the environmental aspect of agricultural and rural change is concerned, it is not the existence of the notion of multifunctionality as such which is controversial, but the nature of the design of farm subsidies and their trade-distorting properties. They highlighted that current policy models (within the EU) envisage "the continued need for multifunctional instruments that support farmers' incomes in marginal areas in order to ensure continued occupancy of rural land and thus the proper [environmental] management of farmed landscapes" (Potter and Burney, 2002: 46). This environmental protection aspect has also been highlighted as an important output and territorial function of multifunctionality by Clark (2003).[55] This environmental component has also been emphasised by agencies and organisations aiming to promote environmental sustainability of the countryside, with the former UK Countryside Agency (amongst many others) suggesting that sustainable land management practices can be encouraged through "multifunctional land management" (Countryside Agency, 2001: 13). The argument here is, therefore, that beyond societal and rural development issues, the *environmental component* has to be another key aspect of multifunctional agriculture.

As Chapters 9 and 10 will highlight, the more 'holistic' view of multifunctionality espoused in these approaches resonates more closely with the notion of multifunctionality this book is suggesting. In particular, it highlights that issues of *environmental, social* and *cultural capital* need to be considered to understand what multifunctionality is about. In Chapters 9 and 10, I will use these more holistic understandings of multifunctionality, rooted in the concept of productivism and non-productivism, as a platform for a reconceptualisation of multifunctionality.

8.4 Cultural interpretations and the spatiality of the multifunctionality concept: neo-liberalism, trade issues and political retrenchment

The above discussion emphasised the importance of different cultural interpretations of multifunctionality and highlighted the spatially heterogeneous adoption and popularisation of the term. In particular, the so-called 'European model of multifunctionality' (Potter and Tilzey, 2005) emerges as a spatially defined rhetoric that is perceived by many non-EU commentators as something that has 'nothing to do with them'. The notion of multifunctionality has, thus, become a *territorially based discursive tool* – or a policy-based *tactical smokescreen* – for European policy-makers, rather than a concept that can be used to fully understand agricultural and rural transitions. In line with Garzon (2005), I argue that this *geographical exceptionalism* – emphasising a European essentialist view – has not greatly aided the wider applicability of the term across diverse agricultural regimes (see also Skogstad, 1998; McCarthy, 2005). Australian colleagues and myself argued that "'multifunctionality' – the notion that farming provides multiple benefits – originated in Europe and has been viewed with suspicion in Australia as a stratagem to permit continued subsidisation of agriculture" (Cocklin *et al.*, 2006: 1; see also Anderson, 2000). Similarly, Hollander (2004) emphasised the difficulties and limitations in the 'geographic transferability' of the Euro-centric notion of multifunctionality to the USA, and suggested that this view represents a 'weak version' of multifunctionality almost purely based on policy discourses attempting to justify continued subsidisation of agriculture. In any discussion of multifunctional agriculture it is, therefore, vital to understand why the term has remained closely associated with a 'European way' of addressing contemporary agricultural processes.

Gorman *et al.* (2001) suggested that the Euro-centrism of the multifunctionality concept can be explained by the specific nature of European farming based largely on relatively small-scale family farming that has always had *multiple functions* (see also Swinbank, 2002; McCarthy, 2005). The main argument here suggests that many economically marginal farming systems responsible for multifunctionality are dependent on CAP subsidies for their

---

[55] Gerowitt *et al.* (2003) made the interesting point that weeds on agricultural land should also be conceptualised as an integral aspect of the multifunctional countryside.

continued existence (Goodman, 2004). This has led authors such as Marsden and Sonnino (2005) to refer to the 'multifunctional character' of the European countryside, and Potter and Burney (2002: 35) to suggest that "multifunctionality is a genuine, and in some respects, unique, feature of European agriculture". Similarly, Losch (2004: 340) emphasised the Euro-centric notion of multifunctionality by arguing that "formulated within the European framework of the early 1990s, the concept of the multifunctionality of agriculture refers to all products, amenities and services created by farming". In this view, Losch (2004) argued, multifunctionality is seen as a concept that helps defend Europe's territorial harmony and that attaches particular value to the protection of the European countryside and food quality (i.e. cultural, culinary and sanitary values). Thus, Gallardo *et al.* (2003: 169; emphasis added) suggested that "the fundamental difference between the European model and that of our main competitors lies in the *multifunctional nature* of agriculture in Europe and in the role it plays in the economy and the environment, in society, and in the conservation of the countryside", or, in the words of Garzon (2005: 18) "multifunctionality [has been] part of the solutions found by European policy makers to renew the social contract over agriculture" (see also Anderson, 2000). As a result, McCarthy (2005) emphasised that it was the EU and a few allied countries that first advanced the notion of multifunctionality. It is little wonder, therefore, that the first policy-related allusion to multifunctional agriculture is found in reports by the European Commission (e.g. CEC, 1988; see Section 8.2).

Delgado *et al.* (2003), therefore, argued that the need for defining a new *social* role for agriculture through the notion of multifunctionality has been particularly crucial in the context of European integration. This is probably most evident in the European Commission's Agenda 2000 formulation that "the fundamental difference between the European model and that of our main competitors lies in the multifunctional nature of agriculture in Europe and in the role it plays in the economy and the environment, in society, and in the conservation of the countryside" (cited in Van Huylenbroek, 2003: xii). It is also evident in EU Agriculture Commissioner Fischler's remarks in the late 1990s that "safeguarding the future of the European model of agriculture, as an economic sector and as a basis for sustainable development, is of fundamental importance because of the multifunctional nature of Europe's agriculture and the part agriculture plays in the economy, the environment and landscape as well as for society" (Fischler, 1999: 1). Clark (2006) suggested that this *opacity* of the notion of multifunctionality has proved a considerable asset in EU policy-making, as it is precisely such broad-brush notions that are needed to engineer alliances among member states and to provide political discretion to domestic elites in addressing the needs of complex stakeholder interests. In other words, EU policy-makers have exploited the vagueness of the term in setting their own multifunctionality agendas.

The role of the USA has been particularly important in 'relegating' the notion of multifunctional agriculture to a European context (Brazier, 2002; McCarthy, 2005). The USA has considered the introduction of the notion of multifunctionality in EU policy discourses as merely an attempt to maintain the distortion of EU internal protection policies on the world market (Delgado *et al.*, 2003). Carpentier *et al.* (2004: 307), therefore, argued that "the term multifunctionality most commonly enters agricultural policy discussions in Europe and Japan, but rarely in North America". Hollander (2004: 301) similarly suggested from a US perspective that "EU agricultural advocates, in their opposition to neoliberal trade policies, have used the ideas of landscape, livelihood, and agroecology, encompassed by the term 'multifunctionality', in defence of domestic agricultural supports". This is an important criticism considering that the EU is the world's second largest agricultural exporter and, as a result, the EU's position on 'multifunctionality' (the 'weak version' in Hollander's, 2004, view) holds particular weight in global agricultural trade negotiations (Clark, 2005; Potter and Tilzey, 2005). According to Freshwater (2002), the US position can be explained because US agriculture is not as socially embedded as in the EU. Thus, Bohman *et al.* (1999), in a report for the US Department of Agriculture entitled *The use and abuse of multifunctionality*, argued that it is particularly countries situated in the 'amber box' of the

URAA (see Chapter 7) – countries under the greatest obligation to change the orientation of their agricultural policies (i.e. most EU member states) – that have defended the concept of multifunctionality particularly vehemently (Blandford and Boisvert, 2002). Freshwater (2002: 6) similarly emphasised from a US perspective that "multifunctionality is a recent term that came out of European discussions of sustainability", and "when multifunctionality is explicitly considered [by US policy-makers] it is seen as a foreign strategy that will harm US interests. The conventional American view of multifunctionality is that it is promoted by countries with no comparative advantage in agriculture to protect their domestic producers from competition" (Freshwater, 2002: 3). As a consequence, the notion of multifunctionality in the USA is inherently linked to *political* issues and, consequently, the USA has (so far) not fully embraced the concept (Bills and Gross, 2005). Although some US policies exist with *implicit* reference to multifunctionality (Freshwater, 2002), there is only limited support for key elements of multifunctionality such as jointness of production (Bohman *et al.*, 1999).

As a result of this *politicisation* of multifunctionality, the role of WTO negotiations in *strengthening* EU-centric multifunctionality discourses has attracted particular attention, as has the role of supra-national bodies such as the OECD. Debates have focused largely on the macro-political role of multifunctionality discourses enshrined in key policy documents such as the OECD's (2001) analytical framework on multifunctionality. It is argued that through the OECD-specific conceptualisation of multifunctionality (see above), the EU succeeded in convincing non-EU countries as varied as South Korea, Japan, Norway or Switzerland (the 'friends of multifunctionality'; cf. Potter and Burney, 2002; Hollander, 2004) of the importance of maintaining multifunctional agriculture *in Europe* during the subsequent Multilateral Round of the WTO (Hollander, 2004; Losch, 2004). McCarthy (2005) suggested that these countries, in particular, have continued to advance stronger and more codified versions of multifunctionality in recent WTO negotiations (see also Goodman, 2004). The USA, meanwhile, has tended to view multifunctionality as a device for the EU to restore production subsidies that have been limited under WTO agreements (Freshwater, 2002), leading Hollander (2004: 310) to argue that the "claims of the case for European agriculture as somehow distinct from the rest of the world do not withstand scrutiny".

The idea that multifunctionality is a key characteristic of European agriculture has reinforced arguments in defence of EU agricultural policy. From the early 2000s, the EU began to mark out the terms of WTO debates by defending its right to use domestic support programmes to uphold the 'European model of agriculture' (Potter and Burney, 2002; Hollander, 2004). This has given a new source of legitimacy to the CAP, justifying the concession of a 'social salary' for farmers for generating positive benefits for society. The key argument here is that agricultural liberalisation in general, and the elimination of 'blue box' payments (i.e. CAP production subsidies) in particular, would threaten the continuation of economically marginal farms that are seen to contribute greatly to the multifunctional character of the EU countryside (Potter and Tilzey, 2005). Thus, Gallardo *et al.* (2003: 171) argued that "until now the change of paradigm and orientation in agricultural and rural matters has served only to legitimize the negotiating strategy of the EU within the WTO". Even more trenchantly, they suggested that "multifunctionality, as it has been considered by the European Commission, was *not* the result of reflection on the European agricultural model, but was born due to external reasons as an *ideological alibi* for opposing our [Europe's] main rivals in the WTO Negotiation Round" (Gallardo *et al.*, 2003: 172; emphasis added). In a similar vein, Garzon (2005: 2) posited that "the European birth of the multifunctionality concept helps synthesize the diversity of national views, is part of the policy change process at European level and has deep roots in the history of the common agricultural reforms, but was primarily used in an international context at a moment where various processes of policy change were colliding". Interestingly, the EU was largely successful in its legitimisation of multifunctionality in the WTO, despite the fact that multifunctionality was interpreted in different ways by each of its member states (Delgado *et al.*, 2003). Yet, although the EU succeeded in getting a mention of non-trade concerns into the draft text of the WTO Seattle meeting, the term 'multifunctionality' did not appear in the final agreed text (Potter and Burney, 2002). This shows that, although the EU has reiterated its support of multifunctionality, it has been careful to couch the argument in terms of WTO-accepted rhetoric of 'non-trade concerns'. It also emphasises the power of the notion of a 'multifunctional European agricultural territory' in macro-policy rhetoric based on specific national (and even regional) farm structures and history.

Approaching the issue from neoclassical trade theory, Potter and Burney (2002) analysed the extent to which the notion of 'multifunctional agriculture' has been used to *justify* continuing EU domestic subsidies to farmers in WTO agriculture trade talks. They highlighted that it has been a premise of supporters of the notion of multifunctionality that trade liberalisation may result in a degree of agricultural restructuring within the EU which, in turn, would threaten biodiversity and landscape values. This suggests a 'green box' that is broad enough to accommodate partially decoupled measures designed to support Europe's multifunctional farming systems (Potter, 2004). Hollander (2004), therefore, argued that multifunctionality provides a *strategic* opening in which to recognise the landscape functions of agriculture and rural settlements, so that the resulting social and ecological complexity can then be defined as 'public goods' and maintained through state policies. The USA, meanwhile, has favoured liberalisation of agricultural trade, thereby counteracting the EU position on multifunctionality by arguing both that high support prices in the EU are leading to distortions in agricultural markets and that liberalisation will result in extensification of production with less pollution and less damage to the environment (Brazier, 2002; Durand and Van Huylenbroek, 2003). The 1996 Federal Agricultural Improvement and Reform Act provided the policy platform for the USA to enter the URAA talks with much less interest in sustaining 'blue box' subsidies (i.e. farm subsidies) and, therefore, with less support for the Euro-centric notion of multifunctionality. As a result, we witness less use of the multifunctionality concept in contemporary US policy rhetoric (Brazier, 2002; Garzon, 2005). Yet, McCarthy (2005) acknowledged that the core ideas of multifunctionality are nonetheless circulating in the USA, albeit often under the narrow label of 'working landscapes' (see also Bills and Gross, 2005). Thus, Dobbs and Pretty (2004: 225) suggested that "the basic concept of multifunctionality is the same on both sides of the Atlantic, although it manifests itself differently", in particular concerning the use of environmental sustainability indicators of farmland in the USA (Hollander, 2004), and based on much less coordinated policy approaches than in Europe (Bills and Gross, 2005). Overall, the US approach to multifunctionality could be interpreted as a cynical attempt at hidden protectionism, as, although committed rhetorically to free trade, it continues to heavily subsidise its own agricultural products (Hollander, 2004; McCarthy, 2005).

The Cairns Group[56] meanwhile have been the most outspoken critics of the European multifunctionality concept (McCarthy, 2005). They advocated a reduction in the size of green box support measures (non trade-distorting) and to use countervailing measures against those countries continuing to deploy trade-distorting support (Randall, 2002; Hollander, 2004). Potter and Burney (2002) argued that this 'radical' position is shared by some developing countries (e.g. members of the so-called 'Like-Minded Group' including Cuba, Pakistan, Kenya and Uganda) who view the green box as an instrument for the protection of developed country interests. Both for the Cairns Group and the Like-Minded Group, the EU is, therefore, using multifunctionality rhetoric as a *smokescreen* to hide continuing productivist subsidy policies (Potter, 2004). McCarthy (2005: 777) argued that most developing countries have been opposed to the notion of multifunctionality as they largely see it as "defined in terms of exceptionalist understandings of European agriculture, which can only work against them" (see also Hollander, 2004; Losch, 2004). McCarthy (2005: 779)

---

[56] This group represents Australia, Canada, New Zealand, Indonesia, Malaysia, Thailand, Philippines, South Africa, Fiji, Costa Rica, Guatemala, Argentina, Bolivia, Brazil, Chile, Colombia and Uruguay. Together, these countries account for one-third of the world's agricultural exports.

even argued that when the different Euro-centric components of multifunctionality "are used to reinscribe ontological differences and perpetuate inequalities between the global North and South, 'multifunctionality' can begin to appear downright reactionary".

Potter and Burney (2002) highlighted that issues over the (mis)use of multifunctionality have also been raised *within* the EU, as it is increasingly clear that for EU Southern member states, for example, the debate about joint production and environmental issues linked to EU policy-makers' conceptions of multifunctionality is less of an issue than concern about rural development implications of complete decoupling (see also Wilson and Juntti, 2005). This latter point may suggest that the contemporary policy notion of multifunctionality may only apply to a small subset of Northern EU member states and that, as outlined above, it fails to provide a robust *policy concept* for most other countries. Lowe *et al.* (2002) have further emphasised substantial divisions within the EU about the meaning of multifunctionality, with the policy-based view of multifunctionality (see above) more closely linked to the French agenda, while the UK has (arguably) continued to pursue a more 'holistic' countryside-based agenda (see also Holmes, 2006).

The discussion so far has highlighted that most of the debates on the 'European' model of multifunctionality have focused largely on policy-based conceptualisations. Yet, some authors have gone beyond such policy-based explanations to explain the Euro-centricity of the multifunctionality concept. Van der Ploeg and Roep (2003), for example, argued that as Europe is increasingly moving away from the agricultural modernisation paradigm towards the rural development paradigm, this has left room for notions of multifunctionality to influence societal discourses (see also Marsden, 2003). In other words, rural development is seen here as reconstructing the eroded economic base of both the European rural economy and farm enterprise. Van der Ploeg and Roep (2003: 38), therefore, see rural development itself as a "multilevel process rooted in historical traditions". These authors, therefore, turn the above-mentioned policy-based concept of multifunctionality partly on its head. Instead of policy developments driving the European multifunctionality process, they acknowledge that there may be European-specific autochthonous social and cultural processes that drive multifunctional pathways and that, in turn, have influenced the CAP policy-making environment to become more 'multifunctional'.

However, Durand and Van Huylenbroek (2003: 10) argued that "multifunctionality is not a European invention", while Delgado *et al.* (2003) highlighted that the notion of multifunctionality is beginning to be profiled as a legitimising argument that can articulate the necessary complementarity between agriculture and the rural world at the *global* scale. They pointed towards multiple treaties and international conventions that have made explicit reference to multifunctional agriculture in order to specify its non-market functions (see also Potter and Burney, 2002). As highlighted above, the most important in this respect have been policy and conference documents produced by the FAO (e.g. 'Quebec Declaration' 1995; 'Rome Declaration and Action Plan' 1996; the FAO conference on the 'multifunctional character of agriculture and the land' 1999) and OECD policy documents that have acknowledged the importance of the multifunctional character of farming (OECD, 2001; Losch, 2004). Similarly, Holmes (2002) highlighted that multifunctional agriculture is not a new concept to Australia and argued that ideas of decision-makers about the best management of Australia's rangelands "have matured from the geographically limited and single-use oriented to the holistic, *multi-use, multi-value* view of rangelands" (Holmes, 2002: 364; emphasis added). Holmes (2006) further argued that in Australia multifunctionality is now receiving some research attention at the farm and catchment scales[57] (see also Martin and Halpin, 1998; Wilson, 2004). Thus, the existence of debates in institutions and countries beyond the EU emphasises that discourses of multifunctional agriculture have not been

---

[57] This includes, for example, a project funded by the Australian Research Council (2005-2007), involving researchers from the School of Geography at Monash University (Melbourne, Australia) and myself, investigating challenges and opportunities for the 'multifunctional countryside' in the state of Victoria.

restricted to the 'European model' of multifunctionality. Nonetheless, the OECD (2001: 10) acknowledged that its member countries "have fundamentally different opinions and positions concerning the definition of multifunctionality, its utility for the agricultural policy debate, and its implications for policy reform". While EU policy-makers are struggling to find common ground over the meaning and purpose of multifunctionality, there is even less consistency in conceptions of multifunctionality beyond Europe. There is one key explanation for the relative reluctance of many New World countries to adopt the notion of multifunctionality. Freshwater (2002) emphasised that the USA has long used the notion of multifunctionality as a tool to manage *public* lands (e.g. in the state forestry sector; see Section 8.2), but that US policy-makers have shied away from applying it to *private* land. The same is true in Australia and New Zealand, where the willingness to adopt notions of multifunctionality has been more closely associated with state-owned lands rather than privately owned farmland. As Wilson and Memon (2005) highlighted, this is largely because in these former New World 'settler societies' ownership of private land was the ultimate goal to gain 'freedom' from the shackles of the Old World. As a result, policy-makers have been reluctant to 'impose' policy frameworks – be they multifunctional or not – that threaten the sanctity of private land and the associated freedom of decision-making for owner occupiers (Wilson, 1993).

A final, but equally important, point relates to the virtual absence of contemporary debates on multifunctionality in a *developing countries* context (see Pretty, 2002, and Wilson and Rigg, 2003, for notable exceptions). In many ways, and similar to debates on 'post-productivism' (see Chapter 7), this reinforces the above notion that multifunctionality is still largely seen as a 'European project'. In Chapter 10, I particularly wish to return to the issue of underrepresentation of developing countries in multifunctionality discourses, and I will propose a framework for understanding multifunctionality applicable in both developed *and* developing countries.

8.5 Multifunctionality and (the lack of) theory

The above discussion has highlighted that, although there are some new and interesting emerging interpretations of multifunctional agriculture, many debates have continued to focus on economistic, policy-based and structuralist interpretations that place great emphasis on the political economy as a key driver of multifunctional transitions. Based on our discussion in Part 2, the reader will recognise here the part-legacy of structuralist interpretations that dominated conceptualisations of agricultural and rural change during the 1990s. It is only in the last few years that wider issues linked to culture, the consumption countryside, and society's changing views of agriculture have been incorporated into conceptualisations of multifunctionality (e.g. Marsden, 2003; Holmes, 2006), while discussions of the importance of grassroots actors' multifunctional identities and aspirations have only recently become the focus of attention (e.g. Burton and Wilson, 2006). In my view, this leaves a relative gap in our understanding of what multifunctional agriculture is, or should be, about. In particular, the above debates have been poorly interlinked with theorisations of the p/pp transition discussed in Part 2 – a body of theory that has greatly influenced academic discourses on agricultural change throughout the 1990s. In addition, the predominance of structuralist interpretations of multifunctional agriculture – especially through the lens of public goods and policy – have led to a relatively narrow view that focuses more on multifunctionality as a *process* rather than as a *concept* that describes *and* explains contemporary agricultural change. This process-based view also neglects the role of agency (as with conceptualisations of the p/pp transition) and other less tangible (and, therefore, less measurable) 'indicators' of change linked to social and cultural capital, such as changing agriculture-society relations or shifts in views about agriculture. In this final section, I wish to explore these issues in greater detail, first by analysing the effects that the early appropriation of the notion of multifunctionality by policy-makers has had for conceptualisations of multifunctionality (Section 8.5.1) and, second, by arguing that this has resulted in weak theorisation underpinning multifunctionality debates (Section 8.5.2).

8.5.1 The early appropriation of multifunctionality by policy-makers

As highlighted above, multifunctional agriculture discourses emerged *synchronously* within both policy-based and academic discourses, a process referred to by McCarthy (2005) as the 'institutionalisation of multifunctionality' and by Garzon (2005) as the 'exogenous character' of multifunctionality. I argue that this has possibly negated the urgency for thorough *scientific theorisation* of the term due to its early 'acceptance' by policy-making circles as a notion with seemingly 'real' policy relevance. In other words, we have witnessed a process where – at least in EU and WTO frameworks – policy-makers have appropriated the notion of multifunctionality early on in the 1990s as a *mechanism* to ward off external pressures for a complete reform of the CAP (Lowe *et al.*, 2002; Potter and Burney, 2002). Thus, Garzon (2005: 2; emphasis added) suggested that "academic discourse has been *influenced* by political discourse rather than influenced it and … the approach taken by international organizations [i.e. WTO] tends to evacuate the very reasons of the emergence of the multifunctionality concept". She further argued that the result was that academic discourse on multifunctionality "appeared after the first policy changes had already generated an active public policy debate" (Garzon, 2005: 18) and that, therefore, "publications [on multifunctionality] followed the policy debate rather than influenced it" (Garzon, 2005: 2).

Durand and Van Huylenbroek (2003: 13) argued that multifunctionality could be seen as a "pretext to defend protectionist policies", and that "the EU has adopted the concept of multifunctional land use as a central principle to legitimate further support of agriculture" (Durand and Van Huylenbroek, 2003: 1). They suggested that, at the international level, multifunctionality has become the topic of negotiation whenever there is a need to reach an agreement about the application of policy tools accepted by all parties (i.e. Hollander's, 2004, notion of 'weak' multifunctionality). Similarly, Clark (2003: 12) suggested that the "inherent ambiguity [of multifunctionality] has enabled its promotion across the EU to suit the different political priorities of Member States". Clark further argued that one of the most important attributes of the multifunctionality concept is precisely its *malleability* in the face of policy considerations, highlighting that the concept has been reconfigured over the 1990s to accommodate alterations in EU political priorities. In a similar vein, Potter and Tilzey (2005: 12) suggested that "the concept of multifunctionality has been recruited to legitimise continued state support", and that "while counter-narratives are available which argue for a continuing social and environmental justification for state support, these have been over-written, and to some extent *appropriated*, in the course of recent policy debates by those seeking to defend policy entitlements in more regressive neo-mercantilist terms" (Potter and Tilzey, 2005: 16; emphasis added). This has been further reiterated by Clark (2003: 285; emphasis added) who suggested that "the dominant legacy of agro-industrial production logics might have resulted in the *appropriation* of multifunctionality to serve existing organisational agendas", further supported by Delgado *et al.* (2003) who argued that the notion of multifunctionality was initially used by the EU as a simple *ideological pretext* in the WTO Millennium Round negotiations. Buller (2005), therefore, posited that the EU has gone to enormous lengths to create, both domestically and internationally, the 'camouflage' of multifunctionality to justify the continuation of farm subsidies, while Losch (2004: 353) saw multifunctionality largely as a conceptual innovation of protectionist rhetoric because "certain industrialized countries invoke multifunctionality to justify their choice of policies favouring their own farmers". Garzon (2005) referred to these processes as the 'tactical positioning' of the notion of multifunctionality for policy-related ends – a positioning that has severely weakened the conceptual utility of the term.

As we saw in Section 8.4, similar tactical positioning has been identified for the macro-politics of multifunctionality. Indeed, the most tangible evidence for (mis)appropriation of the notion of multifunctionality is linked to recent negotiations in WTO negotiation rounds. As the definition of the notion of multifunctionality was not finalised during negotiations of Agenda 2000 (see above), this confirms to some authors that multifunctionality was never aimed to orient Agenda 2000, but was used as a *tool* for the European negotiating strategy with the WTO to gain sufficient time for a new reform of the CAP (Potter and Burney, 2002). Thus, although multifunctionality is now a controversial and somewhat discredited term in WTO negotiations (for reasons highlighted above), Potter and Tilzey (2005: 14) highlighted that it "retains considerable discursive resonance in [EU] Member States". Potter and Tilzey, thus, suggested that European policy-makers have continued to qualify their support for market liberalisation with the need to retain a multifunctional agriculture, translating this need into a *bi-modal* approach to CAP reform that moves to *decouple* support from production on the one hand, while *recoupling* it to agri-environmental and rural development on the other. For WTO agriculture trade talks, therefore, Potter and Burney (2002) highlighted that the notion of multifunctionality has been used as a *smokescreen* for the continuation of protectionist agricultural policies. Garzon (2005: 1), thus, argued that the notion of multifunctionality "seems to have evolved into a central organizing domestic political concept [within the EU] for the purpose of conducting reforms", and that "the concept of multifunctionality was the best synthesis found by European policy makers to respond to the urgency to find a renewed legitimacy of agricultural policy" (Garzon, 2005: 16). As highlighted in Section 8.4, this may have been exacerbated by the fact that powerful countries in the WTO, such as the USA, relegated the notion of 'multifunctional agriculture' to EU *policy* discourses rather than seeing multifunctionality as a *concept* inherently linked to agricultural change. Indeed, we saw that countries in the amber box of the URAA (i.e. most EU countries) have most vehemently lobbied for the promotion of multifunctional policies to address possible shortfalls in productivist subsidies (Bohman *et al.*, 1999; Potter and Burney, 2002). This emphasises that the vague notion of multifunctionality has been (mis)appropriated to plug the threat emanating from gradual withdrawal of EU agricultural subsidies based on external pressure from the WTO (Potter and Burney, 2002). Indeed, total liberalisation of agricultural trade – as envisaged by the WTO – may represent a "threat to the maintenance of the multifunctional character and sustainability of [EU] agriculture" (Durand and Van Huylenbroek, 2003: 2).

This early appropriation of the notion of 'multifunctionality' by policy-makers to suit narrow policy objectives has strengthened *policy-based* conceptualisations of the term. Multifunctionality, therefore, has been used as an *excuse* by policy-makers to continue backdoor subsidisation of farmers without having to refer to the link between production and subsidies criticised by the WTO. This led Hollander (2004: 311) to criticise the notion of multifunctionality embedded in macro-political rhetoric, as "by attempting to pull into a common framework issues such as food security in developing countries and the preservation of stone fences in rural Scotland, the term begins to lose coherence". Gallardo *et al.* (2003: 172; emphasis added), therefore, argued that "understanding multifunctionality in this way is no more than interpreting and using this concept in a *limited* way". As the following will show, the use of multifunctionality as a political smokescreen has left a conceptual vacuum to the detriment of deeper theorisation of the term.

8.5.2 The weakly theorised nature of multifunctionality debates

It is evident that the predominance of policy-driven discourses has greatly influenced conceptualisations of the notion of multifunctional agriculture. This *politicisation* of multifunctionality has led to a theoretically weak concept of multifunctionality, aptly referred to as the "invented discourse of agricultural multifunctionality" by Potter and Tilzey (2005: 1). Thus, Garzon (2005: 1) highlighted that "contrary to previous experiences of policy change, the conceptualisation process [of the notion of multifunctionality] does not stem from academic or experts arena. The concept belongs to the process of policy exchange and change where it finds its roots". Holmes (2006: 145), therefore, lamented that "it would be unfortunate if such a useful, indeed pivotal, term [multifunctionality] should acquire restricted usage through the preemptive discourse of policy-makers and advisers in the EU and OECD". Similarly, French rural researchers also criticised that "the word multifunctionality has many different realities ... but it is often used as if its meaning does not require further questioning", and that "clarification of this concept of multifunctionality, at the intersection between different disciplines, therefore, remains an important challenge for researchers" (Lardon *et al.*, 2004: 15; my translation) – a proposition that partly forms the justification for the reconceptualisation of multifunctionality presented in Chapters 9 and 10.

The discussion in this chapter has shown that a policy-led notion of multifunctionality has resulted in a relatively narrowly circumscribed concept. As a result, the largely European-based notion of multifunctionality has, so far, not required a holistic, broad-based and theoretically well informed vision (Garzon, 2005; Holmes, 2006). Instead, multifunctionality has largely been relegated as a short-term and pragmatic tool and mechanism for shallow policy decision-making – in other words, a *quick-fix notion* that lives an uneasy existence between political practicality and academic unease as a term that has not yet convinced at the conceptual and theoretical level (see also Mather *et al.*, 2006).[58] This is particularly the case as most existing conceptualisations have seen multifunctionality merely as an economic or policy-based *process* rather than an overarching *concept* that both describes *and* explains contemporary agricultural change (see Clark, 2003, and Holmes, 2006, for notable exceptions). Thus, Gallardo *et al.* (2003: 172) suggested that one of the main shortfalls of the term is because "the treatment of multifunctionality and the adoption of support measures are significantly different, depending upon whether multifunctionality is considered to be a characteristic of agriculture [or] ... an objective of agricultural policy". These early appropriations of the notion of multifunctionality by policy-makers and economists may have led to *retrospective* and *partial* academic conceptualisations of multifunctional agriculture. Buller (2005: 19) suggested that "this is understandable ... given the pragmatic focus upon evidence-based assessments that underlies ... evaluation criteria as 'value for money' and efficiency" (see also Wilson and Buller, 2001). Further, as Section 8.3 highlighted, contemporary conceptualisations of multifunctionality have also been marred by sectoral theoretical interests within relatively narrow disciplinary confines and by the predominance of structuralist political economy interpretations that have largely neglected post-structuralist and 'cultural-turn'-related emphases on grassroots stakeholder actions and thought. Although multifunctional agriculture has been approached from various disciplinary and conceptual angles, it continues to suffer from relatively weak theorisation, and no holistic assessment of multifunctionality has yet acknowledged, let alone analysed, the interlinkages between the different conceptual strands of the concept (Goodman, 2004).

There have, nonetheless, been some attempts at linking the notion of multifunctional agriculture to wider theoretical questions, including, in particular, work by Potter and Burney (2002), Clark (2003), Marsden (2003), Knickel *et al.* (2004), Potter and Tilzey (2005), McCarthy (2005) and Holmes (2006). A key example here is Potter and Burney's (2002) attempt to conceptualise the emergence of multifunctionality in the context of the post-Fordist transition in agriculture. Yet, based on our discussion in Part 1 of this book, such a theoretical linkage is problematic, as it suffers from the same assumptions about linearity and spatial homogeneity we encountered when deconstructing the postulated p/pp transition. Nonetheless, Potter and Burney's work offers one of the few promising theoretical windows into a conceptual world that lies beyond the narrow confines of economistic and policy-based

---

[58] It is interesting to note that such theoretical weaknesses have also been highlighted in other multifunctionality debates, with urban theorists, for example, suggesting that 'multifunctional urban land use' is still largely 'an empirical phenomenon' and not one that is yet theoretically well grounded (Rodenburg and Nijkamp, 2004: 274).

interpretations of multifunctionality (see above), and their contribution has been key for engendering further urgency for more thorough conceptualisations of multifunctional agriculture (see also Holmes, 2006) – an issue that Chapters 9 and 10 will investigate in more detail. Yet, this part of the multifunctionality debate is currently only in its infancy, and a major justification for the argument in this book is that the possible linkages between theorisations of multifunctional agriculture and other societal transitional processes are, so far, poorly understood and developed. This is particularly true for the wealth of information on social and cultural capital at grassroots level that has, so far, not been extensively acknowledged in existing conceptualisations of multifunctionality.

In this book, therefore, I attempt to move beyond existing structuralist conceptualisations of multifunctional agriculture by acknowledging the multi-faceted dimensions of the debates. As Chapters 9 and 10 will discuss, I will suggest that the notion of multifunctional agriculture can only be fully understood by taking into account existing debates and theorisations of agricultural change and transition. In particular, by building on the multiple discourses of multifunctionality highlighted in Chapter 8, I will highlight that only by theoretically anchoring the concept of multifunctional agriculture in the revised p/np spectrum of agricultural decision-making outlined in Chapter 7 will it be possible to provide a holistic understanding of what multifunctionality should be about.

8.6 Conclusions

This chapter has provided a synthesis of current understandings of 'multifunctional agriculture'. I discussed the genesis of the notion of 'multifunctionality' and highlighted that the term has been applied to many different arenas from several disciplinary vantage points. This showed that multifunctionality has only relatively recently (i.e. from the early 1990s) been appropriated by agricultural and rural researchers and decision-makers as a key concept both for policy-making and for describing the complexity of contemporary agricultural change. Current understandings of multifunctionality have been relatively reductionist and based largely on narrow *economic* and *policy-based* approaches predicated on structuralist interpretations of agricultural and rural change that have only provided partial answers. In addition, many contributions only make implicit allusions to multifunctionality, while only few have engaged with in-depth critical analysis of the concept, let alone attempted to (re)conceptualise what 'multifunctional agriculture' could mean.

In recent years a more holistic view of multifunctionality has emerged that places more emphasis on the interlinkages of the concept with rural development, culture, the consumption countryside, societal needs, agency-led patterns and processes of agricultural and rural change, as well as environmental issues. As Freshwater (2002: 16) emphasised, "as the interest [in the notion of multifunctionality] increases there is greater chance that more of the basic elements that define multifunctionality will become more relevant". Yet, the relatively narrow structuralist interpretations of multifunctionality have suffered from *discursive insularity* that has confused rather than clarified what multifunctionality could be about. As Clark (2006: 334) emphasised, multifunctionality has largely remained a "high-level discursive construct" that has been used, abused and (mis)appropriated by policy-makers (Bohman *et al.*, 1999; Garzon, 2005). As a result, the notion of multifunctional agriculture is still largely understood as a policy-led *process describing* current agricultural trends, rather than as a *normative concept explaining* agricultural change.

Problems surrounding contemporary notions of multifunctionality also relate to 'cultural' and 'territorial' questions. The Euro-centric notion of multifunctionality, interpreted by many as a 'smokescreen' to defend the continuing productivist subsidy culture of European agriculture, has emerged as a particularly problematic issue. Many have questioned the applicability of the term beyond Europe, because multifunctionality is perceived by many global commentators as a 'European policy project' with little relevance to non-European (especially non-EU) regions. Yet, problems surrounding current conceptualisations of multifunctionality are also linked to the early (mis)appropriation of the term by European policy-makers eager to use it as a tool for *shallow* policy decision-making (or in Hollander's, 2004, term 'weak multifunctionality' contexts) to further productivist subsidy-oriented CAP goals in light of mounting WTO pressures. This gave early 'legitimacy' to the notion of multifunctionality before it had been thoroughly conceptualised at academic level.

Many authors have also suggested that agricultural multifunctionality is a *new* process, i.e. something that did not exist before popularisation of the term by policy-makers. The result has been a notion of multifunctionality largely predicated on its relevance for policy-based decision-making, characterised largely by short-termism, weak theorisation and pragmatism. McCarthy (2005: 778), therefore, argued that "most authors seem to accept that articulations of and struggles over multifunctionality all turn on the rapid revaluation of rural natures in the context of trade liberalization during the neoliberal area". Similarly, Freshwater (2002: 17) suggested that with reference to the linkages between multifunctionality and social capital we need "to recognize that the concept is more than a device for creating trade barriers and that it is a way to think about balancing the multiple outputs of agriculture in a way that increases aggregate social welfare". Ironically, this gap in current conceptualisations may have, in turn, lessened the need for thorough conceptual and theoretical investigations of what multifunctional agriculture could mean, as the term was already part of established 'mainstream' political and economic discourses by the early 1990s. This has been much in contrast with the comparable notion of 'sustainability', first theorised at academic level during the 1980s and then gradually appropriated by policy-makers from the 1990s onwards *after* it had been exposed to thorough academic debate. The result of this anachronistic process has been an understanding of multifunctionality that has not yet gelled into a *coherent workable framework* for fully *understanding* contemporary agricultural and rural change.

These shortfalls highlight that we urgently need a *new* concept of multifunctionality that is conceptually and theoretically better anchored in current debates on agricultural change. In particular, we need a model that draws on existing *holistic* debates of multifunctionality (see Section 8.3.4), that goes beyond mere economic and policy-based understandings, and that is applicable not only in a European context but also *globally* for explaining agricultural change in *any* agricultural area. This chapter, thus, has highlighted that there is an urgent need to adopt a more *inclusive* concept of multifunctionality that will also find applicability in a developing areas context. Academic and scientific debate needs to re-appropriate the notion of multifunctionality from policy-makers, expose it to thorough theoretical and conceptual analysis which has, so far, been lacking, and reconceptualise it into a *normative* concept that can be used to *explain* what is happening in the countryside. This critique forms the justification for the revised model of multifunctionality anchored in notions of productivism and non-productivism discussed in Chapters 9 and 10.

Chapter 9

(Re)conceptualising multifunctionality

9.1 Introduction

Chapter 8 highlighted that some useful discussions have addressed the concept of 'multifunctionality', but that none of these debates has shed sufficient light on *what* the notion of multifunctionality implies, *who* the beneficiaries should be and *how* it ought to be implemented into practice – in other words, the notion of multifunctionality has remained undertheorised, process-oriented and poorly linked to wider debates in the social sciences. The aim of this book is to conceptualise multifunctionality in the context of debates on the postulated shift from productivism to post-productivism. Parts 1 and 2 showed that applying transition theory to better understand the p/pp transition has highlighted that little evidence can be found for a wholehearted shift towards post-productivist action and thought. Instead, Chapter 7 suggested that we should keep the notion of 'productivism' as a useful notion characterised by relatively well defined indicators and parameters, but that we should replace the notion of 'post-productivism' with *non-productivism* as the 'true' opposite of productivism. I further argued that productivist and non-productivist action and thought occur simultaneously, both temporally and spatially, and that they form the extreme ends of a *spectrum* of agricultural decision-making.

In this chapter I will suggest that the concept and territory of multifunctionality is situated between these extreme ends of the decision-making spectrum – in other words, that the notion of multifunctionality can be *theoretically anchored* in a revised p/np spectrum of decision-making. This new conceptualisation of *multifunctional agriculture* is, inevitably, complex, and will rely on a series of conceptual and logical steps that will gradually unravel the different components of multifunctionality. To set the wider framework, Section 9.2 will briefly outline the link between multifunctionality, transition theory and the p/np boundaries of decision-making. Section 9.3 will highlight that we can use this new conceptualisation of multifunctionality to *define* 'agriculture' itself. The normative issue of different *types of multifunctionalities* ('weak', 'moderate' and 'strong') is discussed in Section 9.4, with particular reference to the fact that assessing multifunctionality inevitably implies value judgements. Section 9.5 will then discuss issues of *scale* and multifunctionality by investigating at what spatial level multifunctionality should apply and why. Here, I will particularly touch upon ideas of 'nested hierarchies' and the territorial importance of multifunctional decision-making. Concluding comments will be given in Section 9.6.

9.2 Multifunctionality and the productivist/non-productivist boundaries of decision-making

9.2.1 Reconceptualising multifunctionality

The application of transition theory has allowed us to reconceptualise the p/pp transition into a revised model characterised by agricultural and rural decision-making bounded by the extreme pathways of 'productivism' and 'non-productivism' (the p/np spectrum; see Section 7.4). I suggest here that the conceptual territory of *multifunctionality* can be theoretically anchored in this p/np model and that ***multifunctionality occupies the (conceptual and real) space between the extreme pathways of productivism and non-productivism*** (Fig. 9.1.). I will refer to this as the *multifunctionality spectrum*. This spectrum should be seen as comprising multiple *Deleuzian* transitional pathways open to various actors and stakeholder groups at various scales, or, as Dear (1986: 373) argued, as "an infinity of overlapping realities ... [characterised by] the simultaneous appearance of two [or more] objects at the same chronological moment".

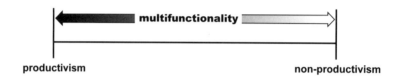

productivism   **non-productivism**

Fig. 9.1. Multifunctionality and the productivism/non-productivism spectrum of decision-making

Let us briefly consider how the extreme ends of the productivist/non-productivist spectrum shown in Figure 9.1 can be defined, bearing in mind Marsden's (2003) cautionary note that new conceptualisations of agricultural/rural change need to be based on *integrating local* and *external* processes and on the recognition of the importance of local/non-local network configurations. As highlighted in Chapter 7, defining the productivist end of the spectrum is relatively unambiguous and is largely linked to the seven dimensions of productivism discussed in Chapter 5. Based on Chapter 7, and partly building on Knickel *et al.*'s (2004) synthesis of 'multifunctionality indicators', the non-productivist end of the spectrum can be characterised by eight inter-related themes. These are closely associated with conceptualisations of post-productivism but in a temporally and spatially non-linear sense (see also Clark, 2003). They are also building on Durand and Van Huylenbroek's (2003) suggestion that multifunctional agriculture includes three key functions linked to space (stewardship, landscape, environment), production (food security, diversity) and service functions (maintenance of rural areas, biodiversity, rural development); on Holmes' (2006) seven modes of 'multifunctional rural occupance' based on a mix of production, consumption and protection goals; and on 'elements of multifunctionality' used as a conceptual framework in the EU-funded Multagri project (Marsden and Sonnino, 2005).

First, high environmental sustainability plays a key role in the multifunctional territory close to the non-productivist pathway boundary (and conversely low environmental sustainability for farms/rural actors located close to the productivist end of the spectrum). Second, actors close to the non-productivist end of the spectrum will show a tendency for local embeddedness, in particular through horizontally integrated rural/farming communities with close interaction between local communities through reciprocal rural-agricultural relationships (cf. Goodman, 2004). Meert *et al.* (2005) described such embeddedness as the presence of closely related family in the neighbourhood, a number of other befriended farm households, good relationships with direct neighbours (also non-farm), and participation of the farmer/farm couple in local social life and associations. Third, the non-productivist end of the spectrum is characterised by short food chains and, fourth, has a tendency for low farming intensity and productivity. Fifth, agricultural actors on non-productivist development pathways will show high degrees of diversification, in particular away from traditional productivist food and fibre production. Sixth, in a conventional Marxist interpretation, non-productivist action and thought is also more likely to be weakly integrated into the global capitalist market (cf. McMichael, 1995; Goodman and Watts, 1997). Seventh – and in line with Boyer and Saillard (2002) and Knickel *et al.* (2004) who suggested that arguments, ideas and ideology are important during societal transitions – non-productivism is also characterised by attitudinal changes, in particular through open-minded farming and rural populations who see 'farming' and 'agriculture' as processes that go beyond productivist food and fibre production, mass production and profit maximisation. Finally, the non-productivist end of the spectrum is closely associated with open-minded societies who accept that the nature of 'farming' and 'agriculture' is in the process of change.

As these themes highlight, multifunctionality is about *multiple functions* of agricultural land use, practices and thoughts, as well as about how such agricultural processes influence rural areas and communities and societal conceptions of farming and agriculture (see also Clark, 2003; Van der Ploeg and Roep, 2003). In line with Knickel and Renting (2000) and De Groot (2006), multifunctionality is the antithesis of *monofunctionality* which is closely associated with the extreme productivist end of the multifunctionality spectrum in which agriculture is seen to serve relatively 'narrow' functions of food and fibre production (see Chapter 5). As Pretty (2002) suggested, multifunctionality should be seen to create 'diverscapes' as opposed to 'monoscapes'. Section 9.4 will explore the defining characteristics of the multifunctionality spectrum in more detail in the context of a proposed conceptualisation of 'weak', 'moderate' and 'strong' multifunctionality.

I, therefore, agree with Holmes (2006: 143) who argued that "multifunctionality rather than post-productivism is the central dynamic driving rural change", and who suggested several 'occupance modes' in the Australian context that implicitly refer to a spectrum of multifunctional processes[59] where "within each of these modes, markedly divergent resource-use trajectories can be delineated" (Holmes, 2006: 146). Similarly, Marsden (2003: 7) suggested that "the evolution of new alternative food governance systems is not as simple as suggesting that there is a binary contradiction between the industrial [i.e. productivist] and the 'alternative' [i.e. non-productivist] food system", and that such more holistic conceptualisations "take us significantly beyond the dualistic assumptions of rural land development associated with either an agricultural or non-agricultural focus" (Marsden, 2003: 142). The multifunctionality spectrum also concurs with Potter's (2004: 29) suggestion that "an alternative definition of the multifunctional countryside is one in which productivist and post-productivist [read 'non-productivist'] functions are delivered side by side through an enforced segregation of rural space" (see also Wilson, 2001, 2002). Contrary to most conceptualisations of multifunctionality as a *static state* discussed in Chapter 8, this new view of multifunctionality sees it as a *dynamic transitional process* occurring within the boundaries of the productivist and non-productivist decision-making spectrum – a process that can be influenced, shaped and, as Chapter 10 will discuss, orchestrated in different ways.

As with any model attempting to understand complex processes behind human action and thought, I acknowledge that the notion of a multifunctionality spectrum can be criticised as

---

[59] Holmes' (2006) occupance mode spectrum includes a productivist mode, a rural amenity mode, a pluriactivity mode, a peri-metropolitan mode, a marginalised agricultural mode, a conservation mode and an indigenous mode (see also Section 9.4 below).

yet another reductionist attempt at conceptualising complexity. Yet, as Holmes (2006: 157) succinctly argued, "although incomplete and capable of encompassing only one dimension in multidimensional contexts, the spectrum analogy, applied to the relative power or importance of two contending interests [i.e. productivism and non-productivism], can at times provide a useful, partial appraisal of appropriate (or probable) outcomes". The utility of such a spectrum has also been acknowledged by Potter and Tilzey (2005: 15) who conceded that the notion of multifunctional agriculture based on a bi-modal spectrum "does appear to be a defining feature of the neo-liberal, post-Fordist regime which we believe better describes the emerging institutional and policy landscapes of European agriculture" (see also Guillaumin *et al.*, 2004). Similarly, Luttik and Van der Ploeg (2004) argued that multifunctionality should be seen as a *continuum*, not a dichotomy, and there are also parallels with discussions of the 'hybridity' of societal transitions discussed in Part 1 (e.g. Whatmore, 2002; Nederveen Pieterse, 2004), further emphasised by French commentators Barthélémy *et al.* (2004: 124; my translation) who argued that "multifunctionality forms a hybrid entity". A spectrum, therefore, best encapsulates the embedded hybridity in human decision-making processes and pathways. Contrary to the binary notions of productivism and post-productivism, the notion of a spectrum enables us to situate and understand *multiple* actions and processes simultaneously in a non-linear and spatially heterogeneous way, without losing sight of transitional processes affecting agricultural pathways.

The underlying assumption of this new understanding of multifunctionality is that *almost any* agricultural/rural action and thought will contain elements of *both* productivism and non-productivism – in other words few actions, thoughts and development pathways can be described as solely either productivist or non-productivist. Thus, Walford (2003b, 501) emphasised the ability of many farms to "follow contrasting development pathways simultaneously", while Knickel *et al.* (2004: 83; emphasis added) argued that "it is only a small proportion of all farm households that is characterised by a *monofunctional* pattern of resource use and income generation". Indeed, Vatn (2002) emphasised that conceptualisations of multifunctionality should not lose sight of continuing productivist goals linked to food security and safety issues that will always be at the forefront of human needs. I, therefore, agree with Swagemakers (2003: 189) who suggested that "agriculture has always been multifunctional"; with Belletti *et al.* (2003a: 59) who argued that "all farms have some degree of multifunctionality"; and with Knickel *et al.* (2004) who suggested that most farms in Europe are multifunctional. Marsden (2003: 101) further emphasised that rural space is increasingly conceived of as "a highly elastic phenomenon, constructed out of combinations and layers of social, political and economic relations, traversing different physical spaces at any one time". He argued that we need to conceive of differentiating agricultural and rural spaces that are caught up in different webs of local, regional, national and international supply chains, networks and regulatory dynamics. Further, McCarthy (2005: 774) emphasised that the notion of multifunctionality is also "inherently sensitive to spatial and social differentiation, the fact that different rural areas clearly can and will produce very different, even unique, combinations of use values", while Walford (2003b: 496) emphasised that the "multifunctional agricultural regime may be represented by farms' differential positioning with respect to … separate dimensions of change".

In Chapter 7, I referred to this as the *territorialisation* of productivist and non-productivist actor spaces. I highlighted that there are overlaps with various conceptualisations of agricultural change, including Ward's (1993) notion of the 'two-track countryside', the 'agrarianism-environmentalism' spectrum by Buller (1998), the 'production-consumption spectrum' by Marsden (1996), or Marsden's (2003) notion of the 'bifurcation' of farmer roles between production and consumption spaces. Similarly, Potter and Tilzey's (2005) notion of a 'bi-modal productivist/post-productivist policy strategy' implicitly assumes a spectrum of decision-making choices, as does the 'super-productivism-rural idyll spectrum' by Halfacree (1999), the 'intensification-extensification spectrum' by Caraveli (2000), Potter's (2004) suggestion of an increasing integration of production (i.e.

productivism) and consumption (i.e. non-productivism) spaces, or Delgado *et al.*'s (2003) notion of the 'dualism of European agriculture' (see also Goodman, 2004). There are also conceptual overlaps with the notion of 'multifunctionality as an integrative concept' suggested by Knickel *et al.* (2004), the notion of 'joint production' discussed in Chapter 8 (see Nowicki, 2004; Winter and Morris, 2004), and with Lang and Heasman's (2004) concept of 'food wars' that are fought at the interface between productivist forces and the 'ecologically integrated' paradigm (the latter echoing conceptualisations of non-productivism). Similarly, there are overlaps with Lowe *et al.*'s (2002) notion of multifunctionality as the 'third way', and with Van der Ploeg and Roep's (2003) concept of the three-dimensional farm in which the arenas of the 'rural area', the 'mobilisation of resources' and the 'agro-food supply chain' lead to a complex positioning and repositioning of agricultural and rural actors along a spectrum of decision-making opportunities. Finally, synergies can also be found with the 'competitive/non-competitive farm spectrum' suggested by Gallardo *et al.* (2003) in which non-competitive farms can be seen to be situated closer to the non-productivist end of the decision-making spectrum (see also Chapter 10).

Yet, the spectrum of multifunctional decision-making suggested here goes well beyond mere issues of *processes* and *production*. In my view, multifunctionality should mean that elements of *both* productivism and non-productivism will *at any time and in any space* influence action on the ground that will result in multiple development pathways for agricultural and rural areas along a spectrum of different decision-making opportunities (Murdoch and Marsden, 1995). Holmes (2006: 145), therefore, rightly argued that "multifunctionality is increasingly recognised as a characteristic of all rural holdings, even those outwardly in pursuit of monofunctional production or consumption goals". Similarly, Van der Ploeg (2000) argued that what is vaguely referred to as 'normal agriculture' should also be situated within models that acknowledge the multifunctional contribution of such systems. Marsden and Sonnino (2005) referred to this as 'bifurcation' of multifunctional spaces where farmers reconnect with their conventional markets on the one hand, and become part of 'alternative' food supply chains on the other. I, therefore, agree with Clark (2003: 247; emphasis added) that "there is ... no *single* [farm] business characteristic that could be ... identified as definitely contributing to the initiation and sustenance of multifunctional business practice", and that "the complex multifaceted notion of multifunctionality is effectively outside any single organisation's competence" (Clark, 2003: 296). I also agree with Goodman's (2004: 12) suggestion that "analytically, this re-labelling of long-standing or 'old' practices shifts the centre of gravity away from 'vanguardism', 'rupture' and paradigm change towards continuity and incrementalism". Indeed, as the discussion below will show, it is this complex combination of *multiple* old and new factors that make up multifunctional agriculture and that influence multifunctional agricultural pathways. Parallels can be drawn with Storper's (1997: 255; emphasis added) view of the societal transitional 'puzzle' where "the modern economy can ... be conceived as a complex organisational puzzle, consisting of *multiple* and *partially overlapping* worlds".

While the notion of post-productivism implied a *directionality* of action and thought towards a specific goal (i.e. that *all* actors aim at moving towards post-productivism), the multifunctionality spectrum allows for multidimensional coexistence of *both* productivist and non-productivist action and thought and is, in my view, a more accurate depiction of the multi-layered nature of past, present and future rural and agricultural change. Holmes (2006: 145) also argued that "the term MRT [multifunctional rural transition] is to be preferred over the descriptor PPT [post-productivist transition], not only because of its accuracy, conciseness and generality in conceptualizing the current rural transition, but it avoids the historicist myopia attached to the prefix 'post'". The multifunctionality spectrum allows, for example, an understanding of the synchronous coexistence of pressures to maintain productivist competitiveness of European agriculture on the one hand, and the consolidation of an agricultural model based more on non-productivist principles on the other. Inevitably, this conceptualisation of multifunctionality challenges existing understandings of multifunctional agriculture that see it merely as an economic or policy-based *process* (see Section 9.2.2 below). Instead, the notion of multifunctionality suggested here sees it as an overarching *concept* that both describes *and* explains contemporary agricultural change.

As relativists tell us, there are 'multiple realities' – an issue not sufficiently recognised by those advocating that a post-productivist transition has taken place. As Evans *et al.* (2002: 327) emphasised, "farming continues to be dominated by production and rural space remains devoted to agricultural production" – in other words, productivist and non-productivist spaces of decision-making lie side-by-side. This echoes debates on 'hybridity' (see Chapter 4), where authors such as Nederveen Pieterse (2004: 67) argued that "hybridisation and the mélange of diverse modes of organisation give rise to a pluralisation of forms of cooperation and competition as well as to novel mixed forms of cooperation" – patterns not dissimilar to those that occur within the multifunctionality spectrum (see also Whatmore, 2002). Holmes (2002: 381; emphasis added), in his critique of the p/pp model, similarly argued that "an understanding of the post-productivist transition in any context, whether Western Europe or the Australian rangelands, requires systematic scrutiny of the [key] driving forces, recognizing *variability* in origins, impulses, processes, actors, ideologies and outcomes, leading to more *heterogeneous, multifunctional* rural regions" (see also Smailes, 2002; Winter, 2005). Knickel and Renting (2000) also highlighted that rural development consists of a wide variety of new *multidimensional* and *integrated* activities which fulfil several functions not just for the farm, but also for the region and society as a whole, while Hoggart and Paniagua (2001: 42; emphasis added) argued that in agricultural/rural research we need an "appreciation of the *multiplicity* of causal forces that impact on change forces, with recognition that these forces can *coalesce* in dissimilar ways in different places".

Actors in the multifunctional agricultural regime could be seen to be imbued with a sense of reflexivity in that – similar to notions of reflexive modernisation (Beck, 1992; Ray, 1998) – there may be a recognition (and acceptance?) of actors about their relative 'location' in the p/np spectrum. In other words, multifunctional agriculture may be a regime where the boundaries between the different dynamics (Marsden, 2003) or rural structured coherences (Cloke and Goodwin, 1992) may become more distinct, thereby allowing for productivist action and thought to co-exist alongside non-productivist modes of agricultural and rural change. This new conceptualisation of multifunctionality, therefore, addresses Marsden's (1999a: 504) call for a "more integrated, holistic and spatial rather than sectoral approach to understanding rural transformation".

9.2.2 How does the multifunctionality spectrum differ from other conceptualisations of multifunctionality?

It is evident that this new conceptualisation of multifunctionality both differs from, and builds on, other conceptualisations discussed in Chapter 8. We saw that only limited communication exists between the different strands of multifunctionality conceptualisations and that the many isolated discourses have not yet gelled into a coherent whole. The multifunctionality spectrum suggested here *incorporates* many of the different approaches discussed in Chapter 8 in a holistic way, and enables the positioning of any policies, economic processes, consumption spaces and rural development processes within a wide array of decision-making opportunities bounded by productivist and non-productivist action and thought. For example, it builds on the OECD (2001) definition of multifunctional agriculture that argued that the term should be interpreted as a characteristic of agriculture as an economic activity which produces multiple and interconnected effects – effects that may be positive or negative, intentional or unintentional, synergetic or conflictive, and valued on the market or not (see Chapter 8). The multifunctionality spectrum, however, suggests that all productivist and non-productivist action and thought is in some ways *intentional* (see also Chapter 10) and places much less emphasis on economistic indicators than the OECD. In particular, the multifunctionality spectrum acknowledges the multiple and interconnected nature of agricultural transition, and emphasises the *synchronous* positive/negative and synergetic/conflicting effects of farming.

The notion of a spectrum with multiple *Deleuzian* parallel (and not so parallel) pathways of multifunctional decision-making particularly challenges the dualism inherent in traditional (agricultural) transition models. Although bounded by two 'extreme' pathways (productivism/non-productivism), the myriad of possible pathways in between these spaces suggests anything but dualistic processes. Most importantly, the new concept of multifunctionality suggests that agriculture has *always* been multifunctional in some way or another (for similar viewpoints see also FAO, 1999; Larsen, 2005; Holmes, 2006). In other words, contrary to most existing conceptualisations of multifunctionality that see it as a relatively *new* process (Freshwater, 2002; Hollander, 2004; Mather *et al.*, 2006), or as a *byproduct* of productivist agricultural activities, the multifunctionality spectrum suggests that multifunctionality is an *inherent* aspect of any agricultural activity – referred to in the following as the *multifunctional agricultural regime*.[60] Indeed, multifunctionality has been present since the emergence of the first agricultural societies, albeit with a highly differentiated spectrum of multifunctional 'quality' (see Section 9.4). I, therefore, agree with Goodman (2004: 12) that we have lost conceptual clarity and empirical understanding based on exiting models of "what distinguishes the 'alternative' from the 'normal'", but that this is "compensated by a fuller appreciation of the spatio-temporal disjunctures, complexities and continuities of contemporary rural change" – issues that the multifunctionality spectrum directly addresses. As a result, I disagree with Mather *et al.* (2006: 451) who argued that multifunctionality "is not an obvious improvement on post-productivism", or with Vandermeulen *et al.* (2006) who suggested that 'traditional' farming systems should be *transformed* into 'multifunctional' ones (see also Lowe *et al.*, 2002). Indeed, as this chapter will highlight, all 'traditional' farming systems are multifunctional in one way or another. I also disagree with those who suggest that multifunctionality exemplifies a *shift away* from productivism (e.g. Walford, 2003b; Kristensen *et al.*, 2004; Potter, 2004) or a shift from *post-productivism* to multifunctionality (e.g. McCarthy, 2005; Clark, 2006). In my view, both *productivism* and *non-productivism* form *inherent components* of what can be described as multifunctional agricultural processes. Barthélémy *et al.* (2004) also argued that multifunctionality *simultaneously* comprises complementary and opposing processes without either ever able to substitute the other.

It is evident that the multifunctionality spectrum shows parallels with Marsden's (2003) spatiality of the p/pp transition and with Marsden and Sonnino's (2005) acknowledgement of the coexistence of what they term 'neo-productivist' and multifunctional pathways. Similarly, it echoes Sonnino and Marsden's (2005) conceptualisation of 'multi-layered' 'alternative' and 'conventional' rural spaces, while also drawing on Clark's (2003) useful conceptualisation of multifunctionality into different conceptual 'levels' including multifunctional policy goals, outputs and territorial functions. Further, it also builds on Murdoch *et al.*'s (2003) notion of the spatially and conceptually 'differentiated countryside' in which multiple and parallel rural and agricultural development pathways are increasingly acknowledged as the norm, as well as on debates on 'alternative rural development' that some suggest have succeeded the now outdated notion of post-productivism (e.g. Goodman, 2004). As Section 9.4 will discuss, there are also some conceptual overlaps with Hollander's (2004) notion of 'strong' and 'weak' multifunctionality – i.e. a differentiation of multifunctional quality inherent in all agricultural systems.

As Section 7.3.4 highlighted, Marsden (2003) also suggested an approach that acknowledges the spatial heterogeneity inherent in agricultural trajectories by identifying

---

[60] While multifunctional agriculture can, in this sense, be interpreted as a 'regime' (i.e. as a system of accepted norms and institutionalised conventions, rules and processes), our deconstruction of the transition to post-productivism has shown that it would be problematic to refer to productivist or post-productivist *regimes*, as this would imply clearly defined sets of structured coherences and epochs – assumptions that, as we saw in Chapter 7, are difficult to substantiate.

three broader heuristic categories of territorialisation including the productivist *agro-industrial dynamic*, the *post-productivist dynamic* and the *rural development dynamic*. While Marsden (2003: 235; emphasis added) argued that all three dynamics can occur together, he nonetheless posited that the rural development dynamic "seeks to move *beyond* the post-productivist countryside", thus suggesting an element of temporal linearity and spatial homogeneity in the transition towards a new rural development regime. The new concept of multifunctionality suggested here goes beyond such heuristic terminology, as it enables the positioning of all three of Marsden's dynamics *within* temporally and spatially complex productivist and non-productivist decision-making pathways. In particular, it challenges Marsden's notions that the 'rural development dynamic' finds its most tangible expression at the regional level, or that non-productivist action and thought is largely driven by urban and suburban calls for re-orientation of the food chain. As we will see in Section 9.5, the multifunctionality spectrum enables us to conceptualise productivist and non-productivist action and thought at *any scale* and in *any territorial* and *temporal* context, and across various stakeholder group interests. Contrary to Marsden who sees inherent conflict between the agro-industrial and rural development models, the multifunctionality spectrum acknowledges that these two dynamics have and will always *coexist*, albeit with different positions along the p/np spectrum. The framework suggested here, therefore, echoes Marsden and Sonnino's (2005: 28) recent acknowledgement that "there has not been enough critical research in examining how the three models of rural development … actually play out in different rural regions". While the agro-industrial dynamic includes agricultural development trajectories situated close to the productivist end of the multifunctionality spectrum, and while the post-productivist dynamic is more closely related to *non*-productivist action and thought, Marsden's 'rural development dynamic' is more difficult to position as it overlaps both with agricultural and rural spaces that may go beyond notions of 'agricultural' multifunctionality. As Marsden (2003: 18) acknowledged, this dynamic is currently marginalised in mainstream political and policy-oriented thinking and is "a much broader and diffuse church". The multifunctionality spectrum enables us to bring together Marsden's three dynamics within *one unifying concept* that is both spatially and conceptually robust. In particular, it enables the conceptualisations of multiple pathways (see Chapter 10) that bring agriculture 'back in' as a significant shaper of the countryside and rural areas, both for productivist and non-productivist purposes.

It is evident that the multifunctionality spectrum differs markedly from economistic and policy-based conceptualisations of multifunctionality discussed in Chapter 8. Thus, Durand and Van Huylenbroek's (2003: 11) suggestion that "the principal problem of multifunctionality is how to remunerate farmers for providing non-market goods or in other words how the provision can be stimulated in the context of the present market system", in my view only addresses a small subset of processes driving multifunctional agricultural pathways. Thus, the concept developed here differs from economistic interpretations that only associate multifunctionality with *non-commodity* or *non-competitive* forms of agriculture (e.g. Gallardo *et al.*, 2003), and also challenges the notion that multifunctionality could be interpreted as an *innovation*. Similarly, conceptualisations of multifunctionality driven by policy-based approaches that place heavy emphasis on the policy environment as a *process* for guiding multifunctionality action can be equally criticised for only providing partial answers (see Chapter 8). In particular, the multifunctionality spectrum goes well beyond conceptualisations that place great emphasis on neo-liberal agricultural trade discourses and the WTO as global policy drivers for the seemingly *new emergence* of multifunctional agriculture (e.g. Hollander, 2004; Potter and Tilzey, 2005).

The multifunctionality spectrum particularly acknowledges that highly productivist action and thought can still be 'multifunctional' (see particularly discussion below on 'weak', 'moderate' and 'strong' multifunctionality). Thus, I also partly disagree with Clark (2003: 44) who suggested that multifunctionality "represents the first alternative business 'logic' to productivism", and with Losch (2004: 356) who argued that multifunctionality "represents the potential of a genuine and complete paradigm change for farming". Similarly, Holmes' (2006: 154; emphasis added) assertion that "the increasingly differentiated territorial expression of the multifunctional transition [is] *displacing* the formerly near-ubiquitous productivist agricultural mode" can be questioned because productivism, in my view, forms an *important part* of multifunctionality. Finally, the multifunctionality spectrum challenges the Euro-centric paradigm that claims that European farming landscapes are more multifunctional than others (Hollander, 2004; McCarthy, 2005). It thereby also addresses Losch's (2004) suggestion that the notion of multifunctionality can not simply be absorbed into a new recipe for policy-making, and even less to a 'ready-made' package directly transferred from the European experience. The new framework, therefore, enables acknowledgement of a much wider set of criteria, driving forces, and patterns influencing multifunctional agricultural decision-making that range across productivist and non-productivist decision-making pathways over time.

9.2.3 The multifunctionality spectrum: conceptual implications

The advantages of the new multifunctionality spectrum are manifold. The concept allows the exact positioning of any actor, actor space or process along a spectrum of decision-making that is based on the various dimensions of productivist and non-productivist spaces – i.e. including all of Marsden's above-mentioned dynamics while simultaneously addressing notions of the 'synergy in rural and agricultural development' suggested by Van der Ploeg and Roep (2003). In particular, conceptualising multifunctionality as part of a wide spectrum of decision-making opportunities acknowledges the *complexity* and *multi-causality* underpinning agricultural change, as it takes into account the *multiple* dimensions of existing conceptualisations of multifunctionality based on policy change (Potter and Burney, 2002), economic factors (Van Huylenbroek and Durand, 2003), and other approaches highlighted in Chapter 8.

In both Chapters 9 and 10, we will explore the nature of this complexity and the multiple dimensions of multifunctionality, where economic and policy dimensions are but two of multiple processes defining multifunctional agricultural decision-making pathways. This helps us move away from the problems associated with the 'European model of multifunctionality'. Indeed, the new multifunctionality concept can be applied *in any location around the world*, as it sees policy change as only one of many components of multifunctionality and is, therefore, largely independent of the history of policy-making trajectories specific to a given territory (e.g. the EU) – a problem that has marred the exporting of the 'European multifunctionality concept' to non-European geographical contexts. As Hollander (2004: 302-303) argued, the 'conventional' view of multifunctionality was "articulated as an anti-development or alternative development discourse emanating from the 'North'". The concept of a multifunctionality spectrum turns this argument on its head, as it can be invoked to explain processes in any farming system (both with productivist and non-productivist tendencies) and is not limited to 'traditional' forms of agriculture. Most importantly, and contrary to the proposition that we may witness different types of 'multifunctionality' in the developed and developing world (e.g. Bresciani *et al.*, 2004), multifunctionality conceptualised as a decision-making spectrum no longer needs to be associated with a Euro- or Northern-centric policy discourse restricted by anti-development rhetoric. As Chapter 10 will discuss, it can be applied equally to *both* the developed and the developing world.

Four key points emerge from this. First, the multifunctional agricultural regime, therefore, *exists everywhere* and has *always* existed in some form or another (see Section 10.4). It is here that the new concept differs most markedly from other conceptualisations as I argue that *every* farm holding on Earth is in some way or another multifunctional (albeit with varying multifunctional 'quality'; see Section 9.4). The artificial division made between 'conventional' and 'multifunctional' agricultural businesses by some authors (e.g. Clark,

2003; Hollander, 2004) is, therefore, no longer valid. Second, contrary to other concepts, multifunctionality is not restricted to a 'European model' of managing agricultural change, nor is it solely linked to policy and policy change, to economistic dimensions, or uniquely associated with joint production at the field scale. Third, conceptualising *multifunctionality* along the p/np spectrum enables us to acknowledge decision-making pathways that are most often characterised by non-linearity, heterogeneity and complexity – conceptual issues that, as Chapter 7 illustrated, have often marred conceptualisations of agricultural change. The new multifunctionality concept is not merely a new compromise that falls into the trap of traditional Western *dualistic imagination* (see Chapter 3), but a clear and workable alternative to, and critique of, contemporary unilinear and directional conceptualisations of agricultural change. Echoing recent work by Sayer (1991), Murdoch (1997a, b) and Gerber (1997), I argue that the multifunctionality spectrum emphasises the non-dualistic but symmetrical perspective on agricultural change that is increasingly gaining ground in agricultural and rural research (see in particular Evans *et al.*, 2002; Marsden, 2003)[61]. Fourth, the multifunctionality spectrum also has important theoretical repercussions, as there remains the question of theorising in a way that rejects large-scale universalising theory on the one hand (i.e. our deconstruction of the p/pp transition in Chapter 7), while, on the other hand, 'replacing' this framework with an alternative – but arguably equally universalising – framework of a multifunctional agricultural spectrum of decision-making. I will argue in Section 9.5 that the strength of the new multifunctionality spectrum is that it has an *in-built localising tendency* that allows greater temporal and spatial portability than the p/pp model, and that it is more sensitive to local geographies while also enabling analysis of macro-scalar patterns and processes.

How does this discussion interlink with transition theory? Transitional pathways based on the multifunctionality spectrum most closely resemble the *Deleuzian transition model*. In the multifunctional agricultural regime we need to acknowledge the existence of manifold, parallel and not so parallel, and at times highly complex development pathways over time, bounded by productivism and non-productivism. In particular, the multifunctionality spectrum can only be fully appreciated within the framework of similarly evident 'territorialisations' of Fordism/post-Fordism, modernism/post-modernism, with deconstructions of the transition towards post-socialism and post-colonialism, as well as with debates on environmentalism, technological and even evolutionary transitions (see Chapter 4). Building on Figure 2.1.e in Chapter 2, Figure 9.2 shows how transitional pathways in the multifunctional agricultural regime, bounded by the extreme pathways of productivist and non-productivist action and thought, can be hypothetically conceived for any actor or actor space. The figure shows Deleuzian pathways that vary over time, depending on the 'mix' of productivist and non-productivist drivers. These trajectories can vary substantially over time, shifting from action and thought that may be close to the productivist end of the spectrum at certain times, while close to non-productivist action and thought at others. As Potter and Tilzey (2005: 13) argued, "farm businesses and farming families are moving simultaneously along trajectories that are integrating some operators more closely within the industrial agro-food system while at the same time taking others further away". Only a few selected pathways are depicted in the figure. In theory, one line can be drawn for each actor involved in, or influencing, the multifunctional agricultural regime, and one has to imagine myriads of other lines occupying the conceptual space between non-productivism and productivism.

Two important points need to be emphasised here. First, there will be very few actors whose actions and thoughts will be *exclusively* productivist or non-productivist. The multifunctionality spectrum implies that any agricultural action will *always* contain some productivist *and* non-productivist elements. For any agricultural action (be it on a farm or in

---

[61] See also Deleuze and Guattari (1987), who criticised the use of 'single signifiers' in processes with multiples and hybrids transforming themselves among independent and interlinked multiple transitional pathways, as well as Bookchin's (1982) critique of reductionist thinking and antagonistic dualism that tends to neglect diversity, complexity and variety.

the national policy realm), we, therefore, witness the *multidimensional coexistence* of productivist and non-productivist tendencies, albeit with a multitude of different, and highly individual, expressions of such coexistence of what can be, at times, highly contradictory pressures (see Section 10.3). As Section 9.3 will highlight, a stakeholder situated exactly at the non-productivist end of the spectrum or 'beyond agriculture' will, by definition, no longer be involved in farming or agricultural activities and will, therefore, be 'outside' the multifunctional *agricultural* regime. This means that all 'agricultural' actors included within the multifunctionality spectrum shown in Figure 9.2 will still somehow be involved with *food and fibre production* (see Section 9.3). Second, opportunities for multifunctional action and thought by agricultural actors are likely to *vary over time*, with some actors moving towards non-productivist action and thought, and some moving towards the productivist end of the spectrum. Others may, indeed, move *beyond* non-productivism and leave farming altogether. Chapter 10 will discuss factors influencing transitional pathways and constraints and opportunities faced by actors when adopting pathways tending towards non-productivist action and thought.

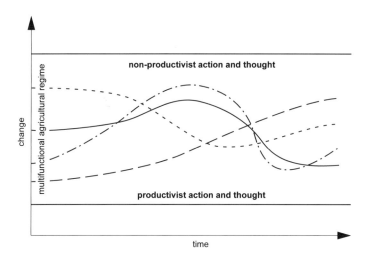

Fig. 9.2. Deleuzian transitional pathways and the multifunctionality spectrum

9.3 Multifunctionality and the boundaries of 'agriculture'

The new conceptualisation of multifunctionality – embedded within the extreme pathways of productivist and non-productivist action and thought – has important repercussions for our understanding and definition of 'agriculture'. As Marsden (2003: 180) emphasised, we have now reached a point where "we need to reconceptualise the role of 'the farm' as a site of production, reproduction and exchange". Similarly, Garzon (2005: 11) argued that "the concept of multifunctionality [has been] applied indifferently to agriculture as well as to rural areas". It is important to distinguish between multifunctional *agricultural* space (as part of our conceptualisation of a multifunctional agricultural regime) and multifunctional *rural* space – or as Kneafsey *et al.* (2004) suggested, 'multifunctional rurality'[62] (see also Lowe *et al.*, 2002). The discussion of multifunctionality in this book focuses largely on

---

[62] Some commentators argue that we should also conceptualise multifunctional *food regimes* (e.g. Buller, 2004). I argue here that food regimes are encapsulated both within agricultural and rural multifunctional processes.

multifunctional *agricultural* space (see Chapter 1), but understanding the concept is impossible without also considering what implications the revised p/np model has for conceptualising multifunctional *rural* space. Multifunctional agricultural space is, therefore, *embedded* in a less well delineated concept of multifunctional rural space (Fig. 9.3). As Marsden (2003: 19) emphasised, "rural areas are more than the sum of their agricultural parts". Multifunctional agricultural action and thought only applies to actors and spaces that are still concerned with *food and fibre* production (fibre in the context of agricultural fibre plants such as cotton, flax or linseed, not in the context of forestry; Schakel, 2003; Robinson, 2004) – i.e. the 'traditional' definition of agriculture. As highlighted below, this also includes farms that have entirely switched towards production of *biofuels* (e.g. Knickel and Renting, 2000). Only actors directly or indirectly concerned with food and fibre production are part of multifunctional *agricultural* space. As Section 9.2 emphasised, such actors may be very close to the non-productivist boundary of decision-making, but even these actors retain some link to food and fibre production.

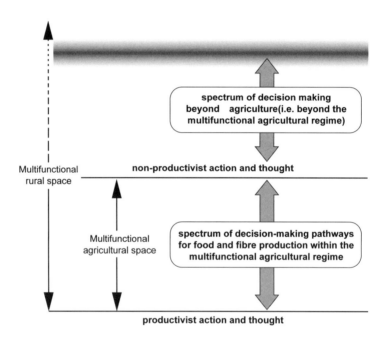

Fig. 9.3. Multifunctional agricultural and rural spaces

Multifunctional agricultural space, therefore, also includes farms that have largely opted out of food and fibre production by adopting non-agricultural diversification activities (e.g. establishment of golf course on farm, tourist accommodation), but that still produce *some* food or fibre from activities on the farm (Rapey *et al.*, 2004b). As Bell (2004: 248) emphasised, "growing food is only one dimension of ... the purpose of agriculture: cultivation". In other words, farms where marginalisation of the farm as a profitable enterprise is occurring (based on Bowler's, 1992, farm diversification classification; see below) are still situated within the multifunctional *agricultural* regime. According to Meert *et al.* (2005), such farms would include both diversification 'within agriculture' (e.g. innovative marketing of produce) and 'outside agriculture' (e.g. on-farm tourism, pluriactivity, golf courses, etc.). These actors would most likely continue to see themselves as 'farmers', although income from *agricultural* activities may be negligible (Rapey *et al.*, 2004b; Burton and Wilson, 2006). Based on a case study in Belgium, Meert *et al.* (2005: 93) mentioned the example of a farmer/worker who argued that "from Monday until Friday I am first and foremost a farmer, but during the weekend I go to work in the [factory]". Part-time farmers are, therefore, firmly embedded within the multifunctional *agricultural* regime, as they continue food and fibre production, although other non-farming activities may form the bulk of their income. Multifunctional agricultural space also includes 'hobby farms' that, although often operating only for self-sufficiency purposes largely outside the capitalist global market – and, therefore, close to the non-productivist boundary of decision-making – nonetheless continue to produce some food and fibre (Gasson, 1988).

The situation is less clear if we consider other (usually very small) actor spaces that may be part of food and fibre production. Urban allotments or vegetable plots in private gardens, for example, fall into this 'fuzzy' category. I would argue that these examples are useful illustrations of multifunctional spaces that often *straddle* the non-productivist boundary of multifunctionality but that, most often, remain embedded within the agricultural territory – in other words, they could be interpreted as representing an important form of food and fibre production that does not differ *conceptually* from a small subsistence farming plot (cf. Pretty, 1995). Indeed, at times vegetable production from private gardens can be a highly productivist activity, as the contribution of private vegetable plots to the overall food production in the former GDR testified. Although occupying only a small share of the GDR agricultural land area, the contribution of these private plots to total production was substantial (e.g. most of the national production of eggs, honey, chicken and rabbit meat) (Wilson and Wilson, 2001).

A territory that lies firmly beyond the non-productivist boundary of multifunctional agriculture relates to *parks* and other 'manicured' landscapes (including most private gardens) whose main roles are not food and fibre production, but recreation, amenity and visual appeal (Schama, 1995; but see also Bell, 2004, for an opposing viewpoint). In addition, Harvey (2004) suggested that today's remnant farm landscape features (e.g. hedges) could be conceptualised as situated 'outside' the multifunctional farm landscape. He argued that we may have witnessed a change in the use of landscape features, such as hedges of farm woodlands, from 'endogenous' uses to farming (i.e. embedded with other farming activities) towards 'exogenous' use, where a hedge can be seen as a non-productivist feature solely used for biodiversity protection. However, I argue that such non-productivist spaces on farms are part of the wide-ranging spectrum of multifunctional activities found on most farms and should, therefore, be seen as part of the multifunctional *agricultural* landscape. However, the extraction of plant or animal-based materials from natural uncultivated areas in the developing world (e.g. hunter-gathering, rubber tapping of naturally grown trees in tropical forests), meanwhile, lies *beyond* the boundary of the multifunctional agricultural regime (i.e. the highly environmentally sustainable practice of rubber tapping cannot be regarded as an agricultural activity as such; cf. Grigg, 1974). Yet, the latter example highlights that the definition of the non-productivist pathway boundary should not be taken too literally and that classifications – as usual – depend on local context. Thus, size of a spatial unit should not be used to categorise spaces as either within or beyond multifunctional agriculture, as the example of (often productivist) very small-scale intensive wet paddy rice fields illustrates.

Multifunctional *rural* space, meanwhile, comprises a wider spectrum of decision-making that *includes* multifunctional agricultural space, but that may also include actors who are not (or no longer) involved in food and fibre production. As Buller (2005: 1) emphasised, "agriculture is a vital component of rural economic diversification and environmental management but not the sole component", while Lowe *et al.* (2002: 15-16) highlighted that "farmers are [now only] one set of economic, social and environmental actors among others [in rural areas]". Figure 9.3 highlights that there is, therefore, a substantial multifunctional rural territory that lies 'beyond agriculture' (see also Marsden, 1999b). This territory, for example, includes former farmers who have chosen to completely leave food and fibre production behind and who are using their land entirely for non-farming purposes (Bohnet *et al.*, 2003; Cayre *et al.*, 2004) – referred to as 'effaced rurality' by Halfacree (2004) in which agriculture only retains a 'ghostly' presence. Examples include 'farms' where activities such as horse grazing, shooting or forestry (or combinations thereof) form the *only* use of the land. Schakel (2003: 232) painted an extreme view of a future countryside where "farming will be disconnected from the rural area and the countryside will only be the domain for other activities such as nature conservation, recreation, landscape planning, urbanization and so on".

The issue of *forestry* is particularly interesting, as on-farm woodlands (whether for commercial or non-commercial use) often form an integral part of *agricultural* land use (e.g. for livestock shelter; fodder) and are, therefore, often firmly embedded within multifunctional *agricultural* space. However, a farm that has been completely converted to forestry (commercial or non-commercial) has to be classified as 'beyond agriculture' (in the traditional definition of *agricultural fibre* production) and, therefore, at best embedded within the multifunctional *rural* regime (or, indeed, the multifunctional forestry regime; see Chapter 8). The owner of a forest no longer engaged in agriculture would also most likely not, or no longer, see him/herself as a 'farmer', even if the person's background was in farming (Burton, 2004; Burton and Wilson, 2006). This notion of multifunctional rural space resonates most closely with Marsden's (2003) *rural development dynamic* in which actor spaces intersect both with the agricultural and rural in complex ways (see also Marsden and Sonnino, 2005).[63] With recent counter-urbanisation trends placing increasing pressure on rural housing (see Chapters 6 and 7), much former agricultural land is taken over by ex-urban populations who have no intention of using this land for agriculture. Urbanites purchasing such parcels of land are, therefore, *shifting* such territories (possibly irreversibly) from multifunctional agricultural to rural space.

Recent dramatic technological developments in food production are adding another interesting philosophical dimension to the debate. New research suggests, for example, that meat may be able to be produced *without* livestock rearing through the growth of meat from animal muscle tissue alone (e.g. Bethge, 2005). This suggests that, in future, the rearing and slaughter of animals may no longer be needed, effectively removing one aspect of food production from the 'agricultural' and 'farming' domain (see also notions of the 'virtual farmer' suggested by Van der Ploeg, 2003). Where would such a development fit into the conceptualisation of agricultural and rural multifunctional spaces in Figure 9.3? I would argue that artificial production of meat in the laboratory lies conceptually 'beyond agriculture' (although it involves food production), and that our concept of the multifunctional agricultural regime needs to maintain somehow the link between food and fibre production and the *land/soil* (i.e. the Greek 'agros' means 'field') in order to remain a conceptually robust model. Thus, the *land-based nature of farming* should be seen as an integral component in conceptualisations of the multifunctional *agricultural* regime (see also Section 9.5). If future 'agricultural' production takes place without physical recourse to soils and land, then an entirely new concept of what multifunctionality entails may be needed. Indeed, it may be questionable whether such food production would even fall into the

---

[63] To contextualise this discussion, I wish to briefly highlight a parallel arena of investigation where similar classification issues have come to the fore. The classification of national parks by the International Union for the Conservation of Nature (IUCN) takes into account agricultural land use within parks (e.g. in the UK) and territories 'beyond agriculture' usually characterised by relatively untouched wilderness. The various IUCN grades (from IA to VI) accord different roles to protected areas, some of which are clearly 'beyond agriculture' (e.g. Grade 1A 'strict nature reserve'), while others allow social activities (such as farming and settlements) within designated areas (e.g. Grades V [protected landscape] and VI [managed resource protected area]). We see similarities here between 'multifunctional' national park aims that include protection of cultural *and* social values (e.g. most national parks in Europe) and 'monofunctional' parks (Grades 1A and 1B) where protection of wilderness and nature are the *only* purpose (e.g. core wilderness areas in the USA or New Zealand where public access is restricted).

category of multifunctional *rural* space, as such food production could become a largely urban phenomenon! This highlights that future technological change in food production may lead to deeper philosophical questions underlying conceptualisations of multifunctionality.

Leaving such 'extreme' possible developments aside, Figure 9.3 (above) shows that while the boundaries of multifunctional agricultural space are clear(er) – based on our conceptualisation of the extreme boundaries of productivism and non-productivism in Chapter 7 and Section 9.2 – the 'upper' boundary of multifunctional rural space remains fuzzy. There are two reasons for this. First, the boundaries of what constitutes the 'rural' are in themselves predicated on fuzzy notions of rurality (Hoggart, 1990; Woods, 2005; see Chapter 1), and the boundary is becoming increasingly blurred where it intersects with urban spaces (Murdoch and Lowe, 2003; Woods, 2005). Second, multifunctional rural *actor* spaces also become increasingly less clearly defined as we move away conceptually from the non-productivist boundary of multifunctional agricultural space – i.e. the classification of some actors as 'rural' is difficult in an increasingly urbanised world (Murdoch and Lowe, 2003).

I will refer to aspects of multifunctional rural spaces in the following discussion, but most of the analysis will focus on the multifunctional *agricultural* regime. Yet, we need to bear in mind that – as argued in Chapter 1 – terms such as the 'multifunctional agricultural regime' continue to be far from ideal, particularly as the emphasis on 'agriculture' may be seen to indicate a continuing emphasis on *agricultural* production and, therefore, on traditional *productivist* notions of rurality. We will need to keep this caveat in mind when interpreting multifunctionality in the remainder of this book.

9.4 Weak, moderate and strong multifunctionality: a normative view

Let us investigate in more detail how the *quality* of the multifunctional agricultural regime can be more clearly defined. So far, we have established that multifunctionality is situated along a *broad spectrum* of decision-making opportunities (see Sections 9.2 and 9.3). Can the notion of multifunctionality be narrowed down further so that it may become a tangible concept? At this juncture, it is important to recognise that the multifunctionality spectrum inevitably implies *normative judgements* about the 'value' and 'direction' of multifunctionality. In other words, multifunctionality *can* and *should not* be a neutral term, as Chapter 8 amply demonstrated. In particular, to make multifunctionality a relevant term that does not just highlight conceptual and academic issues, but that also has relevance for tangible decision- and policy-making 'on the ground', multifunctionality has to be accorded a certain normative *value* and *positionality* along the p/np decision-making spectrum. In other words, defining multifunctionality is about accepting that *priorities* need to be set for specific actions and processes influencing agricultural (and rural) development pathways – priorities largely absent from existing conceptualisations of multifunctionality (see Chapter 8). As Knickel and Renting (2000: 512) emphasised, we need to acknowledge the "different levels of multifunctionality" (see also Hollander, 2004; Potter and Tilzey, 2005).

This section will, first, address how different conceptual levels of multifunctionality can be identified and what the specific 'ingredients' of these different *multifunctionalities* should be (Section 9.4.1). Section 9.4.2 then discusses how specific farming systems can be categorised based on different types of multifunctionality, while Section 9.4.3 looks at interactions between policy and different levels of multifunctionality. This will form an important basis for our discussion of how to 'manage' the multifunctional agricultural regime in Chapter 10.

9.4.1 Conceptualising 'weak', 'moderate' and 'strong' multifunctionality

I propose that the multifunctionality spectrum comprises three conceptual levels: 'weak', 'moderate' and 'strong' multifunctionality (Fig. 9.4).[64] The use of these value-laden terms is deliberate. *Weak multifunctionality* is the 'worst' type of multifunctionality close to the productivist decision-making pathway of decision-making, while strong multifunctionality is the 'ideal' model that all societies should be striving for.[65] Indeed, and as the discussion below will highlight, *strong multifunctionality* will be conceptualised as engendering *synergistic mutual benefits* between individual actions and processes (e.g. landscape protection and food quality; direct marketing and rural community embeddedness; etc.) and, therefore, as good for the environment, good for farmers, good for rural social relationships and governance structures, good for food quality, and good for agriculture-society interactions in general (see also Knickel *et al.*, 2004). I will argue that strong multifunctionality is, therefore, predicated on strong *natural, social, moral* and *cultural capital* (Bourdieu, 1983; Putnam, 1993; Bryant, 2005). Further, strong multifunctionality is closely associated with the emergence of consumption processes in rural areas that focus not only on consumption of food and fibre, but also on the NCOs of farming such as protection of landscape, habitats and biodiversity (Marsden, 2003; Potter, 2004).

Adopting such a quality-based approach is not new. Potter and Tilzey (2005), for example, also referred to 'weak' and 'strong' versions of multifunctionality, albeit in the context of a policy-based conceptualisation (see Chapter 8), while Guillaumin *et al.* (2004: 54; emphasis added; my translation) suggested that "multifunctionality is a response to diverse drivers requiring a multi-faceted and variable conceptualisation". For debates on 'sustainable development' many commentators have suggested an equally value-laden spectrum, ranging from weak to strong sustainability, where strong sustainability is (often implicitly) expressed as the ultimate societal goal (e.g. Redclift, 1987; Gibbs *et al.*, 1998). Similarly, as Chapter 8 highlighted, urban theorists have also used notions of multifunctional 'quality' in their attempts to map urban multifunctionality across spatial and temporal scales, with some acknowledging that multifunctionality is not always associated with 'positive' processes (Rodenburg and Nijkamp, 2004).

I argue, therefore, that agricultural systems characterised by ***weak multifunctionality*** have weak environmental sustainability, in particular through external-chemical-input-heavy 'neotechnic' farming practices predicated on high farming intensity and productivity often characterised by relative *uniformity* of crops or animals (cf. Harris, 1978). Lifran *et al.* (2004) suggested that the more intensive agricultural systems are, the less multifunctional they are likely to be. Arguably, such systems are also likely to be based on GM crops, although heated debates continue about the relative ecological/social positionality (i.e. technological treadmill) of such crops in the context of controlled seasonality, generational sequencing, reproduction, maturing, ripening, ageing and decaying (e.g. Adam, 1999; Pretty, 2002). Thus, Lang and Heasman (2004: 180) argued that "despite the widespread introduction of GM crops, many research and scientific questions remain to be satisfactorily answered". Marsden (2003: 168) argued that in such systems, "the disconnection of food production from nature is mediated by two interrelated processes: first appropriation, that is the attempt to replace previously natural production processes by industrial activities; and second, substitution, or the attempt of industrial capitals to replace natural products in the food system with industrially produced substitutes". Both appropriation and substitution suggest a productivist philosophy linked to a weakly multifunctional agricultural dynamic.

---

[64] Based on our discussion in Section 9.3, Figure 9.4 also shows the territory 'beyond agriculture' that lies beyond our *agricultural* multifunctionality spectrum.

[65] In this way it differs from Hollander's (2004) notion of 'weak' and 'strong' multifunctionality that is purely defined from a policy analysis perspective.

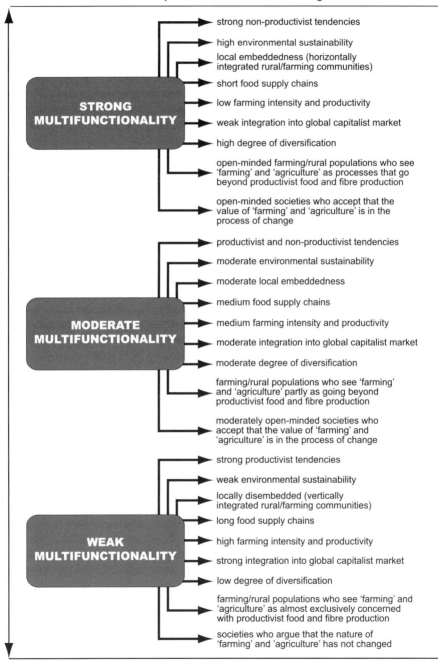

Fig. 9.4. Weak, moderate and strong multifunctionality

Weakly multifunctional agricultural systems also depict locally disembedded and vertically integrated rural/farming communities that show little internal socio-political cohesion, long food chains (i.e. little redistribution of knowledge and locally produced food within the immediate locality),[66] and tendencies of disempowerment of local actors and stakeholder groups linked to the rising power of corporate retailers (Goodman, 2004; Lang and Heasman, 2004). Pretty (1995) referred to such processes as the 'suffocation of local institutions', while Bell (2004: 39) associated this with the experience of a "disappearing rural culture". The weak multifunctionality regime will, therefore, also show tendencies of strong integration into the global capitalist market (especially with large transnational bioscience companies or retailing clusters), with a strong agricultural export orientation that allows agricultural actors to largely bypass their immediate rural neighbourhood for sale of agricultural commodities (Lang and Heasman, 2004). As Marsden (2003) emphasised, this regime is associated with the globalised production of standardised products, highlighting the increasing role of 'downstream' sectors in shaping at distance the farm sector. According to Marsden, strong integration into the global capitalist food markets, therefore, can be seen to tie agro-food centrally into an industrial dynamic, treats natural food products as merely industrial products, and tends to see CAP reform as a preface for further concentration and global competition. Most crucially, globalisation leads to a progressive lack of self-reliance and the loss of local social embeddedness of farmers (see also Swyngedouw, 1997; Goodman, 2004).

Agricultural regimes characterised by weak multifunctionality also have low degrees of diversification into non-agricultural activities, and, from an occupational identity perspective, landholders would see production of food and fibre as their foremost goal as a 'farmer'. It is useful to link the depth of diversification activities with Bowler's (1992) model showing different 'degrees' of diversification commitment. Thus, weakly multifunctional farm holdings are likely not to diversify at all or, if diversification is taking place, to diversify into diversification pathways described by Bowler (1992) as 'maintaining a viable agricultural enterprise' by either following a productivist industrial model pathway or by embarking on *agricultural* diversification (i.e. diversification activity still situated in the field of agricultural production; e.g. alternative crops or animals; organic farming) (Fig. 9.5). In a similar vein, pluriactivity (as a type of income diversification; cf. Ilbery, 1991) is only weakly developed (Robinson, 2004; Clark, 2005). In addition, weak multifunctionality may also be encouraged by action and thought beyond the farm gate, in particular linked to how rural populations see the role of agriculture in societal development. In this regime farming/rural populations will see 'farming' and 'agriculture' as almost exclusively concerned with productivist food and fibre production (in other words, the hegemonic role of agriculture is not questioned), while society as a whole will argue that the nature of both farming and agriculture have not changed or, more importantly, are in no need for change (Guillaumin *et al.*, 2004). In Pretty's (2002: 53) words, modern agriculture often predicated on the weak multifunctionality model "has brought a narrow view of farming". In line with debates on the positive and negative sides of agricultural change outlined in Chapters 5 and 6, weak multifunctionality, therefore, comes closest to what Marsden has termed the 'agro-industrial' dynamic.

Geographical examples of weakly multifunctional agricultural systems include areas where maximum production of food and fibre for an export-driven rural economy is the norm – or, in Harris' (1978) agricultural production systems classification, they include most 'neotechnic' agricultural systems predicated on intensive, biodiversity-poor and technologically heavy farming practices. Such areas include, for example, large parts of the highly productivist arable areas of the Ukraine, the American Midwest, East Anglia in the UK, the Paris Basin, Australian cotton and rice farming areas, the Börde area in central

---

[66] Marsden (2003) highlighted that in the UK, for example, an alarming 70% of the British public have no idea what food farmers in their local area produce.

Germany or the Emilia Romagna in northern Italy, to name but a few. Marsden and Sonnino (2005: 25) argued for the UK that "in the more industrialized agricultural areas of Eastern England ... where farmers increasingly have to accommodate the private-interest models of regulation and governance led by the corporate retailers ... there is very little (if any) potential for the development of a more re-embedded and multifunctional model of agriculture". Some would also argue that most of the Netherlands – one of the most intensively used agricultural areas in the world – would largely fall under the weak multifunctionality label (e.g. Knickel *et al.*, 2004). Similarly, highly intensive agricultural areas in developing countries, such as arable areas in China, intensive maize farming areas in Mexico or cash crop plantations in sub-Saharan Africa are characterised by weak multifunctionality (Robinson, 2004; Mazoyer and Roudart, 2006).

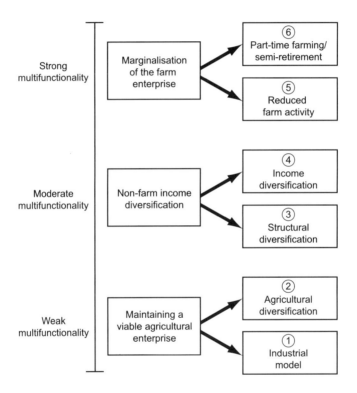

Fig. 9.5. Multifunctional quality and farm diversification pathways (Source: adapted from Bowler, 1992)

In particular, and as Parayil (1999) discussed in the context of technological transition, many areas planted with the help of Green Revolution seeds and technology are often characterised by weak(ening) multifunctionality, in particular due to high levels of vertical integration with, and dependence on, multinational agri-business corporations (Kloppenberg, 1988; Pretty, 1995). Thus, Parayil (1999: 177-178) argued for India that "the Green Revolution ... evolved as the 'traditional' agricultural system gave way to a new one with a different dynamic. In the old case, the boundary of the sociotechnological system did not stretch beyond the village. With the new one, it stretched far beyond the village". Farming systems characterised by rice terraces (wet rice) are often interesting examples of weak multifunctionality. Although these systems have often been in existence for thousands of years in relatively unchanged form and, therefore, suggest some level of sustainability (e.g. Philippines), these systems are usually intensive, show low levels of biodiversity, often rely on Green Revolution plant technologies, and are geared towards maximum production of a uniform plant species. As Pretty (2002: 36) argued, through Green Revolution technologies "in the blink of an eye, rice modernization during the 1960s and 1970s shattered ... social and ecological relationships by substituting pesticides for predators, fertilizers for cattle and traditional land management, [and] tractors for local labour groups" (see also examples of Green Revolution Indonesian rice farming or Mexican wheat and maize farming discussed in Pretty, 1995).

However, while earlier conceptualisations of agricultural change made it difficult to nuance such rather crude categorisations, the multifunctionality spectrum enables us to provide important caveats for above-mentioned examples. Thus, as Section 9.2 emphasised, there will be virtually no agricultural area in the world where productivist action and thought *exclusively* predominates. Indeed, even in the 'worst' agricultural landscapes (in environmental and social terms in particular), some evidence of action and thought leaning towards non-productivism can be found. As this chapter will highlight throughout, it is, therefore, dangerous to paint all agricultural/rural actors in a given region with the same brush. Thus, the neighbour of a highly productivist agro-business in the American Midwest may be a farmer whose practices can best be described as non-productivist (Bell, 2004). Further, while the German Börde area contains many highly productivist and export-oriented commercial farm businesses, there are also some small-scale, much less productivist, holdings that do not fall into the general mould of the weak multifunctionality model (Wilson and Wilson, 2001). Similarly, while most of the Netherlands is very intensively farmed, there is evidence to suggest that certain aspects of Dutch farming (e.g. local embeddedness of farms; society's view of agriculture) may be close to the non-productivist end of the multifunctionality spectrum (Meert *et al.*, 2005). Finally, while many Australian intensive cotton farmers may have exhausted their fragile soils through over-intensive productivist agricultural practices, some farmers in the same area may be heavily engaged in local Landcare groups attempting to mitigate desertification and soil degradation processes on their farms (Cocklin and Dibden, 2004; Wilson, 2004).

Three points emerge from this. First, weak multifunctionality suggests tendencies towards *monofunctionality*. In other words, monofunctional action and thought tends to predominate in these systems characterised by relatively uniform food and fibre production and mental images of farming as a relatively monofunctional (as opposed to multifunctional) activity. As Pretty (2002: 47) argued, such an agricultural system "is single coded – it does one thing (produces food) and it does it well. It draws on no local traditions; it is placeless, unflexible and monocultural", highlighting that weakly multifunctional agricultural systems are often characterised by agricultural processes that are in *competition* with strongly multifunctional drivers (see below). Second, just as it is difficult to find *purely* productivist systems, *purely* monofunctional agricultural systems are also unlikely to exist. Weak multifunctionality, therefore, implies that productivist tendencies tend to *predominate* in a given system, but not that this system is *exclusively* characterised by productivist action and thought. This means that, third, analysing whether a geographical area or an individual farm is weakly multifunctional needs a research methodology that allows each entity to be investigated on a case-by-case basis, also investigating *relative changes* in multifunctional quality over time. The larger the scale of investigation, the less likely it will be that we can 'accurately' position a given area along the multifunctionality spectrum (see also Section 9.5 on the spatial dimensions of multifunctionality).

Moderate multifunctionality combines elements of both productivist and non-productivist action and thought, that more or less balance each other out. It, therefore, occupies the 'middle ground' of the multifunctionality spectrum shown in Figure 9.4. Farming systems characterised by moderate multifunctionality have higher levels of environmental sustainability than weak multifunctionality systems, but are still linked to environmental degradation caused by agriculture. Further, actors in moderately multifunctional systems have higher levels of local embeddedness than those in weakly multifunctional systems, and show some evidence of horizontally integrated rural/farming communities with close(r) interaction between local rural communities and their farming populations. Yet, in these systems, many farmers continue to operate 'outside' of the rural community and may not necessarily operate within short food chains. Indeed, Marsden (2003) highlighted that the increasingly complex interlinkages between retailers and processors are producing interesting 'mutations' in supply chains where some corporate retailers are, for example, developing home deliveries through internet ordering that can be seen to combine both weak and strong multifunctionality (see below) characteristics. Similarly, Holmes (2006) highlighted how many rural communities in Australia currently straddle the line between strong and weak multifunctional quality (which he terms 'complex multifunctionality'), in particular with reference to small farms that show a mix of production and consumption values, but also regarding marginalised agricultural/pastoral occupance where there is, yet often untapped, potential of integration of productivist and non-productivist processes (see also Holmes, 2002).

Moderately multifunctional farmers generally show tendencies for lower farming intensity and productivity than their weakly multifunctional counterparts, but they will still farm in relatively intensive ways. Some farmers in moderately multifunctional systems will have diversified into activities not linked to productivist food and fibre production, but many others will not and will continue to see food and fibre production as their ultimate goal. Diversification activities are more likely to fall into Bowler's (1992) pathway of 'non-farm income diversification' (see Fig. 9.5 above) that involves both *structural* diversification (redeployment of farm resources into new non-agricultural products or services; e.g. farm gate sales, farm tourism) and *income* diversification through more elaborate forms of pluriactivity (use of non-specific farm household assets for non-agricultural activities unconnected to the farm; e.g. off-farm employment by any member of the farm household) (Evans and Ilbery, 1993; Clark, 2005). Further, on moderately multifunctional farms, integration into the global capitalist market will be less pronounced than in weakly multifunctionality systems, but most farmers will produce for a market that lies well outside their immediate neighbourhood and rural community. In addition, in moderately multifunctional agricultural regimes some mental changes through more open-minded farming and rural populations will have taken place, although few stakeholders will see 'farming' and 'agriculture' as processes that go *entirely* beyond productivist food and fibre production, mass production and profit maximisation. Finally, wider society in the moderately multifunctional regime will have partly accepted that the nature of 'farming' and 'agriculture' is in the process of change, but will still accord a relatively important role to agriculture and agricultural production as key pillars of economic and social development.

Geographical examples of agricultural systems characterised by moderate multifunctionality include farming regions that are relatively well embedded in the global capitalist system, but that also show tendencies of localisation and extensification. In this sense, the middle ground of moderate multifunctionality occupies most agricultural areas in the world. In a European context, most agricultural areas in the lowlands that do not depict weak multifunctionality fall into the moderate multifunctionality category. In the UK, these include, for example, most of the lowland areas characterised by mixed farming or by less intensive arable farming. In the USA, meanwhile, most farming areas along the eastern seaboard and parts of California, Oregon and Washington would be characterised by moderate multifunctionality, while in New Zealand most of the 'hillcountry farming' in both the South and North Island could be seen as moderately multifunctional (Primdahl and Swaffield, 2004). In developing countries, farming areas that are beginning to be embedded within capitalist profit-maximising structures could be included in the moderate multifunctionality model (Pretty, 1995, 2002). This will, for example, include farming areas in India that currently still show high levels of local embeddedness and highly diverse and extensive production systems, but that are in the process of intensification, capitalisation and polarisation of agricultural production (see detailed example mentioned in Chapter 10) – thereby depicting both non-productivist and productivist development trajectories (Parayil, 1999).

Strong multifunctionality, meanwhile, is characterised by action and thought located close to the non-productivist pathway boundary of the multifunctionality spectrum, with a strong emphasis on public good characteristics (Potter and Tilzey, 2005) and strong social, economic, cultural, moral and environmental capital (Bourdieu, 1983; see also Worster's, 1993, principles for good farming). In line with notions of strong multifunctionality expressed by *diversity, functional interweaving* and *spatial heterogeneity* developed by urban multifunctional theorists (e.g. Priemus *et al.*, 2004; see Chapter 8), strong agricultural multifunctionality comprises multiple inter-related dimensions, as well as being closely associated with both the notions of 'diversity' (cf. Vreeker *et al.*, 2004) and 'deepening', 'broadening' and 'regrounding' of farming and non-farming activities (Knickel *et al.*, 2004; see also Bell, 2004, for the US context). This notion of strong multifunctionality, therefore, differs from Hollander's (2004) policy-oriented notion of strong multifunctionality, but shows parallels with Bryden's (2005) 'broad definition' of multifunctionality based on a more holistic interpretation that also includes social and cultural capital.

High environmental sustainability plays a key role in strongly multifunctional systems, based on often highly diverse assemblages of crops and livestock often akin to 'palaeotechnic' types of agricultural production (cf. Harris, 1978). As Belletti *et al.* (2003a: 61) suggested, "greater attention and sensitivity to the farm landscape and to the role it carries in the success of the farm activities frequently characterize [strongly] multifunctional enterprises". Guzman and Woodgate's (1999) notion of agro-ecological 'co-evolutionary pathways' that address the needs of society/nature without jeopardising the integrity of social and natural systems, therefore, shows parallels with strong multifunctionality (see also Altieri, 1987). Knickel and Renting (2000) emphasised how a reduction in yields is often part-and-parcel of farming systems where multifunctional quality is increasing, and, similarly, Holmes (2006) argued that a reduction in 'agricultural overcapacity' should be seen as key to increasing multifunctional quality. Environmental sustainability of strongly multifunctional systems also includes agricultural and farm household activities with a *low carbon footprint* (e.g. energy-efficient machinery) and should be predicated on agro-food chains that reduce the need for long-distance food transport (see also below).

Actors in the strongly multifunctional agricultural regime show strong tendencies for local and regional embeddedness, in particular through horizontally integrated rural/farming communities with very close interaction between local communities through reciprocal rural-agricultural relationships characterised by strong lateral actor linkages and newly empowered stakeholder groups (Goodman, 2004). This resonates with Leyshon *et al.*'s (2003) notion of 'spaces of hope' whereby newly assembled (or rediscovered) horizontal platforms of action are (re)created out of local agricultural/rural systems (see also Knickel *et al.*, 2004), and with Buller's (2004) notion of 'transversal linkages and networks' that permit social and economic capital to be retained within rural communities. Bryden (2005) also referred to the important linkage between strong multifunctionality and 'quality of life' in rural areas. As Pretty (1995: 132) highlighted, "local groups and indigenous institutions have ... long been important in rural and agricultural development", also highlighted by French researchers as a key indicator of strong multifunctionality through preservation of the 'social and cultural tissue' of rural areas and through facilitation of processes of 'cohabitation' of agricultural/non-agricultural actors (Guillaumin *et al.*, 2004; Roux *et al.*, 2004). Focusing on food chains, Marsden (2003: 116) similarly referred to this as "the formation of new food 'circuits' [that] are demonstrating the significance of embedded social networks as a forerunner for local rural economic development".

In line with similar conceptualisations of multifunctional forestry (e.g. Freshwater, 2002; Larsen, 2005), *social* and *cultural capital* are well developed in strongly multifunctional agricultural systems (Daugstad *et al.*, 2006).[67] As Lardon *et al.* (2004: 16; my translation) emphasised from a French perspective, "multifunctionality implies the existence of collective action" (see also Pivot *et al.*, 2003), while Boody *et al.* (2005) highlighted the importance of job creation and maintenance of the social fabric as key pillars of cultural capital in strongly multifunctional US farming systems. Such embedded social networks particularly include activities that will help provide new income and employment opportunities for the agricultural sector (i.e. new or strengthened relationships between the agricultural sector and rural society), as well as so-called 'associational interfaces' that are both informal but highly significant in establishing trust, common understandings, working patterns, and different forms of cooperation and positive interaction between stakeholder groups in the food supply chain (Pretty, 2002; Marsden, 2003). As Meert *et al.* (2005: 88) emphasised, "strong social networks are often a key factor for the successful development of new activities in the farm household". Clark (2003: 29) also emphasised the importance of networks of sub-national actors and agencies, working cooperatively and systematically in the promotion of multifunctional agriculture, by arguing that "researchers are beginning to recognise the tremendous importance of networks of agencies at the sub-national scale in shaping and promoting the new multifunctional agenda for agriculture". Thus, partnerships and close interpersonal linkages of farmers with other grassroots actors in the locality will be particularly well developed (Bennett, 2005; Meert *et al.*, 2005), and strong governance structures will be evident (Wilson, 2004).[68] In countries with ethnic minorities living in rural areas (e.g. Australia, New Zealand, Canada) local embeddedness in strongly multifunctional agricultural regimes will also mean empowerment and inclusion of ethnic groups (e.g. Australian Aborigines) in agricultural/rural decision-making processes (e.g. Argent, 2002; Holmes, 2006).

Strongly multifunctional systems will also be characterised by short *food chains* and high(er) *food quality*[69] associated with more differentiated food demand by consumers (the 'quality turn'; Goodman, 2004), the capacity to re-socialise or re-spatialise food, a demand for food products with high (often regionally based) symbolic characteristics, the creation of additional value for rural regions (Belletti *et al.*, 2003b; Renting *et al.*, 2003), and enlightened visions about food and health focusing on the interlinkages between both human *and* environmental health (Lang and Heasman, 2004). Buller (2005: 15) emphasised the importance of quality foods as a multifunctionality indicator by arguing that "a number of studies have been undertaken … that examine the new multifunctionality roles that farming performs through the provision (and on occasion marketing) of quality foods or local foods". Further, according to Marsden (2003: 152), "quality food markets raise the spectre of local organisation and the potential development of regionally embedded supply chains", while Venn *et al.* (2006) emphasised that 'alternative' and 'conventional' food networks cover a wide (and at times overlapping) spectrum of multifunctional possibilities, with alternative food networks often depicting strongly multifunctional tendencies. Delgado *et al.* (2003: 25) also suggested that "multifunctional agriculture can offer the consumer a range of goods and services of higher quality than those sold on a purely competitive market". As a result, strongly multifunctional food consumption habits can be interpreted as a *countervailing* force to Popkin's (1998) notion of the 'nutrition transition' (rapidly declining dietary quality with

---

[67] Rodenburg and Nijkamp (2004) also argued from an urban multifunctionality perspective that advantages of synergy between individual components of multifunctionality involving multiple actor levels are key for the economic vitality and environmental quality of modern cities.

[68] Parallels can also be drawn here with other 'localisation movements', the best known of which will be Schumacher's (1973) notion of 'small is beautiful' – an idealised vision of human society predicated on small-scale relocalised, horizontally embedded and environmentally sustainable communities.

[69] I acknowledge that the meaning of food quality is a complex notion that is culturally and socially constructed, (Ilbery and Kneafsey, 2000), although there are many commonalities with regard to how different cultures around the world differentiate between low and high quality foods (Lang and Heasman, 2004).

increased obesity and coronary problems) through moves towards strongly multifunctional healthy diets (e.g. Mediterranean foods) predicated on food quality and good nutritional balance combined with increased exercise. Marsden (2003: 17) also emphasised that short food chains "represent a more profound shift ... than the emergence of new products, in that they are being explicitly used to create markets independently of the [weakly multifunctional] industrialised and intensive ones led by the globalised multiple retailers". In addition, strongly multifunctional systems will also depict low farming intensity and productivity (Murdoch *et al.*, 2000; Evans *et al.*, 2002), in most cases characterised by a reluctance to use Green Revolution or GM crops (Adam, 1999; Pretty, 2002), and will also emphasise carbon neutral food production and transport.

There will also be high degrees of *diversification* and *pluriactivity* in strongly multifunctional agricultural systems, in particular away from traditional productivist food and fibre production (Durand and Van Huylenbroek, 2003; Marsden, 2003), leading to a revaluation of existing farm household knowledge (e.g. of women and young people) and the need to develop new skills and professional abilities (Belletti *et al.*, 2003a). Clark (2003: 226) emphasised that "diversification is seen as a key characteristic of multifunctional [farm] business behaviour", although diversification opportunities will vary substantially from region to region (Marsden, 2003; see also Chapter 10). Strongly multifunctional farming systems are particularly likely to embark on Bowler's (1992) diversification pathways that imply the *marginalisation* of the farm enterprise (see Fig. 9.5 above) through reduced farm activity, or, as both Knickel *et al.* (2004) and Turner and Reed (2006) suggested, through the 'deepening' of diversification activities and/or the move towards part-time farming or even semi-retirement which may also involve the sale of pockets of land (Meert *et al.*, 2005).

More controversially, and partly in line with the pessimistic view on agricultural liberalisation taken by powerful UK conservation agencies (e.g. English Nature, 2000; RSPB, 2001), some academics (e.g. McMichael, 1995; Goodman and Watts, 1997; Shiva and Bedi, 2002) or anti-globalisation advocates such as the Confédération Paysanne in France (Coleman and Chaisson, 2002) have argued that strongly multifunctional farms are more likely to be *weakly integrated into the global capitalist market*, as only partial or complete disengagement from global capitalist (productivist) networks and agriculture liberalisation processes will enable on-farm implementation of above dimensions of strong multifunctionality (see also discussion of 'relocalisation' in Chapter 10).[70] This was echoed by McCarthy (2005) who emphasised that strongly multifunctional systems are most threatened by trade liberalisation and globalisation, and by Hollander (2004) who argued that strong multifunctionality may be a form of resistance against global neoliberalism (see also Swyngedouw, 1997, and Goodman, 2004, for more critical views).

Strong multifunctionality will also imply that *substantial mental changes* have taken place among various stakeholder groups, in particular through open-minded farming and rural populations who see 'farming' and 'agriculture' as processes that go well beyond productivist food and fibre production, mass production and profit maximisation. Building on Di Iacovo's (2003, 2006) more inclusive view of multifunctionality (see Chapter 8), strongly multifunctional agriculture can also be seen to act as a bridge between agricultural practices and the wider community (in particular disadvantaged groups in rural society), or – as Clark (2003) emphasised – to address the *social dislocation* of agriculture from other production sectors and from mainstream territorial economic development efforts that have often bypassed farming. It is interesting to note that a recent study in the Netherlands found that "farmers mentioned societal limitations ... as the most important factor for being

---

[70] There are parallels here with Gallardo *et al.*'s (2003) notion of farms that may not be *competitive* on the national and global markets, but that may, nonetheless, be *viable*. Thus, a strongly multifunctional farm deciding to partly or fully disengage from the global capitalist market may continue to be viable, albeit often in an economically marginal sense. Ervin (1999) cautioned that the state of knowledge on environmental effects of withdrawal of farming systems from globalised capitalist networks is 'sketchy', particularly concerning site-specific and regional impacts where the land may need to be farmed to maintain its conservation values.

involved in (or forced into) multifunctional agriculture" (Jongeneel and Slangen, 2004: 201). In addition, Clark (2003) suggested that for farms to be strongly multifunctional they must somehow make a *purposeful* contribution to local or regional rural development. Strongly multifunctional systems will also be closely associated with *open-minded societies* who accept that the nature of 'farming' and 'agriculture' is in the process of change. Overall, the 'strongest' level of multifunctionality can be achieved if all of the above processes and activities occur *simultaneously* (which will rarely be the case; see Chapter 10), based on what Knickel and Renting (2000) called *self-reinforcing multiplier* effects of multifunctional activity, or what Van der Ploeg and Renting (2000) called 'clusters of compatible and mutually reinforcing multifunctional activities'.

Geographical examples for the strong multifunctionality model are manifold and include areas characterised by low intensity, high biodiversity and closely integrated farming communities where agricultural production may not be the main source of income (Mollard *et al.*, 2004; Dijst *et al.*, 2005). In Europe, such areas include many upland and mountainous areas (e.g. Alpine valleys; British moorlands) (e.g. Dax and Hovorka, 2004). In the USA, states such as Montana may contain many elements of strong multifunctionality (Diamond, 2006), while in Australia and New Zealand farming districts that have managed to ensure survival of biodiversity-rich indigenous forests on farms show tendencies of strong multifunctionality (Wilson and Memon, 2005). In the European rural literature, in particular, there is much emphasis on why certain farm regions tend to have shorter food supply chains, with better integration of food production into local sale and consumption (Marsden, 2003; Goodman, 2004). Italy has been hailed as a particularly successful example of this strong multifunctionality indicator, especially through its successful 'slow food' movement (food as pleasure and social integration based on local cuisines and artisanal food), its successful regionally-specific food brands, and its close connection between consumers and producers (Pretty, 2002; Lang and Heasman, 2004).

As I will discuss in more detail in Section 9.4.2 and Chapter 10, the situation among subsistence farmers in the developing world is particularly interesting in this respect, as it could be argued that most agricultural areas in the developing world probably (still) fall into the category of the strong multifunctionality model. Indeed, based on Harris' (1978) classification of many of these agricultural systems as 'palaeotechnic' (i.e. low input, highly biodiverse extensive production for subsistence or local markets) they would meet most of the criteria of strong multifunctionality. Thus, large parts of India and China, for example, would (still) fall under the strong multifunctionality model, with over one billion farmers around the world currently falling under this particular model (Losch, 2004). However, as Section 10.4 will highlight, this situation may be changing due to globalisation pressures.

Six points emerge from this discussion. First, if we accept that conceptualising multifunctionality along a spectrum is linked to *value judgements*, strong multifunctionality should be seen as the 'best' type of multifunctionality – or, indeed, the type of multifunctionality with the best *quality*. Not only is it predicated on ensuring the protection of the environment, healthy farming and rural communities, but it can also be seen as the most 'moral' type of multifunctionality (see also Lowe *et al.*, 1997). Intuitively, there is something 'good' about strong multifunctionality, as most of its dimensions resonate positively with what producers, rural stakeholders and wider society would see as the 'optimum' type of agricultural regime. Indeed, I would argue that social systems over time have always *attempted* (not always successfully) to either maintain strong multifunctional agriculture regimes or to move away from weak or moderate multifunctionality towards strong multifunctionality (see Chapter 10). This was echoed by Belletti *et al.* (2003a) and Knickel *et al.* (2004) who suggested that a *deepening* of farming activities (i.e. creating new resources by way of new links between internal resources and external diversity) is crucial for a shift towards strong multifunctionality, while McCarthy (2005: 774) argued that the notion of strong multifunctionality "offers a positive characterization rather than a negative one". Marsden (2003) similarly suggested a changing *quality* and *morality* surrounding recent relationships between key actors in agri-food networks, with a move from exploitation towards responsibility/interdependency, from a supply chain focus to partnership/communication, from top-down regulation to governance and empowered networks, and from confrontational relationships between stakeholders with varying views towards collaboration – i.e. where the former processes resemble weak multifunctionality structures, while the latter resonate well with notions of strong multifunctionality.

Clark (2003) also emphasised in the context of advanced economies that farm holdings adopting strongly multifunctional pathways also often *benefit financially* from adoption of stronger multifunctional pathways (e.g. through certain diversification activities). This has been reiterated in a US study investigating multifunctionality pathways on different farms that highlighted that net farm income does not need to suffer from adoption of strongly multifunctional farm development pathways (Boody *et al.*, 2005). Belletti *et al.* (2003b: 158) similarly suggested that strongly multifunctional agriculture "can increase or maintain the local common stock of resources, from natural environment to biodiversity and landscape, up to cultural and food tradition characteristics of the territory. In this way, a [strongly] multifunctional agriculture can be the basis for the diversification, economic integration, and rural area differentiation, so as to promote the chances for enterprises in the territory and consequently the whole economic development". Strong multifunctionality is, therefore, the model policy-makers and agricultural/rural stakeholders should strive for. Yet, we need to be careful not to fall into the trap of romanticising one multifunctionality model over another. Historically, for example, it would be wrong to argue that, for most of the time since the inception of farming, agricultural systems operated on the basis of the strong multifunctionality model (see Chapter 10). Virtually every place on the globe has historically seen different phases of weak, moderate and strong multifunctionality, and it is debateable whether the strong multifunctionality model has been the dominant model over time. Indeed, one needs to ask whether the strong multifunctionality model in one area may be *predicated* on the co-existence (temporally and spatially) of moderate and weak multifunctional agricultural regimes in other areas? In other words, is the aim to achieve strong multifunctionality possible and based on a win-win situation? Or is it possible to conceive of a 'zero-sum-game' situation in which the existence of one multifunctionality regime is predicated on the existence of another? Pretty (2002: 95) emphasised from an environmental sustainability perspective that "what we do not yet know is whether a transition to a [multifunctional] agriculture … will result in enough food to meet current food needs". In other words, could global agriculture be based solely on a strong multifunctionality model? This emphasises that farming and agricultural practices alone may not be the best processes to achieve strong multifunctionality, and that many other stakeholder groups and localities need to be involved in helping to engender strong multifunctionality pathways. I will return to these issues when discussing how to 'manage' the transition towards strong multifunctionality in Section 10.4.

Second, we have seen throughout our discussion of multifunctional quality that *environmental sustainability* forms an important component of conceptualisations[71] (Potter and Burney, 2002; Marsden and Sonnino, 2005). As Swagemakers (2003: 191) emphasised, "multifunctional agriculture plays a key role in maintaining the functions of landscape, nature, and biodiversity", and Garzon (2005: 6) similarly suggested that "the debate on multifunctionality … is intimately linked to the search of public policies aiming at a balanced and sustainable development". Yet, the possible confusion surrounding the close association between notions of 'multifunctionality' and 'sustainability' was particularly highlighted during the 1999 FAO Conference entitled 'Multifunctional character of agriculture and land' in which representatives from the Cairns Group (see Chapter 8) rejected (for political reasons) the idea that there was a distinction between the two terms (Doran *et*

---

[71] In many ways, this echoes arguments long held by agro-ecologists who have emphasised the importance of environmental sustainability in conceptualisations of agricultural 'best practice' (cf. Altieri, 1987).

al., 1999). As a result, the complex interplay between notions of 'multifunctionality' and 'sustainability' need to be clear(er). As Section 9.2 emphasised, it is important to reiterate that environmental sustainability is only *one of several* key components of the strong multifunctionality model.[72] Indeed, as Buller (2005: 2) emphasised "multifunctionality is distinct from nature conservation or biodiversity management per se, because it explicitly links these to processes of agricultural production". Similarly, Garzon (2005) emphasised that while sustainability is a resource-oriented concept, multifunctionality is an activity-oriented concept describing the activities of agriculture as a specific economic sector. Although we see parallels between debates on multifunctionality and environmental sustainability, the two terms are not synonymous. Nonetheless, the notion of environmental sustainability is particularly important in the conceptualisation of strong multifunctionality where environmental considerations are paramount close to non-productivist pathways of action and thought (Brouwer, 2004a). Indeed, understanding the role of nature and the environment are crucial in conceptualisations of multifunctionality in general, as strong multifunctionality is particularly embedded in environmental and sustainability discourses. While weak multifunctionality shares many similarities with the notion of 'sustainable *economic* development' (Barbier, 1987) in which productivist action and thought tend to override environmental concerns (i.e. agro-industrial model), strong multifunctionality is more closely associated with the original Brundtland Commission definition of sustainability that also takes into account inter-generational equity issues and, indirectly, criticises the productivist global capitalist model (Wilson and Bryant, 1997). Environmental sustainability, therefore, is a vital component of multifunctionality, but conceptualisation of multifunctional quality also has to be based on complex social, political and economic characteristics that go well beyond environmental sustainability alone.

Third, there are close associations between the concept of weak, moderate and strong multifunctionality suggested here and Bowler's (1992) spectrum of different 'shallow' to 'deep' diversification pathways (see Fig. 9.5 above). Yet, contrary to debates in the current literature that often *equate* the notion of multifunctionality with diversification/pluriactivity (see Chapter 8), it is important to emphasise that farm diversification – just like environmentally sustainable farming – only forms *one component* of what multifunctional agriculture is about.

Fourth, above discussion has highlighted that we need to be cautious not to oversimplify categorisations. The new multifunctionality model highlights that weak, moderate and strong multifunctionality are in themselves *highly complex* categories that include many diverse farming systems and agricultural areas. The 'mix' of productivist and non-productivist action and thought will vary greatly and between case studies. Weak, moderate and strong multifunctionality should, therefore, not be seen as agricultural regimes with clearly delineated boundaries, but as relatively fuzzy conceptual entities that show 'clusters of dimensions' associated with one of the three multifunctionality regimes, but that also, simultaneously, will often include dimensions from one or both of the 'other' multifunctionality regimes. This is also illustrated by the fact that both of Marsden's (2003) post-productivist and rural development dynamics (see Chapter 7) show parallels with components of *both* moderately and strongly multifunctional farming systems. Thus, monocultural farming associated with the weak multifunctionality model will often go hand-in-hand with weak environmental sustainability and long food chains, but some elements of moderate multifunctionality, such as evidence of farm diversification, may also be present. However, it is unlikely that many dimensions characteristic of the strong multifunctionality model would be found in weakly multifunctional systems (see also Section 10.3 on corridors of multifunctional farm development pathways). In addition, all three types of multifunctionality should be seen as flexible and relatively permeable entities. Some

---

[72] This may be different to fledgling debates on *multifunctional fisheries* that are still largely *equated* with environmental sustainability guaranteeing survival of fish stocks (Schmidt, 2003; see also Chapter 8).

agricultural areas may be moving from strong or moderate multifunctionality towards weak multifunctionality (cf. Marsden's, 2003, notion of the ephemerality of agricultural systems), while others may have just gone beyond the weak multifunctionality model by placing greater emphasis on non-productivist agricultural development pathways (see also Section 10.2 on constraints and opportunities for 'moving along' the multifunctionality spectrum).

Fifth, the acknowledgement that conceptualisations of weak, moderate and strong multifunctionality are based on moral value judgements about 'better' and 'worse' agricultural regimes means that the ultimate aim of any agricultural or associated actors should be to 'move' agricultural systems towards the strong multifunctionality model (see also similar arguments by Morris and Potter, 1995, and Wilson, 1996, about 'moving' farmers along the AEP participation spectrum). Indeed, strong multifunctionality may enable both diversity and sustainability in human and natural systems without undermining economic efficiency (Pretty, 2002). In other words, in the long term economic efficiency and survival of farming systems may be *predicated* on development of strong multifunctionality pathways. Multifunctionality is, therefore, about different pathways opportunities (see Section 10.3) at different scales (Section 9.5) that could lead individuals or entire farming regions towards strong multifunctionality and thereby closer to the boundary of non-productivist action and thought. A strongly multifunctional agricultural regime can be conceived as a 'stable' aspirational regime, while regimes characterised by weak and moderate multifunctionality may be relatively 'unstable' systems. Using transition theory, Chapter 10 will discuss opportunities and constraints for moving agricultural regimes from weak and moderate to strong multifunctionality and will highlight that, often, the barriers for moving systems from one state to another can be difficult to overcome.

Finally, the notion of a multifunctionality spectrum opens the question of how to 'measure' different multifunctional quality. Here it is important to acknowledge that multifunctionality, as conceptualised here, does not describe an *absolute* state but a *flexible transitional* process. This means that the use of absolute indicators – an approach that many policy-making organisations such as the European Commission are advocating for ease of data collection (Wilson and Buller, 2001) – is *not* the right approach. As McCarthy (2005: 778) emphasised, "there are tensions inherent in the fact that indicators simplify, standardize, and quantify complex information and relationships ... when much of the point of multifunctionality is to emphasize the heterogeneous and synergistic aspects of [agricultural and rural processes]". Thus, the use of so-called 'expert knowledges' to assess multifunctionality may need to be questioned at the same time as we re-assess notions of multifunctionality, while evaluations of *relative local change* – possibly assessed by both 'experts' and 'non-experts' – are likely to assume greater importance[73] (Wilson and Rigg, 2003). Reconceptualising the differentiated nature of multifunctionality and associated methodologies should, therefore, be able to address Delgado *et al.*'s (2003: 28) call that "multifunctionality therefore, suitably understood and applied, could be the basis from which to approach the processes of change that European agriculture and the rural world needs".

9.4.2 Farming systems and the multifunctionality spectrum

This section investigates how different agricultural systems (e.g. organic farming, subsistence farming, lifestyle farming, etc.) can be classified in the context of the multifunctionality spectrum. I will not discuss all possible farming systems here, but, based on specific examples ranging from the developed to the developing world, illustrate the complexities underlying the conceptualisation of multifunctionality based on a wide spectrum of opportunities. This section will, therefore, provide empirical substance to the

---

[73] See Knickel and Renting (2000) for a good discussion of the constraints and opportunities for 'measuring' multifunctional quality and for assessing the position of individual actors on the multifunctionality spectrum.

conceptual discussion of multifunctionality presented so far. Section 10.2 will take this argument further by analysing multifunctional transitional potential for different farm *types*.

Let us begin by arguing that *intensive farming systems* (e.g. large-scale grain production, ranching, intensive livestock production or intensive wet-rice farming) are more likely to show weak multifunctionality characteristics. These systems often have low environmental sustainability and are usually aiming at maximum food and fibre production for export beyond the region on the basis of long food chains and strong interlinkages with the global capitalist market. As a result, actors in intensive farming systems may have little opportunities for local embeddedness, often resulting in limited interaction between farmers/agro-businesses and local communities. As farmers in intensive agricultural systems are often preoccupied with putting all their energy into maximising food production, they will often not show high degrees of diversification, associated with traditional ideological notions of 'farming' and 'agriculture' that do not go beyond productivist food and fibre production. Holmes (2006: 145) suggested that "it is only with the emergence of commodified, agro-industrial modes that monofunctionality became the norm, not merely in material outputs but, more importantly, in the supremacy of production goals over consumption and protection". Yet, many studies have highlighted the high degrees of variability inherent in intensive farming systems, and in most cases intensive farmers will also show characteristics of moderate or even strong multifunctionality (Potter, 2004; Robinson, 2004). For example, an intensive arable farmer in Europe may farm most of her/his land in productivist ways (weak multifunctionality) but may decide to set-aside a small part of the land for environmental conservation (strong multifunctionality). Similarly, while the farmer on an intensive holding may be largely preoccupied with maximising production (weak multifunctionality), other members of the farm household may devote their energy to non-agricultural diversification activities (moderate or strong multifunctionality depending on the nature of diversification activity). Even the most industrially farmed holdings will show elements of both productivist and non-productivist action and thought, highlighting the importance of a nuanced multifunctionality *spectrum* of decision-making. Further, as Robinson (2004) emphasised, many small family farms may have little opportunity to opt for strong multifunctionality, as weak multifunctionality pathways may be a necessity rather than a choice. As Freshwater (2002: 9) highlighted for the USA, "multifunctionality is not an appealing option to most farm households" (see also Bohnet *et al.*, 2003, for the UK, or Meert *et al.*, 2005, for Belgium).

Inevitably, discussions about the link between intensive farming systems and multifunctionality also need to take into account that in many agricultural regions *lowland* agriculture is characterised by more intensely farmed systems, while *uplands* are often characterised by more extensive 'traditional' farming systems (Dax and Hovorka, 2004; Mollard *et al.*, 2004; see also Chapter 7). Hodge (2000: 260) suggested that some upland agricultural systems "have often co-evolved with the environment over substantial periods of time to the extent that there is a close relationship between the valued characteristics of the environment and certain attributes of agricultural systems". Of course, much care needs to be taken in such generalisations as the inverse may be the case in some regions (e.g. Andes in South America; lowland desert areas; etc.), and recent research in the EU also suggests that differentiations of weak and strong multifunctionality pathways are not always linked to such lowland/upland 'territorialisation' of multifunctionality transitions (e.g. Kantelhardt, 2006). Yet, as Ronningen (1994) convincingly argued for Switzerland, there is often a link between lowland farming systems and weakly multifunctional pathways and upland systems that often have more opportunities for strong multifunctionality. I will explore this in more detail in Chapter 10 when discussing constraints and opportunities for the adoption of different multifunctionality pathways at farm level.

As Part 2 of this book highlighted, discussions on the positionality of *organic farming* in the multifunctionality spectrum are also illuminating, as organic farming has been the subject of heated debates concerning its contribution towards environmental conservation and sustainability (Tovey, 1997; Dabbert *et al.*, 2004). This has been particularly the case as the extent of officially accredited organic farming areas in Europe has increased from 100,000 ha in 1985 to over 3 million ha by 2000 (Pretty, 2002). On the one hand, there appear to be clear arguments for conceptualising organic farming as embedded in the *strong* multifunctionality model (e.g. Knickel *et al.*, 2004; Lang and Heasman, 2004). Swagemakers (2003: 202), for example, argued that "using some form of organic farming is an obvious choice for multifunctional farms", while Marsden (2003: 232) suggested that "organic production does emphasise the multifunctionality of farming systems". As Twyne (2005) and Hopkins (2005) argued, there is undisputed evidence that organic farming contributes towards increased biodiversity due to lack of application of biochemical inputs on farms. Indeed, trials comparing organic and 'conventional' farming showed 50% more abundance of species on organic farms, and there also appear to be clear advantages for soil protection (Swagemakers, 2003; Twyne, 2005). The environmental sustainability dimension of organic farming as part of the strong multifunctional model is, therefore, not greatly disputed (but see below). There is also an argument that organic farming systems can contribute towards safeguarding local embeddedness of organic farmers, with close interaction between local communities through reciprocal rural-agricultural relationships engendered by local sale of organic products (e.g. in farm shops) and by better 'control' of organic farmers over their crops (i.e. no dependency on large agro-chemical companies). Thus, short food chains often characterise organic farming systems. There is also support for the strong multifunctionality model in the way organic farming is linked to *mental changes*, in particular through open-minded farming and rural populations who may see organic farming as a type of land use that goes beyond productivist food and fibre production. Organic farming is also associated with the strong multifunctionality dimension of open-minded consumers who accept that the nature of farming is in the process of change. As a result, Marsden (2003: 212) suggested that "the organic food sector appears to offer, in theory and rhetoric at least, some form of antidote to the concerns of society over food safety and environmental externalities".

Yet, arguments for organic farming fitting the weak multifunctionality model can also be found (Evans *et al.*, 2002; Lang and Heasman, 2004). Organic farmers are not necessarily more likely than 'conventional' farmers to diversify activities away from food and fibre production, and they are not necessarily more weakly integrated into the global capitalist market than their conventional counterparts. Indeed, some argue that organic farming can be as *intensive* as other forms of farming and can, therefore, be a highly productivist system geared towards profit maximisation (Gregory, 2005). Recent evidence from the EU that farmers converting towards organic farming are increasingly doing so largely for financial reasons (organic conversion subsidies and higher product prices) suggests that ideological and 'green' factors are not always the overriding reason for organic conversion (Twyne, 2005; Kantelhardt, 2006). Some have even contested the environmental sustainability component of organic farming. Schoefield (2005), for example, argued that most organic products in the EU are not produced locally, and that high food miles (with negative effects for climate change) often characterise global organic agriculture. Short food chains are, therefore, not always a characteristic of organic farming systems, exemplified by the increasingly important role played by large supermarkets as the main sellers of organic produce (Marsden, 2003; Kantelhardt, 2006). Some organic farmers (e.g. small producers of mange-tout in Kenya produced for UK supermarkets) are, therefore, more firmly embedded in the global capitalist system than many highly productivist 'agri-businesses' (Barrett *et al.*, 1999; Gregory, 2005) and may, therefore, not be as firmly embedded locally as some authors suggest. 'Indirect' environmental disbenefits of organic farming are also often highlighted: as organic farming uses more space than conventional farming (less productivity), it may encourage the further conversion of non-agricultural land (e.g. tropical rainforest) into productive (often non-organic) use. Thus, although organic farming may contribute *locally* to improved environmental sustainability, at the *global scale* it may encourage further environmental degradation (Gregory, 2005). Marsden (2003: 220), therefore, argued that "the jury is still out on whether organic farming always presents the most environment-friendly or safe-food option" and that the notion of organic farming as an 'alternative' pathway needs to be questioned as it is now difficult to argue that "for the majority of organic produce, systems of retailing and distribution can also defend a similar claim to 'alternative' status". Thus, "organics risks being reduced from a distinct philosophy with profound implications for the way we produce, market and consume food, to just another form of product differentiation and category management" (Marsden, 2003: 218).

This discussion highlights that even those farming systems that intuitively appear as seemingly 'straightforward' to classify as *strongly* multifunctional do not necessarily fit the strong multifunctionality concept on all fronts. Although evidence still seems to suggest, on the whole, that organic farming is situated close to strong multifunctionality and non-productivist action and thought, some elements of organic farming can also clearly fall into the weak multifunctionality category. These dichotomies have been emphasised by Evans *et al.* (2002: 322) who suggested that "although ... organic farming cannot be understood as [strongly multifunctional], neither can they be viewed as productivist in the conventional sense". Depending on individual circumstances of organic farming systems, I would, therefore, categorise organic farming as falling into the moderate/strong multifunctional end of the spectrum. This is also an interesting result for the historical evolution of agriculture. As agriculture before the 19th century was organic everywhere in the world, the fact that organic agriculture is not necessarily strongly multifunctional suggests that it would be wrong to assume that *all* farming systems were historically strongly multifunctional. Indeed, as Section 10.4 will highlight, over time we have also witnessed periods of weak multifunctionality before 1800 within 'organic' farming systems.

What is loosely classified as *subsistence farming* is another interesting example, especially as most of these farms can be regarded as strongly multifunctional (see also Chapter 10). Most farmers around the world are still subsistence-oriented, i.e. where agricultural production is largely aimed at feeding the farm family (and at most the local community) with limited market integration (Pretty, 2002; Barthélémy *et al.*, 2004). Pretty (1995) referred to these systems as the 'forgotten agriculture', as most research and economic attention has been focused on the more profit-driven larger productivist agro-businesses (see also Chambers *et al.*, 1989). Yet, subsistence or semi-subsistence farming still supports around two billion people on Earth (about 30% of the global population), and most of the food production in Africa, for example, comes from such low productivity systems (Pretty, 1995; Losch, 2004). Linking to our discussion above, subsistence farms are most often organic (i.e. no use of external inputs), although very few would be part of *accredited* organic farming schemes. However, compared to their organic farming counterparts in the developed world, most subsistence farmers fit closely with the different dimensions of strong multifunctionality highlighted above, based on the predominance of palaeotechnic and agro-ecological farming strategies (Harris, 1978; Guzman and Woodgate, 1999), their strong ties with their local community (Wilson and Bryant, 1997; Pretty, 2002), the importance of local farmers' knowledge systems, and their mental approach to low-key extensive farming practices (Bell, 2004; Roux *et al.*, 2004). In other words, in these farming regimes the 'new associationalism' identified by Marsden *et al.* (2002) and Clark (2005) as crucial for the re-embedding of agricultural businesses with their locality (and, therefore, a key characteristic of strong multifunctionality) is well developed. As Pretty (2002: xii) highlighted, there is also a historical component to this debate as humans have been farming for some 600 generations (largely for subsistence), and "for most of that time the production and consumption of food has been intimately connected to cultural and social systems". As a result, subsistence farming is one of the few farming systems that comes close to the strong multifunctionality model. McCarthy (2005: 775), therefore, suggested that the positive externalities associated with the notion of multifunctionality "are disproportionately produced by more marginal producers – for example, that farms that are less intensive and more diverse are more likely to provide wildlife habitat and valued cultural landscapes".

Similarly, Holmes (2006: 145) argued that "traditional, primarily subsistence modes of agricultural occupance also are multifunctional, not merely through product diversification, but more critically, through having consumption and protection values embedded within production modes of resource use". Losch's (2004) suggestion that most farmers in developing countries have no resources to pay for 'other' functions of agriculture, thus, may be a too simple view of the multifunctional non-farming benefits produced by subsistence farmers. Indeed, subsistence-based agricultural practices produce strong multifunctionality functions often at *no cost* to the farmer (i.e. these functions are part of their day-to-day farming practices).

However, the subsistence farming category covers a wide range of different farmers. On the one hand, it still includes *shifting cultivators* and *nomadic pastoralists* who roam across wide areas of forest or drylands in cycles that have guaranteed sustainable farming for thousands of years (Wilson and Bryant, 1997; Mazoyer and Roudart, 2006). On the other hand, many subsistence farmers are now becoming more embedded into the globalised capitalist system (see Chapter 10) and may, as a consequence, have more opportunities to sell their products on the market and, in some cases, acquire technological equipment or artificial fertilisers that may boost productivity (Goodman and Watts, 1997). Although most subsistence farmers may be strongly multifunctional *out of necessity*, many may choose farm development pathways that take them away from the strong multifunctionality model (Roux *et al.*, 2004). These farmers, therefore, may sacrifice local embeddedness for global linkages, environmental sustainability for intensification, and mental closeness with nature (although one should not romanticise subsistence farming) with productivist outlooks. The *chosen path* of many subsistence farmers may, therefore, be one that takes them rapidly away from strong multifunctionality (see Section 10.4).

Another farming type that has received substantial coverage over the past few decades – particularly in a developed world context – is that of *hobby* or *lifestyle farming* (Gasson, 1988; Burton and Wilson, 2006). Lifestyle farmers adopt farming as a hobby and do not rely on the sale of food and fibre products for economic survival (i.e. they often have a stable income outside of farming) (Holloway, 2002). Although these farms still fall into our definition of the multifunctional *agricultural* spectrum (see Section 9.3), they are situated close to the strongly multifunctional non-productivist end of the spectrum. Indeed, for these farmers one of the key drivers of agricultural intensification does not exist, as they most often do not need to maximise profits to ensure farm survival. Thus, lifestyle farming can be interpreted as a 'luxury' that will, currently, almost exclusively be found in developed countries. Lifestyle farming, therefore, is closely associated with the strong multifunctionality model. Yet, in the European and North American context, lifestyle farmers are frequently wealthy urbanites who have not been brought up in the regions where they bought their farm (Holloway, 2002). The indicator of 'local embeddedness' may, therefore, be relatively weak. Further, as these 'farmers' often do not come from a farming background, the fact that they will often farm in extensive environmentally-sustainable ways does not necessarily mean that their mental image of farming concurs with the strong multifunctionality model (i.e. they could be productivist in their outlook).

Similar to agro-business farming, *plantation agriculture*,[74] meanwhile, can be positioned largely within the weak multifunctionality model, due to its intensive monocultural and often environmentally unsustainable farming practices (e.g. plantations are often created on formerly forested lands), its export-oriented nature and often strong embeddedness into the global capitalist market (often plantations are owned by foreign companies) and its weak ties with local communities (Harrison, 2001; Robinson, 2004). In many parts of the developing world, plantation agriculture (e.g. coffee, cacao, palm oil) has replaced strongly multifunctional subsistence farming systems (e.g. parts of India or Africa) and has, therefore,

---

[74] I acknowledge that the term 'plantation' is increasingly problematic although still widely used in agricultural literature on developing countries (e.g. Grigg, 1974; Robinson, 2004).

contributed towards a weakening of formerly strongly multifunctional agricultural systems (Barthélemy, 2004; Roux *et al.*, 2004). Thus, Robinson (2004: 149) emphasised that "a central feature of the agricultural geography of the South ... is the ongoing retreat of subsistence production in the face of an 'advancing wave' of commercialisation". In this context, it is important to differentiate between plantation agriculture and other, often similar, forms of agriculture. A 'plantation' implies artificial replacement of original vegetation by intensive monocultures and is, therefore, often environmentally unsustainable. Thus, a coconut plantation established 'artificially' for maximum yields (often for export) has different multifunctional quality than a naturally grown coconut grove that may be environmentally sustainable with coconuts used largely for subsistence (Grigg, 1974).

Farming systems based on *Green Revolution* technologies also often show tendencies of weak multifunctionality (Parayil, 1999; Pretty, 2002). As highlighted earlier, the Green Revolution was characterised by the advent of new high-yielding varieties which, when cultivated with modern fertilisers and pesticides, transformed many agricultural systems in the developing world (primarily wheat and rice cultivation). Agricultural scientists bred new varieties of staple cereals that matured quickly, permitting two or three crops to be grown each year. Although output in these systems has often increased dramatically, this intensification has often occurred through the loss of former agricultural systems characterised by strong multifunctionality (Robinson, 2004). It is estimated that over half of the global area of wheat, rice and maize is now using modern varieties based on Green Revolution technologies (Pretty, 1995; Parayil, 1999). Yet, as Pretty (2002) emphasised, Green Revolution agrarian systems also span a wide range of agricultural systems with different farming intensities, different forms of embeddedness in local rural communities, and pronounced differences in interlinkages with the global market. Thus, some farming systems based on Green Revolution crops will also depict elements of moderate or even strong multifunctionality (e.g. some systems can be environmentally and socially sustainable) (Hazell and Ramasamy, 1991).

A similar trend can be seen in many farming systems (both in developing and developed countries) linked to the development of *biofuels*, as an increasingly viable alternative to fossil fuels (Knickel and Renting, 2000; Belletti *et al.*, 2003a). Although the intensity of biofuel 'plantations' varies widely, on the whole these systems are often characterised by weak multifunctionality. In most cases they are produced relatively intensively, may be damaging to the environment (need for biochemical applications), and usually have long supply chains to centrally located refineries. Many farmers are 'switching' from food crop production to non-food biofuel crop planting (e.g. oilseed rape), resulting in a re-intensification of farming at a time when extensification of farming has become the norm (e.g. in the EU) and, consequently, a move away from moderate or even strong multifunctionality towards weaker multifunctionality (Kantelhardt, 2006). However, there are also some interesting *localising* tendencies in operation, where farmers are calling for more locally embedded production, refining and consumption of biofuels (also to address global warming issues linked to long-distance transport), which suggest tendencies for moderate or even strong multifunctionality despite relatively intensive production.[75]

Finally, *mixed farming systems* are another interesting example highlighting both the complexity of the multifunctionality spectrum and the importance of the recognition of a wide range of multifunctional decision-making pathways open to farmers. Mixed farms often show a wide range of farming crops and activities that can span the entire range of the multifunctionality spectrum. They often have one part of their food and fibre production that is relatively intensive (i.e. weak multifunctionality), while other parts may show tendencies of strong multifunctionality. Some mixed farms may, at times, be more closely associated

---

[75] A similar debate can be found on GM crops. As Chapter 6 highlighted, on the one hand these crops can be interpreted as strongly multifunctional (e.g. reduced need for chemical inputs) while, on the other hand, they may also engender intensification and increased vertical embeddedness characteristic of weak multifunctionality (Robinson, 2004).

with productivist action and thought if the business is oriented towards exporting food outside of their immediate locality, while others may be closer to the non-productivist boundary of farm development pathways. However, it is likely that mixed farms occupy a more moderately multifunctional role than agro-businesses (indeed some highly productivist agro-businesses may have once been mixed holdings) and a more moderately multifunctional role than subsistence or lifestyle farmers (although many of the latter could also be classified as 'mixed' in the wider sense of the word) (see also Fig. 10.2). As highlighted above, it is particularly farming systems that occupy the conceptual 'middle ground' of the multifunctionality spectrum that need to be investigated on a case-by-case basis, due to the complex multifunctional pathway opportunities open to these farms (see also Section 10.3).

Several points emerge from this discussion. First, industrial agricultural systems of the developed world (and where these systems have been adopted in developing countries) are more likely to depict weak multifunctionality, and possibly also have the largest need for a shift towards strong multifunctionality (see Section 10.4). Second, farming in the developing world is still largely characterised by strong multifunctionality *out of necessity*, but, for reasons of income maximisation, the chosen path of many farmers may often be to move towards agricultural regimes characterised by moderate or even weak multifunctionality (e.g. plantation farming, Green Revolution crops). Third, agricultural export-oriented countries will tend towards weak multifunctionality as action and thought is likely to tend towards productivist farming practices. Countries or regions dominated by subsistence farming, on the other hand, are more likely to show strong multifunctionality tendencies. Thus, farming regions more firmly embedded into the global capitalist system are generally more likely to be characterised by weak multifunctionality, while 'traditional' farming systems (usually less intensive, small-scale and locally well embedded) are often more strongly multifunctional.[76] Fourth, and closely linked to the notion of spatial heterogeneity discussed in Chapter 7, mountainous and other agriculturally disadvantaged areas are more likely to be strongly multifunctional, while more intensively farmed lowlands usually would fall into the weak/moderate multifunctionality categories (Rapey *et al.*, 2004b). In the UK context, Marsden and Sonnino (2005: 25) confirmed this territorialisation by arguing that "multifunctional agriculture will develop very unevenly in the UK ... it is likely to be successful in regions, such as Wales and the South West of England, that are, to a different extent and in different ways, committed to the ideals of sustainable rural development, whereas in regions such as the East of England ... [strong] agricultural multifunctionality is not yet a development option". Fifth, there will be high geographical diversity of weak, moderate and strong multifunctionality spaces, even within small territories. This spatial heterogeneity suggests that we may also witness juxtaposition of different types of multifunctionality at both small and large scales (see Section 9.5), depending on orientation of farming districts and individual farmers (Van der Ploeg and Roep, 2003; Rapey *et al.*, 2004b). Sixth, weak, moderate and strong multifunctionality are not fixed entities, as individual farms or even entire regions may change their farming orientations relatively quickly. The dynamic nature of the multifunctionality spectrum, therefore, needs to be acknowledged and will be investigated in further detail in Chapter 10. Finally, even farming systems that intuitively appear easy to categorise (e.g. organic farming as strongly multifunctional; agro-business farming as weakly multifunctional) may show elements of both productivism and non-productivism. The latter highlights the danger of broad generalisations and emphasises the need for case-by-case investigations of the positionality of agricultural actors/institutions/holdings along the multifunctionality spectrum.

---

[76] There are substantial debates about 'traditional' farming, and it would be wrong to argue that all traditionally farmed systems tend towards strong multifunctionality. Indeed, some 'traditional' farming systems, such as intensive wet rice cultivation in Asia, will show tendencies of weak or moderate multifunctionality (see above).

9.4.3 Policy and strong multifunctionality

The discussion in Part 2 highlighted that policy change has formed an integral part of conceptualisations of the transition to post-productivism (see Chapters 5 and 6). Further, Chapter 8 highlighted that the policy-based view of multifunctionality has formed one of the key pillars of earlier conceptualisations of the term. The above discussion confirms that policy is intricately linked to the notion of multifunctional agriculture, in particular regarding conceptualisations of strong multifunctionality. In this section, I wish to take this discussion further by analysing how we can better understand the role of policy by linking discussions on policy to the multifunctionality spectrum. The aim is not to provide a holistic analysis of the interlinkages of policy and multifunctionality (see Chapter 8 for detail), but to highlight, through specific policy examples, the utility of the notions of *weak, moderate* and *strong* multifunctionality (see also Section 10.4).

In line with Potter and Burney (2002) and Potter and Tilzey (2005), I argue that policy can be an important *driver* for action and thought along the multifunctionality spectrum, although, as Chapter 10 will discuss, policy may not *always* be the key driving force for changes in multifunctional decision-making. In many agricultural areas of the world, policy has played an important part in influencing action and thought along the multifunctionality spectrum. On the one hand, policy has often acted as a driver for *weak multifunctionality* by encouraging productivist tendencies, leading Pretty (1995: 3; emphasis added) to argue that farming has remained strongly multifunctional in many parts of the world "*despite* existing policy environments". As Chapter 5 highlighted, early CAP policies, for example, were largely predicated on a weak multifunctionality model, encouraging farm development pathways close to the productivist end of the spectrum. On the other hand, Chapters 6 and 8 discussed recent policies that have tended to promote the *strong multifunctionality* model by encouraging non-productivism and activities that encourage improved embeddedness and interlinkages between rural communities and localities. In particular, AEPs and policies for environmental conservation have been described as types of policies attempting to engender strong multifunctionality pathways (Dobbs and Pretty, 2004; Marsden and Sonnino, 2005), while the EU LEADER programme has been highlighted as a successful policy approach with real potential to empower local stakeholder groups in rural communities akin to strongly multifunctional pathways (Ray, 1998; Di Iacovo, 2003).

It could be argued that, over the past 50 years or so, agricultural policy has generally tended to encourage the weak multifunctionality model, in an EU context based on what Potter (2004) referred to as an 'agri-centric view of rural development' (see also Chapters 5 and 8). The ultimate goal of most agricultural policies has been, at least until recently, to provide an enabling framework for productivist maximisation of food and fibre production, to ensure regional and national self-sufficiency (in particular during the period after the Second World War), and to raise incomes of farmers by providing policy-based incentives for intensification, mechanisation, rationalisation and specialisation (Winter, 1996; Ingersent and Rayner, 1999). As Pretty (2002: 3) emphasised, "recent thinking and policy has separated food and farming from nature, and then accelerated the disconnectedness". As Chapter 5 highlighted, until recently little emphasis was placed on environmental protection, local embeddedness, diversification, or non-productivist perceptions of what farming and agriculture are, or should be, about. However, as Chapter 8 emphasised, recent agricultural reforms linked to national (e.g. USA, Australia), trans-national (e.g. EU through the CAP) or global policies (e.g. WTO negotiations) have placed more emphasis on 'holistic' policies that address both productivist and non-productivist issues (Potter and Burney, 2002).

Recent policy reform could be construed as attempting to engender a stronger multifunctionality model. Knickel *et al.* (2004: 97) suggested that "European agriculture and the CAP are moving from a mere monofunctional towards a more holistic and integrated perspective". Yet, even policy dimensions within Agenda 2000 and the CAP Mid-Term Review to 2007 – hailed by some as a multifunctional policy package (see Chapter 8) – do not necessarily match notions of strong multifunctionality. Thus, Lowe *et al.* (2002: 4) argued that "the Agenda 2000 outcome was … deeply compromised and must be judged a missed opportunity to transform the CAP". Gallardo *et al.* (2003: 170) were similarly critical by arguing that "Agenda 2000 … could be in conflict with the defence of multifunctionality. One of these aspects is the emphasis on the competitiveness and productivity of European agriculture". They, therefore, suggested that "the current CAP is not the most suitable policy, either for the defence of the multifunctional character of agriculture or for the integration multifunctionality and competitiveness objectives" (Gallardo *et al.*, 2003: 174). Marsden (2003) similarly emphasised that the current raft of CAP policy measures still focuses too much on the land-based farmer as recipient of public funding, rather than on non-agricultural actors who may be – as Chapter 7 highlighted – those increasingly shaping the countryside and contributing to the vibrancy of the rural economy. Clark (2003: 227; emphasis added), meanwhile, concluded that many of his respondents "saw the effects of the CAP's commodity regimes as an active institutional *constraint* to the emergence of multifunctional agriculture", and that withdrawal (complete or partial) of farmers from the CAP subsidy regime should be seen as a key indicator of an attempt to establish stronger on-farm multifunctionality pathways. Similarly, Potter and Tilzey (2005) were unsure how effective current CAP measures will prove in holding the line against engrained productivist ideologies driven by neo-mercantilist and neo-liberal policy agendas. They suggested that "while the dominant policy trend may be towards the decoupling of agricultural support in line with WTO disciplines, the rolling out of government-funded rural development programmes under the CAP's 2nd pillar appears to be designed to safeguard what can only be described as post-productivist enclaves" (Potter and Tilzey, 2005: 15). It is, therefore, tempting to agree with Marsden (2003: 2) who suggested that "however urgent the calls are for a serious reform of public funds away from the productivist logic, and towards a more 'multifunctional' or orchestral approach to rural policy, the inertia in this tendency will require alternative strategies associated with different types of state, community and market-based action" (see also Chapter 10).

As the discussion throughout this book has highlighted, the issue is not that of a simple shift from policies engendering weak multifunctionality to a policy environment emphasising strong multifunctionality. Indeed, the complexity of the policy environment highlights the importance of recognising a wide *spectrum of Deleuzian decision-making pathways* that spans the entire breadth of possibilities ranging from weak to moderate to strong multifunctionality. As Figure 9.6. highlights, just as different farming systems can highlight different types of multifunctionality (i.e. weak, moderate or strong), the policy environment depicts highly complex processes that contain both weak and strong multifunctionality dimensions.

Some policy arenas may be relatively easy to brand as either strongly or weakly multifunctional. Thus, production subsidies under the CAP (referred to as 'perverse subsidies' by some; cf. Pretty, 2002; de Groot, 2006) have generally engendered weak multifunctionality by encouraging farmers to intensify production with little regard for the environment or socio-cultural cohesion of rural areas[77] (Lang and Heasman, 2004; Potter and Tilzey, 2005; see Chapter 5). EEC farm modernisation grants and less favoured areas policies of the 1970s, aiming to boost the economic profitability of farms in disadvantaged areas, particularly epitomised a shift towards weak multifunctionality in the wake of the rejection of the Mansholt Plan of the late 1960s and the consequent adoption of a policy of 'farm survival' at whatever environmental and financial cost (Potter, 1990; Caraveli, 2000; see also Argent, 2002, for similar policy processes in Australia, or Boody *et al.*, 2005, for the USA). Similarly, policies encouraging new EU member states (especially Mediterranean countries during the 1980s and 1990s) to 'catch up' in agricultural productivity terms with

---

[77] Globally, it is estimated that such 'perverse' agricultural subsidies total between US$ 1 and 2 trillion (De Groot, 2006).

their Northern European counterparts have also reduced multifunctional quality (Delgado *et al.*, 2003; Wilson and Juntti, 2005). Such policies have been closely associated with the branding of the CAP as an 'engine of destruction', based on increasing policy-led uniformity of the farmed landscape, and are seen to have led to a *homogenisation* of agriculture along often *monofunctional trajectories* (Van Huylenbroek and Durand, 2003). Indeed, many of these policies have led to a *shift away* from what may have been relatively strong multifunctional farming systems in the past – for example, extensive livestock systems in Spain or Greece – often with resulting losses in horizontal embeddedness of rural communities and environmental sustainability (e.g. Povellato and Ferraretto, 2005, for Italy) and, simultaneously, increased embeddedness of formerly traditionally farmed extensive agricultural systems into EU and global capitalist agricultural production systems (e.g. Vieira and Eden, 2005, for Portugal).

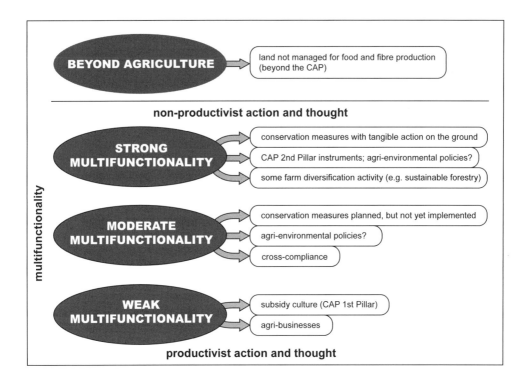

Fig. 9.6. Policy and the multifunctionality spectrum

Figure 9.6 highlights that the same policy arena (e.g. the CAP 2nd pillar since 2000) may have policy components that fall into *both* the strong and weak multifunctionality end of the spectrum (Lowe *et al.*, 2002; Ward and Lowe, 2004). For most agricultural (and environmental) policies, a simple classification into weak, moderate or strong multifunctionality is, therefore, difficult, further exacerbated by the fact that policy does not always lead to *tangible* action on the ground (Wilson and Juntti, 2005). The example of European AEP is a particularly suitable case-in-point due to the multi-faceted and often ambiguous nature of these policies (Buller *et al.*, 2000; see also Chapters 7 and 8). Figure 9.6 suggests that AEP has, on the one hand, helped conservation of the farmed landscape by, for example, encouraging the protection of remnant farm woodlands or by inducing farm extensification practices that have helped reduce environmental pollution (e.g. the new Environmental Stewardship Scheme in the UK, in particular 'Higher Level Stewardship' that requires sophisticated environmental management practices) – referred to as conservation measures with 'tangible' action on the ground in Figure 9.6. AEP, therefore, has to some extent at least contributed towards the *strongly multifunctional* dimension of environmental sustainability (cf. Brouwer, 2004b; Bills and Gross, 2005). The recently introduced policy requirement of *cross-compliance* as part of the CAP 2nd pillar (i.e. farmers will only receive production subsidies if they comply with nationally defined minimum environmental standards) may be a particular case-in-point (Lowe *et al.*, 2002; Gallardo *et al.*, 2003), as is the Conservation Reserve Programme in the USA (since 2002) which pays farmers for greater environmental protection on land kept in production rather than encouraging farmers to take land completely out of production (Boody *et al.*, 2005; McCarthy, 2005). In particular, the need for EU farmers' compliance with Good Agricultural and Environmental Condition (GAEC) requirements as part of cross-compliance linked to the new Single Farm Payments can be interpreted as a step towards stronger multifunctionality (at least in environmental terms) (Buller, 2005; Mather *et al.*, 2006). Thus, some commentators have argued that Single Farm Payments and cross-compliance may *reduce* the need for farmers to further intensify. Osterburg (2005) suggested that the system of 'one-off evaluation' of cross-compliance payments based on *one* reference period (2000-2002) means that already extensive (and arguably more strongly multifunctional) holdings will gain in relative terms, while intensive holdings are not likely to obtain higher payments than their extensive counterparts. This may lead to a gradual lessening of the need for European farmers to embark on weakly multifunctional pathways in the future, with early evidence suggesting that some reductions in production intensity of cereal and beef production are already occurring (see Osterburg, 2005, for the case of Germany).

However, EU AEP has also been negatively branded as a 'backdoor subsidy' that merely adopts the façade of 'green' action, while perpetuating farmers' dependence on subsidies for survival and continuation of conventional food and fibre production (Potter and Burney, 2002; see also Chapter 7). Barthélémy *et al.* (2004) emphasised that the incremental nature of AEP and other 2nd pillar policies has meant that, by default, these policies had to be gradually adjusted to 'fit' emerging European multifunctionality discourses (see Chapter 8), highlighting that the multifunctionality effect of policy also depends on how certain policies are used, interpreted and implemented by various stakeholders (Potter and Tilzey, 2005). In the context of the UK, for example, Buller (2005: ii) referred to "the reluctance of the UK agricultural policy community to embrace multifunctionality as a paradigm for rural policy", while Garzon (2005: 7) argued that the UK was "initially at odds with the 'continental' vision of multifunctionality". That many agri-environmental schemes in the EU merely support *maintenance* of agricultural activities rather than *change* (Buller *et al.*, 2000) lends further credence to the relative lack of enthusiasm for policies encouraging strongly multifunctional farming systems (see also Chapter 7). As Walford (2003b: 501) emphasised, "commentators have identified flaws in interpretations of recent agri-environmental policy changes as representing anything other than superficial tinkering while underlying support for agricultural production persists". Thus, many AEPs have *not* contributed towards substantial changes in farming practices and can, therefore, at most be classified as contributing towards *moderate* multifunctionality (e.g. Kleijn and Sutherland, 2003; Kleijn *et al.*, 2006). Indeed, as Buller (2005: 3) emphasised for most agri-environmental schemes operating in the EU, "the promotion of agricultural multifunctionality [is not] an explicit and stated objective", and "the current implementation of the revised 1st pillar is likely to have significant *positive* and *negative* implications for the delivery of greater agricultural multifunctionality" (Buller, 2005: 9; emphasis added). Further, while cross-compliance can, in principle, be seen to encourage strong multifunctionality (see above), in practice it may result in highly differentiated multifunctionality outcomes on the ground, as EU countries have much leeway about implementation and interpretation of cross-compliance requirements (subsidiarity) (Lowe *et al.*, 2002). This prompted Gallardo *et al.* (2003: 173) to argue that "the weakness with which the possibilities to apply the principle of cross-compliance has been presented ...[and] the contradiction between the new multifunctional aims [of Agenda 2000] ... make the instruments which should in theory defend multifunctionality less efficient" (see also Goodman, 2004).

As Lang and Heasman (2004) emphasised, a similarly ambiguous policy example relates to the recent introduction of improved food quality, branding and assurance schemes linked to designations of, for example, 'appélation d'origine controlée', 'certificates of special character', 'protected geographical indication', 'eat the view' or the 'Scottish and Welsh Quality Assured Beef and Lamb schemes' (see also Ilbery *et al.*, 2000; Marsden, 2003). Such schemes have been encouraged by various policies, such as EU Regulation 2081/92 or the EU LEADER initiative. Intuitively, such designations should be closely associated with strong multifunctionality, as they encourage local embeddedness by emphasising high quality production from one particular region (e.g. wines from the Côtes du Rhône area in France as a region associated with high food quality) and by emphasising food quality over quantity (Moran, 1993; Marsden, 2003). As Belletti *et al.* (2003a: 71) emphasised, "collective valorisation strategies centred on typical products are often strictly related to multifunctionality". Yet, food branding from specific agricultural regions can also promote these products in the global food market, thereby encouraging long food chains (e.g. Côtes du Rhône wines can be bought in most supermarkets of the developed world), and leading to increased interweaving with the global capitalist market of what may have been hitherto locally embedded (and therefore more strongly multifunctional) farm holdings. There is also little evidence that specific regional branding leads to a change in ideological support for productivist farming pathways (Jones and Clark, 2001). In this sense, designations such as 'appellation d'origine controlée' may have highly divergent effects along the multifunctionality spectrum, some of which resonate with strong multifunctionality, while others appear to engender weak(er) multifunctionality pathways. In addition, Evans *et al.* (2002) highlighted that quality is not solely a feature of specialist food policies, and that notions of 'quality' are also part of the mass food market. They argued that "the result is a coexistence of quality and quantity" (Evans *et al.*, 2002: 319) in today's food markets – a coexistence that confirms the existence of a *spectrum* of multifunctional decision-making pathways. Goodman (2004) added to this that global retailers also often source food locally (i.e. individual retail branches make ample use of locally embedded products), thereby further blurring the boundaries between 'global' and 'local' (a process he terms 'market-led competitive territoriality').

Yet, it is for the macro-policy environment that the most challenging questions regarding multifunctionality and policy emerge. The CAP RDR is a key example of the ambiguous positioning of many policies in the multifunctionality spectrum (Lowe *et al.*, 2002; Bryden, 2005). This regulation has undoubtedly placed greater emphasis on agriculture-rural interactions by directing funds not only at agricultural but also rural actors. It, therefore, has the potential to contribute to the strongly multifunctional dimension of improved horizontal embeddedness of agricultural and rural actor spaces (Ward and Lowe, 2004; Clark, 2005). However, the RDR is still part of what many describe as a staunchly productivist CAP and, therefore, may contribute little towards strongly multifunctional *ideological* changes in the way society and rural communities perceive 'agriculture' and 'farming' (Barthélémy *et al.*, 2004; Buller, 2005). That about 90% of current CAP expenditure (in 2006) continues to be targeted at providing production subsidies highlights the continuing predominance of weak multifunctionality ideologies underlying most EU agricultural policy-making. In addition, the flexibility in EU national agricultural policy through subsidiarity means that there is room for significant divergence in national multifunctional policy agendas (Lowe *et al.*, 2002; Goodman, 2004). While CAP subsidies are likely to be phased out by 2013, it remains to be seen whether the weak multifunctionality policy model in the EU will be replaced by a new framework more likely to engender strong multifunctionality (see also Section 10.4).

Similarities can be drawn with the USA where the continuation of productivist agricultural policies led Boody *et al.* (2005: 36) to argue that "rather than support commodity production, US farm policy should support [strongly multifunctional] agricultural diversification to enhance nonmarket ecosystem services". Freshwater (2002: 14) similarly argued from a US perspective that "if multifunctionality is to become the basis for developing farm policy in the US it will have to come from a recognition that the existing way that farm policy is defined and implemented is not adequately serving the broader interests of the American public". In particular, Freshwater emphasised that while the 1996 US Farm Bill adopted a policy stance that would lead to farmers being weaned from government subsidies, the 2002 Farm Bill was a productivist-oriented policy that returned to the 'old' approach of providing counter-cyclical price supports to a small number of major commodities. As a result, Freshwater (2002: 17) argued that there is little link between current US agricultural policy and strong multifunctionality pathways, and that farm policy both in the USA and EU "has remained narrowly focused on the financial returns to farmers from producing specific commodities"[78] (see also Blandford and Boisvert, 2002).

The role of the WTO as a global policy framework that has increasing influence on national, EU and US agricultural policy-making also needs to be considered (Potter and Tilzey, 2005). Chapter 8 highlighted that some authors place great emphasis on changes in global and supra-national policy-making as a basis for conceptualisations of the notion of multifunctional agriculture (e.g. Potter and Burney, 2002). In 1994, the WTO brought agriculture into global discussions for the first time (Bennett, 2005), underpinned by increasing globalisation trends and increasing global pressures for reductions in trade barriers for agricultural products. Undoubtedly, such global policy influences are becoming an increasingly crucial driving force for different multifunctional decision-making pathways in any national, regional or local context (Durand and Van Huylenbroek, 2003; see also Section 9.5). Indeed, as Chapter 8 highlighted, the notion of 'multifunctionality' is used here as a legitimate non-trade concern within the terms of the URAA, justifying special consideration under the negotiations (Potter, 2004). Yet, again, the picture is complex and highlights the importance of acknowledging a wide *spectrum* of multifunctionality along complex decision-making pathways. On the one hand, WTO negotiations fall squarely into the weak multifunctionality dimension, as the ultimate aim is to increase agricultural profit and production by reducing, or abandoning, internal tariffs and thereby increasing the flow of transnational and global trade with agricultural commodities (Potter and Burney, 2002). Thus, Peterson *et al.* (2002: 440) argued that "any resolution to the multifunctionality issue within the current agricultural policy framework would be problematic". Similarly, Blandford and Boisvert (2002) suggested that the current structure of trade policy dispute resolution is fundamentally incapable of dealing with multifunctionality as an argument for agricultural policy. Marsden (2003: 15), therefore, referred to the "new super-productivist, global regimes brought about by the new CAP and WTO agreements", while Potter and Tilzey (2005) highlighted how the WTO framework has forced the EU to agree to a strategy of increased market access. On the other hand, these same processes can be seen to encourage strong multifunctionality by placing pressure on national governments to *withdraw* subsidies and decoupling domestic support, thereby removing key incentives for productivist agricultural intensification (Peterson *et al.*, 2002; Potter and Tilzey, 2005).

Four key points emerge from this. First, as Figure 9.6 (above) highlights, policy influence spans the entire spectrum of multifunctional decision-making, ranging from staunchly productivist policies (e.g. some elements of the CAP) to policies encouraging action and thought close to the non-productivist end of the multifunctionality spectrum (e.g. conservation policies for the countryside that have led to *tangible* environmental protection).

---

[78] Bell (2004) estimated that US farmers receive an average annual subsidy cheque for US$20,000-25,000.

Bryden (2005: 13; emphasis added) argued that "the inter-relationships between farming, farm households, multifunctionality, rural development and policy thus have the characteristics of a dynamic system which will vary between different types of rural area according to characteristics of farming, farm households, institutional and governance structures at local and other levels, the nature and practice of rural development, and policies, and the factors influencing these". Just as different farming systems may depict different levels of multifunctionality (see Section 9.4), the policy process covers an equally wide range of influences ranging from weak, to moderate, to strong multifunctionality. This was reiterated by Clark (2003: 187) who showed how policy interests of regional stakeholders in the UK span the entire spectrum of multifunctionality, with many surveyed stakeholders believing that policies often "impeded the uptake of [strong] multifunctionality".

Second, the more complex a policy, the more likely it is to contain policy elements that span the *entire* multifunctionality spectrum. Thus, 'single' policy issues such as farm modernisation grants of the 1970s can be more easily classified along the multifunctionality spectrum (i.e. as weakly multifunctional) than more complex policies such as Agenda 2000 or the CAP RDR (Lowe *et al.*, 2002; Marsden, 2003). The latter contain various 'policy bundles' with often different expressions and outcomes along the multifunctionality spectrum – exacerbated by highly differentiated transposition of these policies by EU member states based on subsidiarity (Clark, 2003). Policy is, therefore, rarely mono-dimensional as many advocates of the shift towards post-productivism have advocated, but often shows multi-dimensional characteristics along a wide multifunctionality spectrum.

Third, the importance of policy in influencing multifunctional pathways depends on the *political context*. Indeed, policy is closely associated with the political arena and state coercive powers. Based on the notion of the weak and strong state model (Hall, 1993), there are examples at one extreme end of the political coercion spectrum in which the state, and its associated policies, are almost entirely in control of directing multifunctionality pathways (e.g. North Korea; Myanmar; Soviet Union until 1990; Cuba until recently). In countries such as North Korea, for example, weak multifunctionality is likely to predominate, based on state-led goals of maximisation of food and fibre production to satisfy self-sufficiency needs and/or prestige-related maximisation of agricultural exports (Hale, 2005; Woo, 2006). In these autocratic countries, the *weak multifunctionality* regime is, therefore, likely to predominate (see Section 10.2). At the other extreme end of the political coercion spectrum, some agricultural territories remain relatively untouched by state policies (e.g. remote shifting cultivation communities in tropical forests of Amazonia, Cameroon, or Papua) (Parayil, 1999; Mazoyer and Roudart, 2006). Here, tribe-specific 'policies' may continue to regulate agricultural activity as they have done for thousands of years (Wilson and Bryant, 1997) and multifunctional agricultural decision-making will occur largely independently from state influence. Interestingly, such agricultural systems often depict tendencies of *strong multifunctionality* (see Section 9.4). However, such 'state-independent' agricultural territories are becoming rare, and even the remotest corners of the globe are increasingly embedded into the global capitalist system and coming under growing control of nation states with jurisdiction over these tribal territories (e.g. Bryant, 2005, for the Philippines). This highlights that increasing state policy influence over agricultural territories may often herald a shift towards weak(er) multifunctionality pathways (see also Section 10.3).

Finally, although policy can be an important driver for multifunctional decision-making, it is only *one aspect* of multifunctionality (see also Chapter 8). Room needs to be made for the possibility of *policy-independent* autochthonous multifunctional pathway development (e.g. remote shifting cultivator communities). However, we do not necessarily need to invoke examples from far-flung areas to highlight that agricultural change can often occur *parallel* to, or indeed *in opposition to*, the policy process. The inability of EU policy over the past decades to halt the rapid decline in farming, linked to what some refer to as 'accelerated agricultural structural change', is a case in point. Despite concerted efforts of EU policy-makers to find policy solutions to halt the rapid shrinkage in farm numbers since the 1950s, the number of EU farmers selling or abandoning their farms continues to increase (Robinson, 2004). This has potentially disastrous consequences for strong multifunctionality dimensions such as agricultural/rural cohesion, food supply chains, diversification activities and environmental sustainability. As Gallardo *et al.* (2003: 175) argued, "what in other economic sectors would be a process of natural adjustment in search of efficiency has perverse effects in the case of agriculture. Indeed, the termination of the activity of those economic agents incapable of being competitive in the market means not only the abandoning of the productive activity but may also generate loss of joint [i.e. multifunctional] functions". Although policy has had some effect in reducing the acceleration of farm abandonment, autochthonous processes linked to lack of competitiveness, failure of policies to provide sustainable non-agricultural incomes in some rural areas, and general disillusionment of farmers with their own identities as food and fibre producers (in the UK farmers are the income group with the highest rate of suicides!) have taken on a *multifunctional life of their own*. Although policy – or in this case failure or lack of policy – can be blamed for the fate of many of these farmers, this highlights that policy and multifunctionality have to be seen at least as part of a *reciprocal relationship* in which, on the one hand, policy can substantially shape multifunctionality while, on the other hand, autochthonous developments may also lead to changes in multifunctional policy direction (e.g. abandonment of Mansholt Plan in 1968; need for policies relevant to rural communities; policies that help young farmers 'stay on the land'). This may be particularly the case as, in the EU context at least, we may witness a shift from a universalistic CAP-regulated supply system based on productivist production targets towards a more privately regulated supply chain approach which gives precedence to quality control (Marsden, 2003). While the multifunctional agricultural regime has been *partly orchestrated* by policy (at least in a developed world context), autochthonous agricultural processes have also partly shaped the multifunctional policy environment.

9.5 The geography of multifunctionality

The discussion in the previous sections has highlighted that conceptualisations of the multifunctional agricultural regime have to consider spatial issues. This is particularly true for the embeddedness of the multifunctional agricultural regime within the wider sphere of the multifunctional *rural* regime discussed in Section 9.3, as well as regarding multifunctional territories, agricultural systems and the spatiality of multifunctional policies discussed in Section 9.4. Many theorists have emphasised the importance of spatial scale issues in conceptualisations of multifunctionality. For example, Peterson *et al.* (2002: 441) argued that "because of the spatial diversity of the agricultural landscape, some have advanced the notion that many policies aimed at addressing the multifunctional nature of agriculture must be administered at sub-national, regional or local levels". In a similar vein, Priemus *et al.* (2004: 270) argued that "in defining multifunctional land use, it is important to identify time dimensions and scale levels". This was reiterated by Bryden (2005: 8) who emphasised that "there is a continuing debate on the relationship between multifunctionality and territorial development in rural areas".

   Massey (2001) posited that the social relations which constitute space are not organised into scales so much as into *constellations of temporary coherences*. These coherences are set within a social space which is the product of relations and interconnections from the very local to the global. Conceptualisations of agricultural multifunctionality should, therefore, be closely intertwined with issues of scale – the 'geography of multifunctionality' – and we need to know to which specific spatial territories weak, moderate and strong multifunctionality pathways apply. Freshwater (2002: 11), therefore, argued that for "multifunctional policies, it is important to think of the scale at which they are to be operated", reiterated by Durand and Van Huylenbroek (2003: 10) who suggested that "the question must be answered as to what the adequate level of analysis for multifunctionality is and the consequent level of appropriate intervention. Must multifunctionality be analysed at the level of the farm, the surrounding region, or even the entire territory?" (see also Lardon *et al.*, 2004). Similarly, Clark (2003: 13; emphasis added) argued that in the UK "emphasis is placed on promoting an interpretation of multifunctionality that seeks to integrate agriculture with *territorial* economic development efforts" (see also Rapey *et al.*, 2004b, for France). However, as Chapter 8 highlighted, there is confusion about scale-related issues, with some authors suggesting conceptualisations of multifunctionality at the field level (e.g. Winter and Morris, 2004), while others have focused almost exclusively on the global level (e.g. Potter and Burney, 2002).

In line with Knickel and Renting (2000) and Freshwater (2002), I argue that for 'multifunctionality' to become more than a mere *conceptual* issue with potentially limited applicability 'on the ground', we need to know the precise nature, shape and size of this 'ground'. Thus, from a human geography perspective in particular, multifunctionality is ultimately about *territorial expression on the ground* as different actors and stakeholder groups attempt to impose their specific multifunctional strategies within specific *spatial* contexts. In other words, multifunctionality can not be an aspatial concept, but should also have tangible expression rooted in specific localities, in the farmed landscape, and in what has increasingly been termed 'multi-level governance structures' that permeate various scales of decision-making. Thus, at what scale should multifunctionality apply and why, and what is (or should be) the territory of multifunctionality? I argue that multifunctionality has to apply to *multiple* scales and entities. There exist, therefore, different 'multifunctionalities', depending on the aims and objectives of agricultural stakeholders and the nature and scale of investigation. Marsden and Sonnino (2005) similarly argued that England and Wales, for example, have adopted *spatially different* approaches to multifunctional policies, while Marsden (2003) emphasised the importance of understanding the different 'power geometries' operating at different scales of decision-making.

Figure 9.7 shows the multiple spatial scales of multifunctionality based on the multifunctionality spectrum, ranging from the farm level, to the rural community, to regional and national levels and, ultimately, the global level. These scales should be seen as constellations of temporary coherences with relatively open 'boundaries' (Massey, 2001) rather than fixed entities in space and time. The figure highlights several spatial issues. First, the spatiality of multifunctionality can be seen as a *nested hierarchy* in which individual spatial components of multifunctionality interact with each other at different scales and in complex ways, resulting in implementation of manifold multifunctionality actions at the grassroots level 'on the ground'. These different scales may also show substantial levels of spatial overlap (e.g. farm level with rural community level). Second, as we move away from the grassroots level (farm level) – where multifunctionality finds more *direct* expression through tangible actions on the ground – multifunctionality is expressed more *indirectly* (e.g. at national level). Intermediate scales (e.g. regional level), therefore, act as 'filters' for top-down multifunctionality policies/ideas/decision-making for actual implementation of multifunctional processes on the ground. Third, while nested hierarchies of multifunctionality up to and including the nation state level are relatively straightforward to conceptualise (see below), the link between national-level and global-level multifunctionality is less clear. Section 9.5.1 will, therefore, focus on the spatial scales of multifunctionality to the nation state level, while Section 9.5.2 will investigate questions surrounding direct and indirect expression of multifunctionality. The question of global-level multifunctionality is explored in Section 9.5.3.

9.5.1 The spatial scales of multifunctionality

The geography of a particular territory is vital in understanding multifunctionality concerns, and it is important to emphasise the large differences that exist regarding the importance of *agricultural territories* managed at different scales. In most countries, the farm-level scale shown in Figure 9.7 will comprise 100% of agricultural land (i.e. the farm). This is why at this scale multifunctional *agriculture* issues will be at the forefront of concerns. At the rural community level, agricultural land will usually occupy at least half of the territory (otherwise these areas would not be classified as 'rural'; see Woods, 2005). Multifunctional *agriculture* will, therefore, be important although, as highlighted below, *rural* multifunctionality issues also need to be considered. At the regional level, however, we may see large differences in importance of agricultural territory, and in some regions there may be virtually no agricultural land (e.g. most large cities; mountainous areas; etc.). The question of multifunctional agriculture at the regional level will, therefore, not always be relevant. The situation at the national level, meanwhile, is complex. Although there is hardly any country where *no* agricultural territory exists (exceptions include micro-states such as the Vatican State), we need to acknowledge that for some countries the question of multifunctional agriculture may be relatively meaningless due to the small size of their agricultural territory. In some cases, therefore, the multifunctional agriculture concept will only apply to a small proportion of land (e.g. United Arab Emirates [agricultural land occupies less than 3% of land area if we exclude areas used for nomadic pastoralism]; Greenland [ice-free grazing areas occupy less than 1% of land area]). In some countries, the issue of multifunctional agriculture will, therefore, be much more pertinent than in others. In addition, we need to acknowledge that agricultural territory in many countries is shrinking due to urbanisation, land abandonment, infrastructural developments or irreversible land degradation (e.g. desertification; salinisation). This means that the territory to which the notion of multifunctional agriculture applies may be gradually reduced in some areas.

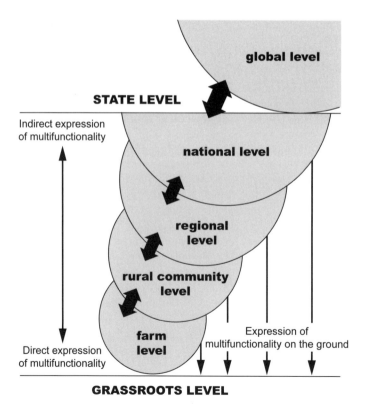

Fig. 9.7. Spatial scales and the nested hierarchies of multifunctionality

Let us begin by investigating *farm level multifunctionality*. In line with Durand and Van Huylenbroek (2003), I argue that the farm level[79] has to be the *smallest spatial unit* of multifunctionality (see also Durand, 2003; Losch, 2004). As Knickel and Renting (2000: 514) emphasised, "farmers constitute the 'centre of gravity' of rural development". Similarly, Buller (2005: 2) highlighted that most policies "seeking to promote multifunctionality … focus predominantly upon the farm". Applying multifunctionality to a smaller scale (e.g. individual fields within a farm; cf. Rapey *et al.*, 2004a) would be relatively meaningless, as, in my view, the smallest common denominator for multifunctionality pathways has to be the level of the *decision-making unit* – in other words, the farm. As Chapter 7 demonstrated, productivist and non-productivist spaces can lie in close vicinity within one farm, but as multifunctionality is about the *totality* of decision-making processes bounded by productivist and non-productivist action and thought, picking out individual spatial subsets of a farm unit (e.g. a field) may result in 'compartmentalised multifunctionality' that may give a distorted picture of wider farm-level multifunctionality pathways (see Chapter 10). Swagemakers (2003: 189), thus, rightly argued that "it is precisely by getting to know the day-to-day work of farming that we can begin to understand the characteristics of multifunctional agriculture".

Winter and Gaskell (1998) emphasised the importance of the 'production unit' of the farm as the basic scale for analysis, as most payments and subsidies to farmers, while no longer tied to the commodity, remain tied to the farm territory. Similarly, Marsden and Sonnino (2005: 5) highlighted that recent agricultural transitional processes "called for the implementation of farm-based … rural policies". This was reiterated by Van der Ploeg (2003) who argued that multifunctionality can only be understood in the context of the farmed landscape where the farm, and the landscape it creates, should become the defining framework informing decisions about agricultural development. Similarly, Durand and Van Huylenbroek (2003: 11) suggested that "at the farm level questions arise about the practical consequences of [multifunctional] policies for farm practices". Thus, Winter and Morris' (2004) recent argument that multifunctionality should only be understood within the narrow confines of a field or, at most, several fields within a farm may be too narrow and, in my view, only forms one small ingredient of what multifunctionality is about. Yet, I agree with Winter's (2005) suggestion that multifunctionality should be seen as a *land-based concept* in which the notion of multifunctionality applies to a specified unit of land (i.e. the farm, the region, etc.).[80]

As research has shown, a farmer may apply specific farming strategies to one part of the farm that may be close to our definition of *non-productivist* action and thought (e.g. withdrawing food and fibre production on one upland field for conservation and biodiversity preservation), while on other parts of the holding that same farmer may apply *productivist* strategies (e.g. maximising food and fibre production on highly productive soils with maximum agro-chemical inputs) (Morris and Potter, 1995; Wilson and Hart, 2001). In other words, farm level multifunctionality is not only about expression of the *physical basis* of food and fibre production, but also about unravelling *actions and thoughts* of the decision-maker for that space – the farmer. Multifunctionality is, therefore, about the link between *human decision-making* and *spatial expression* of these decisions on the ground. Farmers and farm households will adopt productivist or non-productivist pathways, depending on factors exogenous to the farm level (i.e. factors operating at the rural community, regional, national

---

[79] The farm level contains any areas within the boundaries of a farm (either in one block or comprised of several units not necessarily in close vicinity to each other). It is also interesting to note Knickel and Renting's (2000) inclusion of the 'farm household' as a separate level of multifunctionality. I argue that the farm household can be included at the farm-level scale of multifunctionality in which a multitude of individual actor decisions within the farm household take place (particularly with regard to pluriactivity and diversification).

[80] In Chapter 10, I will also discuss how the notion of multifunctionality should be considered as a *person-centred concept* based on multifunctional action and thought by individual agricultural/rural stakeholders.

or global levels highlighted in Figure 9.7), but also based on factors endogenous to the farm such as farmer attitudes, ideologies and identities (Burton, 2004; Burton and Wilson, 2006). Van der Ploeg and Roep (2003: 41), therefore, argued that "the coordination and allocation of family labour between different agricultural and non-agricultural activities in the pluriactive farm household appears to be an important source of [multifunctional] synergy". Using the farm scale as the 'smallest common denominator' allows the bringing together of multiple compartmentalised multifunctionality actions, providing a more complete picture of multifunctionality quality on the *whole* farm.

There will, of course, be many examples where one farmer will not be the only decision-maker (Burton, 2004). As Chapter 10 will discuss, ownership patterns may range from single owners, to family trusts in which multiple farm household members (e.g. father and daughters) make joint decisions about productivist and/or non-productivist farm development strategies, to complex ownership arrangements comprising shareholding companies and multiple farm owners, many of which may not necessarily be from a farming background (a common pattern in the USA, Canada or former East Germany; Potter, 1998; Wilson and Wilson, 2001). Further, many farms are farmed by tenants or farm managers who do not have ownership of the land, and where multifunctional decision-making pathways may be based on a wide spectrum of decision-making constraints and opportunities for those working the land, ranging from complete control over decisions by the landlord to (almost) complete control by the tenant (Mandler, 1997; Woods, 1997). Thus, "it is often difficult to disentangle the work of one farmer from another. In particular, aspects of the farm that contribute to productivity such as soil structure and fertility and field conditions ... are not necessarily laid down by a single ... farmer" (Burton, 2004: 206). Yet, complex ownership (even over longer time scales) still means that the farm should be the smallest unit for conceptualising multifunctionality strategies. While farm ownership patterns have changed considerably over the past few decades in many parts of the world towards a more corporate agri-business model (Marsden, 2003), the notion of a clearly definable farm property, with a distinctive and mappable boundary, has not. In the context of the farm as the main unit of farmer decision-making, Burton (2004: 207; emphasis added), therefore, argued that "in representing the symbolic actions of generations of farmers, *the farm* provides a store of symbolic capital that any new entry to farming ... can draw on". Thus, whatever farm ownership patterns exist, it will still be possible to identify the *territory of the farm* in which multifunctionality strategies are implemented.

The second spatial scale in Figure 9.7 concerns **rural community level multifunctionality**. Van der Ploeg and Roep (2003: 50) emphasised the re-emergence of the locality and community as key in multifunctionality concepts: "while local resources, local town/countryside relations, local initiatives, and local players seemed to be nearly irrelevant [in the past], they are again central to multifunctionality". Similarly, Belletti *et al.* (2003a: 68) highlighted that "the collective level is very important in any analysis of multifunctionality ... [as] multifunctional farm outcomes also have positive effects outside the farm" – referred to as the 'multifunctional local system'. Belletti *et al.* (2003b: 158), therefore, argued that "multifunctional agriculture can not only represent one of the supporting axes for the economic development of rural districts, [but] the rural district may also be the right institution to promote multifunctional agriculture". Blandford and Boisvert (2002: 111) similarly suggested that a reorientation of the multifunctional policy agenda is needed towards "the local or community level", while Knickel *et al.* (2004: 89) emphasised the importance of "the creation of cohesion between activities, not only at farm level but also between different farms, or farms and other rural activities" for the establishment of strongly multifunctional processes. Bell (2004) also highlighted how in the USA the notion of a locally more embedded 'community-supported agriculture' is taking shape that may facilitate stronger multifunctionality pathways. Yet, as Woods (2005) demonstrated, there are many debates about what makes up a 'rural community'. This usually comprises one or several small settlements with a high percentage of people engaged in rural activities, but may also include non-agricultural activities affecting implementation of multifunctional strategies 'on the ground' (e.g. forestry or mining).

Belletti *et al.* (2003a) highlighted that achieving strong multifunctionality at rural community level is predicated on the need to achieve critical mass to build up and sustain a 'multifunctional reputation' for the area to external consumers (e.g. tourists); the possibility to gain from scale economies in providing multifunctional connections between the local system, markets and public institutions; and scope for the successful joining of multifunctional elements in the local area. Adoption of strongly multifunctional pathways is, therefore, not only predicated on reinforcement of on-farm activities, but there also need to be synergies between these reproductive strategies and the imperatives of local or community-based territorial development more generally – a process referred to as the 'new associationalism' between farmers and their community (Marsden *et al.*, 2002). This new associationalism can be interpreted as the vehicle to transform the 'public goods' created by farming and agriculture into human welfare at the community level (Bryden, 2005). Thus, agricultural businesses need "to create and maintain new associations with a whole range of external actors and institutions … [by] constructing and optimising new social networks [with] those involved at different points in the various supply chains, and those significant regional and local actors who are able to facilitate [multifunctional] economies" (Marsden *et al.*, 2002: 814-816). As a result, Belletti *et al.* (2003b) suggested that the rural community level is the *essential dimension* for the implementation of multifunctionality, since a large number of functions connected to agriculture require a territorial concentration of actions and networks (economies of scale) to be successfully implemented – actions that may not have sufficient weight at individual farm level. Thus, "the rural district can allow the reinforcement of [multifunctional] territorial identity and the awareness of this specificity in the mind of the local stakeholders" (Belletti *et al.*, 2003b: 159).

Not all ingredients for successful rural community level multifunctionality will always be in place. That potential for rural community level multifunctionality is often based on tourism development is a particularly contentious issue (Kneafsey *et al.*, 2004; Mollard *et al.*, 2004). As various studies show, not all farm areas are suitable for attracting tourists (Font, 2000; Clark, 2003; see also discussion below on regional level multifunctionality). Nonetheless, Vanslembrouk and Van Huylenbroek (2003) showed that extensification of agriculture – as part of strong multifunctionality trajectories – can have a positive influence on both attracting new tourists and on prices rural tourists are willing to pay. Establishing strongly multifunctional pathways at the rural community level (and beyond) can, therefore, act as a *reinforcing* mechanism for increasing the potential for further multifunctional options through increased tourism opportunities. It is, therefore, at the rural community level that we need to consider the above-mentioned distinction between *agricultural* and *rural* multifunctionality (Rapey *et al.*, 2004a). *Agricultural* multifunctionality strategies at the rural community level will be largely concerned with issues surrounding environmental sustainability (or lack thereof), farming intensity and productivity, the level of integration of farms into the global capitalist market, and diversification activities that may lead farmers away from food and fibre production. *Rural* multifunctionality issues, meanwhile, will be mainly focused on local embeddedness between agricultural and rural actors, where the nature, extent and durability of these agriculture-community relationships will depend on weak or strong multifunctionality pathways adopted by the community as a whole (Belletti *et al.*, 2003a).

Landscape protection and environmental sustainability as key ingredients of the multifunctional agricultural regime will play a particularly important role at the rural community level. Strong multifunctionality pathways are more likely in areas where an integrated and holistic vision exists that helps farmers look beyond their own farm boundaries, and that acknowledges that the choice of individual multifunctionality pathways also has repercussions for the rural community as a whole. Here, leadership offered by individuals in the community often proves crucial for rural community level acceptance of 'innovations' such as strong multifunctionality (Wilson, 1992; Rogers, 1995). The Australian Landcare movement is a particularly apt example for this. Landcare operates at the rural community level (usually at river catchment scale) and has been relatively successful in bringing farmers from one or several districts together to discuss issues of environmental degradation and advantages and disadvantages of adopting different multifunctionality pathways (Wilson, 2004). The key issue here is for farmers to recognise that implementation of strong multifunctionality is virtually impossible at the farm level alone and that the *entire community* has to provide an enabling environment. Van der Ploeg and Roep (2003: 41), therefore, argued that at the rural community level "as a well-defined social and geographical space ... new forms of [multifunctional] articulation are developing". The rural community level also has a crucial role for multifunctional food supply chain issues (Clark, 2003). Indeed, (re)localisation of agro-commodity chains often depends on opportunities for local sale of farm products at the rural community level. Thus, demand for local production will be key in making local farm shop ventures a success, while community support (practical and psychological) for farmers choosing strong(er) multifunctionality pathways is crucial for successful farm management transitions.

The third spatial scale in Figure 9.7 shows the importance of **regional level multifunctionality**. At the regional level, *rural* multifunctionality issues assume greater importance than *agricultural* issues (see Fig. 9.3 above), as agriculture will, in most cases, be only one of many economic sectors present. Nonetheless, the regional level also has important repercussions for enabling or disenabling multifunctional agriculture pathways 'on the ground', not least due to recent trends of the 'regionalisation of rurality'. Clark (2003, 2006) and Marsden and Sonnino (2005) emphasised how regionalisation in the EU has been a particularly important driver for different national interpretations of 'multifunctional' policies (see Chapter 8), while Ward *et al.* (2003: 21) highlighted recent shifts away from a "national, sectoral and individualized notion of agriculture and agricultural competitiveness to a regional, territorial and collective notion". Clark (2003) referred to this process as the 'sub-national promotion of multifunctional agriculture', and highlighted that for England, for example, "a whole tier of English regional governance now has direct or indirect involvement in promoting multifunctionality" (Clark, 2006: 333). Marsden and Sonnino (2005) also emphasised how European Structural Funds have contributed towards a *regionalisation* of multifunctionality opportunities since 1988 (i.e. largely through regionally targeted flows of money), while the EU LEADER programme (since 1991) has greatly aided in reinforcing regionally based territorial identities (Ray, 1998, 2000). In countries such as the UK recent political devolution has provided an additional platform for political empowerment of the regions and, as a consequence, for more regionally oriented multifunctional pathway opportunities (Marsden and Sonnino, 2005). Clark (2006) also highlighted how the development of Regional Development Agencies in the UK has reinforced the importance of regional-level multifunctionality decision-making. Clark (2003: 21), therefore, suggested that "a whole new tier of sub-national agencies has come into being in England ... arguably improving the organisational and institutional conditions for the promotion of the new multifunctional conception of agriculture". Murdoch *et al.* (2003) further suggested that the regionalisation of rurality has helped harness local variety in line with broader policy goals. It is likely, therefore, that regional level multifunctionality pathways may gain further importance in the near future, at least in an EU context of increasing regionalisation (Allen *et al.*, 1998; Durand and Van Huylenbroek, 2003).

Knickel and Renting (2000) emphasised the particular significance of the regional level in conceptualisations of multifunctionality, especially for job creation and options for pluriactivity. However, positive synergies are only possible at regional level where favourable linkages between two or more multifunctional activities or entities can be created. As a result, Benz and Furst (2002: 518) emphasised that strong multifunctionality pathways are particularly dependent on regional-level stakeholder interactions, as "the resulting synergy leads to a [multifunctional] impact that is greater than the sum of the effects produced by the same activities taking place in isolation from each other". These authors, therefore, highlight that the development of strong multifunctionality pathways can only be fully understood if we take into account the *synergies* and *multiplier effects* that exist between the different geographical scales of the multifunctionality spectrum.

At the regional level, five synergistic components will be particularly important, highlighting the existence of manifold tensions within both geographical and actor spaces in which multifunctional differentiation (both in thought and action) are contested through actors attempting to impose their respective representations of multifunctionality over others. First, the regional level is often crucial in providing diversification opportunities to farmers (e.g. based on tourism potential of the region, but see discussion above). If diversification activities are not available to a farm household at the regional level, it is likely that economically marginal farms may give up farming altogether instead of moving their farm business elsewhere or seeking pluriactive income sources far away from the farm. As Clark (2003) emphasised, a buoyant region that places great emphasis on strong multifunctionality should provide farmers with opportunities to embark on strongly multifunctional pathways (see also Vanslembrouk and Van Huylenbroek, 2003). Second, the positionality of a region in local-global processes of interaction (Swyngedouw, 2001) influences the embeddedness of farm holdings in the global capitalist system. Echoing debates on multifunctional quality of different farm systems (see Section 9.4), regions that are weakly embedded into the global capitalist system (e.g. remote tropical forest areas) will often depict strong agricultural multifunctionality. Highly globalised regions, meanwhile, will often contain weakly multifunctional intensive productivist-driven agro-industrial farms geared towards income maximisation and export of food and fibre (e.g. arable farms in the Paris Basin; American and Australian wheat belt farmers; Argentinean cattle ranchers) (Goodman and Watts, 1997). Third, the region also plays a key role in influencing food supply chains available to farmers. Geographically remote regions are more likely to have *locally* integrated food supply and consumption structures (i.e. strongly multifunctional), while centrally located or highly urbanised regions provide farmers with opportunities for *globalised* agro-commodity chains (i.e. often weakly multifunctional) (Marsden, 2003). However, location near urban agglomerations may also open *multiple* and *different* multifunctionality opportunities for farmers due to both higher purchasing power of urban populations (e.g. for more expensive organic products) and more varied consumer demand for agricultural products (Marsden, 2003; Goodman, 2004). Fourth, at the regional level the dimension of multifunctionality concerning open-mindedness of farming and rural populations will be crucial. Implementation of strong multifunctionality 'on the ground' may be difficult for farmers if the regional culture is reluctant to support strongly multifunctional farming strategies. Although there are many examples where individual farmers go 'against the grain' (Potter, 1998), empirical evidence suggests that it is difficult for farmers to implement what may be perceived as 'radical' actions that are not compatible with 'the way things are done' (e.g. Bell, 2004, for the USA). Fifth, the regional level will also be important as the spatial scale in which state-led policy is *mediated* through various 'filter' and 'gatekeeping' processes (Wilson *et al.*, 1999; Wilson and Juntti, 2005). In particular, the region often acts as a vehicle for the street-level-bureaucratic level of policy implementation, in which interpretation of policy as 'weakly', 'moderately' or 'strongly' multifunctional by regional policy stakeholders may have severe repercussions for multifunctional actions on the ground. For example, although some policy intentions at the state level may be strongly multifunctional (e.g. AEP; see Section 9.4), regional street-level bureaucrats may interpret such policies in a different way. Key policy decisions may be made at regional level that *distort* national policy messages or that may lead to *non-implementation* of policy on the ground. The latter is particularly true if certain policies do not 'fit' regional priorities. Although examples of this 'regional policy filter' are manifold, Mediterranean countries exemplify particularly well the importance of this spatial layer through frequent failure of implementation of strong multifunctionality policies perceived as leading to a reduction in regional revenue. In many

Mediterranean regions, implementation of AEPs, for example, is often seen as the imposition of strong multifunctionality that does not fit the weak multifunctionality model adopted by regions aiming at increasing agricultural intensity, exports and profits (Buller *et al.*, 2000; Wilson and Juntti, 2005) (see also Chapter 7).

Finally, conceptualisations of multifunctionality also need to acknowledge the possibility of ***national level multifunctionality***. As with the regional level, national level action and thought can have substantial repercussions for the implementation of strong agricultural multifunctionality on the ground, in particular through national political development pathways and the role that national governments, NGOs and institutions play in policy formulation and implementation (Wilson and Bryant, 1997; Losch, 2004). Four key issues are of particular relevance here. First, the political orientation of a country will define the 'chosen' level of a country's embeddedness into the global capitalist market. While many countries have now adopted neo-liberal free-marketeering ideologies that permeate every aspect of society, including many agricultural development pathways (Australia and New Zealand may be the best examples; Primdahl and Swaffield, 2004), the above-mentioned example of North Korea highlights how political isolation of a country will also substantially influence opportunities for the selection of *different* multifunctionality pathways on farms (Hale, 2005). While it was argued in Section 9.4 that globally well embedded countries often spawn weakly multifunctional agricultural regimes, the globalisation of agricultural trade and production may also open *multiple* and different multifunctionality opportunities for farmers. Countries such as North Korea, meanwhile, may only offer very limited multifunctional opportunities to their farmers (based on a weak multifunctionality model predicated on satisfying national self-sufficiency in food; see also Chapter 10). Opportunities for strong multifunctionality pathways at the farm level are, therefore, intrinsically linked to political decision-making pathways at the national level.

Second, as the discussion of regional level multifunctionality has highlighted, the national level most often provides the framework in which *policy* affecting multifunctional agriculture is formulated. As Clark (2006: 342) argued for the UK, "DEFRA… has responsibility for encouraging the transition to multifunctional agriculture approaches nationally". This is also largely the case in the 'supra-national' EU, as subsidiarity ensures that most policy implementation powers are delegated to the level of the nation state (Jordan, 1999; Buller *et al.*, 2000). As Delgado *et al.* (2003: 31) emphasised, "each country and each territory will have to … propose what they consider most suitable for the future of their rural areas". In the 'strong state' political model (i.e. where the state has much coercive power), in particular, multifunctionality is often *orchestrated* by the state by 'command-and-control' regulatory legislative instruments (Ekins, 1999). Here we can invoke the example of North Korea again where, in what is currently the most extreme example of state control over human decision-making, multifunctionality is equivalent to *national* agricultural multifunctionality (Woo, 2006). Here, the boundaries between the different nested hierarchies in Figure 9.7 are particularly blurred, and it may be difficult to differentiate between farm-level and state-level multifunctionality pathways. Yet, many other examples exist where the nested hierarchies of multifunctionality also remain blurred. In China, for example, while some decision-making powers are granted to the hundred millions of small-scale farmers, the state still retains control over food policy, farming structures and, indeed, agricultural ideologies. At the other extreme are countries such as Australia and New Zealand, where neo-liberal agendas have led to a virtual removal of state influence over individual farm-level multifunctionality decisions (e.g. in New Zealand farm subsidies were abolished in 1984; Wilson, 1994; Primdahl and Swaffield, 2004). There, farmers are essentially left to fend for themselves, exposed to the vicissitudes of the global market, resulting in often weakly multifunctional farm trajectories driven by the need to survive economically in what is often a harsh and unforgiving globalised economic climate (Jay, 2005). Countries in the EU, meanwhile, occupy the 'middle ground' regarding national policy influence on multifunctionality. Here, states provide an enabling policy framework (which focuses on either weak or strong multifunctionality), mediated by regional policy stakeholders (see above), where farmers are often left with substantial flexibility about which multifunctionality pathways they may wish to adopt (see also Chapter 10).

Third, it is at the state level that societal *ideological* multifunctionality dimensions play the most important role – often mediated through the media, art and literature. As Section 9.4 highlighted, the strong multifunctionality model is predicated partly on open-minded societies who accept that the nature of 'farming' and 'agriculture' is in the process of change. While such ideologies may vary considerably from region to region (i.e. a highly urbanised region is likely to have a different view of the importance and required orientation of agriculture than a remote agriculturally based region; see above), it is often ideologies about the position of agriculture in society that define constraints and opportunities for the adoption of strongly multifunctional agricultural pathways *at any scale*. Thus, countries where society continues to see agriculture essentially as a guarantor of self-sufficiency and high export earnings are unlikely to provide an enabling environment for strong multifunctionality pathways. Clark (2003: 19) highlighted that EU member states, for example, "have chosen to implement the RDR, and with it the concept of multifunctional agriculture, in distinctively different ways … in response to the specific political priorities of their rural regions and their agricultural sectors" (see also Barthélemy *et al.*, 2004). Just as farmers will find it difficult to 'swim against the stream' within their own communities (see above), they will also find it hard to implement agricultural pathways that may seem to contradict society's wishes and aspirations (but see also discussion in Section 10.2). In democratic pluralistic societies this will be less of an issue, as manifold expressions of multifunctionality will be allowed and, indeed, encouraged. But in countries with more restrictive socio-political cultures, productivist societal ideologies may be the key factor in dissuading farmers from leaving productivist farm development trajectories.

Finally, when discussing national-level multifunctionality we also need to acknowledge the *global spatiality* of multifunctional processes (see also Section 9.5.3 below). It is possible to conceive of a scenario in which national level multifunctionality of a country may be strong, but where such strong multifunctionality pathways are predicated on weak multifunctionality pathways in other countries. This is particularly the case where a non-self-sufficient country (e.g. Germany) encourages moderate or strong multifunctionality pathways among its own farmers (and even this is questionable; see Section 10.2), while importing large amounts of food from countries or regions where this food has been produced under a weakly multifunctional agricultural regime (e.g. food imports into Germany from the Netherlands or cash crop oriented regions in developing countries). In other words, strong national multifunctionality pathways in one country may only be possible by exploiting weak multifunctionality pathways in another. The increasingly globalised nature of agro-commodity chains (see Chapter 5) makes it less likely that 'self-contained' national multifunctionality (i.e. one in which a country is largely self-sufficient and relies little on food imports) is possible. Fully understanding national ideological strong multifunctionality pathways may, therefore, only be possible by investigating the global embeddedness of a country within a complex web of global weak, moderate and strong multifunctionality farm development pathways.

9.5.2 Nested hierarchies: the 'direct' and 'indirect' expressions of multifunctionality

In the previous section I have outlined the different spatial scales to which the concept of multifunctionality should apply. The four scales highlighted included the farm level, the rural community level, the regional level and the national level (see Fig. 9.7 above), and I argued that application of the multifunctionality concept is ultimately associated with implementing multifunctional pathways *at the farm level*. Yet, the discussion also highlighted that expressions of multifunctionality differ depending on scale issues, as multifunctional action and thought at any level will have repercussions for how multifunctionality is implemented at the grassroots level. As Lardon *et al.* (2004: 15; my translation) emphasised, "multifunctionality does not affect agricultural spaces in a uniform way". These different spatial expressions represent the *nested hierarchies* of multifunctionality. Building on Murdoch and Marsden's (1995) notion of multi-scalar 'actor spaces', and on theories of urban multifunctionality (e.g. Rodenburg and Nijkamp, 2004), the notion of *nested hierarchies* implies that the different multifunctionality scales in Figure 9.7 are interlinked and form *mutually constituted* decision-making spaces that reflect individual national and regional governance structures (see also Cocklin *et al.*, 1997; Jessop, 2002; Hess, 2004).

This addresses Winter's (1998) crucial point that the contemporary emphasis on the local (i.e. farm level) may enhance certain kinds of sensitivities, but may also erase others and thereby truncate rather than emancipate rural research. I also agree with Buller's (2005: 20) suggestion that "the assessment of multifunctionality requires a holistic analytical framework that extends beyond the farm gate and beyond the individual farm" and that "the true multifunctionality of agriculture [needs to be seen] as a set of embedded and socially/spatially interlinked activities". Further, as Marsden (2003: 29) emphasised, "the relative significance of food and food networks in and through space is variable, according to the main activities in those networks, and the relative power to capture the social value of food products". Marsden (2003: 115) emphasised that any agricultural region has to be seen as *vertically* embedded with national governments, *horizontally* through its engagement with the wider rural community, and *diagonally* with sub-regional frameworks, and suggested that "rural spaces are constituted and re-made by cross-cutting networks of power and association, with rural restructuring as an outcome of the aggregated network effects". Knickel and Renting (2000: 513) also argued that "the interrelationship between [multifunctionality] levels must also be considered. Although rural development often starts in the farm or farm household, it must also be defined at the level of the region or countryside". Such hierarchies and spatial networks can, therefore, be seen as 'nested' in as much as they each form the basis for the existence and functioning of the other – both in ascending and descending order from the grassroots to state level[81] (see also Long and Van der Ploeg, 1995). This means, for example, that it is futile to conceptualise multifunctionality at the farm level without also considering multifunctionality drivers at spatial levels beyond the farm, i.e. at the rural community, regional and national (and global) levels. Similarly, understanding national level debates about multifunctionality (e.g. expressed through policy formulation; see Section 9.4) will be difficult without understanding the implementation (or non-implementation) of multifunctional ideas at the regional, rural community and farm levels (Wilson and Juntti, 2005). As O'Sullivan (2004: 284) argued from a perspective of complexity theory, "how local interactions 'scale up' to effects at larger scales is a familiar concern".

A key question is whether *expressions* of multifunctionality are similar across all levels highlighted in Figure 9.7? I would argue that they are not, for the following reasons. As highlighted above, multifunctionality has, first and foremost, to be seen about implementation of productivist and non-productivist action and thought *on the ground*. This means that *agricultural* multifunctionality finds its most *direct expression* at the farm level. *Rural* multifunctionality, meanwhile, will find its most direct expression at the rural community level (but also, at times, with strong influence at the farm and regional levels). National level multifunctionality, meanwhile, should be seen as the scale of decision-making for policy formulation and general societal interpretations of what 'multifunctionality' means in a national context. Thus, multifunctional agriculture will have more *direct expression* at the farm level (i.e. where multifunctionality ideas and policies are implemented), while pathways at the national level are characterised by *indirect expression* of multifunctionality.

---

[81] Cilliers (2001) and O'Sullivan (2004) argued that the boundaries of nested hierarchy systems should not be seen as the 'edge' of an entity separating it from everything else, but as mutually connected systems. Cilliers suggested that intricate non-contiguous spatial structures are common and that boundaries of nested hierarchies are often inter-penetrating (see also Gehring, 1997, and Payne, 2000, for discussions of nested hierarchies in European integration).

This is an important difference between the grassroots and state levels, as it highlights that national level policies and ideas on multifunctionality need to be *translated* into multifunctional actions on the ground. As Hoggart and Paniagua (2001: 51) argued, "change experienced at the local level might be driven by broader forces, but this does not mean it is accepted uncritically or unaltered at the local level, nor does it deny the prospect of inter-local disparities in response to the same inputs". Thus, to speak of 'multifunctional policies' (see Section 9.4) only makes sense through the consideration of the effects of such policies 'on the ground'.

The regional and rural community levels can, therefore, be seen as important *mediators* and *conduits* for both the 'trickling down' of national multifunctionality concepts and ideologies to the farm level and, conversely, for relaying action and thought about the direct implementation of multifunctionality at the farm level to national decision-makers (Wilson *et al.*, 1999; Marsden and Sonnino, 2005). Both the discussion above and Chapter 7 highlighted that such mediation between the national and farm levels can take various pathways. On the one hand, we witness frequent *re-interpretation* of national multifunctionality policies at the regional level in most democratic societies through intermediate or street-level-bureaucratic actors (Cooper, 1998). On the other hand, we may see outright *imposition* of national multifunctionality ideologies on the ground in non-democratic countries with strong coercive state powers and intermediate actor spaces closely aligned with state ideologies (e.g. North Korea; China; Russia to some extent). In complex governance structures where many intermediate actors may mediate multifunctionality policies formulated at the national level, the further removed from the farm level decision-makers are the more difficult it may be for them to influence the trajectories of multifunctionality on the ground. Recent empirical evidence from Southern Europe regarding non-implementation of strongly multifunctional agricultural and environmental policies would tend to support this 'distance-decay' distortionary effect of strong multifunctionality at the grassroots level (Buller *et al.*, 2000; Wilson and Juntti, 2005). However, Clark (2003) also highlighted that even regional expression of multifunctionality pathways may vary substantially between regions based on differences in mediation and re-interpretation of national multifunctionality ideals by powerful regional decision-makers. As a result, some regions may adopt more strongly multifunctional development pathways than national-level multifunctionality ideologies (e.g. some of the German regions; cf. Wilson and Wilson, 2001).

It can, therefore, be surmised that the more complex mediating governance structures are, the more likely it will be that multifunctional action 'on the ground' will be highly varied and span the entire spectrum of productivist and non-productivist action and thought (as will be the case in most EU member states, for example). This may also have important repercussions for the *flexibility* in adoption of changing multifunctionality pathways at different levels of nested spatial hierarchies. Farm level actors may not only have more possibilities in the adoption of highly varied multifunctional farm trajectories, but they may also be able to change the direction of these trajectories more quickly than actors at other levels (see Section 10.3). It will be more difficult for regional and national level decision-makers to rapidly change multifunctional ideas and ideologies – an assertion supported by the quick adaptation of many farmers to changing market forces (Robinson, 2004) and the rather lethargic pace of agricultural policy change at national and, indeed, supra-national levels (Wilson and Juntti, 2005). The more complex and multi-scalar a system – in other words, the further removed it is from the grassroots or farm level – the more difficult it will be for that system to change its multifunctionality trajectories. This is probably best highlighted through innovations such as GM crops, which give added influence to the often weakly multifunctional downstream sectors (e.g. multinational agri-businesses) in shaping the nature of local food production and controlling production schedules at 'arms length' (Pretty, 2002; Marsden, 2003). This also explains why the embeddedness of farmers in globalised agro-commodity production chains largely defines whether farmers will be 'behind' or 'ahead' of regional or national-level multifunctionality trajectories.

The notion of direct and indirect expressions of multifunctionality at various scales highlights one of the key strengths of the multifunctionality spectrum, in particular as the spectrum has an *in-built localising tendency* (i.e. focus on the farm territory for direct implementation of multifunctionality action) that allows greater temporal and spatial portability than the p/pp model discussed in Part 2. The notion of a multifunctional decision-making spectrum is, therefore, more sensitive to local geographies while also enabling analysis of macro-scalar processes. As suggested in Section 9.2, the new multifunctionality concept is not merely a compromise that falls into the trap of traditional Western dualistic imagination (see Chapter 3), but a clear and workable alternative to contemporary unilinear and directional conceptualisations of agricultural change.

9.5.3 Global level multifunctionality?

So far, we have only explored scales of multifunctionality to the national scale, largely because it is still at the nation state level that most policy decisions with tangible expressions on the ground are formulated (Johnston, 1996; Wilson and Bryant, 1997). Yet, socio-political developments over the past few decades have highlighted that, increasingly, global-level multifunctionality drivers need to be taken into account when explaining multifunctionality actions at the grassroots level (Knickel and Renting, 2000; Losch *et al.*, 2004). As Marsden (2003: 142) emphasised, "we have to conceive of rural spaces as ensembles of local and non-local connections, of combinations of local actions and actions 'at a distance', situated in regional economies and different institutional contexts". The nation state may be getting weaker regarding policy influence ('hollowing-out' of the nation state), while supra- or international decision-making structures may be gaining ground (Berger, 2001; Decker, 2002). However, as Buller (2005: 18) emphasised, there is a "general absence of any systematic attempt to globally assess the multifunctional impacts of agriculture and agricultural support mechanisms". In this final section, therefore, I wish to explore whether we can conceptualise 'global level multifunctional agriculture' based on the reconceptualisation of multifunctionality outlined in the previous sections.

Four key issues need to be considered in this debate. The first relates to the question of *global-local spatial interactions* (Massey, 2001; Swyngedouw, 2001). Can direct expression of multifunctionality on the ground be entirely understood without acknowledging global level decision-making pathways that affect national, regional and, at times, rural community level development pathways? Notwithstanding the possibility that "global phenomena emerge unbidden from interactions among lower-level entities" (O'Sullivan, 2004: 285), and Urry's (1984) cautionary note that the local cannot be taken as the national or global writ small, increasing globalisation trends around the globe would suggest that the global level is assuming ever increasing importance, not only for day-to-day lives of citizens, but also concerning multifunctional decision-making pathways on farms (Shiva and Bedi, 2002; Losch *et al.*, 2004). Like other dualisms (see Part 1) the boundaries of the global/local are unstable and blurred, with the global and local increasingly understood as embedded within one another rather than as dichotomous categories. As a result, when global and local are inevitably intertwined in practice, approaches to the study of multifunctionality which look at only one spatial scale will miss much of interest. Global-level studies may fail to take into account local outcomes and responses to global processes by individual actors in the multifunctionality spectrum, while local studies may omit analysis of global economic and socio-cultural influences, thereby only providing limited understanding of multifunctional transitional pathways. Gupta and Ferguson (1997: 37), therefore, referred to the emergence of a "transnational public sphere" that has "certainly rendered any strictly bounded sense of community or locality obsolete". They argued that this reterritorialisation of space requires researchers to reconceptualise fundamentally the politics of community, solidarity, identity and cultural difference. Such 'glocalisation' (Swyngedouw, 2001), characterised by mutually constituting sets of practices, will also influence multifunctionality decisions on individual farms. As Lang and Heasman (2004: 258) emphasised, "the tussle between globalization and localization ... is being fought out in food and health", while Marsden *et al.* (1999: 299) suggested that these glocalisation processes are associated with "the conflict between globalized aspatial systems of production and locally situated ecological systems".

There is ample literature that discusses the effects of globalisation on individual farm decision-making pathways (e.g. Goodman and Watts, 1997; Shiva and Bedi, 2002). What is important for multifunctional agriculture conceptualisations is that influences of globalisation affect different agricultural territories in different ways. While decision-making pathways on super-productivist agro-businesses often occur entirely within the realm of the global (only, at times, influenced by national policy-making), there still exist many farming communities that have not, or only barely, been touched by globalisation. Although Section 9.4 highlighted that such 'remote' communities are increasingly rare, the existence of such communities emphasises that global influences on grassroots-level multifunctionality pathways cover a wide spectrum of influences. At the regional level, meanwhile, it will be virtually impossible to find examples where entire regions are shielded from globalisation processes, while at the national level even the most protectionist and isolationist approaches can not shield states from the vicissitudes, fashions and trends of the global market (Shiva and Bedi, 2002). Thus, even multifunctional national policy trajectories of North Korea will be influenced by market and political developments beyond its borders (the recent need for food imports to stave off starvation in North Korea is a case in point) (Hale, 2005).

Second, global level influences play a key role in our conceptualisations of weak, moderate and strong multifunctionality (see Section 9.4). Indeed, I argued above that the interlinkage of a farm into the global capitalist market can be seen as a possible contributing factor towards weaker multifunctionality. Here, we need to briefly revisit the suggestion made above that globally embedded agricultural stakeholders may be more likely to adopt weak(er) multifunctionality pathways. This may occur out of *choice* (e.g. an agro-business in New Zealand opting for farm intensification based on the opportunity of direct sales of live animals to the Middle East) or out of *necessity* (e.g. an Indian farmer caught in the 'globalisation treadmill' by opting for Green Revolution seeds and fertilisers on his/her formerly more strongly multifunctional subsistence-oriented farm) (Shiva and Bedi, 2002; see also examples in Chapter 10). Rural sociologists (e.g. Friedmann and McMichael, 1989; McMichael, 1995) have highlighted how such 'scaling-up' processes to supranational decision-making often entail the loss of democracy and local/regional decision-making powers. Globally weakly embedded farms, on the other hand, often have strongly multifunctional characteristics, as they are often extensively farmed and more likely to produce (healthier) foods for the local market or for subsistence (Pretty, 1995, 2002; but see also Parayil, 1999, for a contrasting view). Marsden (2003: 47), thus, suggested that globalisation processes "increasingly define agriculture as a somewhat detached commodity system, disengaged from its deeper ecological and social base". Global level multifunctionality drivers, thus, have repercussions for multifunctional decision-making on the ground, and often these influences lead towards a weakening of multifunctionality at farm level. Globalisation, yet again, emerges as the 'great leveller' leading to a more uniform, more monofunctional and less socially rich agriculture (Shiva and Bedi, 2002).

Third, global level influences can not be fully understood without considering the rising importance of political and policy-related drivers of agricultural change emanating from a supra-national decision-making framework. As McCarthy (2005) argued, the role of the state as a 'protector' of national interests is increasingly undermined by neoliberal globalisation. The rising importance of GATT/WTO discussions (since 1994 in particular) needs to be particularly acknowledged, as these are increasingly shaping ideological and practical approaches of national agricultural policies. WTO negotiations influence policy decisions at the nation state level (e.g. recent CAP reforms can be partly seen as a response to WTO pressures to reduce and, eventually, abolish farm subsidies) which, in turn, will influence multifunctional decision-making pathways further down the spatial scales (Potter and

Burney, 2002). However, there is a reason why in Figure 9.7 (above) the 'state level' boundary has been drawn as a relatively firm line separating it from the global level. As with the complex issue of 'trickle down' of multifunctionality drivers from the state to the farm level (see Section 9.5.2), so too the complex global level policy influence on national decision-making needs to be acknowledged. Many commentators have highlighted the continuing lack of a 'global Leviathan' that dictates what nation states should do (Wilson and Bryant, 1997). This makes it difficult to conceptualise global level multifunctionality, as it continues to be unclear *whose* global multifunctionality notion is advocated (e.g. that of the EU, the Cairns Group, the WTO, the United Nations, or even just the G8 countries).

Although recent ratification of the Kyoto Protocol on climate change indicates a gradual change in how national environmental policy is increasingly influenced by global level decision-making, enforcement and monitoring of implementation of such 'international policy' remains weak (Bailey and Rupp, 2005). Yet, even the more structured CAP does not ensure implementation of EU-based policies in individual member states. The example of non-implementation of many aspects of agri-environment Regulation 2078/92 in Southern European member states is a case in point (Buller *et al.*, 2000; Wilson and Juntti, 2005; see also Chapter 7). This highlights that the influence of the supra-national and global level on multifunctional decision-making at national level is highly complex and that this influence will vary greatly depending on specific policy contexts (e.g. whether suggested policy change 'fits' broadly with national policy goals), the coercive powers underlying supranational policy suggestions (the failure of the USA to join the Kyoto Protocol is a case in point), and the general willingness and approach of a nation state to work towards 'the greater good' beyond narrow needs and expectations of the nation state and its citizens (i.e. altruistic national behaviour for which only few examples exist; ratification of the Kyoto Protocol or agreements to continue with the environmental protection of Antarctica may be such examples). What is clear from this debate is that there continues to be much flexibility concerning acceptance (or non-acceptance) of WTO guidelines by individual nation states. Although this is likely to change over the next few decades, at the beginning of the 21st century it is too early to argue that the WTO influences multifunctionality actions on the ground of *all* agricultural actors. WTO negotiations can, therefore, at best be seen as an *enabling* factor for implementation of multifunctionality on the ground that will often lead to weaker multifunctionality pathways (see Section 9.4). For the time being, national level multifunctionality decisions – partly mediated by global pressures – may, thus, continue to have more influence on direct expressions of multifunctionality at the grassroots level.

Finally, any discussion of global level multifunctionality needs to take into account *multifunctional interdependencies* – or in Knickel and Renting's (2000) terminology *multifunctional substitution* effects – between countries and territories. As highlighted above, although the nation state still provides the platform for implementation of strong, moderate or weak multifunctionality pathways at the national, regional and local levels, national-level multifunctionality may be highly dependent on the movement of food in and out of a country. Above I referred to this as the *global spatiality* of multifunctional processes. This highlights the importance of conceiving multifunctionality as composed of all individual parts of an agro-food system, including imports and exports of food in and out of a country. Just as calculations about the ecological footprint of an individual or nation state have to take into account food miles, international air travel or the importing of energy that go well beyond the boundaries of the nation state, so too do we need to factor in processes of 'imported' and 'exported multifunctionality' – possibly best expressed through the notion of the *multifunctional footprint*. As highlighted above, this may well mean that a country is able to maintain relatively strong levels of multifunctionality, precisely because this country imports large amounts of food and fibre from weakly multifunctional countries. Indeed, and as Section 10.2 will discuss, these latter countries may follow a weak multifunctionality model precisely because their agricultural systems are predicated on intensive export-oriented food and fibre production. Japan (importer) and New Zealand (exporter) may be interesting examples to invoke in this context (Jay, 2005). The former has been able to maintain a relatively traditional agricultural sector that could be seen to be moderately or even strongly multifunctional[82] (Francks, 2006). However, this national level strong multifunctionality may have only been possible due to massive food imports from less multifunctional countries such as New Zealand or Australia that have partly adjusted their agricultural trajectories to suit consumer needs of countries such as Japan or EU member states (Peterson *et al.*, 2002). Global multifunctionality may, therefore, only be a *zero-sum-game*, characterised by 'multifunctional competition' in which strongly and weakly multifunctional territories 'balance each other out', rather than a win-win scenario in which all territories can simultaneously move towards strong multifunctionality (Knickel and Renting, 2000; Knickel *et al.*, 2004; see also Section 10.4). This suggests that it may only be possible to implement strong multifunctionality pathways in one region or country *at the expense* of other territories.

9.6 Conclusions

Chapter 9 has provided the basic ingredients for a new conceptualisation of multifunctionality. The new multifunctionality model presented here differs from traditional economistic and policy-based concepts of multifunctionality (see Chapter 8) in that it is theoretically anchored in a revised p/np spectrum of agricultural decision-making and thought. This new concept of multifunctionality has allowed us to delve much more deeply into individual components, drivers and processes of multifunctionality than could hitherto be the case. I, therefore, only partly agree with McCarthy's (2005: 774) assertion that "multifunctionality seems poised to succeed postproductivism [sic] as a framework within which to interrogate contemporary rural dynamics". Although I agree that multifunctionality provides a much more robust framework to explain agricultural and rural change, this is only the case in the context of a *re-framed* p/np spectrum of decision-making opportunities.

The multifunctionality spectrum has allowed us to shed more critical light on the *definition of agriculture* itself as characterised by decision-making pathways along a spectrum bounded by productivist and non-productivist action and thought. Section 9.3 argued that the 'territory' beyond non-productivist action and thought has to be seen as the territory 'beyond agriculture'. This, in turn, allowed us to differentiate between *agricultural* and *rural* multifunctionality, with the former being a subset of the latter. The multifunctionality spectrum also enabled us to propose a *normative* view of different *multifunctionality qualities*, and in Section 9.4 I proposed a new classification based on *weak*, *moderate* and *strong* multifunctionality. I acknowledged that any discussion about multifunctionality has to include value judgements about the 'best' model of multifunctionality and suggested that the 'ideal' type of agriculture would usually be situated within the *strong multifunctionality model*, as this model may be best for farmers, the environment, agricultural-rural social cohesion, food quality, and for agriculture-society interactions as a whole. To provide empirical substance for the conceptualisation of different multifunctionalities, I discussed how different farming systems and agricultural policies can be seen to fall into either the weak, moderate or strong multifunctionality models. I highlighted that in some cases classifications are easy (especially at the extreme ends of the spectrum), but that often classification of a policy or farming system into one multifunctionality type is difficult. This highlighted the importance of conceiving multifunctionality as a *complex spectrum* comprised of many *different decision-making pathways* that may simultaneously straddle both the productivist and non-productivist end of the spectrum. Indeed, many farming systems or agricultural policies depict *simultaneous* characteristics of weak, moderate and strong multifunctionality.

---

[82] This would not include the Japanese fishing sector, which is one of the least multifunctional in the world.

This complexity is further enhanced by questions of scale – an issue that many former commentators on multifunctionality have neglected in their conceptualisations. In Section 9.5, I suggested that multifunctionality applies to *different spatial layers*, but that, ultimately, the most direct expression of multifunctionality is found at the *farm level* within the boundaries of individual farm holdings. Yet, I also highlighted that we can not understand grassroots-level multifunctional decision-making without also taking into account multifunctional action and thought at the rural community, regional, national and global levels. Indeed, strong multifunctionality pathways in one territory may, at times, be *predicated* on weak multifunctionality trajectories in another. Although most decisions affecting current multifunctionality at the grassroots level still occur within the framework of the nation state, recent trends suggest that the global level is assuming ever greater importance in influencing grassroots multifunctionality decisions.

Throughout Chapter 9, the importance of understanding multifunctional decision-making pathways has been implicitly highlighted. Yet, we can not understand how the concept of multifunctionality may be applied 'in practice' without understanding what influences the adoption of multiple and often complex multifunctional decision-making pathways by individual actors or stakeholder groups. It is to the issue of such multifunctional *transitions* that the next chapter will turn.

Chapter 10

Multifunctional agricultural transitions

10.1 Introduction

The aim of this chapter is to analyse the interlinkages between multifunctionality and transition. This will be based on both understanding decision-making processes at multiple levels, and on our newly conceptualised notion of multifunctionality as a spectrum bounded by productivist and non-productivist action and thought. While Chapter 9 highlighted the conceptual basis of the multifunctionality concept, Chapter 10 focuses on the 'practicalities' of understanding, implementing and influencing multifunctional agricultural transitions, in particular by investigating the potential for different actors to adopt multifunctional agricultural pathways closely linked to conceptual debates on transition discussed in Part 1. The chapter seeks to answer the following questions. How can we conceptualise constraints and opportunities for transitional potential of different actors and actor spaces for implementation of the strong multifunctionality model at different scales? How can we explain different multifunctional transitional pathways at farm level, and what are the key factors influencing farmers' multifunctional decision-making trajectories? How and by whom can decision-making within the multifunctionality spectrum best be managed, and how may this help managing multifunctional transitions? Finally, what can we learn from past trajectories for the future of multifunctional decision-making?

Section 10.2 will look at the transitional potential of actor spaces by investigating different *constraints and opportunities* for the evolution of multifunctional action and thought within 'funnels' of decision-making pathways. Such pathways will be further explored in Section 10.3, with particular emphasis on multifunctional transitions at farm level. The focus will be on investigating constraints and opportunities for the adoption of strongly multifunctional farm development pathways over time and with reference to notions of *path dependency* and *'nodes' of decision-making* affecting multifunctional transitions. How to best 'manage' such transitions will then be discussed in Section 10.4, with a view to unravelling *who* the beneficiaries of multifunctionality should be and *how* multifunctionality ought to be implemented into practice. The *temporal dimensions* of managing the multifunctional agricultural regime will also be discussed through investigation of past and future transitions. Concluding comments will be given in Section 10.5.

10.2 Exploring transitional potential: constraints and opportunities for multifunctional decision-making pathways

So far, we have conceptualised multifunctionality at a relatively abstract level (Chapter 9). In this section, I wish to provide a framework for understanding implementation of multifunctional action and thought with a focus on *transitional potential* of actor spaces through investigation of constraints and opportunities for the adoption of the strong multifunctionality model.[83] Building on Knickel *et al.* (2004), the focus here will be largely on multifunctionality as a *person-based concept* where individual stakeholders (e.g. farmers) are the ones defining multifunctional pathways based on a specific area of land (e.g. the farm or the rural community; see Chapter 9). This will form the basis for understanding *multifunctional transitions over time* in Sections 10.3 and 10.4.

To understand the complex interrelationships between constraints, opportunities, and specific enabling and disenabling factors for the adoption of the strong multifunctionality model by various actor groups, a specific conceptual framework based on 'funnel diagrams' will be used. Figure 10.1 shows the basic principle of such a diagram. The x-axis shows factors influencing the scope for decision-making, with 'enabling' factors to the left and 'constraining' factors to the right. The y-axis shows our multifunctionality decision-making spectrum bounded by productivist and non-productivist action and thought comprised of weak, moderate and strong multifunctionality pathways. We now need to imagine multiple decision-making pathways that are possible between the productivist and non-productivist ends of the spectrum, highlighted in the figure as the range of constraints and opportunities influencing multifunctional transitional potential. This range is *wide* towards the left of the figure (i.e. there is high multifunctional transitional potential for actor groups), while it is *narrow* towards the right (i.e. transitional potential between productivist and non-productivist decision-making pathways is limited with little margin for decision-making). This shows that strong multifunctionality pathways are not always *evenly accessible* to farmers and other stakeholders. Yet, although the range of opportunities is 'diminishing' towards the right of the figure, the 'quality' of possible multifunctionality pathways open to various actors is the same on the right and left of the figure. In other words, strong multifunctionality for actors on the right is qualitatively the same as for those on the left, although options for decision-making to achieve such strong multifunctionality are more constrained for the former.

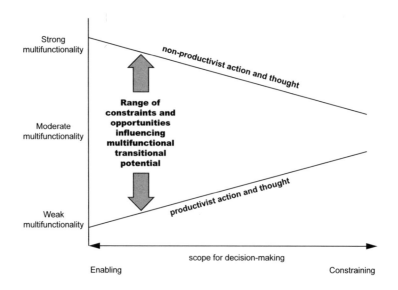

Fig. 10.1. Funnel diagram and the adoption of strong multifunctionality decision-making pathways

---

[83] In Section 9.4 I discussed why the strong multifunctionality model can be seen as the 'best' possible model of multifunctionality.

I wish to begin with a decision-making funnel that analyses constraints and opportunities for multifunctional pathway choices for different *farm types*. The reason for selection of this scale of multifunctionality was outlined in Section 9.5, where I argued that expression of multifunctionality is most important 'on the ground' where it will lead to *tangible* changes in the farmed landscape, agriculture-community interactions, and the quality of food and fibre production. It is at the farm level that we find both the most *direct* expression of multifunctional action and thought and the most important level for mediation of multifunctional influences exerted by other scales in the nested hierarchies of multifunctionality (see Section 9.5). Figure 10.2 shows how different types of farms can be broadly situated within a funnel based on a wide range of enabling factors for adoption of strong multifunctionality pathways for farms on the left of the figure, and a small range of enabling factors on the right. The further to the right a farm is situated, the more constrained the transitional potential of that farm will be. In theory, this model should be applicable anywhere, as most enabling or disenabling factors will be similar irrespective of the political, climatic or socio-economic situation in a given country or region. In most parts of the world large farms will have more transitional potential than small farms, and lowland farms will often have more opportunities for *differentiated* transitional pathways than upland farms (Guillaumin *et al.*, 2004). The figure is illustrative only and 'boundaries' between farm types should not be seen as rigid.

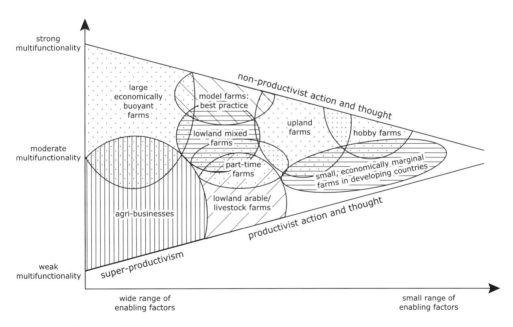

Fig. 10.2. Multifunctionality decision-making funnel for different farm types

If we start at the bottom left corner of the figure, most *agri-businesses* are located towards the weak to moderate end of the multifunctionality spectrum. They usually have a wide range of enabling factors and relatively high transitional potential (spanning a wide spectrum from weak to moderate multifunctionality) as they are often large, vertically integrated (i.e. well embedded into national or global agro-commodity chains), and well capitalised. Walford (2003b: 494) suggested for the UK that "large-scale commercial farmers [are often] viewed as more inclined to adopt innovative approaches, but … may equally be regarded as inveterate agricultural commodity producers". In a developing countries context, this cluster also includes large plantation farms with their often high capitalisation and export-oriented nature (see also Chapter 9). These types of farms have a relatively wide range of decision-making options but, due to their profit-maximising productivist orientation, they tend to be situated towards the weak end of the multifunctionality spectrum (Freshwater, 2002). As Marsden (2003) highlighted, many of these farms have been placed on a technologically driven cost-price squeeze which has led them to race towards reduced prices in the increasingly intensified 'standard' agricultural product sector often predicated on 'disembedded' agro-commodity chains (see also Ward, 1993). This is not to say, however, that agri-businesses may not also embark on moderate or even strong multifunctionality pathways with *some* of their farm management decisions (see also Section 10.3). As Van der Ploeg and Roep (2003: 42) emphasised, "many previously highly specialized, monofunctional farms are being transformed into new, multifunctional enterprises". Similarly, Marsden (2003) suggested that in most intensive agricultural systems, the momentum of change in both mass and quality food markets will be central for providing opportunities for change. Depending on ideological and attitudinal factors, the owner of an agro-business-oriented farm may decide to embark on niche market pathways or set-aside some land for conservation (although any large-scale non-productivist activity would mean that the farm would probably no longer be classified as an agri-business).

At the other side of the multifunctionality spectrum on the left of the figure, *large economically buoyant farms* also have high transitional potential (like agri-businesses), but also often have good opportunities for implementation of strong multifunctionality pathways. Although these farms are, at times, strongly profit-oriented (and therefore only weakly or moderately multifunctional), their economic buoyancy allows them the 'luxury' to embark on qualitatively different non-productivist decision-making pathways. Thus, the wealthy owner of a large farm may decide to take some land out of production for conservation (for example using an agri-environmental scheme), may embark on strongly multifunctional diversification activities to provide employment for the local rural community (e.g. through tourism-related initiatives on the farm), or may decide to partly withdraw from sales of agricultural products associated with high food miles for ideological and environmental reasons (i.e. relocalisation of sale of agricultural products through a farm shop, farmers' market or 'box scheme'), while, at the same time, still continuing with productivist production on the rest of the farm. Whether in India, Germany or Peru, throughout the world it is these large economically relatively well-off farms that do not necessarily rely on the sale of agricultural products for economic survival that can act as important *role models* for implementation of strong multifunctionality pathways.

To the right of these two large clusters of farms, potential is more varied and various drivers for multifunctionality decision-making opportunities become apparent. First, in many parts of the world, *lowland farms* often tend to be farmed along more productivist and weakly multifunctional pathways (Walford, 2003b), while *upland farms* – often characterised by more extensive farming in areas with high conservation value – tend to be situated closer to the strongly multifunctional end of the decision-making spectrum (see Dax and Hovorka, 2004, for Austria, or Louloudis *et al.*, 2004, for mountain areas in Greece). It is estimated that in many European uplands, for example, 80-90% of farms are engaged in pluriactivity (Knickel *et al.*, 2004). Although many exceptions can be found (see Chapter 7), the lowland/upland dichotomy acts as a useful guide for transitional opportunities towards weak or strong multifunctional pathways, particularly as in many upland farming areas multiplier effects of various strongly multifunctional activities are more likely (e.g. landscape protection, tourism, 'deep' diversification; cf. Knickel and Renting, 2000). Yet, as highlighted above, there is a wide spectrum of farm types among both lowland and upland farms, and not all lowland farms will show a propensity for weak multifunctional pathways. A case in point relates to farms in peri-urban fringe areas (usually in the lowlands), where many studies suggest potential for strongly multifunctional pathways involving various multifunctional diversification activities (e.g. golf courses, farm zoos, etc.) that exploit the farm location near large centres of population (e.g. Halfacree, 1997b; Vandermeulen *et al.*, 2006). Luttik and Van der Ploeg (2004: 210) argued that "in the mosaic of agricultural land in near-urban agriculture, there is a special role for multifunctionality", while Gallent *et al.* (2004) referred to the peri-urban fringe as "a zone of transition" with almost unique multifunctional potential. In a similar vein, Fleury *et al.* (2004) pointed towards the increasing diversity of multifunctional opportunities open to peri-urban farms, ranging from farm management pathways close to the productivist end of the spectrum to pathways in which farmers straddle the boundary of non-productivist land use (see also Errington, 1994).

Second, *pure arable or livestock lowland farms* (that may, of course, also be classified as agri-businesses depending on overall farm turnover and integration into the global capitalist system) are often more likely to embark on weak multifunctionality pathways than *mixed lowland farms* (Walford, 2003b). Compared to their monocultural counterparts, the latter are often characterised by less intensive agricultural production practices, higher levels of diversification, better horizontal integration into the local rural community due to more complex product and workforce differentiation, and may have owners who see the role of agriculture in more moderately or strongly multifunctional ways than their more specialised neighbours. As Swagemakers (2003: 198) argued, "the multifunctional farm is, in essence, a mixed farm". Throughout the world, farm specialisation, therefore, may have an important bearing on the position of farms along the multifunctionality spectrum and, in particular, on transitional potential. Third, the level of time commitment of a farmer towards the farm will also often influence multifunctional transitional potential. The example of *part-time farmers* is particularly interesting as these farmers often choose moderately multifunctional development pathways, precisely because they lack the time and commitment to embark on purely productivist trajectories on the one hand (which often require full-time commitment), while also lacking the time and energy needed to move the farm completely towards the non-productivist end of the spectrum. Of course, part-time farmers can just be seen as highly pluriactive farms (with varying degrees of commitment towards the farm), with Knickel *et al.* (2004) highlighting that probably over 50% of European farms are pluriactive in some way or another. Fourth, what could be termed *model farms* occupy an interesting position in the multifunctionality spectrum. In many parts of the world, such farms are established based on a variety of personal or institutional drivers, for example Prince Charles' model farm of Highgrove in the UK, or model farms established by Landcare groups in Australia. These farms are established precisely to show how strong multifunctionality pathways can be established in practice and are seen to serve as a 'model' for other farmers (in the region or nationally) of how productivist pathways can be left behind (although agricultural production may still be highly profitable). For this reason, these farms may be located very close to the non-productivist strong multifunctionality end of the spectrum, albeit with a more limited range of multifunctionality decision-making options than their large agri-business or economically buoyant counterparts.

Although occupying a small conceptual space within Figure 10.2, farm types on the right-hand side of the model arguably contain most farmers on Earth. In particular, the many *small-scale economically marginal farms* in developing countries (subsistence farms) usually occupy the strong multifunctionality end of the spectrum, albeit with an often very limited range of decision-making opportunities for adoption of *different* farm development pathways (i.e. limited transitional potential) (Roux *et al.*, 2004). Pretty (1995) estimated that some two billion people worldwide are still involved in such farming systems (see also Losch, 2004). Freshwater (2002: 14) emphasised the strongly multifunctional social and cultural capital of these farms by arguing that they "have a clear advantage in preserving traditional farm culture and contributing to the broader rural community". These farms are also more likely to be embedded in strongly multifunctional localised food systems. Yet, as Chapter 9 highlighted, subsistence farms are often strongly multifunctional out of *necessity* than choice. It is interesting to contrast these farms with small economically marginal farms in developed countries. Out of economic necessity, the latter are often situated towards the weak or moderate multifunctionality end of the decision-making spectrum, because they have to maximise production in order to stay economically viable – a farm trajectory often aided and abetted by productivist state subsidies (especially in the EU) (see Chapters 5 and 7). This highlights that the drivers for adoption of multifunctionality pathways among small and economically marginal farms still differ between the developed and developing world, while multifunctionality drivers for highly globalised and well capitalised farms are increasingly *converging* around the world (Goodman and Watts, 1997; Parayil, 1999). As Figure 10.2 (above) highlights, although small-scale farms in both the developed and developing world have options to embark on different multifunctionality pathways (ranging from weak to strong multifunctionality), their transitional potential is often based on a relatively narrow range of opportunities. In other words, on these farms it does not take much change to farming practices to 'tip the balance' from strong to weak multifunctionality or vice versa, while on large highly capitalised farms to the left of the figure the transition from one multifunctionality level to another is likely to be more gradual (see also Section 10.3). Yet, as highlighted in Chapter 7, new opportunities for strongly multifunctional pathways may be emerging for small economically marginal farms in the developed world located in the uplands or in areas of high conservation value, as agri-environmental schemes and farm tourism may offer specific non-productivist alternatives to productivist farming trajectories (Ronningen, 1994; Mollard *et al.*, 2004).

The final type of farm shown in Figure 10.2 is *lifestyle* or *hobby farms* – i.e. farmers who adopt farming as a hobby and who do not rely on the sale of food and fibre for economic survival as they often have a stable income outside of farming (see also Chapter 9). This is a most interesting example as hobby farms can be seen to be most closely linked to the non-productivist end of the multifunctionality spectrum[84] (Holloway, 2002; Mather *et al.*, 2006). Often, hobby farmers (mainly in the developed world) have purchased a farm (usually small) to farm in the most strongly multifunctional way – a decision-making pathway made possible as they do not need to maximise profits to ensure survival of the farm, enabling them to focus on agricultural land as a *consumption* good rather than as a *production* asset (Primdahl, 1999). Thus, "able to draw on an income source outside farming, these newcomers are arguably much more 'decoupled' from market trends and agricultural policy decisions than even the most pluriactive mainstream farmers" (Bohnet *et al.*, 2003: 350). Several studies have highlighted how the number of lifestyle farmers is increasing dramatically (e.g. 41% of all UK farms that changed ownership in 2003 were purchased by lifestyle farmers; Mather *et al.*, 2006), which suggests that non-productivist pathways may become much more prominent in the future in some developed countries.

Lifestyle farmers share common elements with small-scale subsistence farmers in developing countries who produce food entirely for their own family or household, although most hobby farmers would be highly pluriactive and not even be self-sufficient in food. Hobby farms can be seen to be the only type of farm that not only straddles the non-productivist boundary of the multifunctional decision-making spectrum, but where some 'agricultural' activities will also lie 'beyond agriculture'. This is particularly true if these 'farmers' are holding livestock as 'pets' (for themselves or their children, for example), or because they engage in other 'agricultural' activities not linked to food and fibre production (Primdahl, 1999; Holloway, 2002). It is for hobby farms, therefore, that the discussion in Section 9.3 on defining the boundaries of agriculture through the p/np spectrum of decision-making making becomes, arguably, most pertinent. As Bohnet *et al.* (2003: 349) emphasised for the UK, we need to start distinguishing "between holdings that are still seen primarily as sites of production by their farming family occupiers and those that are coming to be regarded chiefly as spaces for living by a new category of lifestyle occupiers". Similarly,

---

[84] In this context, I disagree with Evans *et al.*'s (2002: 323) suggestion that the notion of "hobby farming is not envisaged as a challenge to productivism".

Cayre *et al.* (2004: 32; my translation) argued in a French context that "the notion of multifunctional use of rural space highlights the tensions between farmers and lifestylers ['néo-ruraux'] and will lead to a redefinition of the activities, roles and raison d'être of agriculture". Yet, Section 9.4 also highlighted that hobby farming should not be over-romanticised as the 'most' strongly multifunctional farm type. Indeed, as Figure 10.2 shows, hobby farms may also straddle moderate multifunctionality pathways, especially as these farmers are often wealthy urbanites who have not been brought up in the region where they bought their farm. The strong multifunctionality dimension of 'local embeddedness' may, therefore, be relatively weak on these farms. Indeed, "the consequences of … a growing number of residential farms … will be to create a much more fragmented stakeholder community, many of whose members are decoupled from agriculture" (Bohnet *et al.*, 2003: 362). This may be amplified because hobby farmers' mental image of farming may not necessarily concur with the strong multifunctionality model, as these 'farmers' lack a farming background and, therefore, the historical knowledge of how to implement strong multifunctionality pathways (Winter, 1997; Curry and Winter, 2000). As Bohnet *et al.* (2003: 351) emphasised, "the replacement of long-established family farmers with residential occupiers brings with it an erosion of local knowledge and traditions which have been important in sustaining landscape character in the past".

Figure 10.3 takes this discussion further by highlighting the transitional potential of farms based on *ownership patterns*. Across the globe, differences in farm ownership have proven to be key drivers for transitional potential along the multifunctionality spectrum (Sobhan, 1993; Robinson, 2004). Thus, owner occupiers (often in the form of 'family farms') have most control over their multifunctionality decisions as they are not constrained by others in choosing decision-making pathways. As Section 10.3 will also highlight, owner occupiers may also have most flexibility when it comes to rapid changes in multifunctional transitional direction (McReynolds, 1998). There are many examples of family trusts in which multiple household members (e.g. father and daughters) make joint decisions, as well as complex ownership arrangements comprising shareholding companies and multiple farm owners (partnerships), many of which may not necessarily be from a farming background or even from the country within which the farm is located (Pretty, 2002). Here, transitional decision-making potential is constrained by the fact that usually a majority needs to be obtained for decision-making actions to be implemented. In many cases, strong multifunctionality pathway options are constrained by productivist-oriented part-owners (or vice versa). As a result, on farms with multi-member ownership changes to multifunctional transitional pathways will often be slower than on owner-occupied farms, and 'path dependency' also depends largely on the rapidity of common decision-making structures.

Figure 10.3 highlights that it is on tenanted/rented farms (including sharecropping farms) – comprising the largest proportion of farms on Earth – that multifunctional decision-making opportunities are likely to be most restricted. In the USA, for example, about half of all agricultural land is rented or leased (Freshwater, 2002). On these farms, tenants or farm managers do not have ownership of the land, resulting in multifunctional decision-making pathways influenced by *external* decision-making constraints. While some tenants may retain more or less complete control over multifunctional transitional decisions, most often the owner/landlord retains almost complete control. This does not mean, however, that tenanted farms necessarily tend towards weak multifunctionality pathways. As with owner-occupied farms and farms in multi-member partnerships, farm ownership patterns influence the *range* of decision-making opportunities rather than the *quality* of chosen multifunctional pathways. Nonetheless, owner-occupiers may have more flexibility when it comes to *changes* in

multifunctional transitional direction, while on farms with multi-member ownership changes
to multifunctional transitional pathways are likely to be slower[85].

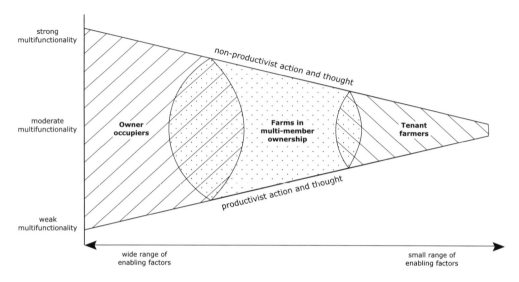

Fig. 10.3. Multifunctionality decision-making funnel for different farm ownership types

Figure 10.4 broadens the discussion of transitional potential by focusing on the regional
and national levels of our nested hierarchies of multifunctionality (see Fig. 9.7). Although
multifunctionality finds its tangible expression on the ground at the farm level, Chapter 9
also highlighted that these decisions are *mediated* through different scales of decision-
making, of which the regional and national scales play a particularly important role for
policy environment, societal ideologies and consumer behaviour (Garzon, 2005). Bresciani *et
al.* (2004: 302) emphasised the "complex web of representations, meanings and values
through which members of national societies conceptualise the contribution of agriculture
and the rural world to the making of their common national identity". Building on Knickel *et
al.* (2004: 91) who argued that "there are important differences in the types and scale of the
multifunctionality of agriculture between countries", Figure 10.4 shows four overlapping
clusters of national/regional multifunctional transitional potential. At the bottom left of the
figure, we find a large cluster of countries that is dominated by agricultural export-oriented
countries, or in Peterson *et al.*'s (2002) terminology 'high-cost agricultural producing
countries'. In most of these countries/regions, transitional potential will be relatively large,
spanning almost the full multifunctionality spectrum. However, national and regional
opportunities would rarely include the extreme non-productivist end of the spectrum due to
encouragement of intensive agricultural practices and policies for export purposes –
processes also often driven by neo-liberalist productivist ideologies (Lifran *et al.*, 2004).
While the domestic role of agriculture may have declined over the past decades, Freshwater
(2002) rightly argued that agricultural exports remain important for a considerable number of
countries. Thus, in these countries the national and regional scales provide a *wide range* of
enabling political and socio-economic processes, but often with an emphasis on weak or

---

[85] Freshwater (2002) also highlighted that public or private ownership of land may influence opportunities for
implementation of strongly multifunctional pathways. Balancing multiple outputs may be easier on public lands
because the public sector is rarely under pressure to maximise profits (see also Wilson and Memon, 2005).

moderate multifunctionality pathways dominated by ideologies of production maximisation and agricultural intensification discourses (Le Cotty *et al.*, 2003). This cluster includes many countries in the developed world where industrial export-oriented agricultural systems predominate, for example the USA, Canada, Australia, New Zealand, the Ukraine and some intensively farmed EU member states such as the Netherlands and Denmark (Brazier, 2002; Jongeneel and Slangen, 2004; Primdahl and Swaffield, 2004). For Australia, Holmes (2006) highlighted that weak multifunctionality pathways are largely predicated on pressures to intensify in already environmentally marginal agricultural areas (agricultural treadmill), continued dependency on agricultural exports, neo-liberal agendas of profit maximisation, and limited opportunities for diversification (see also Cocklin and Dibden, 2004; Wilson, 2004). Freshwater (2002) and Boody *et al.* (2005), meanwhile, suggested that the USA has adopted relatively weak multifunctional pathways based on its industrial and export-oriented agricultural regime (but see also Cochrane, 2003, or Bell, 2004, for opposing viewpoints). Freshwater (2002: 8) further suggested that the desire to preserve a rural way of life predicated on strongly multifunctional pathways "does not fit well with a major part of modern American culture".

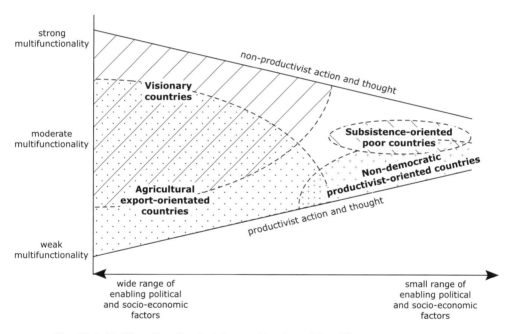

Fig. 10.4. Multifunctionality decision-making funnel for different country clusters

This cluster particularly includes New World countries characterised by highly globalised productivist agro-commodity structures based on colonial and historical agricultural export-oriented ties with the Old World. New Zealand is a classic example, colonised after 1840 largely to provide the UK with a secure supply of livestock-based commodities. Buller (2005) also emphasised that many of these countries have generally adopted more liberal agricultural policy stances, have pushed for greater decoupling, and have sought 'green' (e.g. agri-environmental) rather than 'blue' box solutions, thereby perceiving multifunctionality as a hindrance to free trade and a disguised form of protectionism – perceptions also confirmed by Hollander's (2004) study of the multifunctionality of sugarcane production in Florida (US). Marsden (2003: 168) argued that this productivist model is often reinforced by an alliance of actors "who are still legitimising the intensification and scale economies imported from American and Australasian modernisation theory". In Europe, this cluster also tends to include countries that lack large – and often strongly multifunctional – mountainous areas (e.g. the Netherlands, Denmark) that could have provided a territorial 'sanctuary' for non-productivist agricultural pathways (see Chapter 7). As Andersen *et al.* (2004: 116) emphasised, empirical findings in the EU point towards "a clear differentiation for multifunctionality in relation to environment". Yet, tendencies towards non-productivist action and thought are also beginning to emerge within this country cluster. In Australia, for example, various authors point towards 'new geographies of value', driven by changing expectations and visions of rural areas by middle class urban residents linked to rapidly increasing counter-urbanisation movements (including hobby farming and alternative lifestylers) (Holmes, 2002, 2006; Wilson, 2004). This was echoed by Ilbery *et al.* (1997: 6) who argued that for many countries in the developed world "the agricultural landscape will become even more highly differentiated, with some regions (and within regions, between individual farms) following productivist methods of farming, others following the 'instrumentalist' model of agricultural development and yet others following 'idealist' [i.e. strongly multifunctional] food production methods".

Increasingly, this country cluster includes developing countries aiming to maximise agricultural production through increased embeddedness in globalised agro-food systems (Goodman and Watts, 1997). This includes countries that have embarked on productivist development pathways *out of choice* (e.g. China, Mexico), and those where productivist tendencies have been largely externally enforced through economic and historical (e.g. colonial) factors (e.g. sugarcane plantations in Fiji that superseded traditional subsistence farming; Caribbean countries historically tied to agricultural export markets; developing countries in which large multinationals have exerted control over production and sale of commodities such as Del Monte in Ecuador). This cluster, therefore, includes countries more firmly embedded into the global capitalist system, although there will still be a high geographical diversity of weak, moderate and strong multifunctionality *on the ground*. As highlighted in Chapter 9, the predominance of national weak or moderate multifunctionality ideologies and tendencies does not always result in similar action on the ground, as the latter will still depend on how national and regional influences are *mediated* and *implemented* (or not) by regional actors or farmers, resulting in a wide range of multifunctional grassroots-based farm development pathways that may show a different picture from that of national-scale multifunctionality ideologies (see Fig. 10.2).

At the top left of Figure 10.4 we find a cluster of countries that I term 'visionary countries' – visionary as these countries tend to place more emphasis on moderate or strong multifunctionality pathways. Again, these countries have *large transitional potential*, spanning almost the entire upper end of the multifunctionality spectrum. As Figure 10.4 indicates, we need to acknowledge a large overlap between this country cluster and the cluster of agricultural export-oriented countries discussed above, in particular as moderate multifunctionality pathways may be the outcome of both visionary countries failing to persuade farmers to implement strongly multifunctional pathways, and of agricultural export-oriented countries in which farmers may choose not to heed national ideological and political pressure by adopting less productivist multifunctionality trajectories. This cluster will often include countries where agricultural development pathways may have been slightly (or substantially) different from the productivist-oriented developments that have dominated many parts of the world over the past 50 years (see Chapter 5), where a long tradition of protecting rural landscapes and communities exists, and where consumers exert substantial power over type and quality of food available (cf. Dabbert *et al.*, 2004). Boody *et al.* (2005: 27) referred to this cluster as "countries that value the nonmarket benefits of agriculture". These countries will be largely comprised of developed countries, as only relatively wealthy countries that may have the opportunity to substitute shortfalls in self-sufficiency through expensive agricultural imports can allow themselves the 'luxury' to guide farm development pathways towards the strong multifunctionality model (see Van der Ploeg and Roep, 2003, and Knickel *et al.*, 2004, who place Germany, Ireland and Italy in a 'vanguard' position on multifunctionality). Two key factors are likely to influence strong multifunctionality ideologies and opportunities at national level in this cluster. First, geography matters, as countries with substantial mountainous and other agriculturally disadvantaged areas are more likely to be strongly multifunctional (see Chapters 7 and 9), particularly as agriculture in these areas is often characterised by those dimensions that define the strong multifunctionality model (e.g. opportunities for diversification through tourism in environmentally attractive areas; extensive farming systems; strong local embeddedness of remote agricultural and farming communities; see Section 9.4). As Hall (1998: 277) highlighted, economic and geographical factors matter particularly in relation to options for farm tourism, with the presence of operations closely related to: "the marginality of agricultural activity; the presence of tourism and recreational resources near the farm; and accessibility to major tourist generating regions". In the UK, for example, only about 7% of the total land area is available for recreation by legal right, with only 15% of farmers engaged in tourism-related diversification (Marsden, 2003). Regional and local weak multifunctionality pathways may, therefore, not be a matter of choice but of *necessity* due to lack of alternative non-productivist opportunities.

It is little wonder, therefore, that mountainous countries such as Switzerland, Austria or Slovenia (or the state of Montana in the USA, for example) often show strong multifunctionality tendencies at *both* national and grassroots levels. For example, the Swiss Federal Agricultural Law of 1992 was aimed at actively promoting strong multifunctional pathways throughout its territory (Pretty, 2002), and Ronningen (1994) highlighted how strong multifunctionality pathways in Switzerland have been predicated on 'visionary' agricultural policies and the emergence of a 'green bureaucracy' (see also Schmid and Lehman, 2000). To be sure, many of these countries have the wealth to afford to pay farmers for strongly multifunctional activities – opportunities that will not be available to most other countries (Buller *et al.*, 2000; Losch, 2004). Indeed, Switzerland emerges as the country with the highest rate of financial aid paid to farmers in the world (closely followed by Norway, Korea and Iceland) (Lang and Heasman, 2004), suggesting a close link between state *financial* support and strong multifunctionality (see also Section 10.4).

Historical and cultural factors also play a key role. As Bresciani *et al.* (2004: 302) highlighted, "national cultures' conceptualisation of the role of agriculture may vary according to each country's historical situation and relative degree of development". Countries such as France or Italy that have a long tradition of 'visionary' approaches towards local food production (e.g. 'slow food' movement and high proportion of organically farmed land in Italy; cultural importance of locally produced food in France), strong local agricultural/rural embeddedness and high levels of diversification activities, tend to fall into the visionary country cluster (Jones and Clark, 2001; Lardon *et al.*, 2004). Potter (2004: 23) suggested that "France has moved further than any other Member State in packaging multifunctionality as part of a broader agrarian policy agenda", while Buller (2004: 101) argued that France has been characterised by "a growing acceptance of the multifunctional role of farming within rural space". However, France and Italy also show signs of weak multifunctionality based on their export-oriented rural economies and their 'agrarian' policy agenda (Lowe *et al.*, 2002; Buller, 2004). The latter point re-emphasises how important it is to recognise the *breadth* and *overlapping nature* of multifunctionality decision-making opportunities in all country clusters shown in Figure 10.4. Indeed, Garzon (2005) highlighted through a comparison of southern, northern and eastern Germany how multifunctionality pathways within one country (both at grassroots and regional levels) can span almost the entire spectrum from strong(er) (e.g. Bavaria) to weak(er) multifunctionality (e.g. parts of northern Germany) (see also Wilson and Wilson, 2001; Knickel *et al.*, 2004).

The 'visionary' cluster also includes developing countries in which strong multifunctionality pathways are driven by national ideologies. Bhutan in central Asia is particularly interesting. Recent news coverage from Bhutan suggests that this traditional and technologically still relatively underdeveloped country is attempting to resist globalisation influences, and that its main goal is to maintain traditional economic structures predicated on small-scale farming. Although there are overlaps here with the country cluster comprised of subsistence-oriented poor countries discussed next, it appears that Bhutan has embarked on a strong multifunctionality pathway *out of choice* (at least for the time being). Although information on Bhutan is sparse, it appears that Bhutan is able to feed its population with nationally produced foodcrops (self-sufficiency) and some limited imports. It may, therefore, be in a position – as opposed to many other subsistence-oriented countries discussed below – to choose the best possible multifunctional agriculture pathways and, as a result, fiercely resist a productivist model of intensification and export-orientation (similarly cautious approaches are used in Bhutan for the development of the tourist industry). These processes may be aided by two factors. First, Bhutan was never a colony of European powers. As a result, it was never incorporated into the often weakly multifunctional globalised agro-commodity chains that characterise almost all former colonies. Second, the link between strong multifunctionality pathways and an extremely mountainous terrain again needs to be emphasised, highlighting that the survival of a strongly multifunctional agricultural regime in Bhutan was probably only possible due to its remote and relatively inaccessible nature. Cuba is another interesting country in which a formerly socialist-driven productivist agriculture has been increasingly replaced by more strongly multifunctional pathways based on the Cuban government's declaration of an 'alternative model' of agriculture (Pretty, 2002; Lang and Heasman, 2004). This model has focused on diversification of agriculture (i.e. rediscovering the previous richness of a diversified agriculture), the replacement of 'heavy' technology with 'soft' technology (e.g. oxen for tractors), integrated pest management, and strongly multifunctional cooperation among farmers within and between communities.

The middle right-hand side of Figure 10.4 shows a country cluster comprised of subsistence-oriented poor countries. In this cluster, national and regional policies and incentives will often be aimed at encouraging weak(er) multifunctionality pathways predicated on increasing agricultural exports and intensifying national agriculture. Although Figure 10.2 showed the moderately to strongly multifunctional nature of small subsistence *farms* in these countries, *national* level ideologies in this cluster aim to change the situation of their small-scale and often very poor farming populations. In other words, pressures for *change* are generally strongest within this country cluster. Yet, transitional opportunities are often limited due to economic, political and social constraints, and strong multifunctionality pathways generally predominate *out of necessity* rather than choice. This cluster includes most developing countries, in particular those where subsistence or highly localised farming continues to predominate (e.g. large parts of India and Indonesia; most sub-Saharan African countries; many South American countries). However, although most of these countries still show high levels of local embeddedness and highly diverse and extensive production systems, many are in the process of intensification, capitalisation and polarisation of agricultural production – often based on national and regional policies and incentives predicated on weak(er) multifunctionality models (see also Section 10.4).

The final cluster in Figure 10.4 comprises an increasingly shrinking number of countries with non-democratic systems in which agricultural production is still highly regulated by the state and where sub-national governance structures are weak. These countries are almost exclusively productivist and tend to favour *weak multifunctionality* pathways for ideological and self-sufficiency reasons. These pathways predominate *out of choice* and are not overtly challenged by critical voices. Transitional potential towards moderate or strong multifunctionality is weak or non-existent for ideological, political (and often economic) reasons. Although information is sparse, it is in countries with democratic deficit that local and national multifunctionality goals are most similar (see also Figs 9.7 and 10.2 above). In other words, the state encourages weak multifunctionality through coercive non-democratic means, while farmers in these countries implement weak multifunctionality models based on state indoctrination, political and economic pressure, or for ideological reasons (Forsyth, 2003). The example of North Korea has already been mentioned, but examples of the Soviet Union (until 1990), the GDR (until 1990) or Myanmar can also be invoked here, where state control over agricultural multifunctional pathways was, or continues to be, almost absolute (e.g. Bryant, 1994, 1996; Woo, 2006). The key strong multifunctionality dimensions of changing local and national ideologies towards non-productivist thinking, as well as encouragement of environmentally sustainable farming methods, are likely to be absent in these countries. Unless political structures in these states change substantially, it is unlikely that both the position of these countries within Figure 10.4, as well as their transitional potential towards strong multifunctionality, will change substantially in the near future.

Three important caveats regarding Figure 10.4 need to be highlighted. First, the 'boundaries' between individual clusters have to be seen as *overlapping, flexible* and *permeable*. Most of the country examples highlighted above contain various ideological and policy-oriented goals that often span the entire spectrum from weak to strong multifunctionality. Second, just as individual farms can quickly change their multifunctional transitional trajectories (see above and Section 10.3), entire regions or even countries may change their multifunctional policies (and even ideologies) relatively quickly. Dramatic political upheaval, such as the end of the Soviet Union and the 'post-socialist' transition in many local systems and networks in Eastern European countries (see Chapter 3), often means that multifunctional transitional trajectories can change quickly (Gatzweiler, 2004; see also Section 10.4). Third, we need to be aware of the danger of broad generalisations and that we need further case-by-case (i.e. country-by-country) investigations of the positionality of national and regional agricultural actors/institutions/holdings along the multifunctionality spectrum (see also Chapter 11).

10.3 Multifunctional transitions at farm level

Section 10.2 explored the transitional potential of various actor spaces at different geographical scales. In this section, I wish to extend this discussion by focusing on multifunctional changes over time, in particular by investigating the nature, shape and processes of multifunctional *transitions*. The discussion will be closely interlinked with theoretical and conceptual issues of transitional processes discussed in Part 1. The focus will be on the farm level – the crucial level for implementation of multifunctional action and thought (see Section 9.5) – while Section 10.4 will focus more broadly on understanding how to best manage multifunctional transitions at various time scales and with regard to different, and often competing, actor spaces.

The following discussion will be both conceptual and practical. Section 10.3.1 will first explore conceptual issues linked to multifunctional farm transitions. It will argue that we need to conceive of multifunctional farm development pathways as Deleuzian transitions (see Chapter 2) and will explore the importance of understanding multifunctional path dependency and system memory for conceptualisations of farm-level multifunctionality transitions. Section 10.3.2 will then illustrate farm transitions at a more practical level using two hypothetical examples from the developed and developing world. The purpose of this section will be to both illustrate the complexity of farm-level multifunctional transitions, and to highlight the utility of the multifunctional decision-making spectrum for understanding farmers' multifunctional decision-making.

10.3.1 Path dependency, system memory and farm-level multifunctional transitions: some conceptual issues

In Section 9.2, I highlighted how multifunctional decision-making occurs along Deleuzian transitional pathways bounded by productivist and non-productivist action and thought. Building on Figure 9.2 in Chapter 9 and based on the notion that "multifunctionality [can] be operationalised at the level of the individual farm household" (Knickel *et al.*, 2004: 97), Figure 10.5 shows such Deleuzian pathways for hypothetical individual farm trajectories (farms 'a' to 'h' in the figure). The figure shows examples of farm trajectories and how these may vary over time, ranging from strong multifunctionality pathways to moderate and weak multifunctionality. Two key points need to be highlighted here. First, and building on Van der Ploeg's (2003) notion of different 'farming styles', the figure shows that farm development pathways can span the entire multifunctionality spectrum. Some farms (e.g. farm 'h') are firmly embedded in the strong multifunctionality model over the entire time period, while others (e.g. farm 'c') are closely associated with weak multifunctionality. Other farms, meanwhile, show a steady transition from weak to moderate multifunctionality over time (e.g. farm 'b') or from moderate to strong multifunctionality (e.g. farm 'e'), while others depict highly fluctuating transitional trajectories that shift several times between different multifunctionality levels (e.g. farm 'g'). Others, meanwhile, show relatively slow changes for a length of time and then sudden changes over a relatively short period (e.g. farm 'a' that suddenly shifts from weak to strong multifunctionality; see discussion of 'transitional ruptures' below). Second, the figure shows that it would be rare for a farm to stay at the same level of multifunctionality for any longer time period. Although changes may be small (e.g. farms 'd' or 'f'), subtle changes in the position of a farm in the multifunctionality spectrum will *always* occur (especially the longer the transitional time frame under consideration), based on changing personal, farm-level or external circumstances – pathways often referred to as 'farm survival strategies' (e.g. Meert *et al.*, 2005). As Marsden *et al.* (2002) argued, many agricultural businesses have begun to develop 'new economic logics' (scope economies) through which they exploit the best position (regarding income, job satisfaction, employment creation, etc.) within the multifunctionality spectrum (see also Fairweather and Campbell, 2003; Van der Ploeg, 2003). This means that on-farm multifunctional pathways are never static but always *dynamic*.

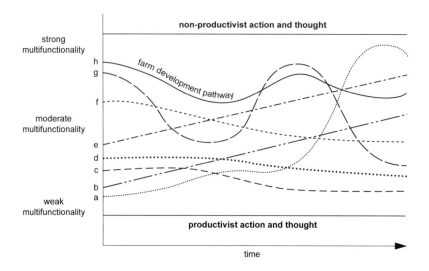

Fig. 10.5. Multifunctional farm-level transitional trajectories

Let us look more closely at the example of a specific farm multifunctional trajectory over time. Figure 10.6 shows the example of hypothetical Farm X and how this farm has, over a period of time, undergone a transition from relatively weak towards strong multifunctionality.[86] Various conceptual issues need to be considered here. First, Farm X is undergoing a transition that can hypothetically be shown as a *line* with a start and end point. Based on our discussion in Chapter 9, we need to conceptualise this hypothetical line as the *sum total* of the different dimensions of multifunctionality pressures acting upon the farm territory (see Sections 9.4 and 9.5). As discussed above, a complex mixture of productivist and non-productivist action and thought will be present on every farm, with, for example, some activities (e.g. intensive agricultural production) situated more closely towards the productivist end of the multifunctionality spectrum, while others (e.g. diversification activities) may be located towards the non-productivist end. It is the 'mixture' of all the activities that make up the multifunctionality pathway of a farm – i.e. the constant tussle between productivist and non-productivist action and thought (cf. Van der Ploeg, 2003). Bohnet *et al.* (2003: 361) aptly described this as "farming careers punctuated by periods of intensification, consolidation and extensification … The dynamics of change are thus complex, location sensitive and subject to various feedback effects" (see also Primdahl, 1999). For illustrative purposes the line in Figure 10.6 is drawn as a clearly definable single line, but in reality it is more likely to resemble a thick and fuzzy line with indistinct edges.

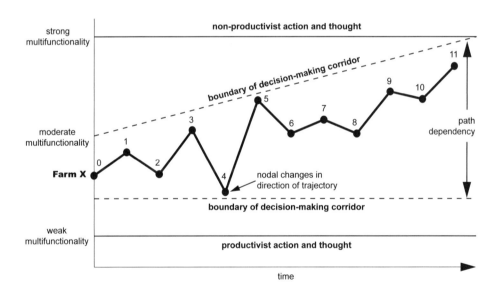

Fig. 10.6. Multifunctional transitional trajectories (Farm X)

Second, if we accept that such a line can be identified, Figure 10.6 suggests that farm transitional trajectories will be characterised by relatively linear transitional patterns for certain periods of time (i.e. little or no change in farm management decisions), interrupted by sudden, at times dramatic, changes in transitional direction. In the figure I term these *nodal changes* in direction of the transitional trajectory, as they can best be visualised as nodes, or pivotal points, in which the direction of a multifunctionality pathway may suddenly change.

---

[86] See Bohnet *et al.* (2003) for similar detailed examples of multifunctional farm biographies from a case study in south-east England.

For ease of explanation and discussion, I have numbered these nodes from '0' (starting point) to '11' (end point) in Figure 10.6. Each farm will have its own 'fingerprint' of successions of linear transitional patterns and nodal changes over time. Indeed, on some farms change may be very gradual over decades, while on others a rapid succession of nodal changes may occur within a short time span. As Holmes (2002, 2006) showed using the example of marginal areas in Australia, agricultural territories can switch quickly from pastoral dominance to 'multifunctional occupance', with the swift emergence of, for example, indigenous, conservation and tourism uses. As a general rule of thumb, one may argue that the more stable and economically secure a farm or rural community is, the more linear the multifunctional transition is likely to be (although this does not say anything about the quality of multifunctionality on a farm), while farms undergoing major upheaval (e.g. change in ownership; impending retirement of owner; drought) may have to adjust their multifunctional strategies through rapid changes in multifunctional transitional direction (see Barthélémy *et al.*, 2004, for Senegalese farmers).

Third, nodal changes to a farm trajectory occur within specific boundaries defined by *path dependency*. As Chapter 2 discussed, transition theorists with roots in both evolutionary political economy and complexity theory highlight that the notion of path dependency means that changes to any system (social or natural) often only occur within specified limits of what is *likely* or *possible*, also referred to as 'lock-in effects' (Grabher and Stark, 1997; O'Sullivan, 2004). Based on multifunctional farm pathways in the UK, Clark (2005: 495) emphasised that "some business managers are 'locked in' to types of agro-food diversification that are framed by the agro-industrial [productivist] model". This highlights that path dependency depends on both the *starting position* of a given system (i.e. Farm X at the beginning of the time axis) and the *history* and *geography* of a system. The starting position has obvious influence on possible transitional trajectories (see also Fig. 10.7 below), as it is difficult for a farm situated at a specific position within a spectrum of possibilities (e.g. Farm X located close to the weak multifunctionality end of the spectrum at point '0') to rapidly move far away from this starting position (Grübler, 1997). Thus, the probability of a system making an *extreme* change away from its starting position is, probabilistically, low. As Casterline (2001) argued, personal choices can be self-reinforcing and, therefore, self-fulfilling, and alternative pathways may often not even be considered.

Further, path dependency of individual farms also depends on path dependency characteristics at the community, regional and national levels (see Fig. 9.7 above) – a notion termed 'rural development clustering' by Marsden (2003) based on farmers' capacity to interface with other local/regional actors. It is here, therefore, that the above-mentioned notions of *social*, *cultural*, *economic* and *moral capital* in a rural community will be of particular importance, as it is only in systems with such well developed capitals that a shift towards strong(er) multifunctionality is likely (Clark, 2005). Thus, Lowe *et al.* (1999) argued that farmers are often embedded in complex techno-economic networks that encourage the development of agriculture along a particular trajectory which gives farmers little option to veer from an established path (self-perpetuating pathways). Clark (2003) also argued that multifunctionality pathways of farms are often highly dependent on *external* drivers such as the policy environment (see Chapter 9), market forces (e.g. for strongly multifunctional diversification activities), or other 'local' obstacles such as rigid planning laws and regulations that may hinder the development of strong multifunctionality pathways. Ward's (1993) notion of the 'productivist treadmill' (see Chapter 5), meanwhile, emphasised the existence of *internal* lock-in effects, particularly for farmers who have decided to embark on strongly productivist pathways. Thus, different farms, farm regions and rural communities will have different possibilities to embark on specific pathways, depending on available social and environmental capital, networks of stakeholder interaction, economic path dependency, ideological path dependency (particularly important at the national level; see Fig. 10.4) and mental path dependency (i.e. the 'mindset' of different actors and actor groups). Lock-in mechanisms can occur at any pathway level and may, at times, severely influence transitional opportunities on individual farms (see example of North Korea above). Thus, understanding multifunctional pathway choices at different scales is vital for understanding multifunctional transitional opportunities at farm level.

Fourth, Figure 10.6 also shows the importance of *system memory*. This notion suggests that a given system (i.e. Farm X) may be at its specific starting location on the spectrum precisely because of the history of decision-making trajectories *preceding* the starting point shown in Figure 10.6. In other words, a system carries with it the memory – or in a more negative sense the 'baggage' – of previous decision-making trajectories (including missed opportunities and wrong pathway choices but, at times, also highly 'positive' choices). O'Sullivan (2004: 285) emphasised that "history matters" and that "path dependence holds that a system's trajectory is a function of past states, not just the current state". At any given point in time, therefore, system memory has a large influence on transitional pathway opportunities in the future (see also Fig. 10.7 below). As Chapter 2 highlighted, an important difference may exist here between human and natural systems. While most natural systems will not be 'encumbered' by memory or historical baggage (i.e. they are more likely to switch from one end of a spectrum of possibilities to another at short notice), *all* human systems are influenced by system memory in one way or another in their transitional processes. In other words, no human system exists without some knowledge (memory) about past transitional trajectories. While this memory may not always be at the forefront of decision-making, it will, nonetheless, often influence current, and therefore, future decision-making pathways. For Farm X in Figure 10.6, system memory can best be shown through the farm being passed on from generation to generation with associated transfer of knowledge systems about what the farm family may see as the 'best' way of farming (Wilson, 1997c; Bell, 2004). Sociological literature suggests that such knowledge systems are often slow to change (i.e. the importance of passed 'wisdom' from parents to children; family-centred lock-in transitional pathways). As Jay (2005) highlighted for New Zealand, if a farm knowledge system has been predicated on the weak multifunctionality model for a considerable length of time, it will be difficult to change this system memory in a short time (i.e. within less than a generation). This is not to say, however, that in some cases system memory can be overridden by more immediate factors that may lead to rapid shifts in decision-making (e.g. whether to abandon farming altogether due to economic hardship; see below).

Fifth, Figure 10.6 also shows that the 'corridor' of path dependency widens over time. The reason for this is because the further one moves from the temporal starting point of a transitional process, the less influence system memory will play (i.e. new influences and factors take on more importance) and the more decision-making options are open to a farmer. This means that, at the end of the temporal period shown in Figure 10.6, the path dependency corridor (i.e. multifunctionality options open to the farmer) is nearly as wide as the entire multifunctionality spectrum itself. The shape and size of the path dependency corridor, therefore, depends on the time scale of the investigation and the temporal starting point from which future transitional development is assessed (see also Chapter 2 and Fig. 10.7 below). Hoggart and Paniagua (2001: 44), thus, cautioned that "there are different 'start positions' for interpreting societal transformations. As each start position is embedded in a narrow analytical lens, once it is opened up to the requirement of holistic vision [e.g. the multifunctionality spectrum], the inadequacies of a restricted field of vision are quickly revealed". Future empirical investigation of path dependency corridors within the multifunctionality spectrum need to take time horizons and the selection of 'starting points' into particular consideration (see Chapter 11). Yet, the shape and size of these corridors of possibilities will also depend on how *risk averse* a farmer may be about embarking on a different trajectory. Thus, corridors of decision-making possibilities are closely linked to a farmers' *risk perception* (Adams, 1995). As Clark (2003: 222) emphasised, "pursuing a [strongly] multifunctional business strategy [has] greater short-term risk attached to it than adhering to the conventional option of commodity production". The p/np boundaries of decision-making possibilities are, therefore, also *risk boundaries* beyond which farmers may be unwilling to step.

Let us investigate the notion of *transitional corridors* in more detail, as understanding the 'boundaries' of these corridors is crucial for understanding constraints and opportunities faced by individual farmers in the multifunctional transitional process. Figure 10.7 shows a conceptual extension of Figure 10.6 and suggests, first, that at each nodal point that demarcates a change in transitional direction, the boundaries of decision-making corridors also shift. Thus, while Figure 10.6 only showed one 'starting point' for the resulting decision-making corridor (at nodal point '0'), in Figure 10.7 each nodal point is seen as the starting point of a *new* decision-making corridor. In other words, over time path dependency decreases compared to the starting point, as new constraints and opportunities act as *cumulative* new drivers for subtle changes in the shape and size of corridors of decision-making possibilities. Further, at each nodal point an individual farmers' decisions have different relations to other farmers' decisions, suggesting that *where* a nodal point is situated in the system has significance for the unfolding multifunctional behaviour of individuals *and* the collective system. At each nodal point, therefore, new options may be available to a farmer to either increase or reduce multifunctional quality (see also Law, 1994, on the theory of ordering, and O'Sullivan, 2004, on theorising complexity). As highlighted above, location of a farm will be crucial, as farms close to the urban periphery, near large population centres, or in areas with high tourism potential will usually have better opportunities for strong multifunctional pathways (Fleury *et al.*, 2004; Mollard *et al.*, 2004). Meert *et al.* (2005: 84), for example, highlighted that for farms in Belgian peri-urban fringe areas the proximity of the urban centre may "stimulate farmers to diversify their activities in order to attract city dwellers to purchase produce direct from farms, or for non-agricultural activities". On the other hand, Walford (2003b) highlighted how many intensive arable and mixed livestock farms in the UK lowlands have opted for a continuation of highly productivist pathways. Ultimately, and as Marsden (2003) suggested, farm strategies have to navigate a pathway through this complexity in their attempts to maintain *reproduction* of the business. Indeed, the end of farm reproduction also means the end of agricultural multifunctionality for a farm family (but not for the farm if purchased by another farmer; see Section 9.3).

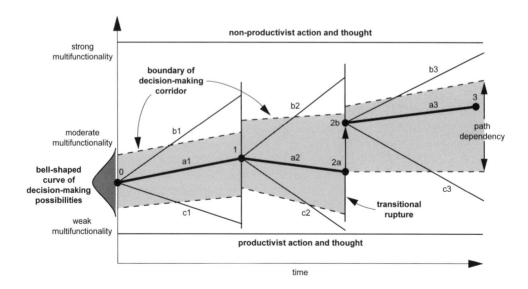

Fig. 10.7. Transitional shifts, decision-making corridors and transitional ruptures

Second, the figure highlights that decision-making options within each of the decision-making corridors can be conceptualised as a *bell-shaped curve* that shows both the multifunctional possibilities open to a farmer at any given nodal point and the probabilities for the adoption of new multifunctional transitional pathways. Conceptualising multifunctional transitional corridors as a bell-shaped curve highlights that most farmers have a *range* of possibilities, but that the most likely pathway is situated in the centre of the corridor. Giddens (1984) referred to such corridors as the 'parameters of the possible' in human decision-making. Unlike Marsden's (2003: 7) recent suggestion that "in the case of the classic 'normal curve' bell-jar, only the minority of [agricultural] producers are clear about where they are going in this system: i.e. the productionists [sic] who are of the scale and integration to compete for world markets; and the growing band of 'multi-functional farmers' [sic] now individually and locationally capable of developing a sustained alternative", I argue that *all* farmers are constrained by decision-making corridors in both their productivist *and* non-productivist actions. Similarly, Meert *et al.* (2005) emphasised that most farms follow a *combination* of two or more different multifunctional pathways, and that a specific farm trajectory is the sum total of multiple, and at times competing, productivist and non-productivist pressures and opportunities. Thus, at nodal point '0' in Figure 10.7 the farm trajectories b_1 and c_1 are both equally unlikely, as they would propel the farm into a multifunctional direction (e.g. weaker multifunctionality for trajectory c_1) that lies beyond the extreme boundaries of the corridor for nodal point '0'. Similarly, trajectories b_2 and c_2 are unlikely from nodal point '1' onwards as they lie beyond the 'possible' corridor boundaries. Thus, rapid multifunctional change from any nodal point that lies beyond the boundaries of risk a farmer is willing to take (i.e. corridor boundaries) is unlikely. Holmes (2006: 155; emphasis added) illustrated this through his example of 'entrenched' Australian pastoral regions that are experiencing economic and environmental stress, "yet remain *trapped* within marginalized occupance, unable to adjust towards long-term viability … or to achieve [strong] multifunctional outcomes". Meert *et al.* (2005) further suggested that *reverse evolution* is also unlikely, especially for farmers who have opted towards strongly multifunctional pathways (e.g. by marginalising agricultural activity). Such farmers may find it difficult to re-embark on moderate or weak multifunctionality pathways because of competition from productivist farms with established economies of scale.

The shape and nature of risk boundaries within transitional corridor opportunities and multifunctional pathways are closely linked to debates on *innovation diffusion theory* discussed in Chapter 2 (e.g. Rogers, 1995). It can be argued that *innovators* (and to some extent early adopters in a rural community) – i.e. 'venturesome' farmers interested in new ideas that may influence their peer networks and who often launch new ideas by importing innovative ideas from outside a system's boundaries (gatekeepers) – may be able to operate in substantially larger transitional corridors than the *late majority* and *laggards* (the latter often being in the majority). Thus, for innovators, risk and uncertainty associated with nodal changes in multifunctional transitional direction may be less restricting than for less venturesome farmers (Rogers, 1995). For an innovator, transitional pathways b_1-b_3 in Figure 10.7 may well be in the realm of possibilities, even if these pathways would take the farm into a very different transitional direction from the original position.

As Guy (2005) highlighted, *new entrants* into farming often act as innovators as they can be more open-minded about multifunctional innovations (i.e. no baggage or system memory) and may, as a result, be more likely to embark on strongly multifunctional pathways. Pretty (2002: 148) argued that "it sometimes takes 'incomers' with a different perspective … to provoke changes in thinking amongst local people who are wedded … to industrialized agriculture because they know no alternatives". Similarly, farmers who have been exposed to new knowledge environments based on agri-environmental scheme participation may also become innovators and be more likely to recognise new (at times non-productivist) pathway options (Wilson, 1997a; Fish *et al.*, 2003). As Swagemakers (2003: 197) argued, "farmers with a good feel for the mood of society can successfully incorporate new activities into their business strategy alongside the traditional food production functions". In a UK context, Clark (2003) listed several innovative and strongly multifunctional farm activities including establishment of a nursery for garden plants, growing willow coppice, production of niche-marketed cheeses, marketing of bottled mineral water from an on-farm spring, collecting locally produced straw to make animal bedding products, setting up an on-farm community archaeology site, establishment of a rural training centre, or establishment of a nature reserve for educational use. As Marsden (2003: 181) reminded us, farmers' responses to the challenge of multifunctionality "will be critically determined by factors such as geographical location, individual business and management skills, practical skills, access to capital, entrepreneurialism, and the availability of support (both institutional and financial)".

Laggards, meanwhile, may only wish to implement new multifunctional transitional steps that are closely linked to 'what they know best' and that will not lead to substantial shifts in transitional direction. Laggards and the late majority are also more likely to be influenced by system memory (past as point of reference, rather than future) – in other words, they will cling to 'established wisdom' as long as possible before being persuaded (if at all) to change the ways they farm. Much work has highlighted that the role of innovators in any farming area is crucial for the pace of adoption of innovations (e.g. Wilson, 2004), and strong multifunctionality pathways, in particular, can be conceptualised as a powerful 'innovation' – especially for farmers who farm in weakly or moderately multifunctional ways. The extent to which innovators may be able to motivate other farmers will often be crucial in determining strong multifunctionality adoption at farm, community and regional levels. In particular, research has shown that farmer *self-concepts* can be influenced by innovators in a farming district. For example, if a laggard farmer started to perform conservationist roles based on peer pressure from innovators or early majority farmers, interactions with these innovators (and also with associated wildlife organisations, government grant bodies and extension services, or other farmers in discussion groups with similar interests) would lead to an increase in the number of social contacts with other 'conservationists' (Ward *et al.*, 1998; Wilson, 2004), an increase in interactive (and eventually affective) commitment of laggard farmers (Coughenour, 1995) and, possibly, an eventual increase in the salience of a strongly multifunctional 'conservationist' identity for that farmer (Burton and Wilson, 2006). Of course, innovators may also be highly influential in *reducing* multifunctional quality by persuading farmers that moderate or weak multifunctionality pathways will yield higher profits through increased integration into global markets and access to new technologies.

As Part 1 amply illustrated, transitions rarely occur along a smooth straight line. Figure 10.7, therefore, also shows a 'transitional rupture' between nodal points 2a and 2b. For farm-level transitions, such a rupture is most likely to occur when a farm is taken over by a new owner not linked to the previous farm family. As Holmes (2006: 155) argued, "rapid change may be propelled by ... agricultural or pastoral redundancy leading to disinvestment, facilitating a shift from agricultural to other occupance modes". Similarly, Bell (2004: 154) argued that many US farmers who had chosen strong(er) multifunctionality pathways experienced "a sudden, disorienting change, in most cases during a period of severe economic stress, in which they had to rethink not only their farming practices but their practices of self". The most extreme transitional rupture occurs when farms have to be sold, abandoned or where land use has to change towards activities 'beyond agriculture' such as forestry.[87] As many authors highlight, such pressures often occur in agricultural areas situated in peri-urban areas that are susceptible to rapid, unpredictable, divergent and dissonant change (Hoggart *et al.*, 1995; Holmes, 2006). Such extreme transitional ruptures are often *forced* rather than voluntary (i.e. previous owner has no choice but to sell the farm).

---

[87] Burton (2004) argued (from a UK context) for a reintegration of woodland management within the farmer's notion of 'farming roles'. Acceptance of woodland management and forestry as integral parts of 'farming' would mean that the perceived extent of transitional ruptures from one land use to another could be minimised.

BSE and foot-and-mouth disease are probably the best recent examples of transitional ruptures severely affecting livestock farming in many European countries, further accelerating the process of farmers leaving agriculture, or, in some cases, providing a platform for partial disconnection from the productivist treadmill towards non-productivist farm development pathways (Hinchcliffe, 2001; Scott *et al.*, 2004). In the case of farm purchase by a new owner, a transitional nodal point may be elevated (or lowered) to a new plain on the multifunctionality spectrum, as the new owner is not encumbered with the memory (baggage) of knowledge systems and farm management approaches used by the previous farm owner(s), and may bring with him/her a new mindset and approach towards farming. In other words, system memory (of the farm system) may be temporarily less important as the new ownership situation creates a 'level playing field' for multifunctionality options. There are substantial overlaps here with our discussion of transition theory in Chapter 2. In particular, the notion of transitional ruptures suggested in Figure 10.7 shows similarities with debates on stepped transition (see Fig. 2.1.b in Chapter 2) in which a system may show relatively uniform transitional pathways for a certain time, interspersed by abrupt and dramatic changes in transitional trajectory. Further, parallels can be drawn here with evolutionary transition, in particular with Eldredge and Gould's (1972) notion of *punctuated equilibrium* or with Dear's (1986) incisive criticism of post-modernism based on 'significant breaks' in societal evolutionary pathways (see Chapter 4).

Figure 10.7 shows the hypothetical case of a sudden elevation of multifunctional quality from nodes 2a to 2b (from a farm trajectory below moderate multifunctionality towards one slightly above) as the new farm owner takes over. A new decision-making corridor opens up for the new owner that has different boundaries to the previous corridors. However, although human system memory may be partly or completely lost through the takeover of the farm (especially if the new owner comes from outside the district and is without social links to the immediate community), path dependency from nodal point 2b is still partly defined by characteristics (and system memory) of the *farm itself* (see also Bell, 2004, for a detailed discussion of the importance of system memory among farmers in the USA). Dear (1986: 374) argued from a post-modern transition theory perspective that "radical breaks between periods do not generally involve complete changes of content but rather restructuration of a certain number of elements already given; features that in an earlier period or system were subordinate now become dominant, and features that had been dominant become secondary". For example, previous farm family generations may have removed all remnant wildlife habitats on the farm for productivist reasons, which means that it would be difficult for the new owner to rapidly increase the environmental capital of the farm towards strong multifunctionality. Research has shown that it may take several farm generations to re-establish lost biodiversity levels (Bohnet *et al.*, 2003). In addition, a farmer may well wish to increase multifunctional quality on his/her farm, but may be impeded by neighbours. For example, super-productivist agriculture such as intensive dairy farming with its associated noise, smells and slurry will limit the options of neighbouring farmers to diversify into strongly multifunctional activities (e.g. farm tourism).

Thus, geography matters as the choice of strong multifunctionality pathways may not be entirely dependent upon the farm decision-maker him/herself (see also discussion in Chapter 9 on tourism potential [or lack thereof] of farming areas) but simply on the locational multifunctional potential of a farm. For these reasons (and many others), there is still considerable overlap in Figure 10.7 between transitional corridors around nodes '1' and '2b', as it is difficult for a new owner to embark on a completely different multifunctional farm trajectory. This re-emphasises the importance of understanding multifunctionality at farm level and not just with regard to the farmer him/herself (see Section 9.5). Indeed, the new owner taking over the farm at nodal point 2b may be strongly multifunctional in his/her farming outlook (although would s/he then have purchased a moderately multifunctional farm?), but may be precluded from implementing a strongly multifunctional farm trajectory (at least for a while) due to system memory inherent in the farm transitional history itself.

Understanding transitional ruptures through farm sale or abandonment is gaining importance, as the pace of agricultural structural change is increasing in most countries with resulting sale and/or abandonment of farms (Woods, 2005).

10.3.2 Farm-level multifunctional transitions: two hypothetical examples

Having outlined key conceptual issues regarding farm-level multifunctional transitions, I wish to highlight the 'practicalities' of farm transitions through two hypothetical examples. The specific focus will be on how individual farmers make multifunctional decisions, what factors influence specific 'nodes' demarcating changes in the direction of multifunctional decision-making pathways, and what transitional pressures and opportunities are faced by farmers both in developed and developing countries. I will also investigate how individual farmers' multifunctional transitional decisions occur within a relatively narrow spectrum of decision-making opportunities constrained by both path dependency and corridors of decision-making bounded by productivist and non-productivist action and thought. I also wish to analyse in more detail debates surrounding risk, constraints and opportunities that farmers face when making multifunctional decisions at a given point in time.

Before analysing two farm examples in more detail, it is worth briefly considering what influences farmers' decision-making processes. The aim here will not be to review the vast literature on this topic, but, through the two examples below, to draw on useful conclusions reached by previous researchers in their attempts to explain why development trajectories of individual farms (and any other human system) are in a constant state of flux. Some general categorisations of factors influencing on-farm decision-making have been produced (e.g. Brotherton, 1991; Wilson, 1992, 1997a; Jay, 2005), and there is general consensus that these factors can be grouped into three distinctive categories: farmer-related factors, farm-related factors and factors external to the farm. *Farmer-related factors* that influence multifunctional transitions include all factors linked to the farmer or the farm household and include, for example, age, gender, household size, type of family partnership and ownership structures, health, education, wealth (available capital), a farmer's social networks or – as Chapter 7 highlighted – a farmer's identity and psychology and his/her views of farming and agriculture (e.g. McDowell and Sparks, 1989; Burton and Wilson, 2006). As mentioned above, farmer-related factors will also include system memory through knowledge systems passed on by various generations of the farm family. *Farm-related factors*, meanwhile, include factors such as farm size, ownership patterns, geographical location (e.g. lowlands/uplands), accessibility of the farm to markets (including local and regional infrastructure and urban networks), soil type, local climate, drainage and availability of water, or surviving on-farm biodiversity (remnant habitats) (Andersen *et al.*, 2004; Kristensen *et al.*, 2004; Meert *et al.*, 2005). *Factors external to the farm* include, for example, regional climate (including issues linked to global warming), the wider economy (national and global), community structures (e.g. rural and urban pressures and opportunities), the availability of farm workers, accessibility to local and regional technological innovations, the regional potential for tourism and other diversification activities, the policy environment and national politics (including institutional thickness and permeability; cf. Amin and Thrift, 1993), and social and cultural factors including support (or lack thereof) of wider society for farmers and farming (Clark, 2003). All these factors (and many more) will combine in complex ways to define the multifunctional transitional trajectory of a farm, at times providing a mixture of enabling factors for the implementation of strong multifunctionality pathways, while at other times compelling farmers to embark on weak(er) multifunctional trajectories. The following two examples will illustrate how some of these factors can influence multifunctional transitional trajectories.

For our first example, let us return to Farm X shown in Figure 10.6 above. Let us assume that this farm is a middle-sized mixed farm located in a touristically attractive area in south-west England. Figure 10.2 highlighted that the *transitional potential* for this type of farm (lowland mixed farm) is medium, with this farm type often tending towards moderate or strong multifunctionality. Let us further assume that Farmer X was a middle-aged farmer when she took over the farm from her father at the start of the temporal transition under consideration. Her father had been a 'traditional' farmer who strongly believed in agriculture as a productivist-oriented endeavour based on his experience of severe food shortages in the UK during and immediately after the Second World War. As a result, Farmer X's father shaped the farm towards a productive relatively intensive mixed farm (arable and livestock) and, knowing that the daughter would take over the farm eventually, strongly influenced Farmer X in her initial views towards the productivist role of agriculture in UK society (system memory). Taking over this farm, Farmer X was, therefore, to a large extent locked into a specific multifunctionality pathway (nodal point '0' in the figure) situated somewhere between weak and moderate multifunctionality (path dependency). Indeed, for the first few years after taking over the farm, it was largely a matter of 'getting on with the business', with little opportunity for major changes (relatively linear transitional change).

However, the first opportunity arose for Farmer X to change the multifunctional transitional pathway after a few years (nodal point '1' in Fig. 10.6) when she (and many other farmers in the region) was approached by a national supermarket chain with an opportunity to sell dairy products directly to the supermarket. This led to a phase of farm intensification and, consequently, to a lowering of multifunctional quality on the farm. Yet, a new opportunity emerged soon after through the availability of the first agri-environmental scheme in the area (Environmentally Sensitive Area scheme [ESA]) in which Farmer X was persuaded by the local extension officer 'to give it a go' (external policy influence) (nodal point '2'). The decision was made easier as Farmer X had just lost her lucrative contract for direct sale to the supermarket. As a result, livestock densities on the farm were reduced to make the farm eligible for maximum green subsidies through the ESA scheme, resulting in an increase in multifunctional quality on the farm (extensification of farm production). At the end of this period the farm had reached moderately multifunctional levels (nodal point '3') and the level of multifunctionality could have been further raised if Farmer X had wished to do so. However, Farmer X had an accident in which she broke her leg with a complicated fracture (at nodal point '3'). This meant that she could not directly manage the farm for a while. As a result, she rented the land on her farm to her (highly productivist) neighbour while staying in the farmhouse herself to convalesce (transitional rupture). Her neighbour was given relatively free reign and immediately intensified production (the ESA agreement had been discontinued after 5 years). This new trajectory put the farm on a steep downwards trend towards weaker multifunctionality (to levels lower than at starting point '0'), resulting in severe nitrate pollution of soils and further reduction in biodiversity.

Farmer X was gradually recovering from her injury and, in the meantime, had inherited money which reduced the need to farm for maximum profit. However, Farmer X wanted to stay in farming as this was 'all she knew' (system memory). At the same time, her long-standing partner moved in with her and, although he showed no direct interest in farming, began contemplating the possibility of converting some unused farm buildings for tourism. Based on her additional financial capital, Farmer X was now in a position to extensify her farm by reducing her livestock herd to one-third of previous stocking density (with a new ESA agreement), and by replacing the original herd with a local rare breed with the aim to produce dairy products for the newly established local farmers' market (changing views of Farmer X about the purpose and role of 'agriculture'; changing opportunities within the local community) (nodal point '5'). However, the transition from a productivist profit-oriented nationally embedded holding to a more locally embedded non-productivist holding was far from smooth. Initially, the reaction from local consumers towards the specialist dairy products (a specialist cheese made from milk from the rare breed) was subdued and Farmer X had to reinvestigate other outlets for her produce (regional supermarket chain) with mixed results (nodal points '6-8'). In the meantime, her partner had completed the conversion of an old barn complex on the farm into tourist accommodation and had begun to convert one of the four fields of the farm into a scenic woodland for biodiversity conservation and tourist enjoyment. This further reduced overall agricultural production of the farm and moved the farm territory closer towards the non-productivist end of the spectrum (nodal point '9'). Although tourism income took a while to provide a substantial part of the income generated from the farm territory (nodal point '10'), a sudden upsurge in tourist interest (linked to better tourism marketing of the region), together with the opening of Farmer X's own farm shop focusing on farm-based low-intensity produce, firmly propelled the farm into the strong multifunctionality realm at the end of the transitional period under consideration. Thus, at nodal point '11' Farm X is still a 'farm' (see Section 9.3), albeit a very different one from the original weakly/moderately multifunctional holding Farmer X took over from her father. Although still within the original boundaries of path dependency at nodal point '0', Farmer X had managed to shift multifunctional trajectories of the farm firmly into the strongly multifunctional end of the spectrum.

Other examples from advanced economies could be discussed in which the multifunctional transitional pathway is 'negative', i.e. in which multifunctional quality is gradually lost over time. We also need to acknowledge that the example has largely compartmentalised decision-making structures into specific more or less clearly defined entities (i.e. one action following on from another), while in reality farmers (as any other actor) often face multiple pressures *simultaneously* that may pull them into contradictory multifunctional directions along Deleuzian transitional pathways. Yet, as highlighted in Chapter 9, the advantage of the multifunctionality spectrum is its ability to accommodate *multiple* and *parallel* influences of multifunctional decision-making, the totality of which define the 'position' of a farm on the multifunctionality spectrum at any point in time.

Figure 10.8 shows another example of a multifunctional transition on a hypothetical farm in the hill areas of Kerala in south-west India (Farm Y). Farm Y is on a 'downward' multifunctional trajectory with gradual loss of multifunctional quality. At nodal point '0', this farm started as a subsistence farm with various characteristics typical for strongly multifunctional farming systems (see Section 9.4). Although at nodal point '0' Farm Y could be categorised as an intensive subsistence farm, it had high environmental sustainability based on palaeotechnic agricultural techniques and farming processes (Harris, 1978), was well embedded in the locality (with only occasional sales of surplus production on the local market), and was not integrated into global agro-commodity chains. As Figure 10.2 highlighted, most subsistence farms are situated close to the non-productivist end of the multifunctionality spectrum but often have limited transitional potential.[88] For the first few years at the beginning of the transitional period under consideration, Farm Y, managed by a farmer and his wife, underwent relatively few changes. However, at nodal point '1' the farm couple had the opportunity to rent additional land which could be relatively intensively farmed. Some remnant forest was cleared to make room for further fields, resulting in a gradual loss of multifunctional quality. However, a severe drought (external driver for changing multifunctionality pathway) forced the wife to seek employment in the local village (nodal point '2'), resulting in a reduction in agricultural intensity for a while with a resulting relative increase in multifunctional quality (nodal point '3').

At the same time, the region within which Farm Y was located underwent gradual changes linked to technological/infrastructural change (e.g. the building of an all-weather road to the local village linking the region with urban centres on the coast) and the rapid growth of international tourism in coastal areas of Kerala. This enabled farmer Y to begin a process of intensification due to new food sale opportunities beyond the local market (nodal point '4') (see also Pretty, 1995, for discussion of a similar example of the gradual disintegration of traditional wet rice cultivation in Bali). Yet, middlemen were the ones

---

[88] This was reiterated by Barthélemy *et al.* (2004) for the case of subsistence-based agriculture on the island of Mayotte (Indian Ocean) where strong multifunctionality pathways are predicated on processes of reciprocity, redistribution, responsibility and the symbolic importance of agriculture-related practices (see also Pretty, 1995).

initially profiting from farm intensification in the region rather than the farmers themselves, leading to temporary stagnation of intensification processes (nodal point '5'). At this stage, Farm Y could still be classified as a subsistence farm (most farm products were consumed by the farmer and his family), although an increasing amount of products was now sold. The farm had begun to embark on a (partly self-chosen) trajectory of moderate multifunctionality.

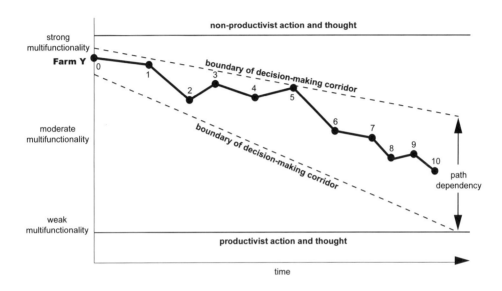

Fig. 10.8. Multifunctional transitional trajectories (Farm Y)

A key change occurred at nodal point '5', when a tourist hotel chain on the coast established direct contact with farmers in the neighbourhood of Farm Y for the direct purchase of fresh horticultural products. For the first time, this enabled farmers in the district to sell their produce directly (and at a reasonable price) to consumers outside the immediate farm district. At this stage, Farm Y moved into what Losch (2004: 349) termed the transition to 'business agriculture', "arising from family farms that had initial advantages [e.g. location and accessibility to markets] or that benefited from direct investment [e.g. direct sale] in the most profitable sub-sectors [e.g. tourism]". On Farm Y, this led to a rapid period of intensification in which the farmers' wife was brought back full-time into agriculture, and other relatives began to work part-time on the farm, helping with the clearing of the last remnant pockets of semi-natural vegetation for the establishment of new fields (nodal point '6') (see also Chaudhri, 1992; Watts, 1996). Profits rose and Farm Y was no longer a 'subsistence farm' but had become a fledgling farm business linked to regional agro-commodity chains through a rapidly increasing regional tourist industry.[89] Additional income also meant new opportunities for the adoption of new technologies leading to further intensification (nodal point '7'). Machines were bought but, most importantly, Farmer Y began to contemplate a deal with a multinational agro-business selling Green Revolution seeds that guaranteed 50% higher yields (see also Parayil, 1999; Pretty, 2002). Farmer Y now had the initial capital to purchase the land he had rented before (thereby changing the geographical scale within which multifunctional farm trajectories of Farm Y should be

[89] A similar trend was described by Marsden (2003) for the island of Barbados (Caribbean) where good returns from the sale of fresh vegetables to local tourist centres provided a stimulus for agricultural change amongst small farms.

assessed), and to change part of the production of the farm to high yielding Green Revolution crops (nodal point '8'). Although this moderately multifunctional strategy was successful for some time (nodal point '9'), an exceptionally strong monsoon led to severe erosion and loss of valuable topsoil on steeper parts of the farm, resulting in irreversible reduction in agricultural potential (external multifunctionality driver; nodal point '10').

Like many former subsistence farms in the developing world, at the end of the transitional period under consideration Farm Y is at a crossroads. By now, it had left the strong multifunctionality model behind and had moved firmly into a moderately multifunctional farm development pathway (largely out of choice). The farm is still relatively strongly embedded within the locality (albeit much less than at nodal point '0') and retains some of its initial strong multifunctional characteristics (e.g. original low-intensity crops are still planted on part of the farm). However, links with globalised agro-commodity chain networks have been established through the tourist industry and Green Revolution crop agro-businesses, and it would not take much further transitional change to propel the farm into a spiral of weakening multifunctional trajectory predicated on further intensification, the complete substitution of traditionally-grown crops with high yielding Green Revolution varieties for sale in global agro-commodity markets, and resulting (irreversible?) environmental degradation (Shiva and Bedi, 2002). As Marsden (2003: 32) highlighted for Brazil, "key actors and agencies [in the locality] become the conduits of globalised knowledges, and internally the gatekeepers of globalised market entry for the numerous and variable producer sector" (see also Rigg, 1989, 1990). Although income for Farm Y has dramatically increased and the quality of life of the farm couple has been substantially improved, this has also been associated with accelerated loss of environmental sustainability (e.g. clearance of protective natural vegetation cover on steep slopes), partial loss of local embeddedness (and, therefore, partial loss of local social networks and safety nets) and a resulting uncertain future increasingly dependent on the vicissitudes of the global market. Marsden's (2003: 71) case study of a farming region in Brazil also found that "new patterns of regional environmental development ... associated with the globalised and intensive food system are creating ... far more immediate types of environmental and social vulnerability".

Although the 'downward' multifunctional transitional trajectory of Farm Y is relatively typical for many farms in the developing world,[90] many examples could be invoked in which strongly multifunctional subsistence farming systems *have stayed* strongly multifunctional over time (see also Losch, 2004). As Pretty (2002: 80) emphasised, "many poorer farmers have made the transition from largely pre-modern agricultural systems directly to sustainable and highly productive systems". Yet, as highlighted above, it cannot be denied that globalisation pressures are increasingly impinging upon even the remotest farming districts (Shiva and Bedi, 2002). For the island of Réunion (Indian Ocean), Roux *et al.* (2004: 146-147; my translation), for example, highlighted the constant tussle between productivist and non-productivist drivers on farms in the developing world by arguing that "opposition between actors advocating 'reasonable' [i.e. strongly multifunctional] models and those who favour productivism is growing". While some countries/regions may choose to actively resist the temptations of globalisation (see example of Bhutan above, but for how much longer?), many farmers may often not have any other transitional alternatives. As Section 10.4 will discuss, transitional pressure for a move away from strongly multifunctional agricultural systems is, therefore, particularly strong in developing countries.

Several key points emerge from these two examples. First, the Indian example highlights that globalisation and technological change are important drivers for *reduction* of multifunctional quality (Shiva and Bedi, 2002). However, as some commentators have argued, globalisation is not necessarily always 'bad' for multifunctional transitions

---

[90] See also Watts (1996) and Marsden (2003) for similar discussions of the loss of multifunctional quality on farms in Brazil and the Caribbean, Altieri (2000a, 2000b) and Higgott and Philips (2000) for Latin America, Pretty (2002) for the transformation of traditional rice cultivation systems, or Robinson (2004) for Malaysia (Sabah) and the Himalayas.

(Nederveen Pieterse, 2004). Globalisation may also bring with it improved access to education and knowledge systems, potentially aiding farmers in the process of viewing agriculture in different ways and thereby influencing farm transitional pathways (partly) towards stronger multifunctionality. As Shiva and Bedi (2002) emphasised for India, individual attitudes of farmers, and their willingness (or ability) to adopt globalised forms of agricultural production, will ultimately define the nature of multifunctional trajectories. Second, the UK example has highlighted that transitional ruptures (e.g. changes in farm ownership; sudden change in farmers' health) can often set in motion large nodal changes in transitional pathways. As highlighted in Figure 10.7, these transitional ruptures may elevate transitional corridors onto a new (higher or lower plain) with new opportunities for multifunctional farm transitions. Although system memory may often stifle *radical* change even after transitional ruptures, a new owner (or a new partner in a farm partnership) may bring with them new ideas that, gradually, may result in changes in multifunctional farm transitions. Third, the above examples also emphasise that although some farmers may be *mentally* willing to embark on a strongly multifunctional trajectory, they may be prevented by physical factors of doing so (e.g. need of Farmer X to rent her land to another farmer due to health reasons). Thus, the mental imagery of multifunctionality in a farmers' mind may be very different from the *actual expression* of farming activities on the ground (see also Chapter 7). It can also be argued that mental changes will be more gradual and that they highlight transitional *possibilities* of a farm within path dependency corridors that may substantially differ from *actual* farm development trajectories (Burton, 2004). Indeed, multifunctional action on the ground may only be possible if the change resulting from this action has been accepted mentally by the farmer (Burton and Wilson, 2006). Here, we may need to envisage mental and practical on-farm processes characterised by a *multifunctionality feedback loop* in which the mental imagery of a farmer is constantly readjusted based on experiences gained on the ground.

10.4 Managing transitions

In this final section I wish to broaden the discussion by investigating issues linked to *how* multifunctional transitions for any actor spaces can be managed, *by whom* and *for whom* they should be managed and, indeed, whether these transitions should be managed at all. It is important to re-emphasise that, throughout Chapters 9 and 10, I have argued that the *strong multifunctionality* model should be seen as the 'ultimate goal' of any agricultural system (for reasons outlined in Section 9.4). This book should, therefore, be partly seen as a *plaidoyer* against the economistic interpretation of multifunctionality simply as an 'externality' issue (see Chapter 8), against the predominant productivist agri-business model that continues to dominate agriculture in advanced economies and that increasingly influences agricultural practices in the developing world (Pretty, 1995; Marsden, 2003), and against the increasing globalisation of agro-commodity chains that tend to weaken global multifunctional agriculture pathways. As Delgado *et al.* (2003: 20) emphasised, globalisation often involves the transfer of power to negotiate and the ability to make political decisions to supranational organisations, and "the loss of the ability to make decisions is not only occurring on a macroeconomic level; it is also apparent on the microeconomic level of individual farms". This means that the suggested notion of the multifunctionality spectrum implies subjective *value judgements* about what type of multifunctionality provides the *best quality* for all actors involved (*normative* interpretation of multifunctionality). As a result, I argued in Section 9.4 that strong multifunctionality should be seen as the most *moral* and *qualitatively best* type of multifunctionality, as it ensures protection of the environment and survival of healthy farming and rural communities, as well as contributing towards better food quality based on *synergistic mutual benefits* between different multifunctionality components. I suggested that most dimensions of strong multifunctionality resonate positively with what producers, rural stakeholders and wider society would see as the 'optimum' type of agricultural system – in other words, it is the only agricultural regime that allows us to put culture back into agri-'culture'. Delgado *et al.* (2003: 26) also argued for a reorientation of agriculture as "agriculture must … be re-oriented towards the production of intangible and non-food outputs", which is "a more profound change seeking to redefine the links which society maintains with nature through agriculture and farmers" (see also Pretty, 1995, 2002; Bell, 2004).

The discussion about *managing* multifunctional transitions, thus, needs to be based on the underlying assumption that agricultural systems over time may have *attempted* (not always successfully) to either maintain strong agricultural multifunctionality regimes or to move away from weak or moderate multifunctionality towards strong multifunctionality. As Di Iacovo (2003: 121) succinctly argued, "multifunctionality of agriculture means reproducing old values in a new form, but one that is suitable for modern society". Clark (2003) further emphasised that strongly multifunctional farm holdings in advanced economies also often have greater flexibility to shape the enterprise mix of their business than weakly multifunctional farmers, i.e. that they may be more in 'control' over their destiny and feel greater 'empowerment' while, simultaneously, also reaping substantial financial rewards from innovative diversification activities and potentially creating employment in rural areas.[91] Similarly, Bohnet *et al.* (2003: 350) emphasised that "it seems increasingly unlikely that … the classical/monoactive [weakly multifunctional] family farms, drawing most of their income from the single activity of farming, will be able to survive in large numbers". This means that strong multifunctionality should, in theory at least, be the model that policy-makers and agricultural/rural stakeholders will most often strive for (given adequate personal motivations and sufficient institutional and economic resources). Indeed, it could be argued that if countries or regions fail to adopt at least *some* aspects of the strong multifunctionality model, then the social and economic survival of agricultural and rural communities in these areas may be severely jeopardised. There are many historical examples that suggest that where societies failed to adopt strongly multifunctional pathways (or at least some elements of strong multifunctionality; see Section 9.4), ultimate catastrophe was the outcome (Spengler, 1992; Diamond, 2006). This means that, in some cases and at certain times, we also have to be willing to acknowledge that strategies that go 'beyond agriculture' may need to be adopted (see Fig. 9.3), akin to Mansholt's bold 1968 plan that suggested the (politically untenable) complete withdrawal of unproductive or environmentally damaging agriculture from specific areas in Europe.

Yet, important questions remain as to *how* the transition from weak to moderate to strong multifunctionality can be encouraged, and *by whom* this should be done. In particular, there are many debates about how the transition towards strong multifunctionality could be implemented, dependent on political, economic and social factors. Indeed, can and should the strongly multifunctional agricultural regime be *orchestrated* at all based on a 'strong policy' model (Hall, 1993), or do multifunctional agricultural transitions have a life of their own? The following will attempt to provide answers to these questions. Section 10.4.1 will investigate past, present and future multifunctional agricultural transitions, with a focus on whether, and how, we can learn from past transitional trends. Building on this debate, Section 10.4.2 will discuss specific constraints and opportunities for the 'best' management of multifunctional transitions.

10.4.1 Conceptualising multifunctional agricultural transitions over time

In this section, I wish to investigate in more detail past, present and future multifunctional trajectories. First, I will analyse broad transitional tendencies in the developed and

---

[91] For his case study of farms in the East Midlands of the UK, Clark (2003) suggested that farms embarking on strongly multifunctional pathways each created on average 1.4 full-time jobs.

developing world. Second, I will investigate possible historical and future transitional trajectories of agricultural regimes and how these trajectories have been influenced by a multitude of factors over time. This second strand of investigation is partly a response to Hoggart and Paniagua's (2001: 57) call that "there is still a critical need in rural studies to ... [explore] change processes, most especially over longer time periods". The discussion so far has highlighted that it is difficult to conceptualise general patterns for multifunctional transitions at the global scale, as key differences continue to exist between developed and developing countries (see Section 9.5). It is, therefore, important to differentiate between transitional processes in the developed and developing world due to different historical/economic trajectories (Losch, 2004). As a result, I will investigate multifunctional transitions over time in the developing and developed world separately.

Figure 10.9 shows a conceptual model of decision-making pathways in the multifunctionality spectrum from past to future in the *developing* world. Several points need to be emphasised here. First, most agricultural regimes in the developing world began as strongly multifunctional systems in the past, based largely on localised subsistence farming and associated strongly multifunctional dimensions (Pretty, 1995, 2002; Sanders, 2006). Losch (2004) emphasised that most agricultural systems in the South have multiple functions, while Becu *et al.* (2004) suggested that traditional rice-based agrarian systems often show multifunctional elements.[92] However, over time the pressure for these systems to *lose* multifunctional quality has increased (Mazoyer, 2001). Thus, Robinson (2004: 147-148) emphasised that subsistence farming "is inexorably being displaced by production for sale, so that even remote parts of the tropics are beginning to farm on more commercial lines". As highlighted above, this is partly linked to the increasing importance of globalised agro-commodity chains that are beginning to influence agricultural practices in even the remotest corners of the globe (Goodman and Watts, 1997). As Losch (2004: 347) emphasised, "the economic environment of agricultural production [in the developing world] is ... directly conditioned by [the] new rules of play in the international game. Market instability and volatility, the growing disparity between different categories of agents, and the absolute necessity to be competitive from very uneven starting points, have profoundly changed the framework of agricultural production". Based on several case studies from developing countries, Pretty (1995: 30-31) argued that "the pressure to increase economies of scale, by increasing field and farm size, has meant that the traditional mixed farm ... has largely disappeared". Similarly, Barthélémy *et al.* (2004: 127; my translation) argued on the basis of agricultural transitions in Senegal, Brazil, New Caledonia and the island of Mayotte (Indian Ocean) that "multifunctionality in these areas is inexorably linked to the issue of agricultural transformation, which has gradually evolved from local to global embeddedness". Losch (2004) added to this the threat of rapidly increasing agricultural populations in many regions of the developing world, with increases over the past 25 years of over 500 million people. Inevitably, this is placing increasing strain on what are already scarce agricultural resources (Wilson and Bryant, 1997) – with associated loss of multifunctional quality particularly in Africa (>50% increase in agricultural population 1975-2000) but also in many regions of Asia (an additional 400 million people living in rural areas since 1975) (Sanders, 2006). Simultaneously, improved technology is beginning to open new opportunities for many farmers in the developing world to intensify agricultural production (and thereby to embark on weaker multifunctional trajectories), albeit with large regional differences (Pretty, 1995; Parayil, 1999; see example of Farm Y above).

---

[92] Chapter 9 already cautioned that virtually every place on the globe has historically seen different phases of weak, moderate and strong multifunctionality, and it is questionable whether the strong multifunctionality model has been the dominant model *throughout* human history in a developing world context. Many examples can be found (e.g. relatively productivist farming practices in China 2000 years ago discussed in Chapter 7; intensive rice cultivation in many parts of Asia, etc.) where agricultural systems in developing countries have been moderately or even weakly multifunctional for a long time.

Second, Figure 10.9 suggests that decision-making opportunities – or 'transitional potential' as highlighted in Section 10.2 – were relatively limited in the past for most agricultural systems in the developing world, but that the range of opportunities has increased and is likely to further increase in the future (Wilson and Rigg, 2003; Sanders, 2006). Although we may witness a gradual loss of strong multifunctionality pathways in many agricultural districts of the South in the future, simultaneously farmers will also gain more opportunities to choose from a wider array of Deleuzian decision-making pathways bounded by productivist and non-productivist action and thought. Third, the future trajectory of decision-making pathways is speculative and based on the current transitional tendency of a weakening strongly multifunctional agricultural regime in most developing countries. Subsistence farming is still the predominant agricultural model in most developing countries at the moment (over one billion subsistence farmers worldwide) but, as Section 10.2 highlighted, these strong multifunctionality pathways may occur *out of necessity* rather than choice. If maintenance of strong multifunctionality is a global goal (see above), then we may risk losing many of these strongly multifunctional farms as they become increasingly embedded into globalised markets, with resulting increases in productivist tendencies (Goodman and Watts, 1997; Robinson, 2004). As a result, *maintaining* strongly multifunctional pathways may *become* an increasingly important goal in developing countries while in developed countries the *rediscovery* of strong multifunctionality pathways may be the first priority (Sanders, 2006).

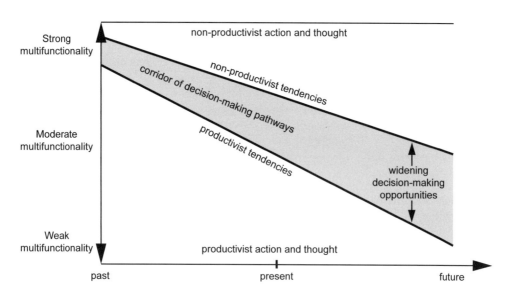

Fig. 10.9. Corridors of multifunctional decision-making pathways over time in the developing world

Figure 10.10 suggests that the situation in advanced economies is different. First, as with most agricultural systems in developing countries in the past, most agricultural regimes in advanced economies began as relatively strongly multifunctional systems (i.e. localised low-intensity systems geared towards subsistence or, at most, towards local markets). However, as discussed in Chapters 5 and 6, increasing productivist pressures for intensification of an increasingly globalised agriculture embedded into large-scale capitalist modes of accumulation have led to a rapid weakening of multifunctional quality over time. As Figure

7.2 highlighted, this led to the 'productivist trough' that has dominated most agricultural systems in advanced economies (Lang and Heasman, 2004).[93] Second, and similar to developing countries, transitional potential may increase in the future for developed countries through a widening of decision-making opportunities including many different multifunctional pathways. Based on Chapters 5-9, we can speculate that the future may offer more opportunities for increasing multifunctional *quality* (Knickel *et al.*, 2004). However, we need to acknowledge that this may occur on the back of relatively *narrow* transitional potential at present, as many agricultural actors in developed countries are still caught in what Ward (1993) termed the 'productivist treadmill'. Lang and Heasman (2004: 285), therefore, cautioned that "although productionism [sic] is weakening in power and public appeal, it is uncertain which ... [other paradigm] ... will replace it for the long term". As I will discuss in more detail below, it is only through positive *transitional management* that a widening of decision-making opportunities for today's farmers may be achieved.

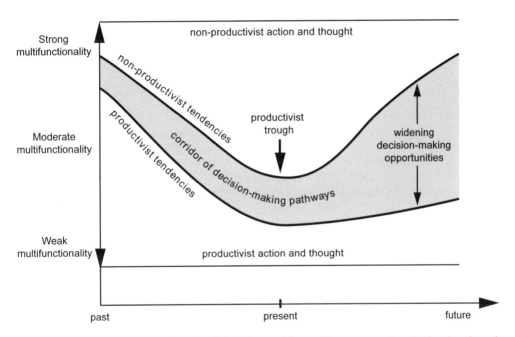

Fig. 10.10. Corridors of multifunctional decision-making pathways over time in the developed world

If we compare Figures 10.9 and 10.10 key differences in transitional tendency become apparent. Most crucially, developed and developing countries are at different temporal points in a *multifunctional transitional cycle* (see also O'Riordan and Voisey, 1998, and Pearce and Barbier, 2000, for similar debates on the transition to a sustainable society). Indeed, most agricultural territories in developing countries have, so far, only undergone the first part of the cycle and may be heading for the 'productivist trough' in the near future. Most agricultural areas in advanced economies, meanwhile, have – due to earlier mechanisation,

---

[93] As with Figure 10.9, Figure 10.10 shows a relatively crude simplification of what are often complex multi-dimensional transitional processes influenced by many interacting factors (see Sections 10.2 and 10.3). Indeed, as Chapter 7 and Figure 10.2 highlighted, many of today's farms operate close to the non-productivist end of the multifunctionality spectrum, although there is little doubt that the *totality* of farming actions in advanced economies has led to a weakening of multifunctional quality culminating in the productivist trough (Robinson, 2004).

capitalisation and embeddedness into the global capitalist system – been firmly embedded in the productivist trough for some time (see Chapter 5) and are now beginning to re-evaluate agricultural approaches based on more moderate or strongly multifunctional concepts (Knickel *et al.*, 2004).

This suggests that there may be an inherent temporality in the seven dimensions that characterise strong multifunctionality (see Section 9.4). While it may be easier to adjust agricultural systems towards stronger multifunctionality pathways based on higher environmental sustainability, better local embeddedness, shorter food chains, improved diversification activities, lower farming intensity and productivity, and weaker integration into the global capitalist market, it will be more difficult to rapidly change the mental images of what 'farming' and 'agriculture' are about – i.e. to create open-minded societies that see 'farming' and 'agriculture' as processes that go beyond productivist food and fibre production. While many agricultural systems in developing countries (and beyond; see example of Mediterranean countries discussed in Chapter 7) still aim to 'catch up' regarding productivist intensification and profit maximisation with resulting loss in multifunctional quality (Dulcire and Cattan, 2002), few societies have made the *mental* adjustments towards strong multifunctionality (see Chapter 7). Indeed, such mental adjustments by both farmers and society as a whole (two key dimensions of strong multifunctionality) may only be a *very recent* feature in agricultural transitional processes (see also below). Only if such mental adjustments towards strong multifunctionality are *completed*, can agricultural systems begin to move out of the productivist trough and 'back' towards strongly multifunctional agricultural pathways. Yet, there is another important point associated with the comparison between the developing and developed world. As Losch (2004) argued, the key multifunctional trajectories in today's world are largely associated with *developing* countries (see Fig. 10.9 above), as it is here that we find over 96% of the world's agricultural population (over two billion). Although loss of multifunctional quality relatively equally impacts all territories of the world for reduction of environmental capital (in particular linked to biodiversity or global environmental degradation), the potential loss of social, cultural and economic multifunctional capital will affect many more people in the heavily populated agricultural regions of the developing world (Pretty, 1995, 2002).

Building on this discussion, Figure 10.11 shows a more nuanced hypothetical depiction of multifunctional agricultural transitions for Western Europe from about AD 1500 to 2150.[94] I have selected Western Europe as a geographical reference point, as there has been relative continuity of agricultural production in this part of the world for this time period (as opposed to the New World, for example, where large-scale agriculture was only introduced after European settlement), and as Western European countries have some homogeneity concerning culture, history and general agricultural trends (Davies, 1997). The figure shows the multifunctionality spectrum ranging from strong to weak multifunctionality on the y-axis and time on the x-axis,[95] and is a conceptual model of key factors that influence transitional changes (or indeed transitional ruptures; see above) within corridors of multifunctional decision-making pathways over time. It highlights a distinctive *corridor* of multifunctional decision-making pathways (fluctuating and characterised by troughs and peaks) bounded by productivist and non-productivist tendencies within which most agricultural decision-making at a given point in time can be situated. Three points emerge from this figure. First, based on Figure 10.10 (above), the figure suggests a gradual decline of the corridor of decision-making pathways from strong/moderate to weak multifunctionality between 1500 and 1980, the 'productivist trough' of the 1950s to the 1990s, and a speculative broadening of transitional possibilities after 1990 with a continuation of future productivist tendencies and, simultaneously, future non-productivist tendencies that show rising potential for the adoption

---

[94] See also Priemus and Hall (2004) for a similar attempt at sketching multifunctional *urban* development over time.
[95] The reader may wish to compare this figure with Marsden's (2003: 184) conceptual model of agriculture/rural development pathways where 'value added' and 'time' are used as reference points on both axes.

of strong(er) multifunctionality pathways. As is still the case in many agricultural regions of the developing world (see above), agriculture in Western Europe around AD 1500 was largely characterised by high environmental sustainability (i.e. almost all systems were 'organic' under current definitions of the term), a tendency for local embeddedness (close interaction with and within local communities),[96] short food chains, low farming intensity and productivity based on low levels of mechanisation and technology, weak or no integration into international markets, and high degrees of diversification activities as 'farmers' were often forced to take on non-agricultural employment to ensure economic survival of their household (Cayre *et al.*, 2004). Thus, most agricultural systems in Western Europe around AD 1500 were strongly multifunctional and, as Priemus *et al.* (2004: 269) emphasised, "in the pre-industrial era, multifunctional land use was commonplace". Yet, it is less certain whether this was also true for how society viewed the role of agriculture at the time. Based on above discussion of *multifunctionality cycles*, it is likely that the two key dimensions of strong multifunctionality – emergence of open-minded farming and rural populations who see 'farming' and 'agriculture' as processes that go beyond productivist food and fibre production and open-minded societies that accept that the nature of 'farming' and 'agriculture' are in the process of change – were largely absent from societal discourses at the time, as agriculture's first and foremost role was seen (or had to be seen) as that of feeding the population of a given territory and ensuring, as much as was possible, self-sufficiency in agricultural production. In other words, while agriculture *on the ground* was mostly strongly multifunctional, *mental images* of agriculture were not. This is the reason why the beginning of the decision-making corridor around AD 1500 is shown in the figure to be situated somewhere between strong and moderate multifunctionality pathways rather than straddling the extreme boundary of non-productivist action and thought.

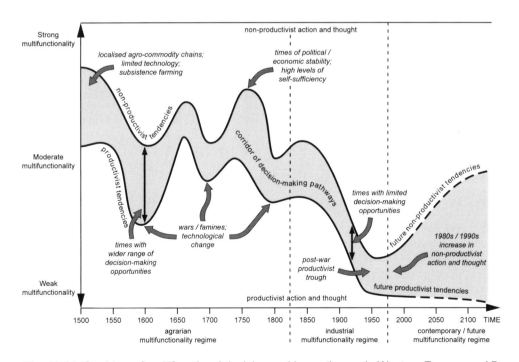

Fig. 10.11. Corridors of multifunctional decision-making pathways in Western Europe *ca*. AD 1500-2150

---

[96] Pretty (1995), for example, highlighted how in the UK the manorial system of integrated farming was sustained for some 700 years by a high degree of cooperation between farmers where local groups established detailed management measures for strongly multifunctional use of agricultural and community resources.

Second, Figure 10.11 suggests that three distinctive multifunctionality regimes can be identified for Western Europe. The first, from AD 1500 (and probably earlier) to the early 19th century I term the 'agrarian multifunctionality regime', characterised by strong to moderate multifunctionality based on pre-Fordist modes of agricultural production and fluctuating productivist and non-productivist corridor boundaries of decision-making pathways. Luttik and Van der Ploeg (2004: 205) argued that "in pre-modern society farming was multifunctional by nature" (see also below). The second regime can be defined as the 'industrial multifunctionality regime' (early 19th century to about 1980), characterised by rapidly weakening multifunctionality (culminating in the 'productivist trough') based on 'Fordist' agricultural production and a relative loss of connection between increasingly urban societies and their rural/agricultural 'hinterland' (Lang and Heasman, 2004; Potter and Tilzey, 2005) – in many ways akin to Friedmann and McMichael's (1989) 'first' and 'second' food regimes. The early 19th century is often seen as an important point of transitional rupture in this respect, due to the rapid increase in farm mechanisation (steam-powered engines) and the development of the first artificial fertilisers, although the timing of this transitional rupture has varied substantially between European countries (Lang and Heasman, 2004; see also Chapter 7). Luttik and Van der Ploeg (2004: 206) suggested that "the decline in the degree [i.e. quality] of multifunctionality started with the industrial modernisation, and continued as this stage developed". During this regime we also witness a narrowing of transitional potential for agricultural actors for reasons outlined above.

The third regime (from about 1980 onwards) could be described as the 'contemporary/future multifunctional regime',[97] characterised by rapid widening of transitional opportunities along the multifunctionality spectrum, a possible continuation of productivist (or even super-productivist; cf. Halfacree, 1997b) action and thought, but also a possible rapid rise in multifunctional quality through new non-productivist opportunities (see also Friedmann and McMichael's, 1989, 'third' food regime). The latter may herald a *rediscovery* of strong multifunctionality, depending on how these new transitional opportunities are 'managed' (see below), but there is insufficient evidence to date as to whether this third multifunctionality regime is characterised by a *transitional rupture* with the former regime or whether it merely depicts a gradual transitional shift with stronger non-productivist tendencies. Pretty (2002) rightly reminded us that the period of the industrial multifunctionality regime only involved six to eight farming generations (compared to about 400 farm generations since the emergence of agriculture in Europe!). Further, as Marsden and Sonnino (2005: 16) argued, "agriculture is still caught very much in between the agri-industrial [productivist in my terminology] and the post-productivist [non-productivist] models", and a similar point was made by Carpentier *et al.* (2004) in the context of continuing super-productivist tendencies in the USA. This emphasises the importance of recognising the constant push and pull factors between the two extreme ends of our multifunctionality spectrum. In particular, these authors highlight the continuing danger of a productivist pathway based on a 'race-to-the-bottom' scenario that dominates many European agricultural regions where it may prove difficult for farmers to leave productivist, or even super-productivist, pathways (Ward, 1993). For the UK, Marsden and Sonnino (2005: 27; emphasis added), therefore, argued that "at central government level … policy development is still stuck in the 'mud' of placating agri-industrial interests on the one hand and post-productivist (environmental and amenity) interests on the other. It has yet to develop a more *autonomous break* or *rupture* with this paradigmatic conundrum". Similarly, Potter and Tilzey (2005: 16) suggested that the relevant question is not how non-productivist European agriculture is likely to evolve, but rather how to find and defend spaces for non-productivism within an inherently *productivist* agriculture.

---

[97] The reader will note that, for reasons outlined in Part 1, I avoid using terms such as 'pre' or 'post' in the characterisation of different multifunctionality regimes.

The third multifunctionality regime is the most speculative, as it is difficult from a presentist perspective to predict the trajectory of future transitional pathways (O'Sullivan, 2004; see also Chapter 7). As Marsden (2003: 211) suggested, we currently witness a "heterogeneous and diverse set of rural development outcomes rather than a clearly generalisable model of convergent spatial development". Buller (2004: 117) argued that this can be explained by the fact that "agriculture is one of the last bastions of the great modernist project of European construction", with all its entrenched pressures to maintain the productivist pathway alive. However, the proposed third regime re-emphasises that contemporary agriculture – at least in a Western European context – may be at an important *crossroads*, as, for the first time, a *wider range* of multifunctional opportunities may be open to agricultural/rural actors than ever before in human history (Wilson, 2001). Thus, Knickel *et al.* (2004: 93) argued that "the overall trend is that all over Europe a very substantial proportion of all farms has been turned into more complex rural [multifunctional] enterprises over the past decades". This echoes Marsden's (1998b) notion of 'optimistic productivism' where future agriculture may not necessarily be a residual category, but where productivist tendencies may be 'reinvigorated' through localisation tendencies focused on agricultural production, while simultaneously enabling non-productivist tendencies to come to the fore. Marsden (2003), therefore, referred to the increased *uncertainties* for different rural spaces, while Delgado *et al.* (2003: 20) argued that "new demands, new customs, and new ways of organizing result in a different social and economic structure in which agriculture occupies a relative position quite different from the traditional one of yesteryear". Yet, if transitional processes over the next decades highlight that mental changes about how society views agriculture will accompany developments that can already partly be witnessed on the ground (see Chapter 6), then we may be able to argue that the turn of the millennium also formed an important *transitional rupture* away from the productivist trough of the late 20th century. If this was the case, the productivist trough could be seen as an *aberration* in a moderately/strongly multifunctional agricultural regime that may be thousands of years old. As Part 1 highlighted, similar questions have been asked about other transitional processes. For the possible transition from socialism to post-socialism, for example, Pavlinek and Pickles (2000: 294) asked whether it may be possible to think of state socialism in Central and Eastern European countries as "some kind of minor interruption in a 'normal' process of capitalist development". Similarly, Dear (1986: 375) suggested that "postmodernism could ultimately prove to be an aberration in the evolutionary cycle".

Third, Figure 10.11 suggests that boundaries of corridors of multifunctional decision-making fluctuate over time. At times, corridors shift towards weaker multifunctionality while at others they may show stronger multifunctionality characteristics. These fluctuations can be largely explained through political and economic factors and, particularly in the past, through levels of self-sufficiency. During times of political and/or economic stability with relatively high levels of self-sufficiency, agricultural systems tend to veer towards stronger multifunctionality, while during times of war, famines, or climate catastrophes, systems tend towards weaker multifunctionality as agriculture has to be intensified to address food shortages. Pretty (2002: xii) argued a similar point over longer time scales by suggesting that over "12,000 years of agriculture, there have been long periods of stability, punctuated by short bursts of rapid change". During 'good times' societies may be able to afford the *luxury* of strong multifunctionality pathways, while during 'bad times' external factors may *force* the adoption of weaker multifunctionality trajectories. Times of extreme fluctuations may also be associated with a widening of multifunctional decision-making opportunities (i.e. when the agricultural system may be 'stretched' in both productivist and non-productivist directions), as not all agricultural actors or regions would necessarily follow broader trends.

It is also interesting to speculate whether *extreme* fluctuations were more a feature of the past, when regional or national self-sufficiency was more important than today (at least in a Western European context). The *globalisation* of agro-commodity chains, in particular, may act as a 'great leveller' of regional or national food production fluctuations (i.e. permanent availability of agricultural commodities irrespective of national circumstances). As a consequence, losses in food production capacity in one territory can be *neutralised* through increased imports of foodstuffs from elsewhere. This debate overlaps with the discussion in Section 9.4 as to whether maintenance of strong multifunctionality pathways in one region occurs at the expense of weakening multifunctionality pathways in other parts of the world. Thus, the more globalised a region's or country's food networks are, the less likely it will be that fluctuations of multifunctionality corridors will be a result of internal factors related to self-sufficiency imperatives. However, as Section 10.2 highlighted, political factors continue to be important drivers for multifunctionality corridors, as political (and associated economic) isolation (e.g. North Korea, Myanmar) places additional pressures on internal multifunctionality opportunities. Indeed, food poverty still affects nearly one billion people on Earth despite progress in mass food production in productivist agricultural systems.

What does the hypothetical model shown in Figure 10.11 tell us about agricultural transitional histories in other parts of the world? Again, we should be cautious about the potential of hypothetical models based on specific cultural, historical and technological contexts of one region (i.e. Europe) to find applicability elsewhere (Said, 1983, 1993) – in particular considering the large differences in affected agricultural populations in the developed and developing world. As above discussion highlighted, transitional corridors in the developing world will, in most cases, be different from those in developed countries, but even non-European developed countries will depict very different (and often much shorter; cf. Australia, USA, Canada) agricultural transitional histories. In the case of Australia (and New Zealand to a lesser extent), for example, right from the start the development of European-style agriculture was predicated on a weak multifunctionality model geared towards productivist exports 'back' to Europe. Although some subsistence-style farming occurred during the first few decades of European settlement in remote forest areas (see Wilson, 1992, for a remote pioneering area in New Zealand), these agricultural systems were rarely strongly multifunctional as they were largely predicated on the removal of original biodiversity-rich forests (Wilson and Memon, 2005). Based on recent incisive critiques of the 'European-based' model of the transition to post-productivism (Holmes, 2002; Jay, 2004; see Chapter 7) it can also be questioned whether a 'productivist trough' can be identified in Australia and New Zealand and, in particular, whether future multifunctional trajectories in these countries may also be characterised by a widening of both productivist and non-productivist opportunities (Cocklin and Dibden, 2004; Holmes, 2006). Indeed, as Evans *et al.* (2002: 316) emphasised, the deregulation of New Zealand agriculture in 1984, in particular, "promoted a development trajectory that has greater alignment with a reformulation of productivism". As Chapter 11 will further emphasise, we should, therefore, see the multifunctional models presented in Figures 10.9-10.11 as initial (hopefully thought-provoking) conceptualisations that await further empirical refinement. Indeed, more studies are now needed to test the robustness of the multifunctionality spectrum (over time) in different geographical and cultural contexts.

10.4.2 Managing multifunctional agricultural transitions

I began Section 10.4 by arguing that the *strong multifunctionality* model should be seen as the 'ultimate goal' of any decision-maker involved in agriculture, as this model can be seen as the most *moral* and *qualitatively best* type of multifunctionality. This means that the discussion about *managing* multifunctional transitions needs to be based on the underlying assumption that the ultimate aim of any agricultural system and associated stakeholders is to either *maintain* strong agricultural multifunctionality or to move away from weak or moderate multifunctionality towards a 'rediscovery' of strong multifunctionality. Empirical support can be given to the need for strong multifunctionality by studies that have highlighted that if countries or regions fail to adopt at least *some* aspects of the strong multifunctionality model, then the social and economic survival of agricultural and rural communities may be severely jeopardised (Diamond, 1998, 2006). Yet, as highlighted above, we need to acknowledge that it is far from established whether multifunctional transitions are a *zero-sum-game* or whether there is a possibility for global-level strong multifunctionality based on a *win-win* situation in which *all* agricultural systems may be able to adopt strong multifunctionality pathways (Lifran *et al.*, 2004; Boody *et al.*, 2005). As Marsden (2003: 27) emphasised, "it would seem that valued consumption spaces need to continuously devalue other spaces in order to reproduce more sensitive capital accumulation ... In this sense globalisation and deregulation tend to redistribute risks and power, making it easier in one place to extract value only at the expense of other places". Indeed, in the dominant consumption spaces of advanced economies, corporate retailers play an increasingly important *gatekeeping* role in translating quality definitions related to strong multifunctionality. Thus, a headlong rush towards strong global multifunctionality may not only be unrealistic, but may defy complex social systems of self-organisation within often only *gradually* changing equilibriums. Echoing Beck (1992), achieving global-level strong multifunctionality may only be possible through a combination of social iteration, pragmatism and, above all, reflexivity across the productivism/non-productivism spectrum. Yet, Chapters 9 and 10 have highlighted that global trajectories of multifunctionality are generally on a *downward* trend with increasing loss of multifunctional quality. While developing countries still have many agricultural systems based on strong multifunctionality pathways, the threats of losing multifunctional quality in the developing world is particularly acute (see above). In developed countries, meanwhile, it remains questionable whether the current situation – with some evidence of increasing non-productivist action and thought – represents a clear transitional *rupture* away from productivist agricultural pathways.

How can the transition from weak to moderate to strong multifunctionality be encouraged and *who* should be encouraging such a transition? Can and should the strong multifunctional agricultural regime be *orchestrated* at all based on a 'strong policy' model (Hall, 1993), or do (should?) multifunctional agricultural transitions have a life of their own? This section will attempt to provide answers to these questions. First, I will discuss contemporary debates on theorisations of transition management, as there is a growing body of literature that is attempting to provide answers about how best to manage increasingly complex societal transitional processes. Second, I will look at the theoretical role of policy in transition management. In democratic societies, policy-led transition management – for better or worse – has played a crucial role in defining transitional corridors and pathways, although the question remains whether such transitions should be orchestrated at all. Third, I will speculate how a transition towards strong multifunctionality can be achieved and by whom it should be implemented. The section will conclude with a critical look at key challenges faced by stakeholders in future implementation of strong multifunctionality trajectories.

Theorising transition management

Current debates in transition theory suggest that it may be possible to develop management strategies that facilitate the shift from one transitional regime to another (i.e. towards strong multifunctionality). The literature on transition theory (see Chapter 2) suggests that various insights can be combined into a management strategy for public decision-makers and other stakeholders. This has been termed *transition management*, an approach developed to guide social, economic and environmental transitions (Rotmans *et al.*, 2002). In line with arguments made in this book, transition management acknowledges that transitions are usually non-linear (see Chapter 4), with an associated need for complex institutional and actor-based responses and policies. As Rotmans and Martens (2002: 113) argued, "transitions are not a law of nature: they do not indicate what is bound to happen. Rather, they represent development pathways ... which provide insight into a range of likely futures, depending on social, economic and environmental circumstances". Although societal transitions are difficult to manage, it should be possible to guide and shape transitions. In other words, although unmanaged transitions can have a *life of their own* (O'Sullivan, 2004), transitional pathways can also be influenced by careful involvement of relevant stakeholders in multi-layered societal decision-making systems (Wilson and Bryant, 1997).

Rotmans *et al.* (2002: 11) suggested that "transition management is based on a ... process-orientated philosophy that balances coherence with uncertainty and complexity" – in the context of this book a transitional complexity highlighted by multifunctional Deleuzian pathway opportunities bounded by productivist and non-productivist actor spaces. According to proponents of transition management, the aim of such management is not so much the realisation of *one* specific transition, as this does not exist (i.e. multiple pathway opportunities), but to address individual components of this Deleuzian transition through *transition-pathway-specific* policies and mechanisms (Rotmans *et al.*, 2002). Just as Figures 10.6-10.11 (above) suggested, transition management should best occur within more or less well defined *corridors* of transitional pathways. It is, therefore, not individual transitional pathways that are of interest to transition managers, but the establishment of the extreme boundaries of transitional *corridors* (in our case productivist and non-productivist pathways) that 'bundle' individual pathways into a manageable whole. As Smith and Pickles (1998) argued, the challenge is to negotiate ways in which we can understand the *diversity of forms* of transition. As long as the general direction of the transitional corridors is agreed upon by stakeholders (see below), specific actions and policies could, therefore, be put into place that may help guide individual 'pathway bundles' into the chosen transitional direction.

This highlights that goals and instruments for transitional management need to be constantly re-evaluated. In other words, there has to be an inherent dynamism in transition management that allows for constant feedback loops and that enables stakeholders to re-evaluate transitional processes at key nodal points of the transition (see Fig. 10.7 above). At each transitional rupture, the extreme boundaries of the transitional corridors need to be re-evaluated and, if needed, readjusted. Rotmans *et al.* (2002) argued that increasing transitional complexity cannot necessarily be *controlled*, but that transition management may help *facilitate* the general direction of transitional pathways. Rotmans and Martens (2002) outlined several points for successful transition management, two of which are particularly pertinent for our discussion. First, *long-term thinking* and attention to *dynamics* as a framework for short-term policies is needed. They suggested that transitional management frameworks of at least 25 years (i.e. one generation) are needed for successful influence of transitional pathways, although 100-year time frames would be much better to address issues of transitional rupture, non-linearity and structure-agency inconsistency. Second, due to the Deleuzian nature of societal transitional corridors, thinking in terms of multiple domains, multiple actors and multiple scales is necessary. As Chapter 9 highlighted, this type of thinking is particularly important for complex actor- and scale-related issues surrounding multifunctional agricultural transitions.

What emerges from these debates is the importance of focusing on transitional corridors (rather than individual transitional pathways that may, at times, be situated outside the main corridors), the acknowledgement that transition management has to be dynamic and flexible, and that it should operate at time scales of at least one generation. As I will discuss in detail below, the latter is particularly problematic if policy is to have an important role to play in transitional management, as in most democratic societies policy cycles are closely linked to *electoral cycles* that only last 3-7 years (Johnston, 1996). Due to the importance of policy as a societal framework for action in democratic contexts, the following will focus specifically on the role of policy in transition management.

Transition management and policy

In democratic pluralistic societies, policy is often seen as the ultimate 'vehicle' with which transition management can be implemented ('strong' policy model; cf. Hall, 1993; Johnston, 1996), and literature on both transition theory and transition management highlights that the magnitude and rate of transitional change can be significantly *influenced* by policy intervention. As Section 9.4 highlighted, policy-led (or, better, policy-influenced) transitional pathways can have positive outcomes if the policy mechanisms have flexible goals and are constantly re-adjusted based on micro-level transitions along Deleuzian multifunctional pathways. Here, we need to return to our concept of multifunctional corridors (see Figures 10.6-10.11). Rotmans and Martens (2002) argued that to mirror the often Deleuzian nature of societal transitions, the 'policy corridor approach' tends to work best. A policy corridor represents a conceptual space within which policy-making can be relatively safe, in our case based on taking into account the multi-dimensional Deleuzian transitional pathways implicit in the multifunctional transitional spectrum. As with individual farmers' transitional decision-making possibilities defined by transitional 'risk boundaries' of productivist and non-productivist action and thought (see Figs 10.6. and 10.7. above), policy corridors can be conceptualised as lying within transitional margins where risks are considered 'acceptable'. Kreileman and Berk (1997) argued that this approach has been used relatively successfully in climate change policy using the 'safe landing' allegory (i.e. an airplane during approach and landing has a certain 'corridor' for error). Similarly, policy using the safe landing approach within multifunctional decision-making corridors can be used to tackle the transition towards strong multifunctionality.

Although, in theory, the policy corridor approach appears as a suitable and logical strategy, a key problem lies with the definition of the 'correct' direction of policy corridors. In particular, *who* is to be in charge of defining the boundaries of the policy corridor? Section 9.4 highlighted that most *state-led* agricultural policies (in both developed and developing countries) have tended to operate within relatively narrow margins of transitional potential, and that state policy has often encouraged weak multifunctionality, partly leading to the current 'productivist trough' that has characterised agricultural transition in many advanced economies (see discussion of impacts of CAP policies in Chapter 7 and Section 9.4). This was also emphasised by Marsden and Sonnino (2005: 16; original emphasis) who suggested that "multifunctional agriculture in the UK, where it has occurred ... has been a kind of *rupture with the State sector*, rather than through policy incentives". This has been exacerbated by global policy pressures (in particular through the WTO; see Section 9.4), suggesting that state-led policy influence on multifunctional agricultural transitions has substantially increased over the past few decades. Thus, 'official' policy formulated and implemented by state-related actors has often been a key driver for weak multifunctionality pathways, rather than a solution for the rediscovery or maintenance of strong multifunctionality (e.g. Wilson and Juntti, 2005, for the Southern European context).

Managing the transition towards strong agricultural multifunctionality: relocalisation and the role of the state

Based on this critical assessment of the possible role of policy in managing multifunctional agricultural transitions, I now wish to discuss whether such transitions should be *orchestrated* at all. As Durand and Van Huylenbroek (2003: 13) rightly asked in the context of multifunctionality, "who will mediate this new contract between society and agriculture? The role of local, regional, national, and international institutions clearly needs to be reviewed". In particular, I wish to analyse how a possible transition towards a strong multifunctionality model can be achieved and by *whom* it should be implemented. It is in this context that my personal preferences for how a future agriculture could/should look like will become most apparent – based on one of the key aims of this book to act as a *plaidoyer* against the economistic interpretation of multifunctionality simply as an 'externality' issue, against the predominant productivist agri-business model that continues to dominate agriculture, and against the increasing globalisation of agro-commodity chains that tends to weaken global multifunctional agriculture pathways.

Let us, therefore, first speculate what would be needed to 'rediscover' strong multifunctionality pathways. Building on O'Riordan (2004) who argued that the transition to a more self-aware environmentally sustainable society requires careful thought and planning based on *localised* instruments such as Agenda 21, and Sonnino and Marsden (2005) who discussed localisation issues surrounding alternative food networks, I argue that achieving strong multifunctionality will only be possible through a complete re-thinking of human/agriculture/food interactions based on *relocalisation* of agro-commodity chains.[98] Relocalisation – or 'reterritorialisation' in Winter's (2005) words – will particularly help contribute towards increased socio-economic 'retention capacity' of rural areas based on strong *social* and *cultural capital* (Oman, 1994; Knickel and Renting, 2000) and will also counteract increasing 'place neutrality' engendered by globalised agro-commodity chains (Blandford and Boisvert, 2002; Freshwater, 2002). Similarly, Knickel *et al.* (2004: 99) suggested that "a characteristic feature of more regional, multifunctional development perspectives is the (re)integration of farming activities into the local economy". Shuman (2000: 46) referred to this as 'community self-reliance', suggesting "personal responsibility, respect for others, and harmony with nature … addition of the word 'community' to self reliance underscores that the ultimate objective is a social and caring one". The importance of the 'local' was also emphasised by Goodman (2004: 5) who argued in the context of the food quality shift that we witness "a transition from the 'industrial world', with its heavily standardized quality conventions and logic of mass commodity production, to the 'domestic world', where quality conventions embedded in face-to-face interactions, trust, tradition and place support more differentiated, localized and 'ecological' [i.e. more strongly multifunctional] products and forms of economic organization".

Clark (2005) termed such relocalisation processes 'neo-endogenous' developments that mobilise local social and economic capital – closely associated with Marsden *et al.*'s (2002) notion of the 'new associationalism' that emphasises the (re)connection of farmers with their local/regional communities, and with Knickel *et al.*'s (2004) suggested potential for a rediscovery of (an often lost) 'cultural repertoire' (see also Pretty, 1995, 2002). There are also parallels with Van der Ploeg *et al.*'s (2000) notion of the 'repeasantisation' of European farming, where the locality and local farm scale is emphasised as a rediscovery of bygone 'peasant' farming styles and virtues. Further, Bell (2004), in his analysis of farming cultures in the USA, makes an interesting allusion to 'farming the self' which can be reinterpreted as a relocalisation of farming priorities in which the farmer and his/her rural community provide the main focus of action and thought. Knickel *et al.* (2004: 88), therefore, argued that "multifunctional agriculture … is not just about 'new things'. It is also about historically rooted realities that are currently reappearing". The notion of relocalisation was also reiterated in a wider economic geography context by Grabher (2001) who suggested that relocalisation processes are usually accompanied by the reframing of businesses (in our case farms) as collective entities, comprising overlapping familial, kinship and wider territorial networks, with managers (i.e. farmers) brokering new relational configurations and acting as a nexus of the 'creative exploitation of embeddedness' for economic, environmental and social gain. In his analysis of the complexity of stakeholder networks involved in 'implementing' multifunctionality at the local and regional level in the UK, Clark (2003, 2005) emphasised that this creative exploitation of embeddedness is the crucial ingredient for 'unlocking' the multifunctional potential of a given rural area. In particular, he emphasised that relocalisation processes are inexorably linked with the relational working and reworking of knowledge and power among actors as a basis for strongly multifunctional pathways (see also Curry and Winter, 2000). Thus, "the new associationalism empowers [farm] business managers to take on new responsibilities and network in ways that are not possible under the [weakly multifunctional] agro-industrial model" (Clark, 2005: 481).

---

[98] See also the associated notion of localised 're-urbanisation' suggested by urban theorists analysing how to improve the multifunctionality of inner city areas (e.g. Priemus and Hall, 2004).

In a broader sense, the notion of relocalisation is also closely associated with the idea of 'foodsheds' based on sustainable agro-food systems within bioregions[99] (Pretty, 2002; Bell, 2004). Kloppenberg (1991) referred to foodsheds as self-reliant, *locally* or *regionally based* food systems comprised of diversified farms using sustainable practices to supply healthier food to small-scale *locally embedded* processors and consumers to whom producers are linked by the *bonds of community* as well as economy. This echoes Van Huylenbroek's (2003: xiii) call for "a more territorial approach of agriculture", or Marsden's (2003: 17) urge for a "reassertion of the value of the local" in food production and consumption as part of what he terms the 'rural development dynamic' (see Chapter 7). This reflects a desire to live and work *within* rather than 'above' or 'in opposition' to natural cycles, where the natural connectedness between people and the land is re-established (Pretty, 2002). Similarly, Guzman and Woodgate (1999: 303) argued that "sustainable societies can only be constructed on the basis of sustainable, locally relevant agricultures ... implying a complete rejection of the homogenising tendencies of the neo-liberal, global modernisation project and the re-direction of co-evolution towards more sustainable ways of living that are based upon the endogenous potential of an infinite diversity of locally relevant agro-ecosystems". Scholten (2004) reported an interesting trend in the USA where in some farm regions there has been a pronounced rise in the number of female farmers, leading to increased organic farming activity and more presence in local farmers' markets. As both Traugher (2004) and Bell (2004) emphasised in the context of women and strongly multifunctional farming in the USA, this may point towards a gender-driven transition towards relocalised, and often strongly multifunctional, farming pathways (see also Robinson, 2004).

In my view, such relocalisation should go well beyond mere shortening of food chains (albeit an important component of strong multifunctionality; see Chapter 9) and should also include the spatial and social *reconnection* between agricultural and rural activities, communities and food consumers. Pretty (1995: 134) suggested that "'local' does have its own special characteristics. It provides the basis for collective action, for building consensus, for undertaking coordination of responsibilities, and for collecting, analysing and evaluating information ... The fact that people know each other creates opportunities for collective action and mutual assistance, and for mobilizing resources on a self-sustaining basis". In this way, relocalisation processes may hold the potential to reinforce the social and environmental role of agriculture as a major agent in sustaining rural economy and culture – in other words, to reintegrate agriculture back into a diversified and successful *local* rural economy. Thus, Lipschutz (1999: 280; emphasis added) argued that "much of the implementation and regulation inherent in regimes *will have to* take place at the regional and local levels, in the places where people live, not where the laws are made". Such calls for relocalisation processes are also evident in Marsden and Sonnino's (2005: 3) conceptualisation of multifunctionality as part of the rural development dynamic which "suggests the potential symbiotic inter-connectedness between farms and the same locale". Similarly, Belletti *et al.* (2003b: 159) aptly suggested that "only a policy focusing on territories and taking serious the bottom up approach and the involvement of territorial actors and specificities, can be coherent with the principle of economic, environmental, and social sustainability". A relocalised agriculture would, therefore, also be in tune with the oft mentioned assumption that "the countryside needs to be occupied by large numbers of farmers in order to retain its multifunctionality" (Potter and Burney, 2002: 39). However, the notion of a strongly multifunctional agriculture predicated on relocalisation processes is not synonymous with *reification* of the 'local' as a spatial configuration that is ontologically given (cf. Swyngedouw, 1997), but is rather a contingent outcome of dynamic processes of

---

[99] The notion if 'bioregionalism' implies the integration of human activities within environmental limits. Pretty (2002) suggested that bioregionalism can be seen as a self-organising concept that connects social and natural systems at a place people can call 'home'.

socio-spatial change necessary for a shift (back) towards a strongly multifunctional agricultural regime.

Pretty (1995, 2002), therefore, suggested that agriculture needs to be *reconnected to the environment* in order to attain higher-level environmental sustainability. This can only be achieved by reducing agricultural *intensity* and by reducing or abandoning *chemical inputs*. In turn, this would necessitate a radical shift in local and global *consumer behaviour* and demand. Globalised agro-commodity chains make it possible for most consumers in the developed world (and increasingly in the developing world) to source any type of food at any time of year, with associated high food miles and a growing (and problematic) role played by supermarkets (Blythman, 2004; Lang and Heasman, 2004). As a result, in a strongly multifunctional agricultural regime global food consumers would need to re-adjust food consumption patterns *away* from global towards local food networks (short food chains), thereby *(re)empowering their locality* and helping globalised farmers to (re)connect with their immediate community (Goodman and Watts, 1997; Pretty, 2002). This would, however, mean acknowledging that *only certain types* of foods would be available at certain times in the year (as was the case over thousands of years), a more *limited range of dietary choices*, and greater *seasonal variation* in the quality of food. This would be equivalent to both a "renewed seasonality of produce, now virtually abolished by continuous availability of most lines in the agro-industrial system" (Marsden, 2003: 18) and a willingness of food consumers to pay the full price of local production instead of opting for cheap imports of weakly multifunctional mass food products (Lang and Heasman, 2004). Ultimately, this would mean *weaker (or no) integration of agricultural systems into the global capitalist market.* Simultaneously, society would also need to accept that farmers may not need to be *full-time* food producers. Strengthening multifunctional pathways would, therefore, also mean providing/enabling more and better opportunities for farmers to find (part-time) employment outside agriculture, especially at times in the year when seasonal agricultural activity is reduced (as is already the case in many rural areas in the European Alps, for example, where many 'farmers' work in the tourist industry in winter). Finally, and most problematically, the shift towards strong multifunctionality would mean that farming and rural populations around the world would need to *rethink* their understanding of 'farming' and 'agriculture' as processes that go beyond productivist food and fibre production, as a key step towards wider acceptance by society that 'agriculture' is in need of change. Van Huylenbroek (2003: xii), thus, argued that strong multifunctionality "not only means a change in support systems to farmers, but also requires a fundamental change in daily practices of farmers, [and] a change in contractual relations between farmers and other stakeholders … If multifunctionality is really taken seriously, it requires a complete rethinking of the institutional system surrounding agricultural production".

A new contract between the farming community and society may have to be established, or, as Watts (1996) suggested, new forms of institutionalisation, regulation and spatialisation become significant in the uneven development of multifunctional agrarian space. Bryant's (2005) notion of 'moral capital' is particularly relevant for the latter, as is the concept of a 'moral contract' guiding wider rural development issues that "recognizes the multifunctionality of agriculture and recommends an integrated and multi-sector approach to the rural economy" (Durand, 2003: 134). Indeed, the rediscovery of strong multifunctionality is an inherently *moral process*. Only if the moral capital of a rural community, region or country is such that envisioning strong multifunctionality becomes a tangible goal, is it likely that action for the implementation of strong multifunctionality will follow. Marsden (2003: 19), therefore, suggested that new conceptualisations of agricultural and rural change need to be able to "assemble legitimate governance and regulatory structures and processes which are integrative and robust". As various studies have highlighted, identification of 'best practice' through mutual farm visits, for example, may be one way to strengthen the moral capital of a rural community, as it is here that actors see positive examples of how strong multifunctionality pathways have been adopted by their neighbours on 'demonstration farms' (e.g. Clark, 2003, for the UK; Wilson, 2004, for the Landcare programme in Australia). It is through these processes that a possible 'snowballing' of moral values underpinning strong multifunctionality may be most effectively promulgated.

This discussion highlights that managing the transition towards strong multifunctionality relies on a series of *interconnected actions* that, due to their complexity regarding temporal characteristics, spatial and scale-dependent processes, and multi-layered actor levels, will be difficult to implement. This was also highlighted by Clark's (2003) in-depth investigation of implementation of multifunctionality in the East Midlands region of the UK that stressed the importance of *communication* between different stakeholder groups in the implementation of strong multifunctionality (see also Wilson and Juntti, 2005). In particular, there is the underlying urge for the *relocalisation of agricultural processes* implicit in the above chain of events that may be particularly difficult to put into action. I, therefore, acknowledge Newby's (1986) caution not to repeat patterns and processes suggested by earlier 'locality studies', and suggest, instead, that the new relocalisation framework proposed here does *not* produce a new descriptivism that reifies exceptionalism. Similarly, and as the discussion above highlighted, we need to be cautious about assuming that relocalisation is necessarily associated with 'positive' outcomes and the global with the often 'negative' universal.

The discussion in the previous section has also questioned whether *state policy regulation* of multifunctional transitions is the best approach, or whether other stakeholders and different forms of transition management and governance need to be found to implement such relocalisation processes? In other words, as Losch *et al.* (2004) and Clark (2006) emphasised, should the state (and its associated polices and institutions) be alone in its role of *orchestrating* multifunctional agricultural transitions? As Marsden and Sonnino (2005: 29; original emphasis) rightly argued, there is an inherent risk that current academic thinking has "led social science researchers into an implicit belief that the State *can* be conducive and empowering with regard to rural development". Similarly, McCarthy (2005: 776) emphasised that "the logic of enacting rural multifunctionality through ... centralized agricultural policies has been questioned" (see also Brouwer, 2004b), while Potter (2004: 29; emphasis added) argued that many studies "map out a more flexible and arms-length pattern of government regulation and intervention in order to unlock what they see as hitherto untapped potential for *endogenous* ... rural development". Marsden and Sonnino (2005, executive summary) equally argued in the context of Wales (UK) that "devolution has turned the State into an orchestrator of agricultural, environmental and rural development networks kept together around a ... multifunctional view of agriculture", but that the UK government, in their view, "has been unable to progress real agricultural multifunctionality". As highlighted, where the state has had large influence over agricultural transitions (e.g. through regulatory top-down policies) the outcome has most often been a weakly multifunctional agricultural regime, with often disastrous consequences for food quantity and quality, environment and biodiversity, and rural/agricultural cohesion (Losch *et al.*, 2004).

Yet, the notion of 'policy' – as a regulatory tool for the management of the transition towards strong multifunctionality – does not necessarily need to imply a top-down state-led structuralist approach (Wilson and Bryant, 1997). Indeed, if we accept that 'policy' can be understood as a broad-based approach that spans the entire spectrum of multi-layered decision-making in society and that, for example, also includes *oral traditions* of subsistence farmers in tropical forests or the *unwritten conveyance of knowledge systems* from one farm generation to another, then 'policy' can also be formulated by individual actors at grassroots level (at least partly independent from state-led policy) and may inform local-level transitional pathways. Thus, the multifunctional transition can also be seen as an *autochthonous* process that, at times, may operate, change and influence multifunctional decision-making outside of the state policy realm (the 'weak' policy model; cf. Hall, 1993). As Van Huylenbroek (2003: vi) emphasised, "rather than organising and prescribing, public authorities should try to stimulate innovation and coordinated action of stakeholders". Similarly, Gerowitt *et al.* (2003: 227) controversially suggested from an environmentally oriented multifunctionality perspective that "the decision as to which ecological goods should be stimulated could be transferred to a decentralized and region-specific public committee in which relevant stakeholders are represented". However, can non-state actors (e.g. at grassroots level) be left alone to guide their own multifunctionality pathways without interference from the state? In many instances the answer would be negative, as individuals will often tend to put personal profit before adoption of strong multifunctionality pathways that may not immediately benefit themselves (Losch *et al.*, 2004). Clark (2003: 185) similarly highlighted that interviewees in his UK case study pointed "to a decisive failure by the region's core agencies to develop 'an overall strategy' for delivering the multifunctional agenda to agricultural businesses ... In particular ... the roles and responsibilities of both peripheral and to a lesser extent core organisations in the delivery of multifunctionality have not been defined with any degree of clarity". In particular, Clark (2006) highlighted the complexity of actor groups involved in implementing multifunctional agricultural and rural agendas at the regional level (his list of actors covers two entire pages!) and, linked to political and knowledge barriers, the often contradictory goals pursued by individual stakeholder groups and the marginalisation of what he termed 'peripheral organisations' (see also Wilson and Juntti, 2005, for similar patterns in Southern Europe).

There are clear links here with debates on the 'tragedy of the commons' (Hardin, 1968), as adoption of strong multifunctionality pathways by individual stakeholders (i.e. based on their own 'policies') would almost always imply a reduction in personal wealth – or, in the case of developing countries, the forsaking of potential increased wealth associated with adoption of weak(er) multifunctionality pathways (see Pretty, 1995, for the case of farming regions in India). Is it, therefore, possible that individuals would be willing to altruistically adopt (or maintain) strongly multifunctional agricultural pathways for the sake of their wider community, food quality and consumer benefits alone? Current evidence would suggest that this is not the case as, given the opportunity (whether state policy-led or not), individuals often choose agricultural trajectories that take them away from strong multifunctionality (see Chapter 5). Clark (2003) further emphasised that some stakeholder groups have clear vested interests that make them actively support weakly multifunctional strategies, which may be particularly true for the growing global power of corporate retailers who increasingly influence and shape food demand, purchasing powers and, ultimately, many farm development pathways (Marsden and Wrigley, 1995; Lang and Heasman, 2004). Here, our discussion overlaps with wider debates about societal transitions in an increasingly globalised capitalist system in which greed often dictates transitional pathways rather than common sense (Lipietz, 1992; Stiglitz, 2002). Thus, the same questions can be asked as to why individuals are not willing to restrict their own energy use and ecological footprint as a move towards tackling global warming. The further intensification and globalisation of individual life, and associated economic pressures for households to seek maximum profits, tends to suggest that non-state-led individual transition management towards strong multifunctionality is unlikely.

This suggests that some *external regulation* of multifunctionality transitions is needed and that the state has to play some (continuing) role in guiding and influencing the transition towards strong multifunctionality (Clark, 2003; Losch *et al.*, 2004). As Lang and Heasman (2004: 303-304) argued in the context of 'hard' policy measures governing the nutrition transition, "the threat of sanction of legal action has to be there, and used appropriately". Similarly, anti-liberalist authors such as Losch (2004: 353) have suggested that to achieve improved multifunctional quality "there is a need for countries to retake control of their own policies. These can only be rebuilt from within each national situation, with its own potential, its own constraints, its own history and internal struggles [i.e. system memory]". Calls for strong state support have also come from those advocating strong sustainability pathways with Pretty (1995: 24), for example, suggesting that "the state should play a supportive role in the development of a more sustainable agriculture". Similarly, Freshwater argued from a US perspective that "by definition multifunctionality demands a more proactive government role because of the existence of non-market goods and joint production". Yet, as Clark (2003) emphasised with reference to the notion of 'agency consensus on multifunctionality', a *combination* between state- and grassroots-led approaches appears to offer the best solution, in particular as a much wider range of actors is now involved in interpreting, producing and consuming the multifunctional countryside (Potter, 2004). This is important regarding the definition of the shape and direction of *policy corridors* mentioned above. Thus, identification of external boundaries of policy corridors and the general direction of this corridor should not be left to state-related actors alone, but should also include non-state actors. In return, grassroots actors should not be the only ones defining transitional trajectories for their localities, but should coordinate their actions and aspirations with the needs of wider society. This means that the *governance* of multifunctional transitional pathways should be inclusive, deliberative and open, involving as many actors and stakeholders in society as possible at any stage of the transitional process. Pretty (1995) emphasised the importance of 'trust' in such transitional processes, both within rural communities and between different segments of society. Thus, for research on agricultural processes, Pretty (1995: 180) argued that "the problem with agricultural science is that it has poorly understood the nature of 'indigenous' and rural people's knowledge". This means that local (or relocalised) processes are still often associated by many in the scientific community with notions of 'primitive' and 'unscientific'.

Agricultural multifunctional transitions are not the only arena where *inclusive governance* and *localisation* of transitional processes are currently called for. Similar calls have, for example, been made by the anti-globalisation movement or by those advocating a transition towards more sustainable environmental management (Wilson and Bryant, 1997). There are also interesting parallels with current and highly pertinent debates on societal transitions away from the dependence on oil as a source of energy (and, in a wider sense, human dependence on carbon-based fuels). Proponents of the 'peak oil' scenario, who argue that global production of oil will soon dramatically decline, also strongly advocate relocalisation processes (i.e. 'small is beautiful'; cf. Schumacher, 1973) as a possible solution for the impending energy (and climate) catastrophe (Shuman, 2000; Heinberg, 2004; Meadows *et al.*, 2004). There is a similar debate here about the respective roles of autochthonous (grassroots-led) and exogenous (state-policy led) policies for the implementation of localised transitional energy-descent strategies, although most critics of spatial reductionism suggest a *mixture* of grassroots- and state policy-led actions (Wilson and Bryant, 1997; O'Riordan, 2004). Crucially, the debates also highlight that the definition of the shape and boundaries of energy-descent transitional corridors should be in the hands of *both* grassroots and state-level actors.

This discussion highlights that a new role may be needed for *state policy*, if it is to facilitate the transition towards both strong(er) multifunctional agriculture in developed countries and the maintenance of strong multifunctionality in developing countries (Bryden, 2005). As Marsden (2003: 190) argued, "now ... is the time for a redefined form of macro thinking based upon the encouragement of diversity and scope of the rural economy, and a re-engaged role for the state in its support". Freshwater (2002) also emphasised that impediments for the implementation of strongly multifunctional policies are often linked to *free market ideologies* which are linked to a 'rolling back' of the state, where interference by government in management decisions of private firms (e.g. farm businesses) is seen as undesirable. Such challenges are also apparent for policy frameworks that operate beyond the boundaries of the state such as CAP and WTO-related policy negotiations (Potter and Burney, 2002). Ideally, and as advocates of ecological modernisation tend to argue (e.g. Marsden, 2004), policy should help in *guiding* individual actions within agreed transitional corridors, and should act as a *facilitator* of strong multifunctionality pathways at different spatial scales of the nested hierarchies of multifunctionality (see Fig. 9.7), rather than as a strongly regulatory framework that, through its lack of inclusivity, has often engendered weak(er) multifunctionality (Freshwater, 2002; McCarthy, 2005).

As the discussion in Section 10.2 highlighted, a *highly differentiated* policy approach for multifunctional agricultural transition management will, therefore, be needed (Freshwater, 2002; Knickel *et al.*, 2004). In some societies, more grassroots-led initiatives may continue to work well (e.g. Switzerland), while in others more regulatory approaches may be needed (Pretty, 1995). The latter will be particularly important if there is evidence of autochthonous state-independent processes towards weaker multifunctionality trajectories. This may, for example, be the case in some Mediterranean countries (Wilson and Juntti, 2005), in some developing countries that are at the cusp of full integration into the global capitalist market (e.g. China, Malaysia, Brazil) with concurrent threatened loss of multifunctional quality (Roux *et al.*, 2004), or in Eastern Europe where commentators have highlighted the inability of the state to control societal transitional processes (Alanen, 1999; Gatzweiler, 2004). More regulatory approaches may also be important in countries with weak governance structures (e.g. current Iraq), or where specific elites control most of the local decision-making powers often in close symbiosis with 'governmental' interests (e.g. some South American or African countries). Inevitably, the solution for identification of the best strategy for implementation of strongly multifunctional agricultural trajectories is in the hands of individual nation states or, depending on political structures, on individual regions (Clark, 2003; Bresciani *et al.*, 2004).

Challenges for the transition towards strong multifunctionality

If we assume that a balance can be struck between state-led orchestration of the transition towards strong multifunctionality and grassroots action helping to define and shape the boundaries of multifunctional transitional corridors, what are the remaining challenges in the quest towards a strongly multifunctional global agricultural regime? Let me return to the *transitional fallacies* identified in Chapters 4 and 7 that formed the basis for our conceptualisation of a multifunctional agricultural spectrum. First, throughout this book I have highlighted the inherent **temporal non-linearity** of agricultural transitional processes. This also needs to be taken into account in transition management, as different temporal pathways will characterise the transition towards strong multifunctionality. I have highlighted above that some of the dimensions of strong multifunctionality (e.g. environmental sustainability, extensification, or the shortening of food chains) could be achieved in relatively short time-spans. Indeed, as Chapters 7 and 10 have highlighted, processes of deagrarianisation, localisation and re-embeddedness are already occurring in many parts of the world. However, the *mental* dimensions of change towards strong multifunctionality thinking will take much longer. Most societies around the globe have, over the past decades, mentally adjusted towards globalised agro-commodity processes. It is likely that mental changes will take more than a generation (possibly even two or more) to readjust towards accepting that weak multifunctionality pathways are *not* the norm.

For *farmers'* mental approaches towards multifunctionality, meanwhile, it is difficult to predict if and how seemingly entrenched productivist self-concepts are likely to change. As Chapters 5 and 6 highlighted, farmers have been under increasing pressure both from a general decline in the fortunes of agriculture and from the substantial effort made by European policy-makers to encourage farmers to adopt new non-farming roles and to abandon production-oriented agriculture (Woods, 2005; Burton and Wilson, 2006). However, because the symbolic beliefs through which identity is maintained are not readily susceptible to revision, identities, as opposed to attitudes, change only very slowly and are, therefore, renowned for their "durability and permanence" (Harvey, 1993: 59). Burton (2004: 211) emphasised how substantial and difficult such shifts will be for farmers, and that if and when farmers have become accustomed to such new strongly multifunctional roles, "they will develop new symbolic meanings behind the 'new' behaviours. For example, learning how to judge when woodland is well managed as a crop, judging a farmer's ability by the number of bird species found on his/her land rather than by the uniformity of the landscape, or even, simply losing the prejudice that diversified farmers represent failed farmers". It can, therefore, be assumed that for farmers to change their predominant self-perception of 'agricultural producer' will require substantial and long-term changes. Thus, "a [non]-productivist farming identity that moves away from contemporary perceptions of farmers' singular role as food producers, and that incorporates farmers' multiple roles as food producers, environmental managers and as producers of 'consumption space' for non-agricultural activities ... remains, therefore, a hypothetical goal for the future rather than a contemporary reality" (Burton and Wilson, 2006: 111).

Ultimately, this will mean that societies in advanced economies will have to change their thousand-year-old notions of both what the term 'farmers' means and what 'farming' and 'agriculture' are about (Pretty, 2002). There are some signs that future agricultural policy will lead to an increasingly differentiated countryside, and it is here that new conceptualisations of the 'good farmer' are likely to emerge. Marsden (2003: 9-10) rightly argued that "the problems of rural development ... have not and cannot be solved by a focus upon the agricultural alone". Argyris and Schon's (1978) notion of 'organisational learning' may be most pertinent here, which suggests that moving towards an acceptance of strong multifunctionality pathways by society will only be possible if there is evidence of changes to the *cognitive* and *normative* propositions held by individuals or stakeholder groups. Durand and Van Huylenbroek (2003: 13), therefore, argued that the implementation of stronger multifunctionality "will probably require not only new instruments but also new institutions and judicial interpretations that recognize farming as a multifunctional activity". Clark's (2003) recent analysis of multifunctionality and the 'learning region' also highlighted which processes would need to be in place for implementation of strong multifunctionality ideas at the regional level (see also Benz and Furst, 2002). Yet, adoption of strong multifunctionality pathways by farmers will also demand new skills and knowledge bases for farmers (Curry and Winter, 2000; Winter, 2003) – knowledge that may take more than one farming generation to be incorporated into farm-based management structures. These studies also suggest that substantial shifts towards non-productivist thinking may be expected in future, as new generations of farmers are brought up in a farming environment already more solidly embedded in strongly multifunctional action and thought. However, there is an inherent danger that many farm families may not adapt to such changes and simply choose to leave agriculture and, thus, "within one generation, decades or even centuries of experience, knowledge and local history could simply be lost" (Burton, 2004: 211).

Second, Parts 2 and 3 of this book have highlighted the **spatially heterogeneous** nature of agricultural transitions. I suggested that there still are differences between transitional pathways in developing and developed countries (albeit with increasing similarities in transitional pressures), and that *one* approach towards strong multifunctionality is unlikely to be successful to address transitional issues in highly geographically, culturally, economically and historically different areas. We have also seen that, even within small geographical territories, large differences in multifunctional transitions can be observed (see discussion of nested hierarchies of multifunctionality in Section 9.5). Inevitably, different understandings and interpretations of multifunctional 'quality' will also exist in different geographical contexts, based on differing world views and differing socio-economic, political and cultural processes underlying agricultural and rural change. This means that *geographically flexible* solutions, policies and pathways will need to be found in transitions towards strong multifunctionality. Each region, and possibly even each farming community, will have to contribute towards its own strategies for tackling both multifunctional quality decline and rediscovery of strong multifunctional trajectories (Clark, 2003). As highlighted above, state-led policy should act as a *facilitator*, rather than as an orchestrator, in these processes.

Third, our discussion above about who should be in charge of managing the transition towards strong multifunctionality interlinks with debates on **structure-agency inconsistencies** (see Chapters 4 and 7) that characterise agricultural transitions. Each region and agricultural community will have different governance structures, with some more inclusive and others more hierarchical. The solution for finding the best pathway towards strong multifunctionality will be to accept that different governance structures exist, and that not one specific transitional strategy can be developed that would suit all multi-layered actor spaces and power structures. The first step in the implementation of strong multifunctionality, therefore, is to accept and use existing governance structures to *maximise the potential* for strong multifunctionality pathways. In a second (future) step, one may then conceive of scenarios where transitional changes towards strong multifunctionality themselves may engender changes in governance structures by empowering actors who currently have little influence in changing the direction of transitional corridors. Clark (2003) emphasised that further encouragement of strong multifunctionality pathways in the UK, for example, will have to be predicated on strengthening and clarifying the role of *sub-national* actors and organisations (see also Marsden and Sonnino, 2005), although in other countries strengthening this governance level may not be the most appropriate strategy (e.g. many developing countries). As highlighted in both Chapters 7 and 9, there are some encouraging signs that this may already be happening in some agricultural regions (Wilson and Juntti, 2005). Yet, as Burton (2004: 211) rightly emphasised, resolving these issues will not be a simple task, and "the primary requirement … is to show greater sensitivity and acknowledgement that changing the role of the farmer does not simply involve structural change, but also a change in the basic social fabric of the community". Pretty (1995), therefore, called for a transition from an 'old' to a 'new professionalism' in which agriculture as a 'profession' will not only have to be rediscovered, but also re-evaluated, reconfigured and, ultimately, remoulded according to strong multifunctionality attributes – or, as Burton and Wilson (2006) emphasised, in which the notion of a 'successful' farmer needs to be recast to also include non-productivist activities.

Additional challenges may jeopardise the possibility for societies to embark on strong(er) multifunctionality pathways in the future. While one no longer needs to invoke the doom-and-gloom population 'explosion' scenarios of the 1960s and 1970s, arguably the most immediate challenge facing agriculture is that of *global warming*. Climate change may, in some regions at least, severely hamper efforts aimed at implementing stronger multifunctionality pathways, even if all stakeholder groups are in favour of such pathways (Olesen and Bindi, 2002; Lang and Heasman, 2004). Indeed, in semi-arid regions, such as the Australian outback or the Sahel zone in Africa, some agricultural activities are already so marginal that they may not be able to survive much longer (Holmes, 2002, 2006; Cocklin and Dibden, 2004). In these cases, strong multifunctionality may become a secondary issue, and farm survival – or possibly moving 'beyond agriculture' (see Fig. 9.3) through complete farm abandonment – may be the only solution. Yet, societal recognition of the problem of global warming may also reinforce strong multifunctionality pathways due to the associated need for shorter agro-commodity chains based on energy-descent scenarios and low carbon emissions, although increased planting of biofuels associated with these processes may contribute towards greater intensification of agriculture with concurrent loss in multifunctional quality (Knickel and Renting, 2000). Simultaneously, global warming may also increasingly threaten global food production through worsening desertification and salinisation (Wilson and Juntti, 2005), with possible concurrent needs to increase agricultural intensity in parts of the world less affected by climate change. However, considering that currently about 60% of food produced globally is wasted (lost along food supply chains through transport, storage, etc.; large amounts of crops fed to livestock for production of meat), there still is *large flexibility* in global agricultural systems which, in turn, may help counter-balance the effects of global warming and allow a continued transition towards strong(er) multifunctionality. Indeed, Pretty (1995) highlighted how farming systems based on strong multifunctionality require less energy (especially fossil fuels) than weakly multifunctional systems. Thus, the future may just mean going back to what humanity has (relatively successfully) done for thousands of years of moderate/strong multifunctionality, but without losing sight of positive lessons that have emerged from productivist pathways (e.g. relative food security). The advantage that current societies may have for a future strongly multifunctional agriculture is that the in-built memory of the global agricultural system should ensure that future strong multifunctionality pathways will be qualitatively, economically and socio-politically *different* – and, therefore, possibly more sustainable – from past experiences.

However, challenges for the transition towards strong multifunctionality do not necessarily need to be global in nature. The repeatedly mentioned 'farm succession crisis' in most advanced economies (but also in the developing world) – where the younger generation is increasingly unwilling to take over farms from their parents – may be an equally great challenge for strong multifunctionality (Meert *et al.*, 2005). Marsden (2003: 78) emphasised that "if conditions are left unheeded we will witness the further marginalisation of small family farms, not least from the progressive out-migration of young farm family members to other occupations and lifestyles". Indeed, the notion of strong multifunctionality is predicated on the *survival* of agricultural communities in which some food and fibre production is still taking place. Although one can envisage scenarios in which entire regions formerly involved in agriculture may move 'beyond agriculture' with associated positive (e.g. some environmental benefits) and negative (e.g. loss of agriculture-community interactions and social networks) outcomes, the issue of multifunctional agriculture becomes meaningless if agriculture *disappears* from an area altogether. Who will manage strong multifunctionality pathways in the future if there are no farmers left? Some respite may be offered by the rapidly increasing numbers of lifestyle farmers in many advanced economies who may, at least partly, reverse possible multifunctional quality loss (but see Chapter 9). As Chapter 5 highlighted, the farm succession crisis and associated agricultural structural change often leads to farm amalgamation with many farms becoming larger, more capitalised, more technologically heavy and – ultimately – more productivist. Pretty (2002) highlighted how in the USA over the past 50 years four million farms have disappeared, equivalent to the loss of over 200 farms every day (see also Wilson and Wilson, 2001, for similar figures for Germany). As Pretty (2002: 107) argued, each of the farm families that leaves the land for good "used to have a close connection with the land, and to other farms in their communities. When they are disconnected, the memories are lost forever" – system memory often linked to strong multifunctionality pathways that may, therefore, be irreversibly lost. Thus, although the *theoretical* aspects of strong multifunctional transition management may be (at least partly) in place, the *practicalities* of implementing this transition are far from clear. It may well be that, in future, the implementation of strong multifunctionality pathways may become more difficult for many farmers on the globe.

10. 5 Conclusions

Building on conceptualisations of multifunctionality in Chapter 9, the focus of this chapter was on understanding multifunctional agricultural transitions. In Section 10.2, I highlighted that it is possible to categorise different farm types along the p/np multifunctionality spectrum, and that transitional potential from weak to strong multifunctionality differs between different categories of farms and between different types of farm ownership. Section 10.2 also showed that we can use the multifunctionality spectrum to understand the transitional potential of individual countries and that – depending on political, economic, social and cultural factors – different nation states show highly varied potential for the implementation of strong multifunctionality pathways. Based on conceptual issues surrounding transitional processes discussed in Part 1, Section 10.3 then focused on multifunctional transitional processes at farm level over time. Here, I introduced the notions of multifunctional *path dependency* and decision-making *corridors*, the latter of which can be understood as 'bundles' of decision-making opportunities bounded by productivist and non-productivist action and thought. I also suggested that *system memory* plays an important role in defining the likelihood of multifunctional actions and argued that *transitional ruptures* – sudden breaks in transitional pathways akin to the notion of punctuated equilibriums in evolutionary transition theory – often characterise farm-level transitions. I used two hypothetical examples to illustrate some of these on-farm multifunctional transitional processes, one from a developed and the other from a developing country. These examples illustrated the complexity, uncertainty and unpredictability of multifunctional transitions, and highlighted the utility of the multifunctionality spectrum developed in Chapter 9 for understanding agricultural transitional processes anywhere in the world.

The final section then broadened the discussion towards past, present and future transitional processes. I focused on differences in the shape and direction of transitional corridors between the developing and developed worlds, but also highlighted the increasing similarities and transitional pressures acting upon all farmers through processes of globalisation, uniformisation of agricultural production and knowledge exchange. I used the example of multifunctional agricultural transitions in Western Europe to illustrate how shifts in the shape and direction of multifunctionality pathways can be better understood, and highlighted the importance of stable economic and political phases for strong(er) multifunctionality tendencies, while catastrophic changes usually tend to shift agricultural pathways towards weak(er) multifunctionality pathways. Linked to discussions in Chapter 7, I discussed the notion of the post-war 'productivist transitional trough' and argued that there is yet insufficient evidence to suggest that societies are firmly moving towards non-productivist multifunctional transitional pathways. I concluded by arguing that theoretical work on transition management may help better target state policy and grassroots action towards stronger multifunctionality, and highlighted the complexity of multifunctional issues facing future decision-makers.

The aim of Chapter 10 was not only to provide a dispassionate and objective assessment of multifunctionality transitions. Throughout Chapters 9 and 10 I argued that, based on a normative approach, we also need to make subjective value judgements about the 'best way forward' for multifunctional transitions. I argued that the strong multifunctionality model is the *qualitatively* best and most *moral* model based on *synergistic mutual benefits* – a model that all societies should strive to *rediscover* (in the case of most developed countries) or to *maintain* (in the case of most developing countries). I used the new conceptualisation of multifunctionality as a framework for criticising the productivist agri-business model that continues to dominate agriculture in advanced economies and that increasingly influences agricultural practices in the developing world, against the increasing globalisation of agro-commodity chains that tends to weaken global multifunctionality and, concurrently as a *plaidoyer* for *relocalised* strong multifunctionality that may, in turn, help re-empower local actors to take more control of multifunctional transitional corridors. Strong multifunctionality should, therefore, not only be seen as an *oppositional* process to continuing productivist tendencies, but as a true *alternative* agricultural regime to the often dominant weak multifunctionality paradigm. Finally, I outlined key barriers for global-level implementation of strong multifunctionality, including, in particular, governance-related issues (who should orchestrate the multifunctional transition?), issues of temporal non-linearity and spatial heterogeneity of transitional processes, the farm succession crisis, and exogenous factors such as climate change. Re-conceptualising the notion of 'multifunctional agriculture' is, therefore, one thing, but it is quite another challenge to operationalise and implement a strongly multifunctional agricultural regime in practice.

Chapter 11

Conclusions

Chapter 11 highlights the key arguments of the reconceptualisation of multifunctionality presented in this book, and briefly discusses how the new conceptualisation of multifunctionality may act as a platform for future research. I will also 'throw down the gauntlet' to future researchers in the hope that theoretical and conceptual ideas presented here will be challenged and, as a result, provide the basis for future critical, and theoretically well informed, debates on agricultural transition.

11.1 What this book has attempted to do

I began this book with the premise that there are two key arenas of concern that have influenced recent research and thinking on conceptualisations of agricultural and rural change. The first was the uncritical and weakly theorised use of 'multifunctionality' in contemporary debates on agricultural change, and that none of the existing debates have shed sufficient light on *what* the notion of multifunctionality implies, *who* the beneficiaries should be, and *how* it ought to be implemented into practice. My second concern was linked to a growing dissatisfaction with the proposed p/pp transition, in particular the notion that agriculture *has* moved from a productivist to a post-productivist era, and that we are now firmly embedded in this new 'post-ism'.

This book has brought these two seemingly separate arenas of concern together by (re)conceptualising agricultural change from a *transition theory* perspective. It also highlighted the difficulties implicit in the attempt to develop a general and comprehensive framework for understanding multifunctional agriculture. The structure of this book into three distinctive analytical steps (Parts 1-3) mirrored the theoretical orientation based around transition theory. This theory was particularly used to inform two analytical processes. First, it helped us deconstruct the notion of a linear p/pp transition, and to argue that, instead, a revised model based on a spectrum of decision-making bounded by productivist and *non-*productivist pathways better encapsulates the temporal non-linearity, spatial heterogeneity, global complexity and structure-agency inconsistency that characterises agricultural transitions. Second, transition theory enabled us to theoretically anchor the notion of multifunctionality as a *concept* characterised by Deleuzian transitional pathways bounded by productivist and non-productivist decision-making trajectories, and, simultaneously, informed our analysis of the dynamics, pace and nature of the transition towards (or away from) a strongly multifunctional agricultural regime. This book, therefore, has particularly challenged existing understandings of multifunctional agriculture based on structuralist approaches that see multifunctionality largely as an economic or policy-based *process*. The book has suggested, instead, that multifunctionality should be understood as a *normative concept* that both describes *and* explains contemporary agricultural change. The step-by-step approach in three distinctive parts, therefore, aimed at highlighting issues, concepts and problems surrounding current conceptualisations of agricultural change through the lens of transition theory.

Part 1 provided the theoretical background for conceptualising transition by introducing 'transition theory'. I began with an overview of what 'transition theory' means and discussed key models of transition from different disciplinary vantage points. I then analysed key debates and theories surrounding the notion of 'transition'. Specific emphasis was placed on analysing different models of transition, and highlighting key debates on possible transitions from 'isms' to 'post-isms'. I concluded by arguing that current debates on the transition to post-productivism in agriculture are likely to share many similarities and problems regarding assumptions about linearity, homogeneity, universality and causality implied in debates about 'other' transitions. This formed the basis for discussion of the postulated transition to post-productivist agriculture in Part 2. Viewing the p/pp model through the lens of transition theory allowed us to *deconstruct* the unilinear assumptions underlying many of the debates surrounding this transition. First, I analysed issues surrounding conceptualisations of 'productivist' and 'post-productivist' agriculture, with particular attention to several dimensions that have formed the basis of these conceptualisations. I then linked this discussion to parallel debates on 'other' transitions discussed in Part 1 and outlined the strengths and weaknesses of the p/pp transition model. I placed particular emphasis on investigating assumptions of linearity, homogeneity, universality and causality underlying the p/pp model. Although I acknowledged that the p/pp transition model has made a vital contribution to current debates, I also suggested that 'post-productivism' should not be seen as the 'end-point' of contemporary agricultural change. Instead, I suggested that a modified transitional model based on the notion of a productivist/*non*-productivist *spectrum* of decision-making is a more appropriate non-linear concept of agricultural change that also enables acknowledgement of both spatially heterogeneous processes and structure-agency inconsistencies. These discussions enabled us to *theoretically anchor* the notion of 'multifunctional' agriculture within the productivist/non-productivist spectrum of decision-making (Part 3). The key argument was that, based on the critique of linear and directional assumptions underlying the post-productivist transition model, the notion of a multifunctional decision-making spectrum bounded by productivist and non-productivist action and thought better encapsulates the complexity that can be observed in agricultural systems.

I then critically examined contemporary conceptualisations of 'multifunctionality' and argued that current notions of the concept have suffered from weak theorisations and relatively narrow conceptualisations. I theoretically anchored the notion of multifunctionality in the context of the productivist/non-productivist decision-making spectrum which, in turn, enabled us to investigate the conceptual boundaries of the multifunctional *agricultural* and *rural* regimes. I also suggested that discussions on multifunctional agriculture have to move away from structuralist theory towards *normative* interpretations which imply using *subjective value judgements*. Based on the conceptualisation of weak, moderate and strong multifunctionality pathways, I argued that the *strong multifunctionality* model is qualitatively and morally the 'best' pathway. As a reaction to weakly conceptualised spatial issues regarding multifunctionality in the current literature, I placed specific emphasis on investigating the spatiality of multifunctionality by analysing at what different scales the notion of multifunctionality should apply. I then investigated issues linked to multifunctional agricultural *transitions over time*. Here, the explicit focus was on *transitional* processes with strong interlinkages to Part 1. I analysed the *transitional potential* of agricultural stakeholders at various scales, and investigated the key issues of path dependency, transitional corridors, system memory and transitional ruptures for the better understanding of multifunctional transitional processes in both the developed and developing world. The discussion focused on *how* the strong multifunctionality model could be implemented, by *whom* this transition should be orchestrated, and highlighted various challenges in the quest towards strongly multifunctional transitional pathways.

The notion of 'multifunctionality', theoretically anchored in a revised p/np spectrum of decision-making, provides a robust conceptual and analytical framework within which to understand and analyse contemporary agricultural change anywhere in the world. The notion of strong multifunctionality, in particular, gives the concept of multifunctional agriculture a framework that, in Delgado *et al.*'s (2003: 28) words will be "politically possible, socially suitable, and economically efficient". It more fully encapsulates agricultural transitions than other recently suggested terminologies such as 'heterogeneous agriculture' (Van der Ploeg, 1990), 'post-rural' (Murdoch and Pratt, 1993), 'post-agricultural' (Marsden *et al.*, 1993), 'differentiated rural spaces' or the 'differentiated countryside' (Lowe *et al.*, 1993; Marsden, 1999b), 'new rural spaces' (Halfacree, 1999; Marsden, 2003), 'sustainable agricultural modernisation' (Marsden, 1999a), or 'neo-productivism' (Evans *et al.*, 2002; Marsden and Sonnino, 2005). I hope, therefore, that by linking these different arenas of investigation together, the revised notion of *a multifunctionality spectrum* suggested here begins to make sense conceptually and theoretically, resulting in a more robust and tangible concept that can be used by decision-makers at various scales to protect, shape and change contemporary agricultural and rural spaces.

11.2 How this book can serve as a platform for future research

This book has been about ideas, debates and conceptual approaches, rather than stating 'facts'. As a result, many subject areas were only briefly mentioned, some arguments had to be left unfinished, and some key points may not have been as fully explained as could be. This leaves many opportunities for future researchers who may want to use arguments made here as a platform for future research. In the following, I will briefly highlight key areas of research linked to multifunctional agriculture that, in my view, would be particularly interesting and challenging, and that would help further refine conceptual and theoretical assumptions underlying arguments made here.

First, this book has provided a theoretical and conceptual analysis of a new model of multifunctionality based on a spectrum of Deleuzian decision-making pathways bounded by productivist and non-productivist actor spaces. Although this argument was partly linked to existing empirical work on agricultural change around the world (the bibliography is testimony to how much work is already available on the issue), what still needs to be done is to find additional *empirical* evidence to further substantiate (or, indeed, refute) the new model of multifunctionality suggested here. Researchers from different disciplinary vantage points now need to: 'swarm out'; get their 'boots dirty'; talk to farmers, rural citizens, policy-makers and other agricultural and non-agricultural decision-makers about multifunctionality-related issues; establish case studies specifically investigating the nature, pace and processes of agricultural multifunctional transitions; and investigate if, and to what extent, the new multifunctionality spectrum suggested here provides a robust framework with which to analyse, and understand, agricultural change in any part of the world. Indeed, McCarthy (2005: 780) highlighted that "we are in urgent need of ethnographies of multifunctional rural areas", while Freshwater (2002) emphasised that the need for more empirical work on multifunctionality is likely to increase as the share of household income spent on food declines with concurrent increase in the importance of social and cultural values associated with agricultural production.

This will need to be associated with a *refinement of methodologies* how to assess multifunctional 'quality', bearing in mind that the use of absolute indicators is most likely *not* the right way forward (see Chapter 9). Thus, I partly disagree with Bryden's (2005: 9; emphasis added) recent assertion that "the degree of co-production [multifunctionality] involved in the production of different commodities by different farm types and farming styles *must be capable of measurement*". Although initial attempts have been made at developing such methodologies (e.g. Holmes', 2006, suggestion of an 'index of multifunctionality' based on production, consumption and protection processes of rural areas; Batty *et al.*'s, 2004, attempt to map different levels of multifunctionality in urban spaces), Knickel and Renting (2000) rightly cautioned that most of the data on agricultural activities and change is still in the form of relatively *productivist* and *positivist* statistics relating to production, while little 'data' exist on less tangible multifunctionality indicators such as *mental* changes or changes in the relative globalised position of agricultural spaces. Further, aggregate data and regional averages may hide significant *individual* multifunctional transitional pathways within a region (Knickel *et al.*, 2004). Randall (2002: 305) attempted to develop a comprehensive methodology, but argued that researchers "have seldom if ever attempted a task so demanding as valuing the outputs of multifunctional agriculture ... Consistency as we move from single to multiple components of multifunctionality, and from local to continental spatial scales [see Section 9.5], is a substantial conceptual and empirical challenge". Andersen *et al.* (2004: 105) further highlighted that "adequate and consistent data and indicators are still not available to monitor the effects of policy changes in the field of multifunctionality and agriculture". Assessment of multifunctional quality will, therefore, only be possible through the use of more *qualitative* and *ethnographic* methods that will enable researchers to engage more closely with individual multifunctional life histories, transitions and development pathways as a basis for developing a *multifunctional farm/rural area typology* based on the multifunctionality spectrum suggested here.

Nonetheless, there are some useful attempts at refining methodologies how to assess multifunction quality, in particular for obtaining data on less tangible qualitative multifunctionality indicators. For example, the study by Bohnet *et al.* (2003), assessing multifunctional trajectories of family farms in a case study in south-east England, offers useful guidance. They emphasised the importance of exploring *life histories* and *biographies* of individual rural actors through conversation and semi-structured interviews (both retrospective and prospective in scope) based on a relatively small sample (i.e. micro-sociological case study analysis). The study by French researchers Cayre *et al.* (2004), who used photographs and ethnographic approaches through an in-depth qualitative survey to elicit reactions by landholders on perceived multifunctional quality of rural landscapes, also offers valuable methodological insights, as does Rapey *et al.*'s (2004b) study that suggested field-parcel-based evaluation of *existing*, *desired* and *potential* multifunctionality functions. Clark's (2005) study of localised multifunctionality networks in a case study in the UK also offers useful guidance, in particular for how multifunctional 'networks' and associated power relations can be identified (see also Wilson and Juntti, 2005, for a similar methodology based on four Mediterranean case studies). Pretty *et al.* (2001), meanwhile, discussed how the 'positive externalities' linked to strong multifunctionality can be assessed methodologically (see also Knickel and Renting, 2000; Knickel *et al.*, 2004), while Andersen *et al.* (2004) provided a typology of multifunctionality for European livestock systems. Hollander's (2004) detailed investigation of the multifunctionality of the sugarcane industry in Florida also emphasised the utility of *ethnographic fieldwork* in unravelling complex multifunctional agricultural pathways. Bell's (2004) detailed socio-ethnographic study of sustainability pathways among Iowa farmers (USA) in my view has succinctly highlighted the qualitative methodological challenges that lie ahead: "A different sustainable [strongly multifunctional] farmer, a different story. A different conventional [weakly multifunctional?] farmer, a different story. Herein lies the humanity of every farmer. There are common features in many of their life stories, in the history of their life conversations ... economic stress, feelings of ill health, doubts about truth, and spiritual doubts, leading to phenomenological rupture and the recognition of social interests and tuning in to a new cultivation of knowledge about the economy and the self – about the structures and practices of the self. But these are not matters for equations, computer programs, and robotic certainties. In this there should be no sociological lament. Quite the contrary; it is cause for celebration. That the outcome of a [farmers' multifunctionality pathways] cannot be predicted is what gives a person her or his unalienable aliveness".

These studies undoubtedly form useful platforms from which to develop more refined methodologies for the analysis of multifunctional 'quality' in future work. Yet, more *empirical research* is needed to substantiate the theoretical and conceptual issues addressed in this book. This was emphasised by Clark (2003: 303) for the UK where currently "the promotion of multifunctionality ... is still at an early stage, and any assessment of the suitability of current strategies must be regarded as provisional". In particular, most information on multifunctionality is at present almost exclusively related to agricultural activities, highlighting that most data currently available do not directly *measure* multifunctionality, but only provide *proxy indicators* of the influence of agricultural activities on the physical, economic and socio-cultural environment (Fry, 2001; Andersen *et al.*, 2004). This is particularly true for non-tangible multifunctionality indicators such as mental changes among actor groups towards weak or strong multifunctionality, 'depth' of diversification activities, or indicators assessing the viability and sustainability of rural communities (Wilson and Buller, 2001). However, future researchers attempting to analyse the holism of multifunctionality will need to avoid the risk of sliding towards empiricism. As past experience has shown, any empirical study – a student dissertation, a PhD, or research by academic staff – will need to continuously interlink empirical work with the wider conceptual and theoretical debates linked to multifunctionality outlined in this book and elsewhere. As Hoggart and Paniagua (2001) cautioned, *non-events* should be as important as events in understanding the character and comprehensiveness of the multifunctionality spectrum. Only through such theoretically grounded empirical work will major contributions towards refinement of the multifunctionality concept suggested here be possible.

Second, more work is also waiting at the conceptual level. As both Chapters 1 and 9 highlighted, conceptualisations of *agricultural* multifunctionality will always need to take into account repercussions of such conceptualisations for our understanding of *rural* processes. Although I have addressed boundary issues related to agricultural and rural multifunctionality, many questions are still left unanswered. Assuming we can somehow agree on a common definition of the 'rural' (which may be impossible), can we apply the notion of a multifunctional decision-making spectrum to *rural multifunctionality*? In particular, what ingredients beyond the p/np spectrum may be needed to provide a comprehensive answer about *global* rural multifunctionality pathways? Although some interesting work is beginning to emerge on these questions (e.g. Clark, 2003, 2005; Holmes, 2006), conceptual issues linked to the interface between 'agricultural' and 'rural' multifunctionality are far from resolved. Beyond this, some commentators are beginning to call for a new research focus on food chains as opposed to considering agriculture as a purely land-based occupation in isolation (Marsden, 2003). Winter (2005), in particular, highlighted the importance in reconnecting research on food, agriculture and nature. Although we can conceive of eating as an 'agricultural act' (Bell, 2004), I admit that the focus on multifunctional *agriculture* in this book may have led to a shift of emphasis away from debates on food networks – debates that, as Goodman (2004) amply illustrated, also have the potential to greatly inform further refinements of the multifunctionality spectrum. I, therefore, agree with Losch (2004: 356) that "it is imperative that farming be removed from the strict sectoral debate in which it has been confined". Much work is, therefore, awaiting future researchers on the question of how the multifunctionality spectrum may apply to food chains and other issues that go 'beyond agriculture'. Is it, for example, possible to conceptualise food chains along notions of weak, moderate or strong multifunctionality? Can Goodman's (2004) assertion about the possible emergence of a new 'multi-tiered food system' in the developed world (differentiated by income and class) be seen as an emerging expression of differing multifunctional pathways along p/np consumption decision-making trajectories beyond the farm gate?

Third, I would also hope that this book will spark a debate about the dimensions, or constituents, of the multifunctionality spectrum. The various dimensions of weak, moderate and strong multifunctionality outlined here are based on current transitional processes.

However, as Figures 10.9-10.11 suggested, we may be at an important crossroads for multifunctional transitional processes in both the developed and developing world, with some uncertainty about the precise future shape of transitional corridors. As the multifunctional transition proceeds over the next few decades, *new* dimensions of multifunctionality may need to be considered, especially at the interface between strong multifunctionality and non-productivist pathways of decision-making. An important question, therefore, is whether the boundary of non-productivist action and thought should be seen as relatively clearly defined, or whether we need even more flexible indicators about what 'agriculture' is? It is likely that discussions about the *boundaries of agriculture* will assume ever greater importance in future, as strongly multifunctional processes of deagrarianisation, diversification and rural gentrification assume greater importance.

Fourth, some of the theoretical assumptions underlying concepts of multifunctionality in general also need further scrutiny. Chapters 3 and 8 highlighted that the notion of multifunctionality goes well beyond agriculture and that the concept (with different conceptual meanings) has also been applied to forestry (where it arguably was first conceptualised) or medicine, among many other fields of enquiry. Although transition theory has given us glimpses into other 'parallel' debates on societal and natural transitions and how these can inform conceptualisations of *agricultural* multifunctionality, the inverse should also be interesting. How can the concept of a multifunctional agricultural spectrum inform conceptualisations of non-agricultural multifunctional pathways (e.g. in forestry or medicine)? I hope, therefore, that researchers interested in broader questions of transition may also find the discussion in this book interesting, as it may aid them to rethink and reconceptualise notions of multifunctionality pertinent in *their fields*. I, therefore, would also like to see this book as a call for more *multi-disciplinary* investigations of the notion of transition in general and multifunctionality in particular.

Fifth, as a human geographer I argue that more work is needed to investigate *spatial issues* of the new multifunctionality concept. A conceptual framework for future analysis has been suggested in Section 9.5, based on nested hierarchies of multifunctionality starting at the spatial level of the farm territory. Future work could focus on the *interrelationships* between the different spatial scales, but also on the complex question of *global-level* multifunctionality. Linked to this is the interesting question (briefly discussed in Chapter 10) whether implementation of strong multifunctionality at the global level can be based on a win-win situation, in which all agricultural territories may be able to adopt strongly multifunctional pathways, or whether it is more akin to a zero-sum-game in which strong multifunctionality in one area is predicated on weak multifunctionality in another? There are, therefore, many fruitful arenas of investigation for those interested in the *geography of multifunctionality*. Particularly challenging questions regarding multifunctionality await research on developing countries where, so far, little work exists on the nature and pace of multifunctional agricultural transitions.

Finally, the notion of multifunctional agriculture also has implications for our construction of knowledge, in particular related to agricultural sciences, rural studies and cognate sub-disciplines such as human geography. In Chapter 8 we saw that approaching multifunctionality from a mono-dimensional and mono-causal perspective (e.g. economic or policy-based perspective) is likely to generate simplistic evaluations of, and solutions for, the challenge of multifunctionality. Only through a multi-disciplinary approach will we be able to fully understand multifunctional agriculture and drive forward constructive agendas for the future. As a strong multifunctional agricultural regime means a relative withdrawal of productivist agriculture, it is evident that 'classical' – often technocentric – agricultural science approaches towards understanding agricultural change may be less relevant in future. Other disciplinary approaches rooted, for example, in rural studies, sociology, psychology, environmental sciences or human geography may take on a more important role in the investigation of future 'agricultural' development pathways that straddle the non-productivist end of the multifunctionality spectrum than has hitherto been the case. As Schakel (2003:

227) emphasised in the context of what he termed a 'post-classical' or even 'post-modern' agricultural science, "multidisciplinary, multifunctional, and integrated knowledge characterizes what may possibly be a new agricultural paradigm". Clark (2003: 203) also emphasised the importance of communication and knowledge transfer for successful implementation of strong multifunctionality, but also conceded "that much still remains to be done to convince agricultural businesses of the … opportunities inherent in multifunctionality". Knickel *et al.* (2004: 101) similarly emphasised that "there is a clear need for more comprehensive multidisciplinary concepts linking the dimensions of agricultural and rural change with the multifunctional character of agriculture". As Chapter 6 highlighted, the recent closure of agricultural colleges in many developed countries can be interpreted as a clear indication that there is now less need for such institutions that are often still embedded in the productivist paradigm and based on outdated productivist teaching structures geared towards training farmers to merely intensify and increase agricultural production. Just as the notion of strong agricultural multifunctionality means a blurring of the traditional boundaries of conventional agriculture, the transition towards strong multifunctionality concurrently necessitates a readjustment in the way academics and scientists will research agricultural and rural issues in the future.

11.3 And finally: throwing down the gauntlet …

This book, therefore, has provided an alternative view of multifunctionality. I have particularly attempted to move beyond simplistic interpretations of multifunctionality by arguing that economic or policy-related drivers of multifunctionality are only a small part of what multifunctionality is, or should be, about. In particular, I have attempted to introduce the notion of multifunctionality as a *transitional* process rather than as a relatively *static* and *compartmentalised* descriptor of agricultural and non-agricultural decision-making at a specific point in time. I have also argued that the notion of multifunctionality should go beyond merely explaining economic and policy-based processes, and that it should also act as a philosophical and theoretically grounded *concept* that both describes and explains agricultural change. In particular, the concept of multifunctionality proposed here challenges the assumption that multifunctionality is a *new* process. While it may be new as an academic or theoretical notion, *all* agricultural systems around the world have *always* depicted some multifunctional tendencies (even the most productivist ones). Thus, this book has gone beyond a simple reconceptualisation of multifunctionality. By arguing that different multifunctionality *quality* exists, I have included *subjective value judgements* about the 'best' and most 'moral' type of multifunctionality all societies should strive for (strong multifunctionality). This has been linked to one of the explicit aims of this book to challenge and criticise processes that are, in my view, currently *weakening* multifunctional quality around the globe, in particular globalisation and the relatively unsustainable productivist agro-industrial dynamic.

Of course, my proposal in Chapter 10 for a relocalised, more locally embedded and more 'moral' agriculture based on strong multifunctionality pathways, will not necessarily resonate positively with some commentators on agricultural change. Indeed, the concept of multifunctionality suggested here may be criticised for arguing a relatively naïve and simplistic case, akin to the 1970s call of 'small is beautiful', that does not sufficiently take into account that globalisation processes now may have a life of their own, irrespective of 'external' regulation and orchestration of multifunctional transitional corridors. I, therefore, agree with Marsden (2003: 116) who argued that the call for relocalised agricultural systems "does not necessarily imply a populistic retreat to localism in the face of the ravages of harsh global forces … [It] is through a more sophisticated analysis of the 'local' that a broader comparative analysis of capitalism and globalisation can be built". I also acknowledge Goodman's (2004) cautionary note that 'local' is not always best (especially when we consider power and gender relations and entrenched ideologies), and that relocalisation processes should always be contextualised within wider processes of both globalisation and, indeed, 'glocalisation'.

I accept this challenge and hope that the new concept of multifunctionality suggested in this book, and the associated call for strong(er) multifunctional agricultural pathways, will provide the basis for future critical, and theoretically well informed, debates about agricultural transition that will, eventually, challenge theoretical and conceptual assumptions made in this book. As Delgado (2003: 34) emphasised, I anticipate that "the debate regarding the interrelations between multifunctionality and the development of [agricultural and] rural areas will continue to grow in intensity". I hope, therefore, that this book will not only form an important reference point for readers interested in multifunctionality, productivism/non-productivism and transition, but that it will also act as a *trigger* for future theoretical, conceptual and empirical work on the nature and pace of global agricultural change.

References

Abdel-Fadil, M. 1989: Colonialism. In: Eatwell, J., Milgate, M. and P. Newmans (eds): *Economic development*. Oxford: Blackwell, pp. 61-67.

Abler, D. 2004: Multifunctionality, agricultural policy, and environmental policy. *Agricultural and Resource Economics Review* 33 (1): 8-17.

Adam, B. 1999: Industrial food for thought: timescapes of risk. *Environmental Values* 8: 219-238.

Adams, J. 1995: *Risk*. London: UCL Press.

Adams, W.M., Hodge, I.D. and N.A. Bourn 1994: Nature conservation and the management of the wider countryside in eastern England. *Journal of Rural Studies* 10 (2): 147-157.

Adedeji, A. (ed.) 1993: *Africa within the world: beyond dispossession and dependence*. London: Zed Books.

Ajzen, I. 1991: The theory of planned behaviour. *Organisational Behaviour and Human Decision Processes* 20: 1-33.

Alanen, I. 1999: Agricultural policy and the struggle over the destiny of collective farms in Estonia. *Sociologia Ruralis* 39: 431-458.

Allen, J., Massey, D. and A. Cochrane 1998: *Rethinking the region*. London: Routledge.

Allen, P., Fitzsimmons, M. and K. Warner 2003: Shifting plates in the agrifood landscape: the tectonics of alternative agrifood initiatives in California. *Journal of Rural Studies* 19: 61-75.

Allison, L. 1996: On planning a forest: theoretical issues and practical problems. *Town Planning Review* 67 (2): 131-143.

Altieri, M.A. (ed.) 1987: *Agro-ecology: the scientific basis of alternative agriculture*. Boulder (Col.): Westview Press.

Altieri, M.A. 2000a: Enhancing the productivity and multifunctionality of traditional farming in Latin America. *International Journal of Sustainable Development and World Ecology* 7 (1): 50-61.

Altieri, M.A. 2000b: Multifunctional dimensions of ecologically-based agriculture in Latin America. *International Journal of Sustainable Development and World Ecology* 7 (1): 62-75.

Altieri, M.A. and B. Rosset 1996: Agroecology and the conversion of large-scale conventional systems to sustainable management. *International Journal of Environmental Studies* 50: 165-185.

Altvater, E. 1993: *The future of the market: an essay on the regulation of money and nature after the collapse of 'actually existing socialism'*. London: Verso.

Altvater, E. 1998: Theoretical deliberations on time and space in post-socialist transformation. *Regional Studies* 32 (7): 591-605.

Amin, A. (ed.) 1994: *Post-Fordism: a reader*. Oxford: Blackwell.

Amin, A. and N. Thrift 1993: Globalisation, institutional thickness and local prospects. *Revue d'Economie Régionale et Urbaine* 3: 405-427.

Amin, S. 1990: *Delinking: towards a polycentric world*. London: Zed Books.

Andersen, E., Henningsen, A. and J. Primdahl 2000: Denmark: implementation of new agri-environmental policy based on Regulation 2078. In: Buller, H., Wilson, G.A. and A. Höll (eds): *Agri-environmental policy in the European Union*. Aldershot: Ashgate, pp. 31-50.

Andersen, E., Elbersen, B. and F. Godeschalk 2004: Assessing multifunctionality of European livestock systems. In: Brouwer, F. (ed.): *Sustaining agriculture and the rural environment: governance, policy and multifunctionality*. Cheltenham: Edward Elgar, pp. 104-123.

Anderson, K. 2000: Agriculture's multifunctionality and the WTO. *The Australian Journal of Agricultural and Resource Economics* 44 (3): 475-494.

Aoki, M. 1990: Intrafirm mechanisms, sharing and employment. In: S. Marglin and J. Schor (eds): *The golden age of capitalism*. London: OUP, pp. 167-179.

Appadurai, A. 1999: Globalization and the research imagination. *International Social Science Journal* 160: 229-238.

Argent, N. 2002: From pillar to post? In search of the post-productivist countryside in Australia. *Australian Geographer* 33 (1): 97-114.

Argyris, C. and D. Schon 1978: *Organisational learning: a theory of action perspective*. Reading (Mass.): Addison Wesley.

Atwood, M. 1988: *Cat's eye*. London: Bloomsbury.

Ausnubel, J.H. and H.D. Langford (eds) 1997: *Technological trajectories and the human environment*. Washington (D.C.): National Academy Press.

Bailey, I. and S. Rupp 2005: Geography and climate policy: a comparative assessment of new environmental policy instruments in the UK and Germany. *Geoforum* 36 (3): 387-401.

Bak, P. 1997: *How nature works: the science of self-organised criticality*. Oxford: OUP.

Bak, P. and K. Sneppen 1993: Punctuated equilibrium and criticality in a simple model of evolution. *Physical Review Letters* 71: 4083-4086.

Bak, P., Flyberg, H. and K. Sneppen 1994: Can we model Darwin? *New Scientist* March 12: 36-39.

Baldock, D. and P. Lowe 1996: The development of European agri-environment policy. In: Whitby, M. (ed.): *The European environment and CAP reform: policies and prospects for conservation*. Wallingford: CAB International, pp. 8-25.

Baldock, D., Cox, G., Lowe, P. and M. Winter 1990: Environmentally Sensitive Areas: incrementalism or reform? *Journal of Rural Studies* 6: 143-162.

Barbier, E.B. 1987: The concept of sustainable economic development. *Environmental Conservation* 14: 101-110.

Barkin, D. 1990: *Distorted development: Mexico in the world economy*. Boulder (Col.): Westview Press.

Barnes, T, 1996: *Logics of dislocation*. New York: Guilford.

Barr, S. 2003: Strategies for sustainability: citizens and responsible environmental behaviour. *Area* 35 (3): 227-240.

Barr, S. 2006: Environmental action in the home: investigating the 'value-action' gap. *Geography* 91 (1): 43-54.

Barrett, H.R., Ilbery, B., Browne, A.W. and T. Binns 1999: Globalisation and the changing networks of food supply: the importation of fresh horticultural produce from Kenya into the UK. *Transactions of the Institute of British Geographers* 24 (2): 159-174.

Barthélémy, D. *et al.* 2004: La multifonctionnalité agricole comme relation entre fonctions marchandes et non marchandes. *Les Cahiers de la Multifonctionnalité* 6: 121-130.

Basalla, G. 1988: *The evolution of technology*. Cambridge: CUP.

Battershill, M. and A. Gilg 1996: Traditional farming and agro-environment policy in southwest England: back to the future? *Geoforum* 27 (2): 133-147.

Batty, M., Besussi, E., Maat, K. and J.J. Harts 2004: Representing multifunctional cities: density and diversity in space and time. *Built Environment* 30: 324-349.

Bauer, P.T. 1976: *Dissent on development*. Cambridge (Mass.): Harvard University Press.

Bauman, Z. 1992: *Intimations of postmodernity*. London: Routledge.

Beck, U. 1992: *Risk society: towards a new modernity*. London: Sage.

Becu, N., Barreteau, O., Perez, P., Walker, A. and P. Garin 2004: Multi-fonctionnalité dans les basins versants du Nord Thailande: entre emergence de points de vue hétérogène et pilotage externe. *Les Cahiers de la Multifonctionnalité* 6: 25-30.

Bell, C. and H. Newby 1974: Capitalist farmers in the British class structure. *Sociologia Ruralis* 14: 86-107.

Bell, D. and G. Valentine 1997: *Consuming geographies*. London: Routledge.

Bell, M.M. 2004: *Farming for us all: practical agriculture and the cultivation of sustainability*. University Park (Penn.): Pennsylvania State University Press.

Belletti, G., Brunori, G., Mascotti, A. and A. Rossi 2003a: Multifunctionality and rural development: a multilevel approach. In: Van Huylenbroek, G. and G. Durand (eds): *Multifunctional agriculture: a new paradigm for European agriculture and rural development*. Aldershot: Ashgate, pp. 55-80.

Belletti, G., Marsecotti, A. and R. Moruzzo 2003b: Possibilities of the new Italian law on agriculture. In: Van Huylenbroek, G. and G. Durand (eds): *Multifunctional agriculture: a new paradigm for European agriculture and rural development*. Aldershot: Ashgate, pp. 143-166.

Bennett, T. 2005: Adapting to change, food, farming and government. Paper presented at the 22nd Rural Futures Farm Management Lecture, University of Plymouth, Plymouth, UK, November 2005.

Benz, A. and D. Furst 2002: Policy learning in regional networks. *European Urban and Regional Studies* 9 (1): 21-35.

Beopoulos, N. and G. Vlahos 2005: Desertification and policies in Greece: implementing policy in an environmentally sensitive livestock area. In: Wilson, G.A. and M. Juntti (eds): *Unravelling desertification: policies and actor networks in Southern Europe*. Wageningen (NL): Wageningen Academic Publishers, pp. 157-178.

Berger, M.T. 1994: The end of the 'Third World'? *Third World Quarterly* 15 (2): 257-275.

Berger, M.T. 2001: The nation-state and the challenge of global capitalism. *Third World Quarterly* 22 (6): 889-907.

Berman, M. 1982: *All that is solid melts into air: the experience of modernity*. New York: Simon and Schuster.

Berman, M. 1992: Why modernism still matters. In: Lash, S. and J. Friedman (eds): *Modernity and identity*. Oxford: Blackwell, pp. 33-58.

Bertens, H. 1995: *The idea of the postmodern*. London: Routledge.

Bethge, P. 2005: Rumpsteak ohne Rind. *Der Spiegel* 2005/34: 124-125.

Bhabba, H.K. 1990: Dissemination: time, narrative and the margins of the modern nation. In: Bhabba, H.K. (ed.): *Nation and narration*. London: Routledge, pp. 2-21.

Bieler, A. and A. Morton 2001: The Gordian knot of agency and structure in international relations: a neo-Gramscian perspective. *European Journal of International Relations* 7 (1): 5-35.

Bignal, E.M. and D.I. McCracken 1996: Low-intensity farming systems in the conservation of the countryside. *Journal of Applied Ecology* 33: 413-424.

Bills, N. and D. Gross 2005: Sustaining multifunctional agricultural landscapes: comparing stakeholder perspectives in New York (US) and England (UK). *Land Use Policy* 22: 313-321.

Bishop, K.D. and A.C. Phillips 1993: Seven steps to market: the development of the market-led approach to countryside conservation and recreation. *Journal of Rural Studies* 9 (4): 315-338.

Blandford, D. and R. Boisvert 2002: Multifunctional agriculture and domestic/international policy choice. *Estey Centre Journal of International Law and Trade Policy* 3 (1): 106-118.

Blythman, J. 2004: *Shopped: the shocking power of British supermarkets*. London: Fourth Estate.

Body, R. 1982: *Agriculture: the triumph and the shame*. London: Temple Smith.

Bohman, M., Cooper, J., Mullarkey, D., Normile, M., Skully, D., Vogel, S. and E. Young 1999: *The uses and abuses of multifunctionality*. Washington (DC): USDA .

Bohnet, I., Potter, C. and E. Simmons 2003: Landscape change in the multi-functional countryside: a biographical analysis of farmer decision making in the English Weald. *Landscape Research* 28 (4): 349-364.

Bongaarts, J. 2001: Fertility and reproductive preferences in post-transitional societies. In: Bulatao, R.A. and J.B. Casterline (eds): *Global fertility transition*. New York: Population Council, pp. 260-281.

Bongaarts, J. and S.C. Watkins 1996: Social interactions and contemporary fertility transitions. *Population and Development Review* 22 (4): 639-682.

Boody, G., Vondracek, B., Andow, D.A., Krinke, M., Westra, J., Zimmermann, J. and P. Welle 2005: Multifunctional agriculture in the United States. *BioScience* 55 (1): 27-38.

Bookchin, M. 1982: *Ecology of freedom: the emergence and dissolution of hierarchy*. Palo Alto (Calif.): CA Cheshire Books.

Bookchin, M. 1992: *Ecology of freedom: the emergence and dissolution of hierarchy* (2nd edn). Montreal: Black Rose Books.

Boserup, E. 1993: *The conditions of agricultural growth: the economics of agrarian change under population pressure*. London: Earthscan.

Bourdieu, P. 1983: Ökonomisches Kapital, kulturelles Kapital, soziales Kapital. In: Kreckel, R. (ed.): *Soziale Ungleichheiten*. Göttingen: Otto Schwarz and Co, pp. 183-198.

Bourdieu, P. 1998: A reasoned utopia and economic fatalism. *New Left Review* 227: 125-130.

Bowes, M. and J. Krutilla 1989: *Multiple-use management: the economics of public forest lands*. Washington (DC): Resources for the Future Press.

Bowie, K.A. 1992: Unravelling the myth of the subsistence economy: textile production in nineteenth century Northern Thailand. *Journal of Asian Studies* 51 (4): 797-823.

Bowler, I. 1992: 'Sustainable agriculture' as an alternative path of farm business development. In: Bowler, I.R., Bryant, C.R. and M.D. Nellis (eds): *Contemporary rural systems in transition* (Vol. 1): *agriculture and environment*. Wallingford: CAB International, pp. 237-253.

Bowler, I.R., Bryant, C.R. and C. Cocklin (eds) 2002: *The sustainability of rural systems: geographical interpretations*. Dordrecht: Kluwer Academic Publishers.

Boyer, R. and Y. Saillard (eds) 2002: *Regulation theory*. London: Routledge.

Boyle, P. and K. Halfacree (eds) 1998: *Migration into rural areas: theories and issues*. Chichester: Wiley.

Bracken, L.J. and J. Wainwright 2006: Geomorphological equilibrium: myth or metaphor? *Transactions of the Institute of British Geographers* 31 (2): 167-178.

Bradbury, M. 1976: The cities of modernism. In: Bradbury, M. and J. McFarlane (eds): *Modernism 1890-1930*. London: Penguin, pp. 96-104.

Bradshaw, M. and A. Stenning (eds) 2001: *The transition economies of East Central Europe and the former Soviet Union*. London: Addison Wesley Longman.

Brassley, P. 2005: The professionalisation of English agriculture. *Rural History* 16 (2): 235-251.

Braun, B. and N. Castree (eds) 1998: *Remaking reality: nature at the millennium*. London: Routledge.

Bray, F. 1984: *Science and civilisation in China: biology and biological technology (Part II: agriculture)*. Cambridge: CUP.

Bray, F. 1986: *The rice economies: technology and development in Asian societies*. Oxford: Basil Blackwell.

Brazier, K. 2002: Ideology and discourse: characterization of the 1996 Farm Bill by agricultural interest groups. *Agriculture and Human Values* 19: 239-253.

Bresciani, F., Dévé, F. and R. Stringer 2004: The multiple roles of agriculture in developing countries. In: Brouwer, F. (ed.): *Sustaining agriculture and the rural environment: governance, policy and multifunctionality*. Cheltenham: Edward Elgar, pp. 286-306.

Brimblecombe, P. 1987: *The big smoke*. London: Methuen.

Brotherton, I. 1991: What limits participation in ESAs? *Journal of Environmental Management* 32 (3): 241-249.

Brouwer, F. (ed.) 2004a: *Sustaining agriculture and the rural environment: governance, policy and multifunctionality*. Cheltenham: Edward Elgar.

Brouwer, F. 2004b: Introduction. In: Brouwer, F. (ed.): *Sustaining agriculture and the rural environment: governance, policy and multifunctionality*. Cheltenham: Edward Elgar, pp. 1-11.

Brouwer, F. and B. Crabtree (eds) 1999: *Environmental indicators and agricultural policy*. Wallingford: CAB International.

Brush, R., Chenoweth, R. and T. Barman 2000: Group differences in the enjoyability of driving through rural landscapes. *Landscape and Urban Planning* 47: 39-45.

Bruun, O. and A. Kalland (eds) 1994: *Asian perspectives of nature: a critical approach*. London: Curzon Press.

Bryant, R.L. 1994: From laissez-faire to scientific forestry: forest management in early colonial Burma 1826-65. *Forest and Conservation History* 38: 160-170.

Bryant, R.L. 1996: The greening of Burma: political rhetoric or sustainable development? *Pacific Affairs* 69: 341-359.

Bryant, R.L. 1997: *The political ecology of forestry in Burma, 1824-1994*. London: Hurst.

Bryant, R.L. 2001: Explaining state-environmental NGO relations in the Philippines and Indonesia. *Singapore Journal of Tropical Geography* 22 (1): 15-37.

Bryant, R.L. 2005: *Nongovernmental organizations in environmental struggles: politics and the making of moral capital in the Philippines*. New Haven: Yale University Press.

Bryant, R.L. and S. Bailey 1997: *Third World political ecology*. London: Routledge.

Bryant, R.L. and G.A. Wilson 1998: Rethinking environmental management. *Progress in Human Geography* 22 (3): 321-343.

Bryceson, D.F. 1996: Deagrarianization and rural employment in sub-Saharan Africa: a sectoral perspective. *World Development* 24 (1): 97-111.

Bryceson, D.F. 1997a: De-agrarianisation in sub-Saharan Africa: acknowledging the inevitable. In: Bryceson, D.F. and V. Jamal (eds): *Farewell to farms: de-agrarianisation and employment in Africa*. Aldershot: Ashgate, pp. 3-20.

Bryceson, D.F. 1997b: De-agrarianisation: blessing or blight? In: Bryceson, D.F. and V. Jamal (eds): *Farewell to farms: de-agrarianisation and employment in Africa*. Aldershot: Ashgate, pp. 237-256.

Bryceson, D.F. 2000: Peasant theories and smallholder policies: past and present. In: Bryceson, D.F., Kay, C. and J. Mooij (eds): *Disappearing peasantries: rural labour in Africa, Asia and Latin America.* London: Intermediate Technology Publications, pp. 1-36.

Bryceson, D.F. 2002: The scramble in Africa: reorienting rural livelihoods. *World Development* 30 (5): 725-739.

Bryceson, D.F. and V. Jamal (eds) 1997: *Farewell to farms: de-agrarianisation and employment in Africa.* Aldershot: Ashgate.

Bryden, J.M. 2005: Multifunctionality, rural development and policy adjustments in the European Union. Paper presented at the Policy Research Institute, Ministry of Agriculture, Fisheries and Forestry, Tokyo, Japan, February 2005.

Buckwell, A., Blom, J., Commins, P., Hervieu, B., Hofreither, M., von Meyer, H., Rabinowicz, E., Sotte, F. and J.M. Sumpsi 1998: *Toward a common agricultural and rural policy for Europe.* Luxembourg: Office for Official Publications of the European Communities.

Bulatao, R.A. and J.B. Casterline (eds) 2001: *Global fertility transition.* New York: Population Council.

Buller, H. 1998: Agrarianism, environmentalism and European agri-environmental policy: toward a new 'territorialisation' of European farm systems. Paper presented at the Annual Conference of the IBG/RGS, Kingston University, Kingston-upon Thames, UK, January 1998.

Buller, H. 1999: Personal comment, Annual Conference of the RESSG, Durham, UK, September 1999.

Buller, H. 2000: *Actors, institutions and attitudes to rural development: the French national report. Research report to the World-Wide Fund for Nature and the Statutory Countryside Agencies.* London: WWF.

Buller, H. 2004: The 'espace productif', the 'théâtre de la nature' and the 'territoires de développement local': the opposing rationales of contemporary French rural development policy. *International Planning Studies* 9: 101-119.

Buller, H. 2005: *Evaluation of policies with respect to multifunctionality of agriculture: observation tools and support for policy formulation and evaluation. (UK national report [WP 6], EU Multagri Project).* Exeter: University of Exeter.

Buller, H. and H. Brives 2000: France: farm production and rural product as key factors influencing agri-environmental policy. In: Buller, H., Wilson, G.A. and A. Höll (eds): *Agri-environmental policy in the European Union.* Aldershot: Ashgate, pp. 9-30.

Buller, H., Wilson, G.A. and A. Höll (eds) 2000: *Agri-environmental policy in the European Union.* Aldershot: Ashgate.

Burawoy, M. 1985: *The politics of production.* London: Verso.

Burawoy, M. 1996: The state and economic involution: Russia through a China lens. *World Development* 24 (6): 1105-1117.

Burton, R.J. 2004: Seeing through the 'good farmer's' eyes: towards developing an understanding of the social symbolic value of 'productivist' behaviour. *Sociologia Ruralis* 44 (2): 195-215.

Burton, R.J. and G.A. Wilson 2006: Injecting social psychology theory into conceptualisations of agricultural agency: towards a post-productivist farmer self-identity? *Journal of Rural Studies* 22: 95-115.

Butler, R. 1980: The concept of tourist area cycle of evolution: implications for management of resources. *Canadian Geographer* 24: 5-12.

Buttel, F.H. 2001: Some reflections on late twentieth century agrarian political economy. *Sociologia Ruralis* 41: 165-182.

Buttel, F.H., Larson, O.F. and G.W. Gillespie 1990: *The sociology of agriculture.* New York: Greenwood Press.

Buttoud, G. and I. Yunusova 2002: A 'mixed model' for the formulation of a multipurpose mountain forest policy: theory vs practice on the example of Kyrgyzstan. *Forest Policy and Economics* 4 (2): 149-160.

Byres, T.J. 1995: Political economy, the agrarian question, and the comparative method. *Journal of Peasant Studies* 22 (4): 561-580.

Byres, T.J. 1996: *Capitalism from above and capitalism from below: an essay in comparative political economy.* Basingstoke: Macmillan.

Cafruny, A. 1989: Economic conflicts and the transformation of the Atlantic order: the US, Europe and the liberalisation of agriculture and services. In: Gill, S. (ed.): *Atlantic relations beyond the Reagan era.* London: St Martin's Press, pp. 179-197.

Caldwell, J.C. 1976: Towards a restatement of demographic transition theory. *Population and Development Review* 2 (3/4): 321-366.

Callero, P. 1985: Role-identity salience. *Social Psychology Quarterly* 48 (3): 203-215.

Callicot, J.B. and R.T. Ames 1989: *Nature in Asian traditions of thought.* Albany (NY): Suny Press.

Campbell, A. 1994: *Landcare: communities shaping the land and the future.* Sydney: Allen and Unwin.

Canevacci, M. 1992: Image accumulation and cultural syncretism. *Theory Culture and Society* 9 (3): 95-110.

Caraveli, H. 2000: A comparative analysis on intensification and extensification in Mediterranean agriculture: dilemmas for LFAs policy. *Journal of Rural Studies* 16: 231-242.

Carpentier, C.L., Ervin, D. and S. Vaughan 2004: Multifunctionality and trade in North America. In: Brouwer, F. (ed.): *Sustaining agriculture and the rural environment: governance, policy and multifunctionality.* Cheltenham: Edward Elgar, pp. 307-328.

Carr, A. 2002: *Grass roots and green tape: principles and practices of environmental stewardship.* Sydney: Federation Press.

Carson, R. 1962: *Silent spring.* Boston: Houghton Mifflin.

Cast, A. and P. Burke 2002: A theory of self-esteem. *Social Forces* 80 (3): 1041-1068.

Castells, M. 1996: *The information age: economy, society and culture* (Vol. 1) – *The rise of the network society.* Oxford: Basil Blackwell.

Casterline, J.B. 2001: The pace of fertility transition: national patterns in the second half of the twentieth century. In: Bulatao, R.A. and J.B. Casterline (eds): *Global fertility transition.* New York: Population Council, pp. 17-52.

Cayre, P., Dépigny, S. and Y. Michelin 2004: Multifonctionnalité de l'agriculture: quelle motivation de l'agriculteur? *Les Cahiers de la Multifonctionnalité* 5: 31-42.

CEC [Commission of the European Communities] 1988: *The future of rural society.* Brussels: CEC [COM (88) 601].

CEC [Commission of the European Communities] 1996: *The Cork Declaration: a living countryside* (European Conference on Rural Development, Cork, Ireland 7-9 November 1996). Brussels: CEC.

CEC [Commission of the European Communities] 1999: *The new rural development policy: elements of the political agreement of the Agriculture Council, 22 February-11 March 1999* (DG Agri press notice, 11 March 1999). Brussels: CEC.

Chakrabarti, A. and S. Cullenberg 2003: *Transition and development in India.* London: Routledge.

Chambers, I. and L. Curti (eds) 1996: *The post-colonial question: common skies, divided horizons.* London: Routledge.

Chambers, R. 1983: *Rural development: putting the last first.* London: Longman.

Chambers, R., Pacey, A. and L.A. Thrupp (eds) 1989: *Farmer first: farmer innovation and agricultural research.* London: Intermediate Technology Publications Ltd.

Chase-Dunn, C. and T.D. Hall 1997: *Rise and demise: comparing world-systems.* Boulder (Col.): Westview.

Chaudhri, D.P. 1992: Employment consequences of the Green Revolution: some emerging trends. *Indian Journal of Labour Economics* 35 (1): 23-36.

Chesnais, J.C. 1992: *Demographic transition: stages, patterns and economic implications.* Oxford: OUP.

Chesnais, J.C. 2001: A march towards population recession. In: Bulatao, R.A. and J.B. Casterline (eds): *Global fertility transition.* New York: Population Council, pp. 255-259.

Childs, P. and W. Williams 1997: *An introduction to post-colonial theory.* London: Prentice Hall.

Chomsky, N. 2000: *Rogue states: the rule of force in world affairs.* Cambridge (Mass.): South End Press.

Cilliers, P. 2001: Boundaries, hierarchies and networks in complex systems. *International Journal of Innovation Management* 5: 135-147.

Clark, J. 2003: Regional innovation systems and economic development: the promotion of multifunctional agriculture in the English East Midlands. Unpublished PhD thesis, Department of Geography, University College London, UK.

Clark, J. 2005: The 'New Associationalism' in agriculture: agro-food diversification and multifunctional production logics. *Journal of Economic Geography* 5: 475-498.

Clark, J. 2006: The institutional limits to multifunctional agriculture: subnational governance and regional systems of innovation. *Environment and Planning C: Government and Policy* 24: 331-349.

Clark, J. and P. Lowe 1992: Cleaning up agriculture: environment, technology and social science. *Sociologia Ruralis* 32: 11-29.

Clark, J., Jones, A., Potter, C. and M. Lobley 1997: Conceptualising the evolution of the European Union's agri-environmental policy: a discourse approach. *Environment and Planning A* 29: 1869-1885.

Cleland, J. 1994: Different pathways to demographic transition. In: Graham-Smith, F. (ed.): *Population: the complex reality*. Golden (Col.): North American Press, pp. 229-247.

Cleland, J. 2001: The effects of improved survival on fertility: a reassessment. In: Bulatao, R.A. and J.B. Casterline (eds): *Global fertility transition*. New York: Population Council, pp. 60-92.

Cloke, P. 1989: Rural geography and political economy. In: Peet, R. and N. Thrift (eds): *New models in geography* (Vol. 1). London: Unwin Hyman, pp. 164-197.

Cloke, P. and M. Goodwin 1992: Conceptualising countryside change: from post-Fordism to rural structured coherence. *Transactions of the Institute of British Geographers* 17: 321-336.

Cloke, P. and J. Little 1990: *The rural state: limits to planning in rural society*. Oxford: Clarendon Press.

Cloke, P. and J. Little (eds) 1997: *Contested countryside cultures: otherness, marginalization and rurality*. London: Routledge.

Clunies-Ross, T. and G. Cox 1994: Challenging the productivist paradigm: organic farming and the politics of agricultural change. In: Lowe, P., Marsden, T. and S. Whatmore (eds): *Regulating agriculture*. London: David Fulton, pp. 53-75.

Coale, A.J. and S.C. Watkins (eds) 1986: *The decline of fertility in Europe*. Princeton (N.J.): Princeton University Press.

Cochrane, W.W. 2003: *Curse of American agricultural abundance: a sustainable solution*. Lincoln (Nebr.): University of Nebraska Press.

Cocklin, C. 1995: Agriculture, society and environment: discourses on sustainability. *International Journal of Sustainable Development and World Ecology* 2: 240-256.

Cocklin, C. and J. Dibden (eds) 2004: *Sustainability and change in rural Australia*. Sydney: UNSW Press.

Cocklin, C., Blunden, G. and W. Moran 1997: Sustainability, spatial hierarchies and land-based production. In: Ilbery, B., Chiotti, Q. and T. Rickard (eds): *Agricultural restructuring and sustainability: a geographical perspective*. Wallingford: CAB International, pp. 25-40.

Cocklin, C., Dibden, J. and G.A. Wilson 2006: From productivism to multifunctionality? Agri-environmental governance in Australia. Paper presented at the International Landcare Conference, Melbourne, Australia, October 2006.

Colas, S. 1994: *Postmodernity in Latin America: the Argentine paradigm*. Durham (N.C.): Duke University Press.

Coleman, W. and C. Chaisson 2002: State power, transformative capacity and adapting to globalisation: an analysis of French agricultural policy 1960-2000. *Journal of European Public Policy* 9: 168-185.

Connell, J. and J. Lea 1995: Distant places, other cities? Urban life in contemporary Papua New Guinea. In: Watson, S. and K. Gibson (eds): *Postmodern cities and spaces*. Oxford: Blackwell, pp. 165-183.

Constant, E.W. 1980: *The origins of the turbojet revolution*. Baltimore: John Hopkins.

Cooper, N. 1998: Street-level bureaucrats and the ESA scheme: the FRCA project officer in the UK. Unpublished PhD thesis, Department of Geography, King's College London, UK.

Coughenour, C. 1995: The social construction of commitment and satisfaction with farm and nonfarm work. *Social Science Research* 24: 367-389.

Countryside Agency 2001: *A strategy for sustainable land management in England*. Cheltenham: CA.

Courtenay, P.P. 1988: Farm size, out-migration and abandoned padi land in Mukim Melekek, Melaka (Peninsular Malaysia). *Malaysian Journal of Tropical Geography* 17: 18-28.

Cox, G. and M. Winter 1987: Farmers and the state: a crisis for corporatism. *Political Quarterly* 58: 73-81.

Cox, G., Lowe, P. and M. Winter 1986: From state direction to self regulation: the historical development of corporatism in British agriculture. *Policy and Politics* 14: 475-490.

Cox, G., Lowe, P. and M. Winter 1988: Private rights and public responsibilities: the prospects for agricultural and environmental controls. *Journal of Rural Studies* 4: 232-237.

Crush, J. 1995: *Power of development*. London: Routledge.

Culler, J. 1983: *On deconstruction: theory and criticism after structuralism*. London: Routledge and Kegan Paul.

Curry, N. and M. Winter 2000: The transition to environmental agriculture in Europe: learning processes and knowledge networks. *European Planning Studies* 8 (1): 107-121.

Curtis, A. and M. Lockwood 2000: Landcare and catchment management in Australia: lessons for state-sponsored community participation. *Society and Natural Resources* 13: 61-73.

Dabbert, S., Häring, A.M. and R. Zanoli 2004: *Organic farming: policies and prospects*. London: Zed Books.

Darwin, C. 1859: *On the origin of species by means of natural selection*. London: Murray.

Daugstad, K., Ronningen, K. and B. Skar 2006: Agriculture as an upholder of cultural heritage? Conceptualizations and value judgements: a Norwegian perspective in international contexts. *Journal of Rural Studies* 22: 67-81.

Davies, N. 1997: *Europe: a history*. London: Pimlico.

Davies, P. 1995: *About time*. London: Penguin.

Davis, K. 1945: The world demographic transition. *The Annals of the American Academy of Political and Social Science* 237: 1-11.

Dawkins, R. 1986: *The blind watchmaker*. London: Longman.

Dawkins, R. 1995: *River out of Eden: a Darwinian view of life*. New York: Basic Books.

Dax, T. and G. Hovorka 2004: Integrated rural development in mountain areas. In: Brouwer, F. (ed.): *Sustaining agriculture and the rural environment: governance, policy and multifunctionality*. Cheltenham: Edward Elgar, pp. 124- 143.

Day, L.H. 1992: *The future of low-birthrate populations*. London: Routledge.

Dean, M. 1999: *Governmentality: power and rule in modern society*. London: Sage.

Dear, M. 1986: Postmodernism and planning. *Environment and Planning D: Society and Space* 4: 367-384.

Decker, F. 2002: Governance beyond the nation-state: reflections on the democratic deficit of the European Union. *Journal of European Public Policy* 9 (2): 256-272.

DeGregori, T.R. 1985: *A theory of technology: continuity and change in human development*. Ames (Iowa): Iowa State University Press.

De Groot, R. 2006: Function-analysis and valuation as a tool to assess land use conflicts in planning for sustainable, multi-functional landscapes. *Landscape and Urban Planning* 75: 175-186.

Deleuze, G. and F. Guattari 1987: *A thousand plateaus: capitalism and schizophrenia*. Minneapolis: University of Minnesota Press.

Delgado, M. del Mar, Ramos, E., Gallardo, R. and F. Ramos 2003: Multifunctionality and rural development: a necessary convergence. In: Van Huylenbroek, G. and G. Durand (eds): *Multifunctional agriculture: a new paradigm for European agriculture and rural development*. Aldershot: Ashgate, pp. 19-36.

Dennett, D.C. 1995: *Darwin's dangerous idea: evolution and the meanings of life*. London: Penguin.

Denslow, J.S. and C. Padoch (eds) 1988: *People of the tropical rainforest*. Berkeley (Calif.): University of California Press.

Der Derian, J. 1988: Introducing philosophical traditions in international relations. *Millennium* 17 (2): 189-193.

Derrida, J. 1994: *Spectres of Marx: the state of the debt, the work of mourning and the New International*. London: Routledge.

Deschamps, J.-C. and T. Devos 1998: Regarding the relationship between social identity and personal identity. In Worchel, S., Morales, J., Páez, D. and J.-C. Deschamps (eds): *Social identity*. London: Sage, pp. 1-12.

Devall, B. and G. Sessions 1985: *Deep ecology: living as if nature mattered*. Salt Lake City: Peregrine Smith.

Diamond, J. 1998: *Guns, germs and steel: a short history of everybody for the last 13,000 years*. London: Vintage.

Diamond, J. 2006: *Collapse: how societies failed to succeed*. London: Penguin.

Dicken, P. 1998: *Global shift: transforming the world economy* (3rd edn). London: Paul Chapman.

Dietrich, V. 1953: *Forstwirtschaftspolitk*. Hamburg: Parey.

Di Iacovo, F. 2003: New trends in the relationship between farmers and local communities in Tuscany. In: Van Huylenbroek, G. and G. Durand (eds) 2003: *Multifunctional agriculture: a new paradigm for European agriculture and rural development*. Aldershot: Ashgate, pp. 101-125.

Di Iacovo, F. 2006: Re-generating contemporary rurality. Paper presented at the Conference 'The rural citizen: governance, culture and wellbeing in the 21st century', University of Plymouth, Plymouth, UK, April 2006.

Dijst, M., Elbersen, B. and K. Willis 2005: The challenge of multi-functional land use in rural areas. *Journal of Environmental Planning and Management* 48 (1): 3-6.

Dobbs, T.L. and J.N. Pretty 2004: Agri-environmental stewardship schemes and 'multifunctionality'. *Review of Agricultural Economics* 26: 221-237.

Dobson, A. 1995: *Green political thought* (2nd edn). London: Routledge.

DOE [Department of the Environment] 1990: *This common inheritance*. London: DOE.

DOE [Department of the Environment] 1994: *Sustainable development: the UK strategy*. London: DOE.

DOE [Department of the Environment] 1995: *Rural England*. London: DOE.

Doran, P., Schmidt, K. and C. Spence 1999: Summary of the conference on the multifunctional character of agriculture and land. *Sustainable Development* 32 (5): 1-10.

Dosi, G. 1982: Technological paradigms and technological trajectories. *Research Policy* 11: 147-162.

Drummond, I., Campbell, H., Lawrence, G. and D. Symes 2000: Contingent and structural crisis in British agriculture. *Sociologia Ruralis* 40: 111-128.

Dulcire, M. and P. Cattan 2002: Monoculture d'exportation et développement agricole durable: cas de la banane en Guadeloupe. *Cahiers de l'Agriculture* 11: 313-321.

Dunford, M. 1998: Differential development, institutions, modes of regulation and comparative transitions to capitalism: Russia, the Commonwealth of Independent States and the former German Democratic Republic. In: Pickles, A. and A. Smith (eds) 1998: *Theorising transition: the political economy of post-communist transformations*. London: Routledge, pp. 76-114.

Duram, L.A. 1997: Great Plains agroecologies: the continuum from conventional to alternative agriculture in Colorado. In: Ilbery, B., Chiotti, Q. and T. Rickard (eds): *Agricultural restructuring and sustainability: a geographical perspective*. Wallingford: CAB International, pp. 153-166.

Durand, G. 2003: The French experiences with Territorial Farming Contracts. In: Van Huylenbroek, G. and G. Durand (eds): *Multifunctional agriculture: a new paradigm for European agriculture and rural development*. Aldershot: Ashgate, pp. 129-141.

Durand, G. and G. Van Huylenbroek 2003: Multifunctionality and rural development: a general framework. In: Van Huylenbroek, G. and G. Durand (eds): *Multifunctional agriculture: a new paradigm for European agriculture and rural development*. Aldershot: Ashgate, pp. 1-18.

Durning, A.T. 1993: Supporting indigenous peoples. In: Brown, L.R. (ed.): *State of the world*. London: Earthscan, pp. 80-100.

Duxbury, D. 2005: Lab instruments provide multifunctionality. *Food Technology* 59 (9): 80-83.

Eco, U. 1984: *Postscript to the Name of the Rose*. New York: Harcourt Brace Jovanovich.

Eder, J.F. 1999: *A generation later: household strategies and economic change in the rural Philippines*. Honolulu: University of Hawaii Press.

Edlin, H.L. 1970: *Trees, woods and man*. London: Collins.

Edwards, C.A., Grove, T.L., Harwood, R. and C.J. Colfer 1993: The role of agroecology and integrated farming systems in agricultural sustainability. *Agriculture, Ecosystems and Environment* 46: 99-121.

Egoz, S., Bowring, J. and H. Perkins 2001: Tastes in tension: form, function, and meaning in New Zealand's farmed landscapes. *Landscape and Urban Planning* 57: 177-196.

Ehrlich, P.R. 1970: *The population bomb*. New York: Ballantine.

Ehrlich, P.R. and A.H. Ehrlich 1990: *The population explosion*. New York: Doubleday.

Eisenstadt, S.N. (ed.) 2002: *Multiple modernities*. New Brunswick (N.J.): Transaction Books.

Ekins, P. 1999: European environmental taxes and charges: recent experience, issues and trends. *Ecological Economics* 31: 39-62.

Eldredge, N. and S.J. Gould 1972: Punctuated equilibria: an alternative to phyletic gradualism. In: Schopf, T.J. and J.M. Thomas (eds): *Models in paleobiology*. San Francisco: Freeman Cooper, pp. 82-115.

Eliade, M. 1959: *Cosmos and history: the myth of the eternal return*. New York: Harper and Row.

Ellin, N. 1996: *Postmodern urbanism*. Cambridge (Mass.): Blackwell.

Ellis, F. 2000: *Rural livelihoods and diversity in developing countries*. Oxford: OUP.

Elson, R.E. 1997: *The end of the peasantry in Southeast Asia: a social and economic history of peasant livelihood, 1800-1990s*. Basingstoke: Macmillan.

English Nature 2000: *International trade rules, agriculture and the environment (position statement)*. Peterborough: English Nature.

Ermann, U. 2005: *Regionalprodukte: Vernetzungen und Grenzziehungen bei der Regionalisierung von Nahrungsmitteln*. Wiesbaden: Franz Steiner Verlag.

Errington, A. 1994: The peri-urban fringe: Europe's forgotten rural areas. *Journal of Rural Studies* 10 (4): 367-375.

Ervin, D. 1999: Towards GATT-proofing environmental programmes. *Journal of World Trade* 33: 63-82.

Escobar, A. 1995: *Encountering development: the making and unmaking of the Third World*. Princeton (N.J.): Princeton University Press.

European Commission 1996: *The Cork Declaration: a living countryside*. Brussels: European Commission.

Evans, N. and B. Ilbery 1993: the pluriactivity, part-time farming and farm diversification debate. *Environment and Planning A* 25: 945-959.

Evans, N., Morris, C. and M. Winter 2002: Conceptualizing agriculture: a critique of post-productivism as the new orthodoxy. *Progress in Human Geography* 26 (3): 313-332.

Fairweather, J. and H.R. Campbell 2003: Environmental beliefs and farm practices of New Zealand farmers: contrasting pathways to sustainability. *Agriculture and Human Values* 20: 287-300.

Falconer, K. and N. Ward 2000: Using modulation to green the CAP: the UK case. *Land Use Policy* 17: 269-277.

FAO [Food and Agriculture Organisation] 1999: Cultivating our futures: taking stock of the multifunctional character of agriculture and land. Paper presented at the FAO Conference 'The multifunctional character of agriculture and land', Maastricht, the Netherlands, September 1999.

Featherstone, M. 1995: *Undoing culture: globalization, postmodernism and identity*. London: Sage.

Feeney, G. and F. Wang 1993: Parity progression and birth intervals in China: the influence of policy in hastening fertility decline. *Population and Development Review* 19 (1): 61-101.

Felipe, J. 2000: Convergence, catch-up and growth sustainability in Asia: some pitfalls. *Oxford Development Studies* 28 (1): 51-69.

Fennell, R. 1987: *The Common Agricultural Policy of the European Community* (2nd edn). Oxford: BSP Professional Books.

Fieldhouse, D. 1981: *Colonialism 1870-1945*. London: Weidenfeld and Nicolson.

Fischler, F. 1999: *CAP reform 1999: a crisis in the making?* Brussels: DG VI.

Fish, R., Seymour, S. and C. Watkins 2003: Conserving English landscapes: land managers and agri-environmental policy. *Environment and Planning A* 35: 19-41.

Fitzpatrick, T. 2004: A post-productivist future for social democracy. *Social Policy and Society* 3: 213-222.

Fleury, A., Moustier, P. and J.J. Tolron 2004: Multifonctionnalité de l'agriculture périurbaine: diversité des formes d'exercice du métier d'agriculteur, insertion de l'agriculture dans l'aménagement des territoires. *Les Cahiers de la Multifonctionnalité* 6: 107-120.

Font, A.R. 2000: Mass tourism and the demand for protected natural areas: a travel cost approach. *Journal of Environmental Economics and Management* 39: 97-116.

Foreman, D. and B. Haywood (eds) 1988: *Ecodefense: a field guide to monkeywrenching* (2nd edn). Tucson (Ariz.): Ned Ludd Books.

Forestry Commission 1991: *Forestry policy for Great Britain*. Edinburgh: Forestry Commission.

Forsyth, T. 2003: *Critical political ecology: the politics of environmental science*. London: Routledge.

Francks, P. 1995: From peasant to entrepreneur in Italy and Japan. *The Journal of Peasant Studies* 22 (4): 699-709.

Francks, P. 2006: Rural economic development in Japan from the 19th century to the Pacific War. London: Routledge.

Francks, P. with Boestal, J. and C.H. Kim 1999: *Agriculture and economic development in East Asia: from growth to protectionism in Japan, Korea and Taiwan.* London: Routledge.

Frank, A.G. and B.K. Gills 1993: *The world system: five hundred years or five thousand?* London: Routledge.

Frank, G. 1969: *Capitalism and underdevelopment in Latin America.* New York: Monthly Review Press.

Freeman, C. 1987: *Technology, policy and economic performance: lessons from Japan.* London: Pinter.

Freshwater, D. 2002: *Applying multifunctionality to US farm policy* (Agricultural Economics Staff Paper No 437). Lexington (Kentucky): University of Kentucky.

Friedberg, S. 2003: Cleaning up down south: supermarkets, ethical trade and African horticulture. *Social and Cultural Geography* 4: 27-43.

Friedman, J. 1999: The hybridisation of roots and the abhorrence of the bush. In: Featherstone, M. and S. Lash (eds): *Spaces of culture: city-nation-world.* London: Sage, pp. 230-255.

Friedmann, H. and P. McMichael 1989: Agriculture and the state system: the rise and decline of national agricultures, 1870 to present. *Sociologia Ruralis* 28: 93-118.

Frigg, R. 2003: Self-organized criticality: what it is and what it isn't. *Studies in History and Philosophy of Science A* 34: 613-632.

Froud, J. 1994: The impact of ESAs on lowland farming. *Land Use Policy* 11 (2): 107-118.

Fry, G.L. 2001: Multifunctional landscapes: towards transdisciplinary research. *Landscape and Urban Planning* 57 (3-4): 159-168.

Fukuyama, F. 1992: *The end of history and the last man.* New York: Free Press.

Fuller, A. 1990: From part-time farming to pluriactivity: a decade of change in rural Europe. *Journal of Rural Studies* 6: 361-373.

Futuyama, D.J. 1998: *Evolutionary biology* (3rd edn). Sunderland (Mass.): Sinauer Associates.

Gallardo, R., Ramos, F., Ramos, E. and M. del Mar Delgado 2003: New opportunities for non-competitive agriculture. In: Van Huylenbroek, G. and G. Durand (eds): *Multifunctional agriculture: a new paradigm for European agriculture and rural development.* Aldershot: Ashgate, pp. 169-188.

Gallent, N., Oades, R. and C. Tudor 2004: Visioning multi-functional urban fringe environments. Paper presented at the 2nd meeting of the Anglo-German Rural Geographers 'Rural multifunctionality: perspectives from policy-making, implementation and practice', University of Exeter, Exeter, UK, July 2004.

Gaonkar, D.P. 2001: *Alternative modernities.* Durham (N.C.): Duke University Press.

Garrido, F. and E. Moyano 1996: The response of the member states: Spain. In: Whitby, M. (ed.): *The European environment and CAP reform: policies and prospects for conservation.* Wallingford: CAB International, pp. 86-104.

Garzon, I. 2005: *Multifunctionality of agriculture in the European Union: is there substance behind the discourse's smoke?* San Franciso: University of California (Institute of Agriculture and Resource Economics).

Gasson, R. 1974: Goals and values of farmers. *Journal of Agricultural Economics* 24: 521-537.

Gasson, R. 1988: *The economics of part-time farming.* London: Longman.

Gatzweiler, F.W. 2004: Institutions for sustainable agriculture in Central and Eastern Europe. In: Brouwer, F. (ed.): *Sustaining agriculture and the rural environment: governance, policy and multifunctionality.* Cheltenham: Edward Elgar, pp. 247-265.

Gehring, T. 1997: Governing in nested institutions: environmental policy in the European Union and the case of packaging waste. *Journal of European Public Policy* 4: 337-354.

Gellner, E. 1992: *Postmodernism, reason and religion.* London: Routledge.

Gerber, J. 1997: Beyond dualism: the social construction of nature and the natural and social construction of human beings. *Progress in Human Geography* 21: 1-17.

Gerowitt, B., Bertke, E., Hespelt, S.K. and C. Tute 2003: Towards multifunctional agriculture: weeds as ecological goods? *Weed Research* 43: 227-235.

Gertler, M.S. 1988: The limits of flexibility: comments on the post-Fordist vision of production and its geography. *Transactions of the Institute of British Geographers* 13: 419-432.

Gibbs, D., Longhurst, J. and C. Braithwaite 1998: 'Struggling with sustainability': weak and strong interpretations of sustainable development within local authority policy. *Environment and Planning A* 30: 1351-1365.

Giddens, A. 1979: *Central problems in social theory: action, structure and contradiction in social analysis*. London: Macmillan.

Giddens, A. 1984: *The constitution of society: an outline of the theory of structuration*. Cambridge: Polity Press.

Giddens, A. 1991: Structuration theory: past, present and future. In: Bryant, C.G. and D. Jary (eds): *Gidden's theory of structuration: a critical appreciation*. London: Routledge, pp. 201-221.

Giddens, A. 1997: *Modernity and self-identity: self and society in the Late Modern Age*. Cambridge: Polity Press.

Giddens, A. 1998: *The third way*. Cambridge: Polity Press.

Gilg, A. 1991: Planning for agriculture: the growing case for a conservation component. *Geoforum* 22 (1): 75-79.

Glacken, C. 1967: *Traces on the Rhodian Shore*. Berkeley (Calif.): University of California Press.

Gleick, J. 1988: *Chaos: making a new science*. London: Cardinal Books.

Goodin, R.E. 2001: Work and welfare: towards a post-productivist welfare state. *British Journal of Political Science* 31: 13-40.

Goodkind, D.M. 1995: Vietnam's one-or-two-child policy in action. *Population and Development Review* 21: 85-111.

Goodman, D. 1999: Agro-food studies in the 'age of ecology': nature, corporeality, bio-politics. *Sociologia Ruralis* 39: 17-38.

Goodman, D. 2004: Rural Europe redux? Reflections on alternative agro-food networks and paradigm change. *Sociologia Ruralis* 44: 3-16.

Goodman, D. and M.R. Redclift (eds) 1989: *The international farm crisis*. London: Macmillan.

Goodman, D. and M. Redclift 1991: *Refashioning nature: food, ecology and culture*. London: Routledge.

Goodman, D. and M. Watts 1997: *Globalising food: agrarian questions and global restructuring*. London: Routledge.

Goodwin, M. 1998: The governance of rural areas: some emerging research issues and agendas. *Journal of Rural Studies* 14: 5-12.

Goodwin, M. and J. Painter 1996: Local governance, the crises of Fordism and the changing geographies of regulation. *Transactions of the Institute of British Geographers* 21 (4): 635-648.

Goodwin, M., Cloke, P. and P. Milbourne 1995: Regulation theory and rural research: theorising contemporary rural change. *Environment and Planning A* 21: 1245-1260.

Gordon, C. 1976: Development of evaluated role identities. *Annual Review of Sociology* 2: 405-433.

Gorman, M., Mannion, J., Kinsella, J. and P. Bogue 2001: Connecting environmental management and farm household livelihoods: the Rural Environment Protection Scheme in Ireland. *Journal of Environmental Policy and Planning* 3: 137-147.

Goss, J. and Burch, D. 2001: From agricultural modernisation to agri-food globalisation: the waning of national development in Thailand. *Third World Quarterly* 22 (6): 969-986.

Gould, S.J. 1982: Darwinism and the expansion of evolutionary theory. *Science* 216: 380-387.

Gould, S.J. 1989: *Wonderful life: the Burgess Shale and the nature of history*. New York: Norton.

Gould, S.J. 1992a: The confusion over evolution. *New York Review of Books* (19.11.1992): 47-54.

Gould, S.J. 1992b: Life in a punctuation. *Natural History* 101: 10-21.

Grabher, G. 2001: Locating economic action: projects, networks, localities and institutions. *Environment and Planning A* 31: 329-341.

Grabher, G. and D. Stark 1997 (eds): *Restructuring networks in post-socialism: legacies, linkages and localities*. Oxford: OUP.

Grabher, G. and D. Stark 1998: Organising diversity: evolutionary theory, network analysis and post-socialism. In: Pickles, A. and A. Smith (eds): *Theorising transition: the political economy of post-communist transformations*. London: Routledge, pp. 54-75.

Graham, S. 1998: The end of geography or the explosion of place? Conceptualising space, place and information technology. *Progress in Human Geography* 22 (2): 165-185.

Gramsci, A. 1971: *Selections from the prison notebooks*. New York: International Publishers.

Grandstaff, T. 1988: Environment and economic diversity in Northeast Thailand. In: Charoenwatana, T. and Rambo, A.T. (eds): *Sustainable rural development in Asia*. Khon Kaen (Thailand): Khon Kaen University, pp. 11-22.

Grandstaff, T. 1992: The human environment: variation and uncertainty. *Pacific Viewpoint* 33 (2): 135-144.

Gray, T. and J. Hatchard 2003: The 2002 reform of the Common Fisheries Policy's system of governance: rhetoric or reality? *Marine Policy* 27 (6): 545-554.

Greenhalgh, S. 1988: Fertility as mobility: sinic transitions. *Population and Development Review* 14 (4): 629-674.

Gregory, D. 1994: City/commodity/culture: spatiality and the politics of representation. In: Gregory, D. (ed.): *Geographical imaginations*. Oxford: Blackwell, pp. 214-256.

Gregory, D. 1998: Power, knowledge and geography. *Geographische Zeitschrift* 86: 70-93.

Gregory, P. 2005: Organic farming as a problem? Paper presented at the workshop 'Organic agriculture: a solution rather than a problem?', IGER North Wyke Research Station, North Wyke, Devon, UK, November 2005.

Grigg, D.B. 1974: *The agricultural systems of the world: an evolutionary approach*. Cambridge: CUP.

Grolle, J. 2004: Warum ist nicht nichts? *Der Spiegel* 2004/39: 190-194.

Grübler, A. 1997: Time for a change: on the patterns of diffusion of innovation. In: Ausubel, J.H. and H.D. Langford (eds): *Technological change and the human environment*. Washington (D.C.): National Academy Press, pp. 14-32.

Grübler, A. 1998: *Technology and global change*. Cambridge: CUP.

Guillaumin, A., Bousquet, D., Perrot, C., Tchakerian, E., Teffene, O., Gallot, S., Hennion, B. and J.L. Demars 2004: Formalisation de connaissances et de methodes pour favoriser la multifonctionnalité de l'agriculture. *Les Cahiers de la Multifonctionnalité* 6: 51-56.

Gupta, A. and J. Ferguson 1997: Culture, power, place: ethnography at the end of an era. In: Gupta, A. and J. Ferguson (eds): *Culture, power, place: explorations in critical anthropology*. Durham (N.C.): Duke University Press, pp. 1-29.

Guthey, G., Gwin, L. and S. Fairfax 2003: Creative preservation in California's dairy industry. *The Geographical Review* 93: 171-192.

Guy, A. 2005: Personal comment, Thorney Abbey Farm, Southwell, East Midlands, UK, June 2005.

Guzman, E. and G. Woodgate 1999: Alternative food and agriculture networks: an agro-ecological perspective on the responses to economic globalisation and the 'new agrarian question'. Paper presented at the EU COST Workshop, Brussels, Belgium, September 1999.

Habermas, J. 1981: New social movements. *Telos* 49: 33-37.

Habermas, J. 1987: *The philosophical discourse of modernity*. Cambridge: Polity Press.

Habermas, J. 1994: *The past as future*. Lincoln (Nebraska): University of Nebraska Press.

Hägerstrand, T. 1952: *The propagation of innovation waves*. Lund (Sweden): Lund Studies in Geography.

Hägerstrand, T. 1967: *Innovation diffusion as a spatial process*. Chicago: Chicago University Press.

Hale, C.D. 2005: Real reform in North Korea? The aftermath of the July 2002 economic measures. *Asian Survey* 45 (6): 823-842.

Halfacree, K.H. 1994: The importance of 'the rural' in the constitution of counter-urbanisation: evidence from England in the 1980s. *Sociologia Ruralis* 34: 164-189.

Halfacree, K.H. 1997a: British rural geography: a perspective on the last decade. In: Lopez Ontiveros, A. and F. Molinero Hernando (eds): *From traditional countryside to post-productivism: recent trends in rural geography in Britain and Spain*. Murcia: Associacion de Geografos Españoles.

Halfacree, K.H. 1997b: Contrasting roles for the post-productivist countryside: a postmodern perspective on counterurbanisation. In: Cloke, P. and J. Little (eds): *Contested countryside cultures: otherness, marginalisation and rurality*. London: Routledge, pp. 70-91.

Halfacree, K.H. 1999: A new space or spatial effacement? Alternative futures for the post-productivist countryside. In: Walford, N., Everitt, J. and D. Napton (eds): *Reshaping the countryside: perceptions and processes of rural change*. Wallingford: CAB International, pp. 67-76.

Halfacree, K. 2004: Radical ruralities and rural structured coherences. Paper presented at the School of Geography lunchtime seminars, University of Plymouth, Plymouth, UK, March 2004.

Halfacree, K.H. and P. Boyle 1998: Migration, rurality and the post-productivist countryside. In: Boyle, P. and H. Halfacree (eds): *Migration into rural areas: theories and issues*. Chichester: Wiley, pp. 1-20.

Hall, C. 1998: *Introduction to tourism: development, dimensions and issues* (3rd edn). Melbourne (Australia): Addison Wesley Longman.

Hall, P.A. 1993: Policy paradigms, social learning and the State: the case of economic policymaking in Britain. *Comparative Politics* 25: 275-296.

Hall, S. 1996: When was 'the post-colonial'? Thinking at the limit. In: Chambers, I. and L. Curti (eds): *The post-colonial question: common skies, divided horizons*. London: Routledge, pp. 242-260.

Hallinan, M.T. 1997: The sociological study of social change. *American Sociological Review* 62 (1): 1-11.

Hanley, N., Whitby, M. and I. Simpson 1999: Assessing the success of agri-environmental policy in the UK. *Land Use Policy* 16: 67-80.

Hardin, G. 1968: The tragedy of the commons. *Science* 162: 1243-1248.

Harper, S. 1993: *The greening of rural policy: international perspectives*. London: Belhaven.

Harris, D.R. 1978: The environmental impact of traditional and modern agricultural systems: In: Hawkes, J.G. (ed.): *Conservation and agriculture*. London: Duckworth, pp. 61-70.

Harrison, C., Limb, M. and J. Burgess 1986: Recreation 2000: views of the country from the city. *Landscape Research* 11: 19-24.

Harrison, M. 2001: *King sugar: Jamaica, the Caribbean and the world sugar economy*. London: Latin America Bureau.

Hart, K. and G.A. Wilson 1998: UK implementation of Agri-environment Regulation 2078/92/EEC: enthusiastic supporter or reluctant participant? *Landscape Research* 23: 255-272.

Harvey, D. 1989: *The condition of postmodernity*. Oxford: Blackwell.

Harvey, D. 1993: Class relations, social justice and the politics of difference. In: Keith, M. and S. Pile (eds): *Place and the politics of identity*. London: Routledge, pp. 41-66.

Harvey, D. 1996: *Justice, nature and the geography of difference*. Oxford: Blackwell.

Harvey, D. 2004: Historical geography and public policy: learning from the past to construct the future. Paper presented at the South West Rural Research Network Seminar, University of Exeter, Exeter, UK, December 2004.

Harvey, G. 1997: *The killing of the countryside*. London: Jonathan Cape.

Hassard, J. 2001: Commodification, construction and compression: a review of time metaphor in organisational analysis. *International Journal of Management Reviews* 3 (2): 131-140.

Hawkins, K. 2002: Potential growth and implications for patterns of retailing: a food retailer's view. Paper presented at the IBG/RGS Conference 'The alternative food economy: myths, realities and potential', London, UK, March 2002.

Hayami, Y. and M. Kikuchi 2000: *A rice village saga: three decades of Green Revolution in the Philippines*. Basingstoke: Macmillan.

Hazell, P. and C. Ramasamy 1991: *The Green Revolution reconsidered: the impact of high-yielding rice varieties in South India*. Baltimore: Johns Hopkins University Press.

Heinberg, R. 2004: *Powerdown: options and actions for a post-carbon world*. Boston: New Society Publishers.

Herrick, C.B. 2005: 'Cultures of GM': discourses of risk and labelling of GMOs in the UK and EU. *Area* 37 (3): 286-294.

Hess, M. 2004: 'Spatial' relationships? Towards a reconceptualisation of embeddedness. *Progress in Human Geography* 18: 165-186.

Hettne, B. 1990: *Development theory and the three worlds*. Harlow: Longman.

Higgott, R. and N. Phillips 2000: Challenging triumphalism and convergence: the limits of global liberalization in Asia and Latin America. *Review of International Studies* 26: 359-379.

Hinchcliffe, S. 2001: Indeterminacy in-decisions: science, policy and politics in the BSE (Bovine Spongiform Encephalopathy) crisis. *Transactions of the Institute of British Geographers* 26 (2): 182-204.

Hinrichs, C. 2003: The practice and politics of food system localization. *Journal of Rural Studies* 19: 33-45.

Hirschman, C. 1994: Why fertility changes. *Annual Review of Sociology* 20: 203-233.

Hirschman, C. 2001: Globalisation and theories of fertility decline. In: Bulatao, R.A. and J.B. Casterline (eds): *Global fertility transition*. New York: Population Council, pp. 116-125.

Hjalager, A.-M. 1996: Agricultural diversification into tourism. *Tourism Management* 17 (2): 103-111.

Hobbs, R.J. and A.J. Hopkins 1990: From frontier to fragments: European impact on Australia's vegetation. *Proceedings of the Ecological Society of Australia* 16: 93-114.

Hodge, I. 2000: Agri-environmental relationships and the choice of policy mechanism. *The World Economy* 23: 257-273.

Hogarth, W.T. 2005: Keeping our fisheries sustainable. *Benthic Habitats and the Effects of Fishing* 41: 11-17.

Hoggart, K. 1990: Let's do away with rural. *Journal of Rural Studies* 6: 245-257.

Hoggart, K. and A. Paniagua 2001: What rural restructuring? *Journal of Rural Studies* 17: 41-62.

Hoggart, K., Buller, H. and R. Black 1995: *Rural Europe: identity and change*. London: Arnold.

Hollander, G.M. 2003: Re-naturalizing sugar: narratives of place, production and consumption. *Social and Cultural Geography* 4: 59-74.

Hollander, G.M. 2004: Agricultural trade liberalization, multifunctionality, and sugar in the south Florida landscape. *Geoforum* 35: 299-312.

Holloway, L. 2002: Smallholding, hobby-farming, and commercial farming: ethical identities and the production of farming spaces. *Environment and Planning A* 34: 2055-2070.

Holmes, J. 2002: Diversity and change in Australia's rangelands: a post-productivist transition with a difference? *Transactions of the Institute of British Geographers* 27: 362-384.

Holmes, J. 2006: Impulses towards a multifunctional transition in rural Australia: gaps in the research agenda. *Journal of Rural Studies* 22: 142-160.

Hopkins, A. 2005: Organic farming as a solution: further comments. Paper presented at the workshop 'Organic agriculture: a solution rather than a problem?', IGER North Wyke Research Station, North Wyke, Devon, UK, November 2005.

Hörschelmann, K. 2002: History after the end: post-socialist difference in a (post)modern world. *Transactions of the Institute of British Geographers* 27: 52-66.

Hoskins, W.G. 1955: *The making of the English landscape*. London: Hodder and Stoughton.

Hösle, V. 1992: The Third World as a philosophical problem. *Social Research* 59 (2): 227-262.

House of Lords 1990: *The future of rural society: report from the Select Committee on the European Communities*. London: HMSO.

House of Lords 1995: *Sustainable development*. London: HMSO.

Hoy, D. 1986: Nietzsche, Hume, and the genealogical method. In: Yovel, Y. (ed.): *Nietzsche as affirmative thinker*. Dordrecht: Martinus Nijhoff, pp. 20-38.

Hudson, R. 2000: *Production, places and environment: changing perspectives in economic geography*. London: Prentice Hall.

Hughes, T.P. 1976: The development phase of technological change. *Technology and Culture* 17 (3): 423-431.

Hulse, M., Wischmann, S. and R. Pasemann 2005: The role of non-linearity for evolved multifunctional robot behaviour. *Lectures Notes on Computer Science* 3637: 108-118.

Ilbery, B. 1991: Farm diversification as an adjustment strategy on the urban fringe of the West Midlands. *Journal of Rural Studies* 7: 207-222.

Ilbery, B. and I. Bowler 1998: From agricultural productivism to post-productivism. In: Ilbery, B. (ed.): *The geography of rural change*. Harlow: Longman, pp. 57-84.

Ilbery, B. and M. Kneafsey 2000: Producer construction of quality products and services in the lagging regions of the European Union. *Journal of Rural Studies* 16: 217-230.

Ilbery, B., Chiotti, Q. and T. Rickard 1997: Introduction. In: Ilbery, B., Chiotti, Q. and T. Rickard (eds): *Agricultural restructuring and sustainability: a geographical perspective*. Wallingford: CAB International, pp. 1-10.

Ilbery, B., Kneafsey, M. and M. Bamford 2000: Protecting and promoting regional speciality food and drink products in the European Union. *Outlook in Agriculture* 29: 31-37.

Imrie, R. and M. Raco 1999: How new is the new local governance: lessons from the United Kingdom. *Transactions of the Institute of British Geographers* 24: 45-63.

Ingersent, K. and A. Rayner 1999: *Agricultural policy in Western Europe and the United States*. London: Edward Elgar.

Inglehart, R. 1977: *The silent revolution: changing values and political style among Western publics*. Princeton (N.J.): Princeton University Press.

Jacob, M. 1994: Sustainable development and deep ecology: an analysis of competing traditions. *Environmental Management* 18 (4): 477-488.

Jacobs, J. 1996: *Edge of empire: postcolonialism and the city.* London: Routledge.

Jay, M. 2004: Productivist and post-productivist conceptualisations of agriculture from a New Zealand perspective. In: Kearsley, G. and B. Fitzharris (eds): *Glimpses of a Gaian world: essays in honour of Peter Holland.* Dunedin (New Zealand): University of Otago Press, pp. 151-170.

Jay, M. 2005: Remnants of the Waikato: native forest survival in a production landscape. *New Zealand Geographer* 61: 14-28.

Jencks, C. 1984: *The language of post-modern architecture.* New York: Rizzoli.

Jencks, C. 1987: *What is postmodernism?* New York: St Martin's Press.

Jenson, J. 1989: 'Different', but not 'exceptional': Canada's permeable Fordism. *Canadian Journal of Sociology and Anthropology* 26: 69-94.

Jessop, B. 1998: The rise of governance and the risks of failure: the case of economic development. *International Social Science Journal* 155: 29-45.

Jessop, B. 2002: Liberalism, neoliberalism, and urban governance: a state-theoretic perspective. *Antipode* 34: 452-472.

Johnston, R.J. 1996: *Nature, state and economy: a political economy of the environment* (2nd edn). Chichester: Wiley.

Johnston, R.J., Gregory, D., Pratt, G. and M. Watts 2000: *The dictionary of human geography* (4th edn). Oxford: Blackwell.

Jones, A. and Clark, J. 2001: *The modalities of European Union governance: new institutionalist explanations of agri-environmental policy.* Oxford: OUP.

Jongeneel, R. and L. Slangen 2004: Multifunctionality in agriculture and the contestable public domain in the Netherlands. In: Brouwer, F. (ed.): *Sustaining agriculture and the rural environment: governance, policy and multifunctionality.* Cheltenham: Edward Elgar, pp. 183-203.

Jordan, A. 1999: The construction of a multilevel environmental governance system. *Environment and Planning C: Government and Policy* 17: 1-17.

Kalb, D., Svasek, M. and H. Tak 1999: Approaching the 'new' past in East-Central Europe. *Focaal* 33: 9-23.

Kaltoft, P. 2001: Organic farming in late modernity: at the frontier of modernity or opposing modernity? *Sociologia Ruralis* 41 (1): 146-158.

Kantelhardt, J. 2006: Exploring multifunctional agriculture in southern Germany. Paper presented at the School of Geography lunchtime seminars, University of Plymouth, Plymouth, UK, May 2006.

Kaplan, C. 1996: *Questions of travel: postmodern discourses of displacement.* Durham (N.C.): Duke University Press.

Karshenas, M. 1994: Environment, technology and employment: towards a new definition of sustainable development. *Development and Change* 25: 723-756.

Kates, R.W. 1997: Population, technology, and the human environment: a thread through time. In: Ausubel, J.H. and H.D. Langford (eds): *Technological trajectories and the human environment.* Washington (D.C.): National Academy Press, pp. 33-55.

Kato, T. 1994: The emergence of abandoned paddy fields in Negeri Sembilan, Malaysia. *Southeast Asian Studies* 32 (2): 145-172.

Kauffman, S. 1993: *The origins of order: self-organisation and selection in evolution.* New York: Oxford University Press.

Kay, J. and E.D. Schneider 1995: Embracing complexity: the challenge of the ecosystem approach. In: Westra, L. and J. Lemons (eds): *Perspectives on ecological integrity.* Dordrecht: Kluwer, pp. 49-59.

Kelly, P. 1999: Everyday urbanization: the social dynamics of development in Manila's extended metropolitan region. *International Journal of Urban and Regional Research* 23 (2): 283-303.

King, A.D. 1995: The times and spaces of modernity (or who needs postmodernism?). In: Featherstone, M., Lach, S. and R. Robertson (eds): *Global modernities.* London: Sage, pp. 108-123.

Kinsella, J., Wilson, S., de Jong., F. and H. Renting 2000: Pluriactivity as a livelihood strategy in Irish farm households and its role in rural development. *Sociologia Ruralis* 40 (4): 481-496.

Kirk, D. 1996: Demographic transition theory. *Population Studies* 50 (3): 361-387.

Kirkby, R. and Z. Xiaobin 1999: Sectoral and structural considerations in China's rural development. *Tijdschrift voor Economische en Sociale Geografie* 90 (3): 272-284.

Kirwan, J. 2004: Alternative strategies in the UK agro-food system: interrogating the alterity of farmers' markets. *Sociologia Ruralis* 44 (4): 395-415.

Kleijn, D. and W.J. Sutherland 2003: How effective are European agri-environment schemes in conserving and promoting biodiversity? *Journal of Applied Ecology* 40: 947-969.

Kleijn, D. *et al.* 2006: Mixed biodiversity benefits of agri-environment schemes in five European countries. *Ecology Letters* 9: 243-254.

Kloppenberg, J.R. 1988: *First the seed: the political economy of plant biotechnology, 1492-2000.* Cambridge: CUP.

Kloppenberg, J.R. 1991: Social theory and the de/construction of agricultural science: a new agenda for rural sociology. *Sociologia Ruralis* 32 (1): 519-548.

Kneafsey, M. 2002: Regional branding: creating associations between product and place. Paper presented at the IBG/RGS Conference 'The alternative food economy: myths, realities and potential', London, UK, March 2002.

Kneafsey, M., Saxena, G. and B. Ilbery 2004: Multifunctional rurality? Policy perspectives on integrated tourism in the English-Welsh borders. Paper presented at the 2nd meeting of the Anglo-German Rural Geographers 'Rural multifunctionality: perspectives from policy-making, implementation and practice', University of Exeter, Exeter, UK, July 2004.

Kneale, J., Lowe, P. and T. Marsden 1992: *The conversion of agricultural buildings: an analysis of the variable pressures and regulations toward the post-productivist countryside* (ESRC Countryside Change Initiative Working paper No 29). Newcastle University: Department of Agricultural Economics.

Knickel, K. 1990: Agricultural structural change: impact on the rural environment. *Journal of Rural Studies* 6: 383-393.

Knickel, K. and H. Renting 2000: Methodological and conceptual issues in the study of multifunctionality and rural development. *Sociologia Ruralis* 40 (4): 512-528.

Knickel, K., Renting, H. and J.D. Van der Ploeg 2004: Multifunctionality in European agriculture. In: Brouwer, F. (ed.): *Sustaining agriculture and the rural environment: governance, policy and multifunctionality*. Cheltenham: Edward Elgar, pp. 81-103.

Kreileman, G.J. and M.M. Berk 1997: *The safe landing analysis: users' manual*. Bilthoven (NL): National Institute of Public Health and the Environment.

Kristensen, L.S., Thenail, C. and S.P. Kristensen 2004: Landscape changes in agrarian landscapes in the 1990s: the interaction between farmers and the farmed landscape – a case study from Jutland, Denmark. *Journal of Environmental Management* 71: 231-244.

Kuhn, T.S. 1970: *The structure of scientific revolutions* (2nd edn). Chicago: University of Chicago Press.

Kuijsten, A. 1996: Changing family patterns in Europe: a case of divergence? *European Journal of Population* 12: 115-143.

Lall, S. 1992: Technological capabilities and industrialisation. *World Development* 20 (2): 165-186.

Lamarck, J.-B. 1809: *Philosophie zoologique*. Paris: Société Zoologique.

Lang, T. 2004: Food wars: the nutrition transition. Paper presented at the 21st HSBC Rural Futures lecture, University of Plymouth, Plymouth, UK, November 2004.

Lang, T. and M. Heasman 2004: *Food wars: the global battle for mouths, minds and markets*. London: Earthscan.

Lanjouw, P. 1999: Rural nonagricultural employment and poverty in Ecuador. *Economic Development and Cultural Change* 48 (1): 91-122.

Lansing, S. 1991: *Priests and programmers: engineering the knowledge of Bali*. Princeton (N.J.): Princeton University Press.

Lardon, S., Dobromez, L. and E. Josien 2004: Traductions spatiales de la multifonctionnalité de l'agriculture. *Les Cahiers de la Multifonctionnalité* 5: 5-16.

Larsen, J.B. 2005: Functional forests in multifunctional landscapes: restoring the adaptive capacity of landscapes with forests and trees. *Proceedings of the European Forest Research Institute* 53: 97-102.

Latour, B. 1993: *We have never been modern*. London: Prentice Hall.

Latour, B. 1996: *Aramis, or the love of technology*. Cambridge (Mass.): Harvard University Press.

Law, J. 1994: *Organizing modernity*. Oxford: Blackwell.

Le Cotty, T., Voituriez, T. and A. Aumand 2003: Multifonctionnalité et cooperation internationale: une analyse du coût de fourniture de biens publics par l'agriculture. *Economie Rurale* 273-274: 91-102.

Leete, R. 1996: *Malaysia's demographic transition: rapid development, culture and politics*. Oxford: OUP.

Le Heron, R. 1993: *Globalised agriculture – political choice*. Oxford: Pergamon Press.

Leonard, H.J. 1988: *Pollution and the struggle for the world product: multinational corporations, environment and international comparative advantage*. Cambridge: CUP.

Leontidou, L. 1993: Postmodernism and the city: Mediterranean versions. *Urban Studies* 30: 949-965.

Lesthaeghe, R. and J. Surkyn 1988: Cultural dynamics and economic theories of fertility change. *Population and Development Review* 14 (1): 1-45.

Levidow, L. 2005: GM herbicide-tolerant crops: modelling farmers as environmental stewards. Paper presented at the IBG/RGS Conference 'Flows and spaces in a globalised world', London, UK, September 2005.

Leyshon, A., Lee, R. and C. Williams (eds) 2003: *Alternative economic spaces*. London: Sage.

Li, L. 1989: Theoretical theses on 'social modernisation'. *International Sociology* 4 (4): 365-378.

Liepins, R. 1995: Women in agriculture: advocates for a gendered sustainable agriculture. *Australian Geographer* 26: 118-126.

Liepins, R. 1998: The gendering of farming and agricultural policies: a matter of discourse and power. *Australian Geographer* 29: 371-388.

Lifran, R., Tiball, M., Le Cotty, T., Prieur, F., Muro, S. and T. Voituriez 2004: La prise en compte de la multifonctionnalité de l'agriculture dans les politiques nationales de commerce et d'environnement. *Les Cahiers de la Multifonctionnalité* 6: 149-153.

Linnaeus, C. 1751: *Philosophia Botanica*. Uppsala (Sweden): Imperial Academy.

Lipietz, A. 1987: *Mirages and miracles: the crises of global Fordism*. London: Verso.

Lipietz, A. 1992: *Towards a new economic order: postfordism, ecology and democracy*. New York: Oxford University Press.

Lipschutz, R.D. 1999: From local knowledge and practice to global environmental governance. In: Hewson, M. and T.J. Sinclair (eds): *Approaches to global governance theory*. Albany (N.Y.): Suny Press, pp. 259-283.

Little, J. 2001: New rural governance? *Progress in Human Geography* 25 (1): 97-102.

Littlefair, J. 2005: Personal comment, UK farmer, BBC News 24 interview, 19.5.2005

Liverman, D. 2004: Who governs, at what scale and at what price? Geography, environmental governance, and the commodification of nature. *Annals of the Association of American Geographers* 94: 734-738.

Lobley, M. and C. Potter 1998: Environmental stewardship in UK agriculture: a comparison of the Environmentally Sensitive Area Programme and the Countryside Stewardship Scheme in south east England. *Geoforum* 29 (4): 413-432.

Lomborg, B. 2005: The truth about the environment. In: Dryzek, J.S. and D. Schlosberg (eds): *Debating the Earth: the environmental politics reader*. Oxford: OUP, pp. 74-79.

Long, N. and J.D. Van der Ploeg 1995: Reflections on agency, ordering the future, and planning. In: Frerks, G. and J.H. den Ouden (eds): *In search of the middle ground: essays on the sociology of planned development*. Wageningen: Wageningen Agricultural University, pp. 64-78.

Loomis, J.B. 1993: *Integrated public lands management: principles and applications to national forests, parks, wildlife refuges and BLM lands*. New York: Columbia University Press.

Losch, B. 2004: Debating the multifunctionality of agriculture: from trade negotiation to development policies by the South. *Journal of Agrarian Change* 4 (3): 336-360.

Losch, B., Perraud, D., Laurent, C. and P. Bonnal 2004: Régulation sociale et régulation territoriale de l'agriculture dans les pays du Groupe Cairns et de l'ALENA. *Cahiers de la Multifonctionnalité* 6: 97-106.

Louloudis, L., Beopoulos, N. and G. Vlahos 2000: Greece: late implementation of agri-environmental policies. In: Buller, H., Wilson, G.A. and A. Höll (eds): *Agri-environmental policy in the European Union*. Aldershot: Ashgate, pp. 71-94.

Louloudis, L., Vlahos, G. and Y. Theocharopoulos 2004: The dynamics of local survival in Greek LFAs. In: Brouwer, F. (ed.): *Sustaining agriculture and the rural environment: governance, policy and multifunctionality*. Cheltenham: Edward Elgar, pp. 144-161.

Lovelock, J.E. 1995: *Gaia: a new look at life on Earth* (3rd edn). New York: Oxford University Press.

Lowe, L. 1991: Heterogeneity, hybridity, multiplicity: marking Asian American differences. *Diaspora* 1 (1): 24-44.

Lowe, P. (ed.) 1992: Industrial agriculture and environmental regulation. *Sociologia Ruralis* 32 (1): 1-188.

Lowe, P. and J. Goyder 1983: *Environmental groups in politics*. London: Allen and Unwin.

Lowe, P., Cox, G., MacEwen, M., O'Riordan, T. and M. Winter 1986: *Countryside conflicts: the politics of farming, forestry and conservation*. Aldershot: Gower.

Lowe, P., Murdoch, J., Marsden, T., Munton, R. and A. Flynn 1993: Regulating the new rural spaces: the uneven development of land. *Journal of Rural Studies* 9: 205-222.

Lowe, P., Murdoch, J. and N. Ward 1995: Networks in rural development: beyond exogenous and endogenous models. In: Van der Ploeg, J.D. and G. Van Dijk (eds): *Beyond modernisation: the impact of endogenous rural development*. Assen (NL): Van Gorcum, pp. 83-105.

Lowe, P., Clark, J., Seymour, S. and N. Ward 1997: *Moralizing the environment: countryside change, farming and pollution*. London: UCL Press.

Lowe, P., Ward, N. and C. Potter 1999: Attitudinal and institutional indicators for sustainable agriculture. In: Brouwer, F.M. and J.R. Crabtree (eds): *Environmental indicators and agricultural policy*. Wallingford: CAB International, pp. 263-278.

Lowe, P., Buller, H. and N. Ward 2002: Setting the next agenda? British and French approaches to the Second Pillar of the Common Agricultural Policy. *Journal of Rural Studies* 18: 1-17.

Luczak, H. 2001: Chinesische Medizin: die Suche nach Qi. *Geo* 2001/6: 72-99.

Lunn, E. 1985: *Marxism and modernism*. London: Verso.

Luttik, J. and B. Van der Ploeg 2004: Functions of agriculture in urban society in the Netherlands. In: Brouwer, F. (ed.): *Sustaining agriculture and the rural environment: governance, policy and multifunctionality*. Cheltenham: Edward Elgar, pp. 204-222.

Lyon, D. 1994: *Postmodernity*. Minneapolis: University of Minneapolis Press.

Lyotard, J. 1984: *The postmodern condition*. Minneapolis: University of Minneapolis Press.

MacFarlane, R. 2000: Managing whole landscapes in the post-productive rural environment. In: Benson, J.F. and M.H. Roe (eds): *Landscape and sustainability*. London: Spon Press, pp. 129-156.

MacKinnon, D. 2000: Managerialism, governmentality and the state: a neo-Foucauldian approach to local economic governance. *Political Geography* 19: 293-314.

MacKinnon, N., Bryden, J., Bell, C., Fuller, A. and M. Spearman 1991: Pluriactivity, structural change and farm household vulnerability in Western Europe. *Sociologia Ruralis* 31: 58-71.

Macnaughten, P. and J. Urry 1998: *Contested natures*. London: Sage.

MAFF (Ministry of Agriculture, Fisheries and Food) 1979: *Farming and the nation*. London: HMSO.

Mandler, P. 1997: *The fall and rise of the stately home*. New Haven: Yale University Press.

Mannion, A.M. 1995: *Agriculture and environmental change: temporal and spatial dimensions*. London: Wiley.

Manser, R. 1993: *Failed transitions: the Eastern European economy and environment since the fall of Communism*. New York: New Press.

Marchand, M. and J. Parpart (eds) 1995: *Feminism/postmodernism/development*. London: Routledge.

Marglin, S. and J.B. Schor (eds) 1991: *The golden age of capitalism*. Oxford: Clarendon Press.

Marsden, T. 1995: Beyond agriculture? Regulating the new rural spaces. *Journal of Rural Studies* 11: 285-297.

Marsden, T. 1996: Rural geography trend report: the social and political bases of rural restructuring. *Progress in Human Geography* 20: 246-258.

Marsden, T. 1998a: Economic perspectives. In: Ilbery, B. (ed.): *The geography of rural change*. London: Longman, pp. 13-30.

Marsden, T. 1998b: New rural territories: regulating the differentiated rural spaces. *Journal of Rural Studies* 14: 107-117.

Marsden, 1999a: Rural futures: the consumption countryside and its regulation. *Sociologia Ruralis* 39: 501-520.

Marsden, T. 1999b: Beyond agriculture? Toward sustainable modernisation. In: Redclift, M., Lekakis, J.N. and G.P. Zanias (eds): *Agriculture and world trade liberalisation: socio-environmental perspectives on the Common Agricultural Policy*. Wallingford: CAB International, pp. 238-259.

Marsden, T. 2003: *The condition of rural sustainability*. Assen (NL): Van Gorcum.

Marsden, T. 2004: The quest for ecological modernisation: re-spacing rural development and agri-food studies. *Sociologia Ruralis* 44: 129-146.

Marsden, T. and R. Sonnino 2005: *Setting up and management of public policies with multifunctional purpose: connecting agriculture with new markets and services and rural SMEs (UK national report [WP 5], EU Multagri Project)*. Cardiff: Cardiff University.

Marsden, T. and N. Wrigley 1995: Regulation, retailing, and consumption. *Environment and Planning A* 27: 1899-1912.

Marsden, T., Murdoch, J., Lowe, P., Munton, R. and A. Flynn 1993: *Constructing the countryside*. London: UCL Press.

Marsden, T., Munton, R., Ward, N. and S. Whatmore 1996: Agricultural geography and the political economy approach. *Economic Geography* 72 (4): 361-376.

Marsden, T., Murdoch, J. and K. Morgan 1999: Sustainable agriculture, food supply chains and regional development: editorial introduction. *International Planning Studies* 4 (3): 295-301.

Marsden, T., Banks, J. and G. Bristow 2002: The social management of rural nature: understanding agrarian-based rural development. *Environment and Planning A* 34: 809-825.

Marsden, T., Eklund, E. and A. Franklin 2004: Rural mobilization as rural development: exploring the impacts of new regionalism in Wales and Finland. *International Planning Studies* 9 (2-3): 79-100.

Marsh, G.P. 1864/1965: *Man and nature*. Cambridge (Mass.): Harvard University Press.

Martens, P. and J. Rotmans (eds) 2002: *Transitions in a globalising world*. Lisse (NL): Swets and Zeitlinger.

Martin, P. and D. Halpin 1998: Landcare as a politically relevant new social movement? *Journal of Rural Studies* 14 (4): 445-457.

Marx, K. 1977: *Capital: a critique of political economy* (Vol. 1). New York: Vintage Books.

Marx, K. and F. Engels 1848/1972: *Werke*. Berlin: Karl Dietz Verlag.

Mason, K.O. 1997: Explaining fertility transitions. *Demography* 34 (4): 443-454.

Mason, K.O. 2001: Gender and family systems in the fertility transition. In: Bulatao, R.A. and J.B. Casterline (eds): *Global fertility transition*. New York: Population Council, pp. 160-176.

Masselos, J. 1995: Postmodern Bombay: fractured discourses. In: Watson, S. and K. Gibson (eds): *Postmodern cities and spaces*. Oxford: Blackwell, pp. 199-215.

Massey, D. 2001: Talking of space-time. *Transactions of the Institute of British Geographers* 26: 257-261.

Mather, A.S. 1991: Pressures on British forest policy: prologue to the postindustrial forest. *Area* 23: 245-253.

Mather, A.S. 2001: Forest of consumption: post-productivism, post-materialism and the post-industrial forest. *Environment and Planning C* 19: 249-268.

Mather, A.S., Hill, G. and M. Nijnik 2006: Post-productivism and rural land use: cul de sac or challenge for theorization? *Journal of Rural Studies* 22: 441-455.

Maynard Smith, J. 1986: Structuralism versus selection: is Darwinism enough? In: Rose, S. and L. Appignanesi (eds): *Science and beyond*. Oxford: Blackwell, pp. 39-46.

Maynard Smith, J. and E. Szathmary 1995: *The major transitions in evolution*. Oxford: Freeman/Spectrum.

Mayr, E. 2002: *What evolution is*. London: Phoenix.

Mazoyer, M. 2001: *Protecting small farmers and the rural poor in the context of globalization*. Rome: FAO.

Mazoyer, M. and L. Roudart 2006: *A history of world agriculture: from the Neolithic to the current crisis*. London: Earthscan.

McAllister, I. 1994: Dimensions of environmentalism: public opinion, political activism, and party support in Australia. *Environmental Politics* 3 (1): 22-42.

McCarthy, J. 2005: Scale, sovereignty and strategy in environmental governance. *Antipode* 37 (4): 731-753.

McCarthy, J. 2005: Rural geography: multifunctional rural geographies – reactionary or radical? *Progress in Human Geography* 29 (6): 773-782.

McClintock, A. 1992: The angel of progress: pitfalls of the term 'post-colonialism'. *Social Text* 31: 84-92.

McCormick, J. 1995: *The global environmental movement* (2nd edn). Chichester: Wiley.

McDowell, C. and R. Sparks 1989: The multivariate modelling and prediction of farmers' conservation behaviour towards natural ecosystems. *Journal of Environmental Management* 28: 185-210.

McEachern, C. 1992: Farmers and conservation: conflict and accommodation in farming politics. *Journal of Rural Studies* 8: 159-171.

McHenry, H. 1996: Farming and environmental discourses: a study of the depiction of environmental issues in a German newspaper. *Journal of Rural Studies* 12 (4): 375-386.

McMichael, P. (ed.) 1995: *The global restructuring of agro-food systems*. Ithaca (N.Y.): Cornell University Press.

McReynolds, S.A. 1998: Agricultural labour and agrarian reform in El Salvador: social benefit or economic burden. *Journal of Rural Studies* 14: 459-474.

Meadows, D.H., Randers, J. and D.L. Meadows 2004: *Limits to growth: the 30-year update*. London: Earthscan.

Meadows, D.L., Randers, W. and W. Behrens 1972: *The limits to growth: a report to the Club of Rome's project on the predicament of mankind*. New York: Potomac Associates.

Meert, H., Van Huylenbroek, G., Vernimmen, T., Bourgeois, M. and E. Van Hecke 2005: Farm household survival strategies and diversification on marginal farms. *Journal of Rural Studies* 21: 81-97.

Mercer, I. 2005: Personal comment, Dartmoor Farmers' Association, Princetown, Devon, UK, May 2005.

Meurs, M. and R. Begg 1998: Path dependence in Bulgarian agriculture. In: Pickles, A. and A. Smith (eds): *Theorising transition: the political economy of post-communist transformations*. London: Routledge, pp. 243-261.

Milton, K. 1996: *Environmentalism and cultural theory*. London: Routledge.

Mingay, G. (ed.) 1989: *The rural idyll*. London: Routledge.

Mitsch, W.J. and J.G. Gosselink 1993: *Wetlands* (2nd edn). New York: Van Nostrand Reinhold.

Mohan, G. and K. Stokke 2004: *Post-colonial geography*. London: Sage.

Mollard, A. *et al.* 2004: Multifonctionnalité, externalités et territoire: evaluation, jeu du marché et gouvernance locale. *Les Cahiers de la Multifonctionnalité* 6: 131-140.

Moran, W. 1993: The wine appellation as territory in France and California. *Annals of the Association of American Geographers* 83: 694-717.

Morgan, M.S. and M. Morrison (eds) 1999: *Models as mediators: perspectives on natural and social science*. Cambridge: CUP.

Morris, C. and N. Evans 1999: Research on the geography of agricultural change: redundant or revitalised? *Area* 31 (4): 349-358.

Morris, C. and C. Potter 1995: Recruiting the new conservationists. *Journal of Rural Studies* 11: 51-63.

Morris, C. and M. Winter 1999: Integrated farming: the third way for European agriculture? *Land Use Policy* 16: 193-205.

Morris, C. and C. Young 2000: 'Seed to shelf', 'treat to table', 'barley to beer' and 'womb to tomb': discourses of food quality and quality assurance schemes in the UK. *Journal of Rural Studies* 16: 103-116.

Motoki, Y. 2002: Transformation of rice farming in the process of modernization: Japan's experience and trends in Southeast Asia. Paper presented at the International Symposium Agrarian Transformation and Areal Differentiation in Globalizing Southeast Asia, Rikkyo University, Tokyo, Japan, November 2002.

Munton, R. 1990: Farming families in upland Britain: options, strategies and futures. Paper presented to the Association of American Geographers Conference, Toronto, Canada, April 1990.

Munton, R. 1995: Regulating rural change: property rights, economy and environment – a case study from Cumbria, UK. *Journal of Rural Studies* 11: 267-284.

Munton, R., Marsden, T. and S. Whatmore 1990: Technological change in a period of agricultural adjustment. In: Lowe, P., Marsden, T. and S. Whatmore (eds): *Technological change and the rural environment*. London: David Fulton, pp. 114-132.

Murdoch, J. 1997a: Inhuman/non-human/human: actor network theory and the prospects for a non-dualistic and symmetrical perspective on nature and society. *Environment and Planning D: Society and Space* 15: 731-756.

Murdoch, J. 1997b: Towards a geography of heterogeneous associations. *Progress in Human Geography* 21: 321-337.

Murdoch, J. 2002: What is 'alternative food economy'? Paper presented at the IBG/RGS Conference 'The alternative food economy: myths, realities and potential', London, UK, March 2002.

Murdoch, J. and P. Lowe 2003: The preservationist paradox: modernism, environmentalism and the politics of spatial division. *Transactions of the Institute of British Geographers* 28: 318-332.

Murdoch, J. and T. Marsden 1994: *Reconstituting rurality: class, community and power in the development process.* London: UCL Press.

Murdoch, J. and T. Marsden 1995: The spatialisation of politics: local and national actor-spaces in environmental conflict. *Transactions of the Institute of British Geographers* 20: 368-380.

Murdoch, J. and A. Pratt 1993: Rural studies: modernism, postmodernism and the 'post-rural'. *Journal of Rural Studies* 9: 411-427.

Murdoch, J., Marsden, T. and J. Banks 2000: Quality, nature and embeddedness: some theoretical considerations in the context of the food sector. *Economic Geography* 76: 107-125.

Murdoch, J., Lowe, P., Ward, N. and T. Marsden 2003: *The differentiated countryside.* London: Routledge.

Murray, R. 1992: Flexible specialisation and development strategy: the relevance for Eastern Europe. In: Ernste, H. and V. Meier (eds): *Regional development and contemporary industrial response: extending flexible specialisation.* London: Belhaven, pp. 67-79.

Naess, A. 1989: *Ecology, community and life-style: outline of an ecosophy.* Cambridge: CUP.

Nassauer, J. 1997: Cultural sustainability: aligning aesthetics and ecology. In: Nassauer, J. (ed.): *Placing nature: culture and landscape ecology.* Washington (D.C.): Island Press, pp. 67-83.

Nederveen Pieterse, J. 2004: *Globalization and culture: global mélange.* New York: Rowman and Littlefield.

Newby, H. 1985: *Green and pleasant land? Social change in rural England* (2nd edn). London: Hutchinson.

Newby, H. 1986: Locality and rurality: the restructuring of rural social relations. *Regional Studies* 20: 209-215.

Newby, H., Bell, C., Rose, D. and P. Saunders 1978: *Property, paternalism and power.* London: Hutchinson.

Newman, C. 1985: *The post-modern aura: the act of fiction in an age of inflation.* Evanston (Illinois): Northwestern University Press.

Nietzsche, F. 1887/1967: *On the genealogy of morals.* New York: Vintage.

Norgaard, R. 1994: *Development betrayed.* London: Routledge.

Nowicki, P.L. 2004: Jointness of production as a market concept. In: Brouwer, F. (ed.): *Sustaining agriculture and the rural environment: governance, policy and multifunctionality.* Cheltenham: Edward Elgar, pp. 36-55.

O'Connor, J. 1989: Political economy of ecology of socialism and capitalism. *Capitalism, Nature, Socialism: a Journal of Socialist Ecology* 3: 93-107.

OECD [Organisation for Economic Cooperation and Development] 2001: *Multifunctionality: towards an analytical framework.* Paris: OECD.

Offe, C. 1985: New social movements: challenging the boundaries of institutional politics. *Social Research* 52 (4): 817-868.

Ogden, P.E. and R. Hall 2004: The second demographic transition, new Household forms and the urban population of France during the 1990s. *Transactions of the Institute of British Geographers* 29: 88-105.

Olesen, J.E. and M. Bindi 2002: Consequences of climate change for European agricultural productivity, land use and policy. *European Journal of Agronomy* 16 (4): 239-262.

Olsson, G. 1991: *Lines of power – limits of language.* Minneapolis: University of Minneapolis Press.

Oman, C. 1994: *Globalisation et régionalisation: quels enjeux pour les pays en développement.* Paris: OECD.

Oñate, J.J. and B. Peco 2005: Desertification and policies in Spain: from land abandonment to intensive irrigated areas. In: Wilson, G.A. and M. Juntti (eds): *Unravelling desertification: policies and actor networks in Southern Europe.* Wageningen (NL): Wageningen Academic Publishers, pp. 73-100.

O'Riordan, T. 1981: *Environmentalism.* London: Pion.

O'Riordan, T. 1989: The challenge for environmentalism. In: Peet, R. and N. Thrift (eds): *New models in geography* (Vol. 1). Boston: Unwin Hyman, pp. 77-101.

O'Riordan, T. 1995: *Environmental science for environmental management.* Harlow: Longman.

O'Riordan, T. 2004: Environmental science, sustainability and politics. *Transactions of the Institute of British Geographers* 29: 234-247.

O'Riordan, T. and H. Voisey 1998: The political economy of the sustainability transition. In: O'Riordan, T. and H. Voisey (eds): *The transition to sustainability: the politics of Agenda 21 in Europe*. London: Earthscan, pp. 3-30.

Osterburg, B. 2005: Modellierung der ökonomischen, ökologischen und sozialen Auswirkungen der reformierten Gemeinsamen Agrarpolitik. Paper presented at the symposium 'Agrarförderung: Motor oder Hemmnis für Innovation im ländlichen Raum?', Berlin-Brandenburgische Akademie der Wissenschaften, Berlin, Germany, October 2005.

O'Sullivan, D. 2004: Complexity science and human geography. *Transactions of the Institute of British Geographers* 29 (3): 282-295.

Parayil, G. 1999: *Conceptualising technological change: theoretical and empirical explorations*. Lanham (Maryland): Rowman and Littlefield.

Passmore, J. 1980: *Man's responsibility to nature* (2nd edn). London: Duckworth.

Patnail, U. 1987: *Peasant class differentiation: a study in method with reference to Haryana*. Delhi: Oxford University Press.

Pavlinek, P. and J. Pickles 2000: *Environmental transitions: transformation and ecological defence in Central and Eastern Europe*. London: Routledge.

Payne, D. 2000: Policy-making in nested institutions: explaining the conservation failure of the EU's Common Fisheries Policy. *Journal of Common Market Studies* 38 (2): 303-324.

Pearce, D. and E. Barbier 2000: *Blueprint for a sustainable economy*. London: Earthscan.

Peck, J. and A. Tickell 1994: Searching for a new institutional fix: the after-Fordist crisis and the global-local disorder. In: Amin, A. (ed.): *Post-Fordism: a reader*. Oxford: Blackwell, pp. 280-315.

Peco, B., Suárez, F., Oñate, J.J., Malo, J.E. and J. Aguirre 2000: Spain: first tentative steps toward an agri-environmental programme. In: Buller, H., Wilson, G.A. and A. Höll (eds): *Agri-environmental policy in the European Union*. Aldershot: Ashgate, pp. 145-168.

Peet, R. 1991: *Global capitalism: theories of societal development*. London: Routledge.

Pepper, D. 1984: *The roots of modern environmentalism*. London: Croom Helm.

Pepper, D. 1996: *Modern environmentalism*. London: Routledge.

Perkins, R. 2003: Environmental leapfrogging in developing countries: a critical assessment and reconstruction. *Natural Resources Forum* 27: 177-188.

Peterson, J.M., Boisvert, R.N. and H. de Gorter 2002: Environmental policies for a multifunctional agricultural sector in open economies. *European Review of Agricultural Economics* 29 (4): 423-443.

Philo, C. 1992: Neglected rural geographies: a review. *Journal of Rural Studies* 8: 193-207.

Philo, C. 1993: Postmodern rural geography? A reply to Murdoch and Pratt. *Journal of Rural Studies* 9: 429-436.

Pickles, J. 2000: Ethnicity, violence, and the production of regions: introduction to the special issue. *Growth and Change* 31 (2): 139-150.

Pickles, J. and R. Begg 2000: Ethnicity, state violence, and neo-liberal transitions in post-communist Bulgaria. *Growth and Change* 31 (2): 179-210.

Pickles, J. and A. Smith (eds) 1998: *Theorising transition: the political economy of post-communist transformations*. London: Routledge.

Pickles, J. and T. Unwin 2004: Transition in context: theory in post-socialist transformations. In: Van Hoven, B. (ed.): *Europe: lives in transition*. Harlow: Pearson, pp. 9-28.

Pivot, J.M., Caron, P. and P. Bonnal 2003: Coordinations locales, actions collectives, territories et multifonctionnalité de l'agriculture: éclairages et perspectives. *Les Cahiers de la Multifonctionnalité* 3: 5-16.

Polanyi, K. 1944: *The great transformation*. Boston: Beacon Press.

Ponte, S. 2002: The 'latte revolution'? Regulation, markets and consumption in the global coffee chain. *World Development* 30 (7): 1099-1122.

Popkin, B. 1998: The nutrition transition and its health implications in lower-income countries. *Public Health Nutrition* 1 (1): 5-21.

Possemeyer, I. and F. Killmeyer 2005: Gesellschaft in Zeitnot: die Diktatur der Uhr. *Geo* 2005/8: 80-109.

Poster, M. 1989: *Critical theory and post-structuralism: in search of a context*. Ithaca: Cornell University Press.

Potter, C. 1990: Conservation under a European farm survival policy. *Journal of Rural Studies* 6 (1): 1-7.

Potter, C. 1998: *Against the grain: agri-environmental reform in the United States and the European Union*. Wallingford: CAB International.

Potter, C. 2004: Multifunctionality as an agricultural and rural policy concept. In: Brouwer, F. (ed.): *Sustaining agriculture and the rural environment: governance, policy and multifunctionality*. Cheltenham: Edward Elgar, pp. 15-35.

Potter, C. and J. Burney 2002: Agricultural multifunctionality in the WTO: legitimate non-trade concern or disguised protectionism? *Journal of Rural Studies* 18: 35-47.

Potter, C. and M. Tilzey 2005: Agricultural policy discourses in the European post-Fordist transition: neoliberalism, neomercantilism and multifunctionality. *Progress in Human Geography* 29 (5): 1-20.

Potter, J.E. 1999: The persistence of outmoded contraceptive regimes: the cases of Mexico and Brazil. *Population and Development Review* 25 (4): 703-739.

Potter, R.B. and G.M. Dann 1996: Globalization, postmodernity and development in the Commonwealth Caribbean. In: Yeung Y-M. (ed.): *Global change and the Commonwealth*. Hong Kong: Chinese University of Hong Kong Press, pp. 103-129.

Povellato, A. and D. Ferraretto 2005: Desertification policies in Italy: new pressures on land and 'desertification' as rural-urban migration. In: Wilson, G.A. and M. Juntti (eds): *Unravelling desertification: policies and actor networks in Southern Europe*. Wageningen (NL): Wageningen Academic Publishers, pp. 101-130.

Pratt, A. 1996: Rurality: loose talk or social struggle. *Journal of Rural Studies* 12: 69-78.

Pretty, J.N. 1995: *Regenerating agriculture: policies and practice for sustainability and self-reliance*. London: Earthscan.

Pretty, J. 1998: *The living land: agriculture, food and community regeneration in rural Europe*. London: Earthscan.

Pretty, J.N. 2002: *Agri-culture: reconnecting people, land and nature*. London: Earthscan.

Pretty, J.N. 2005: Agriculture, biodiversity and economic development: can sustainability in the food system make a difference? Paper presented at the Andrew Errington Memorial Lecture, Rural Futures Seminar Series, University of Plymouth, Plymouth, UK, March 2005.

Pretty, J.N., Brett, C., Gee, D. Hine, R., Mason, C., Morison, J., Rayment, M., Van der Bijl, G. and T. Dobbs 2001: Policy challenges and priorities of internalizing the externalities of modern agriculture. *Journal of Environmental Planning and Management* 44 (2): 263-283.

Priemus, H. and P. Hall 2004: Multifunctional urban planning of mega-city-regions. *Built Environment* 30: 338-349.

Priemus, H., Rodenburg, C. and P. Nijkamp 2004: Multifunctional urban land use: a new phenomenon? A new planning challenge? *Built Environment* 30: 269-273.

Primdahl, J. 1999: Agricultural landscapes as places of production and for living in: owner's versus producer's decision making and the implications for planning. *Landscape and Urban Planning* 46: 143-150.

Primdahl, J. and S. Swaffield 2004: Segregation and multifunctionality in New Zealand landscapes. In: Brouwer, F. (ed.): *Sustaining agriculture and the rural environment: governance, policy and multifunctionality*. Cheltenham: Edward Elgar, pp. 266-285.

Princen, T. and M. Finger 1994: *Environmental NGOs in world politics: linking the local and the global*. London: Routledge.

Przeworski, A. 1995: *Sustainable democracy*. Cambridge: CUP.

Putnam, R.D. 1993: *Making democracy work*. Princeton (N.J.): Princeton University Press.

Randall, A. 2002: Valuing the outputs of multifunctional agriculture. *European Review of Agricultural Economics* 29 (3): 289-307.

Rangan, H. and M. Lane 2001: Indigenous peoples and forest management: comparative analysis of institutional approaches in Australia and India. *Society and Natural Resources* 14 (2): 145-160.

Rantamäki-Lahtinen, L. 2004: On-farm non-agricultural diversification in Finland from a farm management perspective. Paper presented at the South-West Rural Research Network seminar, University of Plymouth, Plymouth, UK, January 2004.

Rapey, H., Lardon, S., Josien, E., Servière, G., Fiorelli, C., Klingenschmidt, F. and E. Matter 2004a: Multifonctionnalité de l'espace agricole d'un territoire: premières conclusions issues d'une demarche de recherche. *Les Cahiers de la Multifonctionnalité* 5: 71-86.

Rapey, H., Josien, E., Lardon, S. and G. Serviere 2004b: Variabilité spatiale et temporelle de la multifonctionnalité de l'agriculture sur un territoire en régions d'élevage: liens avec les dynamiques des exploitations. *Les Cahiers de la Multifonctionnalité* 6: 43-50.

Rawski, T.G. and R.W. Mead 1998: On the trail of China's phantom farmers. *World Development* 26 (5): 767-781.

Ray, C. 1998: Territory, structures and interpretation: two case studies of the European Union's LEADER 1 Programme. *Journal of Rural Studies* 14 (1): 79-87.

Ray, C. 2000: The EU LEADER Programme: rural development laboratory. *Sociologia Ruralis* 40 (2): 163-171.

Reason, P. and B. Goodwin 1999: Towards a science of qualities in organizations: lessons from complexity theory and postmodern biology. *Concepts and Transformations* 4: 281-317.

Redclift, M. 1987: *Sustainable development: exploring the contradictions*. London: Methuen.

Redclift, M., Lekakis, J.N. and G.P. Zanias (eds) 1999: *Agriculture and world trade liberalisation: socio-environmental perspectives on the Common Agricultural Policy*. Wallingford: CAB International.

Regulska, J. 1998: 'The political' and its meaning for women: transition politics in Poland. In: Pickles, A. and A. Smith (eds): *Theorising transition: the political economy of post-communist transformations*. London: Routledge, pp. 309-329.

Reid, A. 1988: *Southeast Asia in the age of commerce 1450-1680: Volume 1 – the lands below the winds*. New Haven: Yale University Press.

Renting, H., Marsden, T. and J. Banks 2003: Understanding alternative food networks: exploring the role of short food supply chains in rural development. *Environment and Planning A* 35: 393-411.

Reynolds, L.T. 2002: Wages for wives: renegotiating gender and production relations in contract farming in the Dominican Republic. *World Development* 30 (5): 783-798.

Rhodes, R. 1997: *Understanding governance*. Buckingham (UK): Open University Press.

Ridley, M. 1996: *Evolution* (2nd edn). Cambridge (Mass.): Blackwell Science.

Rigg, J. 1989: The new rice technology and agrarian change: guilt by association. *Progress in Human Geography* 13: 374-399.

Rigg, J. 1990: Developing World: the Green Revolution 25 years on. *Geography Review* 4 (1): 32-34.

Rigg, J. 1993: Rice, water and land: strategies of cultivation on the Khorat Plateau, Thailand. *South East Asia Research* 1 (2): 197-221.

Rigg, J. 1998: Rural-urban interactions, agriculture and wealth: a Southeast Asian perspective. *Progress in Human Geography* 22 (4): 497-522.

Rigg, J. 2001: *More than the soil: rural change in Southeast Asia*. Harlow: Pearson.

Rigg, J. and S. Nattapoolwat 2001: Embracing the global in Thailand: activism and pragmatism in an era of de-agrarianisation. *World Development* 29 (6): 945-960.

Rigg, J. and M. Ritchie 2002: Production, consumption and imagination in rural Thailand. *Journal of Rural Studies* 18 (4): 359-371

Rip, A. and R. Kemp 1998: Technological change. In: Rayner, S. and E.L. Malone (eds): *Human choice and climate change: resources and technology*. Columbus (Ohio): Battelle Press.

Ritchie, M.A. 1996a: Centralization and diversification: from local to non-local economic reproduction and resource control in Northern Thailand. Paper presented at the 6th International Conference on Thai Studies, Chiang Mai, Thailand, October 1996.

Ritchie, M.A. 1996b: *From peasant farmers to construction workers: the breaking down of the boundaries between agrarian and urban life in Northern Thailand, 1974-1992*. Ann Arbor: UMI.

Ritson, C. and D.R. Harvey (eds) 1997: *The Common Agricultural Policy* (2nd edn). Wallingford: CAB International.

Robinson, G. 2004: *Geographies of agriculture: globalisation, restructuring and sustainability*. Harlow: Pearson.

Roche, M. 2003: Rural geography: a stock tally of 2002. *Progress in Human Geography* 27 (6): 779-786.

Roche, M. 2005: Rural geography: a borderland revisited. *Progress in Human Geography* 29 (3): 299-303.

Rodenburg, C. and P. Nijkamp 2004: Multifunctional land use in the city: a typological overview. *Built Environment* 30: 274-288.

Rodney, W. 1972: *How Europe underdeveloped Africa*. London: Bogle.

Rogers, E. 1995: *Diffusion of innovations* (4th edn). New York: Free Press.

Romstad, E., Vatn, A., Rorstad, P.K. and V. Soyland 2000: *Multifunctional agriculture: implications for policy design*. As (Norway): Agricultural University of Norway.

Ronningen, K. 1994: *Multifunctional agriculture in Europe's playground?* Trondheim (Norway): University of Trondheim (Department of Geography).

Rosero-Bixby, L. and J.B. Casterline 1993: Modelling diffusion effects in fertility transition. *Population Studies* 47: 147-167.

Rotmans, J. and P. Martens 2002: Transitions in a globalising world: what does it all mean? In: Martens, P. and J. Rotmans (eds): *Transitions in a globalising world*. Lisse (NL): Swets and Zeitlinger, pp. 111-126.

Rotmans, J., Kemp, R. and M.B. Van Asselt 2001: More evolution than revolution: transition management in public policy. *Foresight* 3 (1): 15-31.

Rotmans, J., Martens, P. and M.B. Van Asselt 2002: Introduction. In: Martens, P. and J. Rotmans (eds): *Transitions in a globalising world*. Lisse (NL): Swets and Zeitlinger, pp. 1-16.

Roux, B., Bonnal, P. and R. Maluf 2004: Analyse au niveau local de la prise en compte de la multifonctionnalité de l'agriculture au Brésil à partir d'un dispositif comparatif régional: rapprochement du cas français. *Les Cahiers de la Multifonctionnalité* 6: 141-148.

RSPB [Royal Society for the Protection of Birds] 2001: *Eat this: fresh ideas for the WTO Agreement on Agriculture*. Sandy (UK): RSPB.

Sachs, J. 1990: Eastern Europe's economies: what is to be done? *Economist* January 13: 21-26.

Sachs, J. 1992: Building a market economy in Poland. *Scientific American* March: 34-40.

Sahal, D. 1985: Technological guideposts and innovation avenues. *Research Policy* 14: 61-82.

Said, E. 1983: Travelling theory. In: Said, E. (ed.): *The word, the text and the critic*. Cambridge (Mass.): Harvard University Press, pp. 226-247.

Said, E. 1993: *Culture and imperialism*. New York: Knopf.

Sanders, R.A. 2000a: Political economy of Chinese ecological agriculture: a case study of seven Chinese eco-villages. *Journal of Contemporary China* 9: 349-372.

Sanders, R.A. 2000b: *Prospects for sustainable development in the Chinese countryside*. Aldershot: Ashgate.

Sanders, R.A. 2006: A market road to sustainable agriculture? Ecological agriculture, green food and organic agriculture in China. *Development and Change* 37: 201-226.

Sanderson, S.E. 1986: *The transformation of Mexican agriculture: international structure and the politics of rural change*. Princeton (N.J.): Princeton University Press.

Santow, G. and M.D. Bracher 1999: Traditional families and fertility decline: lessons from Australia's southern Europeans. In: Leete, R. (ed.): *Dynamics of values in fertility change*. Oxford: OUP, pp. 51-77.

Sarup, M. 1993: *Introductory guide to post-structuralism and postmodernism*. Athens (Georgia): University of Georgia Press.

Sayer, A. 1989: The 'new' regional geography and problems of narrative. *Environment and Planning D: Society and Space* 7: 253-276.

Sayer, A. 1991: Beyond the locality debate: deconstructing geography's dualisms. *Environment and Planning A* 23: 283-308.

Sayer, A. 1995: *Radical political economy: a critique*. Oxford: Blackwell.

Schakel, J. 2003: Trandisciplinarity and plurality of the consequences of multifunctionality for agricultural science and education. In: Van Huylenbroek, G. and G. Durand (eds): *Multifunctional agriculture: a new paradigm for European agriculture and rural development*. Aldershot: Ashgate, pp. 225-233.

Schama, S. 1995: *Landscape and memory*. London: Fontana Press.

Schmid, H. and B. Lehmann 2000: Switzerland: agri-environmental policy outside the European Union. In: Buller, H., Wilson, G.A. and A. Höll (eds): *Agri-environmental policy in the European Union*. Aldershot: Ashgate, pp. 185-202.

Schmidt, C.C. 2003: Fisheries and Japan: a case for multiple roles? Paper presented at the International Symposium 'Multiple roles and functions of fisheries and fishing communities', Aomori, Japan, February 2003.

Schoefield, D. 2005: Organic farming as a problem: further comments. Paper presented at the workshop 'Organic agriculture: a solution rather than a problem?', IGER North Wyke Research Station, North Wyke, Devon, UK, November 2005.

Scholten, B. 2004: A grass ceiling is haunting America's rural economy: comparing a US alternative agro-food network to those in the UK and Germany. Paper presented at the 2nd meeting of the Anglo-German Rural Geographers 'Rural multifunctionality: perspectives from policy-making, implementation and practice', University of Exeter, Exeter, UK, July 2004.

Schumacher, E.F. 1973: *Small is beautiful: a study of economics as if people mattered*. London: Abacus.

Scott, A., Christie, M. and P. Midmore 2004: Impact of the 2001 foot-and-mouth disease outbreak in Britain: implications for rural studies. *Journal of Rural Studies* 20: 1-14.

Scott, J.C. 1976: *The moral economy of the peasant: rebellion and subsistence in Southeast Asia*. New Haven: Yale University Press.

Seabrook, M. and C. Higgins 1988: The Role of the farmer's self concept in determining farmer behaviour. *Agricultural Administration and Extension* 30: 99-108.

Selby, J. and L. Petäjistö 1995: Attitudinal aspects of the resistance to field afforestation in Finland. *Sociologia Ruralis* 35 (1): 67-92.

Sen, A. 1981: *Poverty and famines*. Oxford: Clarendon.

Sen, K. 1994: *Ageing: debates on demographic transition and social policy*. London: Zed Books.

Sessions, G. (ed.) 1994: *Deep ecology for the 21st century*. Boston: Shambhala Press.

Sheingate, A. 2000: Agricultural retrenchment revisited: issue definition and venue change in the United States and European Union. *Governance* 13: 335-363.

Shiva, V. and G. Bedi 2002: *Sustainable agriculture and food security*. London: Sage.

Shohat, E. and R. Stam 1994: *Unthinking Eurocentrism: multiculturalism and the media*. London: Routledge.

Short, D. 1997: Traditional-style farming and values for sustainable development. *Tijdschrift voor Economische en Sociale Geografie* 88 (1): 41-52.

Shucksmith, M. 1993: Farm household behaviour and the transition to post-productivism. *Journal of Agricultural Economics* 44: 466-478.

Shuman, M. 2000: *Going local: creating self-reliant communities in a global age*. London: Routledge.

Sidaway, J.D. 2003: Sovereign excesses? Portraying postcolonial sovereigntyscapes. *Political Geography* 22 (1): 157-178.

Simmons, I.G. 1989: *Changing the face of the Earth: culture, environment, history*. Oxford: Blackwell.

Simon, D. 1998: Rethinking (post)modernism, postcolonialism, and posttraditionalism: South-North perspectives. *Environment and Planning D: Society and Space* 16: 219-245.

Simon, D. 2006: Development theory and post-colonialism. Paper presented at the School of Geography lunchtime seminars, University of Plymouth, Plymouth, UK, February 2006.

Simon, J.L. 1981: *The ultimate resource*. Princeton (N.J.): Princeton University Press.

Simon, J.L. and H. Kahn 2005: Introduction to 'The resourceful Earth'. In: Dryzek, J.S. and D. Schlosberg (eds): *Debating the Earth: the environmental politics reader*. Oxford: OUP, pp. 51-73.

Singh, S. 2002: Contracting out solutions: political economy of contract farming in the Indian Punjab. *World Development* 30 (9): 1621-1638.

Singh, Y. 1989: *Essays on modernisation in India*. New Delhi: Manohar.

Singhanetra-Renard, A. 1999: Population mobility and the transformation of the village community in Northern Thailand. *Asia Pacific Viewpoint* 40 (1): 69-87.

Skogstad, G. 1998: Ideas, paradigms and institutions: agricultural exceptionalism in the European Union and the United States. *Governance* 11: 463-490.

Smailes, P. 2002: From rural dilution to multifunctional countryside: some pointers to the future from South Australia. *Australian Geographer* 33 (1): 79-95.

Smith, A, 1995: Regulation theory, strategies of enterprise integration and the political economy of regional economic restructuring in central and eastern Europe. *Regional Studies* 7: 761-772.

Smith, A. 1996: From convergence to fragmentation: uneven development, industrial restructuring, and the 'transition to capitalism' in Slovakia. *Environment and Planning A* 28: 135-156.

Smith A. and J. Pickles 1998: Introduction: theorising transition and the political economy of transformation. In: Pickles, J. and A. Smith (eds): *Theorising transition: the political economy of post-communist transformations*. London: Routledge, pp. 1-24.

Smith, A. and A. Swain 1998: Regulating and institutionalising capitalism: the micro-foundations of transformation in Eastern and Central Europe. In: Pickles, A. and A. Smith (eds): *Theorising transition: the political economy of post-communist transformations*. London: Routledge, pp. 25-53.

Smith, F. 2000: 'Multifunctionality' and 'non-trade concerns' in the agriculture negotiations. *Journal of International Economic Law* (2000): 707-713.

Smith, J.K. 1993: Thinking about technological change: linear and evolutionary models. In: Thomson, R. (ed.): *Learning and technological change*. New York: St Martin's Press, pp. 65-78.

Sobhan, R. 1993: *Agrarian reform and social transformation: preconditions to development*. London: Zed Books.

Sonnino, R. and T. Marsden 2005: Beyond the divide: rethinking relationships between alternative and conventional food networks in Europe. *Journal of Economic Geography* 5: 1-19.

Sonoda, S. 1990: Modernisation of Asian countries as a process of 'overcoming their backwardness': the case of modernisation in China. Paper presented at the 12th World Congress of the International Sociological Association, Madrid, Spain, September 1990.

Spencer, H. 1870: *The principles of psychology* (2nd edn). London: Williams and Norgate.

Spengler, O. 1992: *The decline of the West*. New York: Knopf.

Stark, D. 1992: Path dependence and privatization strategies in East Central Europe. *East European Politics and Societies* 6 (1): 17-51.

Stiglitz, J. 2002: *Globalization and its discontents*. London: Penguin.

Storper, M. 1997: Regional economies as relational assets. In: Lee, R. and J. Wills (eds): *Geographies of economies*. London: Edward Arnold, pp. 249-261.

Strassmann, W.P. 1959: Creative destruction and partial obsolescence in American economic development. *Journal of Economic History* 14 (3): 335-349.

Stryker, S. 1968: Identity salience and role performance: the relevance of symbolic interaction theory for family research. *Journal of Marriage and the Family* 30: 558-564.

Stryker, S. 1994: Identity theory: its development, research base, and prospects. In: Denzin, N.K. (ed.): *Studies in symbolic interactionism*. London: JAI Press, pp. 9-20.

Stryker, S. and R. Serpe 1982: Commitment, identity salience, and role behaviour: theory and research examples. In: Ickes, W. and E.S. Knowles (eds): *Personality roles and social behaviour*. New York: Springer, pp. 199-219.

Swagemakers, P. 2003: Novelty production: new directions for the activities and role of farming. In: Van Huylenbroek, G. and G. Durand (eds): *Multifunctional agriculture: a new paradigm for European agriculture and rural development*. Aldershot: Ashgate, pp. 189-207.

Swain, A. 2006: Soft capitalism and a hard industry: virtualism, the 'transition industry' and the restructuring of the Ukrainian coal industry. *Transactions of the Institute of British Geographers* 31 (2): 208-223.

Swinbank, A. 2002: Multifunctionality: the concept and its international acceptability. *Journal of the Royal Agricultural Society of England* 163: 141-148.

Swyngedouw, E. 1997: Neither global nor local: 'glocalization' and the politics of scale. In: Cox, K. (ed.): *Paces of globalization*. London: Guildford Press, pp. 137-166.

Swyngedouw, E. 2001: Elite power, global forces, and the political economy of 'glocal' development. In: Clark, G.L., Feldman, M.P. and M.S. Gertler (eds): *A reader in economic geography*. Oxford: OUP, pp. 541-558.

Symes, D. 1992: Agriculture, the state and rural society in Europe: trends and issues. *Sociologia Ruralis* 32: 193-208.

Szreter, S. 1993: The idea of the demographic transition and the study of fertility change: a critical intellectual history. *Population and Development Review* 19 (4): 659-701.

Tangermann, S. 1996: Implementation of the Uruguay Round Agreement on agriculture: issues and prospects. *Journal of Agricultural Economics* 47: 315-337.

Taylor, P.J. 1996: *The way the modern world works: world hegemony to world impasse*. Chichester: Wiley.

Thompson, W. 1929: Population. *American Journal of Sociology* 34: 959-975.

Thoreau, H.D. 1859: *Walden – or life in the woods*. Boston: Ticknor and Fields.

Thornes, J.B. and D. Brunsden 1977: *Geomorphology and time*. London: Methuen.

Thrift, N. 1995: A hyperactive world. In: Johnston, R.J., Taylor, P.J. and M. Watts (eds): *Geographies of global change: remapping the world in the late twentieth century*. Oxford: Blackwell, pp. 18-35.

Thrift, N, 1996a: Inhuman geographies: landscapes of speed, light and power. In: Thrift, N. (ed.): *Spatial formations*. London: Sage, pp. 256-310.

Thrift, N. 1996b: New urban eras and old technological fears: reconfiguring the goodwill of electronic things. *Urban Studies* 33: 1463-1493.

Thrift, N. 1999: The place of complexity. *Theory Culture and Society* 16: 31-69.

Tilzey, M. 2000: Natural areas, the whole countryside approach and sustainable agriculture. *Land Use Policy* 17: 279-294.

Tominaga, K. 1990: A theory of modernisation of non-Western societies: towards a generalisation from historical experiences of Japan. Paper presented at the 12th World Congress of the International Sociological Association, Madrid, Spain, September, 1990.

Tovey, H. 1997: Food, environmentalism and rural sociology: the organic farming movement in Ireland. *Sociologia Ruralis* 37 (1): 21-38.

Tovey, H. 2000: Agricultural development and environmental regulation in Ireland. In: Buller, H. and K. Hoggart (eds): *Agricultural transformation, food and the environment*. Basingstoke: Ashgate, pp. 109-129.

Toynbee, A. 1962: *A study of history*. Oxford: OUP.

Traugher, A. 2004: 'Because they can do the work': women farmers in sustainable agriculture in Pennsylvania. *Gender Place and Culture* 11: 289-307.

Trippel, K. and M. Davenport 2003: Ein Streitfall namens 'el veneno'. *Geo* 2003/7: 148-163.

Turner, M. and M. Reed 2006: Farm diversification and the new rural paradigm. Paper presented at the Conference 'The rural citizen: governance, culture and wellbeing', University of Plymouth, Plymouth, UK, April 2006.

Twyne, J. 2005: Organic farming as a solution to environmental problems? Paper presented at the workshop 'Organic agriculture: a solution rather than a problem?', IGER North Wyke Research Station, North Wyke, Devon, UK, November 2005.

UNCCD [United Nations Convention on Desertification] 1994: *United Nations convention to combat desertification*. Geneva: UN Environmental Programme.

UNEP [United Nations Environmental Programme] 1992: *World atlas of desertification*. London: Edward Arnold.

United Nations 2003: *World population prospects: the 2002 revision*. New York: UN.

Urry, J. 1984: Capitalist restructuring, recomposition and the regions. In: Bradley, T. and P. Lowe (eds): *Locality and rurality*. Norwich: Geo Books, pp. 45-64.

Usher, A.P. 1954: *A history of mechanical inventions*. Cambridge (Mass.): Harvard University Press.

Van de Kaa, D. 1987: Europe's second demographic transition. *Population Bulletin* 42 (1): 7-9.

Van de Kaa, D. 2001: Postmodern fertility preferences: from changing value orientation to new behaviour. In: Bulatao, R.A. and J.B. Casterline (eds): *Global fertility transition*. New York: Population Council, pp. 290-331.

Van der Meulen, H., De Snoo, G. and G. Wossink 1996: Farmers' perception of unsprayed crop edges in the Netherlands. *Journal of Environmental Management* 47: 241-255.

Vandermeulen, V., Verspecht, A., Van Huylenbroek, G., Meert, H., Boulanger, A. and E. Van Hecke 2006: The importance of the institutional environment on multifunctional farming systems in the peri-urban area of Brussels. *Land Use Policy* 23 (4): 486-501.

Van der Ploeg, J.D. 1990: *Labour, markets and agricultural production*. Boulder (Col.): Westview Press.

Van der Ploeg, J.D. 1997: On rurality, rural development and rural sociology. In: De Haan, H. and N. Long (eds): *Images and realities of rural life*. Assen (NL): Van Gorcum.

Van der Ploeg, J.D. 2000: Revitalizing agriculture: farming economically as starting ground for rural development. *Sociologia Ruralis* 40 (4): 497-511.

Van der Ploeg, J.D. 2003: *The virtual farmer: past, present and future of the Dutch peasantry*. Assen (NL): Van Gorcum.

Van der Ploeg, J.D. and H. Renting 2000: Impact and potential: a comparative review of rural development practices. *Sociologia Ruralis* 40 (4): 529-543.

Van der Ploeg, J.D. and D. Roep 2003: Multifunctionality and rural development: the actual situation in Europe. In: Van Huylenbroek, G. and G. Durand (eds): *Multifunctional agriculture: a new paradigm for European agriculture and rural development*. Aldershot: Ashgate, pp. 37-53.

Van der Ploeg, J.D., Renting, H., Brunori, G., Knickel, K., Mannion, J., Marsden, T., de Roost, K., Sevilla-Guzman, E. and F. Ventura 2000: Rural development: from practices and policies towards theory. *Sociologia Ruralis* 4 (4): 391-408.

Van Hoven, B. 2004a: Understanding transition. In: Van Hoven, B. (ed.): *Europe: lives in transition*. Harlow: Pearson, pp. 29-43.

Van Hoven, B. 2004b: Looking back on 'Lives in transition'. In: Van Hoven, B. (ed.): *Europe: lives in transition*. Harlow: Pearson, pp. 161-163.

Van Hoven, B., Unwin, T. and A. Jansen 2004: Introduction. In: Van Hoven, B. (ed.): *Europe: lives in transition*. Harlow: Pearson, pp. 1-8.

Van Huylenbroek, G. 2003: Preface. In: Van Huylenbroek, G. and G. Durand (eds): *Multifunctional agriculture: a new paradigm for European agriculture and rural development*. Aldershot: Ashgate, pp. xii-vx.

Van Huylenbroek, G. and G. Durand (eds) 2003: *Multifunctional agriculture: a new paradigm for European agriculture and rural development*. Aldershot: Ashgate.

Vanslembrouck, I. and G. Van Huylenbroek 2003: The demand for landscape amenities by rural tourists. In: Van Huylenbroek, G. and G. Durand (eds): *Multifunctional agriculture: a new paradigm for European agriculture and rural development*. Aldershot: Ashgate, pp. 83-99.

Vargas, V. 1992: The feminist movement in Latin America: between hope and disenchantment. In: Nederveen Pieterse, J. (ed.): *Emancipation, modern and post-modern*. London: Sage, pp. 195-214.

Vatn, A. 2002: Multifunctional agriculture: some consequences for international trade regimes. *European Review of Agricultural Economics* 29 (3): 309-327.

Venn, L., Kneafsey, M., Holloway, L., Cox, R., Dowler, E. and H. Tuomainen 2006: Researching European 'alternative' food networks: some methodological considerations. *Area* 38 (3): 248-258.

Vickers, J. and G. Yarrow 1991: Economic perspectives on privatization. *Journal of Economic Perspectives* 5 (2): 111-132.

Vieira, M. and P. Eden 2005: Desertification and policies in Portugal: landuse changes and pressures on local biodiversity. In: Wilson, G.A. and M. Juntti (eds): *Unravelling desertification: policies and actor networks in Southern Europe*. Wageningen (NL): Wageningen Academic Publishers, pp. 131-156.

Virilio, P. 1993: The third interval: a critical transition. In: Andermatt-Conley, V. (ed.): *Rethinking technologies*. London: University of Minnesota Press, pp. 3-10.

Vreeker, R., De Groot, H.L. and E.T. Verhoef 2004: Urban multifunctional land use: theoretical and empirical insights on economics of scale, scope and diversity. *Built Environment* 30: 289-307.

Walford, N. 2003a: A past and a future for diversification on farms? Some evidence from large-scale commercial farms in south-east England. *Geografiska Annaler* 85B: 51-62.

Walford, N. 2003b: Productivism is allegedly dead, long live productivism: evidence of continued productivist attitudes and decision making in south-east England. *Journal of Rural Studies* 19: 491-502.

Walker, K.J. 1989: The state in environmental management: the ecological dimension. *Political Studies* 37: 25-38.

Wallerstein, I. 1979: *The capitalist world-economy*. Cambridge: CUP.

Wallerstein, I. 1991: *Geopolitics and geoculture*. Cambridge: CUP.

Wapner, P. 1995: Politics beyond the state: environmental activism and world civic politics. *World Politics* 47: 311-340.

Ward, N. 1993: The agricultural treadmill and the rural environment in the post-productivist era. *Sociologia Ruralis* 33: 348-364.

Ward, N. 1995: Technological change and the regulation of pollution from agricultural pesticides. *Geoforum* 26: 19-33.

Ward, N. and P. Lowe 1994: Shifting values in agriculture: the farm family and pollution regulation. *Journal of Rural Studies* 10: 173-184.

Ward, N. and P. Lowe 2004: Europeanizing rural development? Implementing the CAP's Second Pillar in England. *International Planning Studies* 9 (2-3): 121-137.

Ward, N. and R. Munton 1992: Conceptualising agriculture-environment relations: combining political economy and socio-cultural approaches to pesticide pollution. *Sociologia Ruralis* 32 (1): 127-145.

Ward, N., Lowe, P., Seymour, S. and J. Clark 1995: Rural restructuring and the regulation of farm pollution. *Environment and Planning A* 27: 1193-1211.

Ward, N., Clark, J., Lowe, P. and S. Seymour 1998: Keeping matters in its place: pollution regulation and the reconfiguration of farmers and farming. *Environment and Planning A* 30: 1165-1178.

Ward, N., Lowe, P. and T. Bridges 2003: Rural and regional development: the role of the Regional Development Agencies in England. *Regional Studies* 37 (2): 201-214.

Warren, M. 2006: The cloud in the silver lining: exploring links between social disadvantage and digital exclusion in rural areas. Paper presented at the Conference 'The rural citizen: governance, culture and wellbeing', University of Plymouth, Plymouth, UK, April 2006.

Watkins, S.C. 1987: The fertility transition: Europe and the third world compared. *Sociological Forum* 2 (4): 645-673.

Watts, M. 1996: Development III: the global agro-food system and late twentieth century development (or Kautsky redux). *Progress in Human Geography* 20: 230-245.

Webber, M.J. and D. Rigby 1996: *The golden age illusion*. New York: Guilford Press.

Welford, R. (ed.) 1996: *Corporate environmental management: systems and strategies*. London: Earthscan.

Whatmore, S. 1986: Landownership relations and the development of modern British agriculture. In: Cox, G., Lowe, P. and M. Winter (eds): *Agriculture: people and policies*. London: Allen and Unwin, pp. 41-60.

Whatmore, S. 1995: From farming to agribusiness: the global agro-food system. In: Johnston, R.J., Taylor, P.J. and M.J. Watts (eds): *Geographies of global change: remapping the world in the late twentieth century*. London: Blackwell, pp. 36-49.

Whatmore, S. 2002: *Hybrid geographies: natures, cultures, space*. London: Sage.

Whatmore, S., Munton, R. and T. Marsden 1990: The rural restructuring process: emerging divisions of agricultural property rights. *Regional Studies* 24: 235-245.

Whitby, M. (ed.) 1996: *The European environment and CAP reform: policies and prospects for conservation*. Wallingford: CAB International.

Whitby, M. and P. Lowe 1994: The political and economic roots of environmental policy in agriculture. In: Whitby, M. (ed.): *Incentives for countryside management: the case of Environmentally Sensitive Areas*. Wallingford: CAB International, pp. 1-24.

White, L. 1967: The historical roots of our ecological crisis. *Science* 155: 1203-1207.

Williams, M. 1989: *Americans and their forests: a historical geography*. Cambridge: CUP.

Williamson, O.E. 1996: *The mechanisms of governance*. Oxford: OUP.

Wilson, A.G. 1981: *Catastrophe theory and bifurcation: applications to urban and regional systems*. London: Croom Helm.

Wilson, G.A. 1992: *The urge to clear the 'bush': a study of native forest clearance on farms in the Catlins District of New Zealand, 1861-1990*. Christchurch (New Zealand): University of Canterbury Press.

Wilson, G.A. 1993: The pace of indigenous forest clearance on farms in the Catlins District, SE South Island, New Zealand, 1861-1990. *New Zealand Geographer* 49 (1): 15-25.

Wilson, G.A. 1996: Farmer environmental attitudes and ESA participation. *Geoforum* 27 (2): 115-131.

Wilson, G.A. 1997a: Factors influencing farmer participation in the Environmentally Sensitive Areas scheme. *Journal of Environmental Management* 50: 67-93.

Wilson, G.A. 1997b: Selective targeting in Environmentally Sensitive Areas: implications for farmers and the environment. *Journal of Environmental Planning and Management* 40 (2): 199-215.

Wilson, G.A. 1997c: Assessing the environmental impact of the Environmentally Sensitive Areas scheme: a case for using farmers' environmental knowledge? *Landscape Research* 22 (3): 303-326.

Wilson, G.A. 2001: From productivism to post-productivism … and back again? Exploring the (un)changed natural and mental landscapes of European agriculture. *Transactions of the Institute of British Geographers* 26 (1): 77-102.

Wilson, G.A. 2002: 'Post-Produktivismus in der europäischen Landwirtschaft: Mythos oder Realität? *Geographica Helvetica* 57 (2): 109-126.

Wilson, G.A. 2004: The Australian Landcare movement: towards 'post-productivist' rural governance? *Journal of Rural Studies* 20: 461-484.

Wilson, G.A. 2005: Towards a post-productivist countryside in the UK and Germany? In: Schmied, D. and O.J. Wilson (eds): *The countryside in the 21st century: Anglo-German perspectives.* Bayreuth (Germany): Naturwissenschaftliche Gesellschaft Bayreuth, pp. 111-120.

Wilson, G.A. and R.L. Bryant 1997: *Environmental management: new directions for the 21st century.* London: UCL Press.

Wilson, G.A. and H. Buller 2001: The use of socio-economic and environmental indicators in assessing the effectiveness of EU agri-environmental policy. *European Environment* 11: 297-313.

Wilson, G.A. and K. Hart 2000: Financial imperative or conservation concern? EU farmers' motivations for participation in voluntary agri-environmental schemes. *Environment and Planning A* 32 (12): 2161-2185.

Wilson, G.A. and K. Hart 2001: Farmer participation in agri-environmental schemes: towards conservation-oriented thinking? *Sociologia Ruralis* 41 (2): 254-274.

Wilson, G.A. and M. Juntti (eds) 2005: *Unravelling desertification: policies and actor networks in Southern Europe.* Wageningen (NL): Wageningen Academic Publishers.

Wilson, G.A. and P.A. Memon 2005: Indigenous forest management in 21st century New Zealand: towards a 'postproducticvist' indigenous forest-farmland interface? *Environment and Planning A* 37: 1493-1517.

Wilson, G.A. and J. Rigg 2003: 'Post-productivist' agricultural regimes and the South: discordant concepts? *Progress in Human Geography* 27 (5): 605-631.

Wilson, G.A. and O.J. Wilson 2001: *German agriculture in transition: society, policies and environment in a changing Europe.* Houndmills: Palgrave.

Wilson, G.A., Petersen, J.E. and A. Höll 1999: EU member state responses to Agri-environment Regulation 2078/92/EEC: toward a conceptual framework? *Geoforum* 30: 185-202.

Wilson, O.J. 1994: 'They changed the rules': farm family responses to agricultural deregulation in South Island, New Zealand. *New Zealand Geographer* 50 (1): 3-13.

Winter, M. 1996: *Rural politics: policies for agriculture, forestry and the environment.* London: Routledge.

Winter, M. 1997: New policies and new skills: agricultural change and technology transfer. *Sociologia Ruralis* 37 (3): 363-381.

Winter, M. 1998: Embeddedness, the new food economy and defensive localism. *Journal of Rural Studies* 19: 23-32.

Winter, M. 2001: *Multifunctionality: final report to OECD.* Cheltenham: Countryside and Community Research Unit.

Winter, M. 2003: Embeddedness, the new food economy and defensive localism. *Journal of Rural Studies* 19: 23-32.

Winter, M. 2005: Geographies of food: agro-food geographies – food, nature, farmers and agency. *Progress in Human Geography* 29 (5): 609-617.

Winter, M. and P. Gaskell 1998: *The effects of the 1992 reform of the Common Agricultural Policy on the countryside of Great Britain.* Cheltenham: Countryside Community Press.

Winter, M. and C. Morris 2004: Multifunctionality and farming systems: the non-commodity outputs of organic and integrated systems of agriculture. Paper presented at the 2nd meeting of the Anglo-German Rural Geographers 'Rural multifunctionality: perspectives from policy-making, implementation and practice', University of Exeter, Exeter, UK, July 2004.

Wolf, E. 1982: *Europe and the people without history.* Berkeley: University of California Press.

Woo, M.J. 2006: North Korea in 2005: maximising profit to save socialism. *Asian Survey* 46 (1): 49-55.

Wood, W. 2000: Attitude change: persuasion and social influence. *Annual Review of Psychology* 51: 539-570.

Woods, M. 1997: Discourses of power and rurality: local politics in Somerset in the 20th century. *Political Geography* 16: 453-478.

Woods, M. 2005: *Rural geography: processes, responses and experiences in rural restructuring.* London: Sage.

Worden, N. and E. Van Heyningen 1996: Signs of the times: tourism and public history at Cape Town's Victoria and Albert Waterfront. *Cahiers d'Etudes Africaines* 141: 215-236.

Wormell, P. 1978: *The anatomy of agriculture: a study of Britain's greatest industry*. London: Harper and Row.

Worster, D. 1993: *The wealth of nature: environmental history and the ecological imagination*. New York: Oxford University Press.

Yarwood, R. 2005: Beyond the rural idyll: images, countryside change and geography. *Geography* 90 (1): 19-32.

Yearley, S. 1988: *Science, technology, and social change*. London: Unwin Hyman.

Young, R. 1990: *White mythologies: writing history and the West*. London: Routledge.

Zoomers, A.E. and Kleinpenning, J. 1996: Livelihood and urban-rural relations in Central Paraguay. *Tijdschrift voor Economische en Sociale Geografie* 87 (2), 161-174.

Index